LIBRARY
NORTH HIGHLAND COLLEGE
ORMLIE ROAD
THURSO
CAITHNESS
KW14 7EE

Library

D1357715

LIBRARY
THE NORTH HIGHLAND COLLEGE
ORMLIE ROAD
THURSO
CAITHNESS
KW14 7EE

UHI Millennium Institute

10301392

Changing Wildlife of Great Britain and Ireland

LIBRARY
THE NORTH HIGHLAND COLLEGE
ORMLIE ROAD
THURSO
CAITHNESS
KW14 7EE

1974 saw the publication of *The Changing Flora and Fauna of Britain*. This, the last major report on the state of all kinds of British wildlife stemming from a conference in 1973, became a benchmark for wildlife conservationists. Twenty-five years after the initial meeting, *The Changing Wildlife of Great Britain and Ireland* was instigated to assess the changes in biodiversity that have occurred over that time and also looks to the future to identify what we need to do to preserve our natural heritage.

The contributions, written by some of the most eminent researchers across the British Isles, several of whom contributed to the original 1973 meeting, cover most groups of organisms: from viruses, diatoms, protozoa, fungi, lichens, mites and nematodes; through butterflies, dragonflies, flies, slugs and snails; to flowering plants, ferns, mammals, birds and fish. The state of knowledge and data capture systems in different groups are assessed. Losses and declines, increases and additions are addressed. The effectiveness of statutory and other measures taken to safeguard wildlife is also assessed.

The picture is mixed: reduced sulphur dioxide levels have benefited sensitive lichens and mosses in a dramatic way, water quality improvement has been beneficial and there have been rediscoveries of species thought to have been lost, and few definite extinctions. Biodiversity Action Plans have also benefited targeted species, but habitat restoration and management for some is not always good for others. There are worrying trends in declining populations, with an increasing number of species being regarded as threatened or endangered, especially in agricultural areas. Introduced species are a major cause of disruption and a cause for concern, and may be encouraged by climate change. Many smaller organisms remain poorly known, a situation likely to remain as a result of scant and declining expertise.

This stock-check and look to the future, the most authoritative statement yet of our current knowledge of diverse kinds of British and Irish wildlife, will be a key resource for conservationists, naturalists, government agencies, taxonomists and other professional biologists for many years to come.

Professor David L. Hawksworth CBE is a past President of the International Union of Biological Sciences, Council Member of English Nature 1996–99, and was Director of the International Mycological Institute 1983–97. He is an internationally recognised authority on biodiversity issues. His previous books include *The Changing Flora and Fauna of Britain* (Academic Press 1974), *Biodiversity: Measurement and estimation* (Chapman & Hall 1994) and contributions to *The Global Biodiversity Assessment* (Cambridge University Press 1995).

The Systematics Association Special Volume Series

Series Editor

Allan Warren
Department of Zoology, The Natural History Museum,
Cromwell Road, London SW7 5BD, UK.

The Systematics Association provides a forum for discussing systematic problems and integrating new information from genetics, ecology and other specific fields into taxonomic concepts and activities. It has achieved great success since the Association was founded in 1937 by promoting major meetings covering all areas of biology and palaeontology, supporting systematic research and training courses through the award of grants, production of a membership newsletter and publication of review volumes by its publishers Taylor & Francis. Its membership is open to both amateurs and professional scientists in all branches of biology who are entitled to purchase its volumes at a discounted price.

The first of the Systematics Association's publications, *The New Systematics*, edited by its then president Sir Julian Huxley, was a classic work. Over 60 volumes have now been published in the Association's 'Special Volume' series often in rapidly expanding areas of science where a modern synthesis is required. Its *modus operandi* is to encourage leading exponents to organise a symposium with a view to publishing a multi-authored volume in its series based upon the meeting. The Association also publishes volumes that are not linked to meetings in its 'Special Volume' series.

Anyone wishing to know more about the Systematics Association and its volume series are invited to contact the series editor.

Forthcoming titles in the series

Morphology, Shape and Phylogenetics
Edited by Norman Macleod and Peter L. Forey

Other Systematics Association publications are listed after the index for this volume.

The Systematics Association Special Volume Series 62

The Changing Wildlife of Great Britain and Ireland

Edited by David L. Hawksworth

LIBRARY
THE NORTH HIGHLAND COLLEGE
ORMLIE ROAD
THURSO
CAITHNESS
KW14 7EE

London and New York

First published 2001
by Taylor & Francis Limited
11 New Fetter Lane, London EC4P 4EE

Simultaneously published in the USA and Canada
by Taylor & Francis, Inc.
29 West 35th Street, New York, NY 10001

Taylor & Francis is an imprint of the Taylor & Francis Group

©2001 Systematics Association

Typeset in Sabon by Bookcraft Ltd., Stroud, Gloucestershire
Printed and bound in Great Britain by TJ International, Padstow, Cornwall

All rights reserved. No part of this book may be reprinted or reproduced or utilised in any form or by any electronic, mechanical, or other means, now known or hereafter invented, including photocopying and recording, or in any information storage or retrieval system, without permission in writing from the publishers.

Every effort has been made to ensure that the advice and information in this book is true and accurate at the time of going to press. However, neither the publisher nor the authors can accept any legal responsibility or liability for any errors or omissions that may be made. In the case of drug administration or the use of technical equipment mentioned within this book, you are strongly advised to consult the manufacturer's guidelines.

British Library Cataloguing in Publication Data
A catalogue record for this book is available from the British Library.

Library of Congress Cataloguing in Publication Data
A catalogue record for this book has been requested.

ISBN 0–7484–0957–2

Contents

LIBRARY
THE NORTH HIGHLAND COLLEGE
ORMLIE ROAD
THURSO
CAITHNESS
KW14 7EE

List of contributors   viii

Foreword   xi
SIR JOHN BURNETT

1 Fifty years of statutory nature conservation in Great Britain   1
EARL OF CRANBROOK

2 Flowering plants   23
TIMOTHY C. G. RICH

3 Ferns and allied plants   50
CHRISTOPHER N. PAGE

4 Mosses, liverworts and hornworts   78
ANTHONY J. E. SMITH

5 Larger fungi   103
ROY WATLING

6 Microscopic fungi   114
PAUL F. CANNON, PAUL M. KIRK, JERRY A. COOPER AND DAVID L. HAWKSWORTH

7 Lichens   126
BRIAN J. COPPINS, DAVID L. HAWKSWORTH AND FRANCIS ROSE

8 Terrestrial and freshwater eukaryotic algae   148
DAVID M. JOHN, ALLAN PENTECOST AND BRIAN A. WHITTON

9 Cyanobacteria (blue-green algae)   150
BRIAN A. WHITTON AND S. J. BRIERLEY

10 Diatoms   152
ELIZABETH Y. HAWORTH

11 Viruses   164
ROGER T. PLUMB AND J. IAN COOPER

12 Protozoa   175
BLAND J. FINLAY

13 Freshwater invertebrates   188
JOHN F. WRIGHT AND PATRICK D. ARMITAGE

14 Nematodes   210
BRIAN BOAG AND DAVID J. HUNT

15 Mites and ticks   230
ANNE S. BAKER

16 Flies   239
ALAN E. STUBBS

17 True bugs, leaf- and planthoppers, and their allies   262
PETER KIRBY, ALAN J. A. STEWART AND MICHAEL R. WILSON

18 Butterflies and moths   300
RICHARD FOX

19 Grasshoppers, crickets and allied insects   328
JUDITH A. MARSHALL

20 Dragonflies and damselflies   340
STEPHEN J. BROOKS

21 Land slugs and snails   355
ROBERT A. D. CAMERON AND IAN J. KILLEEN

22 Birds   367
DAVID W. GIBBONS AND MARK I. AVERY

23 Mammals   399
GORDON B. CORBET AND D. W. YALDEN

24 Fishes   410
ALWYNE WHEELER

25 Tracking future trends: the Biodiversity Information Network   422
KEITH PORTER

26 Prospects for the next 25 years   435
DAVID L. HAWKSWORTH

    Subject index   447
    Systematics Association publications   451

Contributors

Armitage, Patrick D., CEH Dorset, Winfrith Technology Centre, Winfrith Newburgh, Dorchester, Dorset DT2 8ZD, UK (formerly Institute of Freshwater Ecology, River Laboratory, East Stoke, Wareham, Dorset BH20 6BB, UK).

Avery, Mark I., Royal Society for the Protection of Birds, The Lodge, Sandy, Bedfordshire SG19 2DL, UK.

Baker, Anne S., Department of Entomology, The Natural History Museum, Cromwell Road, London SW7 5BD, UK.

Boag, Brian, Scottish Crop Research Institute, Invergowrie, Dundee DD2 5DA, UK.

Brierley, S. J., Environment Agency, Phoenix House, Global Avenue, Leeds LS11 8PG, UK.

Brooks, Stephen J., Department of Entomology, The Natural History Museum, Cromwell Road, London SW7 5BD, UK.

Burnett, Sir John, National Biodiversity Network Trust, 13 Field House Drive, Oxford OX2 7NT, UK.

Cameron, Robert A. D., Department of Animal and Plant Sciences, University of Sheffield, Sheffield S10 2TN, UK.

Cannon, Paul F., CABI Bioscience, UK Centre (Egham), Bakeham Lane, Egham, Surrey TW20 9TY, UK.

Cooper, J. Ian., Institute of Virology and Environmental Microbiology, Mansfield Road, Oxford OX1 3SR, UK.

Cooper, Jerry A., Landcare Research, Canterbury Agriculture and Science Centre, Gerald Street, Lincoln, New Zealand.

Coppins, Brian J., Royal Botanic Garden, 20A Inverleith Row, Edinburgh EH3 5LR, UK.

Corbet, Gordon B., Little Dumbarnie, Newburn, Upper Largo, Leven, Fife KY8 6JG, UK.

Cranbrook, Earl of, Glemham House, Great Glemham, Saxmundham, Suffolk IP17 1LP, UK.

Finlay, Bland J., CEH Windermere, Ferry House, Far Sawrey, Ambleside, Cumbria LA22 0LP, UK.

Fox, Richard, Butterfly Conservation, Manor Yard, East Lulworth, Wareham, Dorset BH20 5QP, UK.

Gibbons, David W., Royal Society for the Protection of Birds, The Lodge, Sandy, Bedfordshire SG19 2DL, UK.

Hawksworth, David L., MycoNova, 114 Finchley Lane, Hendon, London NW4 1DG, UK.

Haworth, Elizabeth Y., Freshwater Biological Association, Ferry House, Far Sawrey, Ambleside, Cumbria LA22 0LP, UK.

Hunt, David J., CABI Bioscience, UK Centre (Egham), Bakeham Lane, Egham, Surrey TW20 9TY, UK.

John, David M., Department of Botany, The Natural History Museum, Cromwell Road, London SW7 5BD, UK.

Killeen, Ian J., Malacological Services, 163 High Road West, Felixstowe IP11 9BD, UK.

Kirby, Peter, 21 Grafton Avenue, Netherton, Peterborough PE3 9PD, UK.

Kirk, Paul M., CABI Bioscience, UK Centre (Egham), Bakeham Lane, Egham, Surrey TW20 9TY, UK.

Marshall, Judith A., Department of Entomology, The Natural History Museum, Cromwell Road, London SW7 5BD, UK.

Page, Christopher N., Gillywood Cottage, Trebost Lane, St Stythians, Truro, Cornwall TR3 7DW, UK.

Pentecost, Allan, Division of Life Sciences, King's College London, Manresa Road, London SW3 6LX, UK.

Plumb, Roger T., IACR-Rothamsted, Harpenden, Hertfordshire AL5 2JQ, UK.

Porter, Keith, English Nature, Northminster House, Peterborough PE1 1UA, UK.

Rich, Timothy C. G., Department of Biodiversity and Systematic Biology, National Museum and Galleries of Wales, Cathays Park, Cardiff CF10 3NP, UK.

Rose, Francis, Rotherhurst, 36 St Mary's Road, Liss, Hampshire GU33 7A, UK.

Smith, Anthony J. E., 5 Queens Gardens, Llandudno, Conwy LL30 1RU, UK.

Stewart, Alan J. A., School of Biological Sciences, University of Sussex, Falmer, Brighton, Sussex BN1 9QG, UK.

Stubbs, Alan E., 181 Broadway, Peterborough PE1 4DS, UK.

Watling, Roy, Caledonian Mycological Enterprises, 26 Blinkbonny Avenue, Edinburgh EH4 3HU, UK.

Wheeler, Alwyne, 14 Theydon Park Road, Theydon Bois, Epping, Essex CM16 7LP, UK.

Whitton, Brian A., Department of Biological Sciences, University of Durham, South Road, Durham DH1 3LE, UK.

Wilson, Michael R., Department of Biodiversity and Systematic Biology, National Museum and Galleries of Wales, Cathays Park, Cardiff CF10 3NP, UK.

Wright, John F., CEH Dorset, Winfrith Technology Centre, Winfrith Newburgh, Dorchester Dorset DT2 8ZD, UK (formerly Institute of Freshwater Ecology, River Laboratory, East Stoke, Wareham, Dorset BH20 6BB, UK).

Yalden, D. W., School of Biological Sciences, 3.239 Stopford Building, University of Manchester, Manchester M13 9PL, UK.

Foreword

There are two good reasons, among many others, why I welcome and commend this invaluable book, a more than worthy successor to a similar volume published in 1974. First, it is especially timely. Second, like the earlier volume also edited by Professor Hawksworth, it is comprehensive, covering as it does most of the major groups of organisms found in the British Isles apart from the marine invertebrates.

It is timely not only because a new millennium is always an appropriate time to take stock but, more importantly, because increasing numbers of the population are becoming conscious of the importance of wildlife to them, their children and their children's children. Much of the planning and thinking that has taken place since the UK signed the Biodiversity Treaty at Rio in 1996 will come to fruition in the next few years. Globally, 2000–2001 is International Biodiversity Observation Year when attention will be focused worldwide on the state of the world's biodiversity. In Britain, the National Biodiversity Action Plan is beginning to take shape and, perhaps of greater importance to the book's potential readers, Local and Species Action Plans are a major contemporary challenge at the start of the twenty-first century. For all of these, *The Changing Wildlife of Great Britain and Ireland* provides just the kind of background that field naturalists, volunteer or professional, need.

Its comprehensive nature is a remarkable achievement. It is never easy to find individuals who possess a sufficiently wide knowledge of the current status of any particular group of organisms, even if it is their chosen group. Once found, it is often even more difficult to persuade them to take time to review the situation in a manner comprehensible to non-specialist readers! That Professor Hawksworth has been so persuasive and that the contributors have met so successfully the challenge that a combined specialist and non-specialist audience poses is greatly to the credit of all involved. Consequently, this book is eminently readable, up-to-date and informative. It is bound to provoke serious thought for the future of our wildlife.

The message the book conveys is not as depressing as might have been expected, although it gives no cause for complacency. Although the numbers of species that have become extinct in the last quarter-century is not high, more careful and widespread observations suggest that the number of species at risk has greatly increased. Improved analysis of the causes of reduction in numbers, or extinction, in almost all groups of organisms indicates that species loss arises from a variety of different causes. That is one of the important messages of the book. There is neither a universal cause for species decline or even extinction in the UK, nor any universal panacea for their recovery and sustainability. Despite the concern such new knowledge engenders the good news, echoed in many of the contributions, is that the last 25 years has seen a transformation in the state of knowledge of the distribution, abundance and ecology of many species. Such knowledge provides a basis on which sounder conservation action can now be based.

Loss or reduction is not the only kind of change noted. In many groups there has been an increase in new species, the numbers of particular species have increased, or their distribution has been found to be wider than had been realized. Additions to species lists stem from three sources: the recognition of overlooked species, taxonomic splitting of species into smaller taxa and immigration of species new to the British Isles. Unfortunately, increase in numbers of a native species is not always to be welcomed if it reflects an imbalance, however caused, in an ecologically balanced situation that results in the serious decline or even loss of other species. In many cases, however, the native species are declining in numbers as a result of increase in the numbers of alien species. Two well known situations demonstrate this clearly, namely the increase in grey squirrels and American mink with corresponding declines both in numbers and sites of the native red squirrel and water vole, respectively.

The common denominator in virtually all these changes is human activity of various kinds. So far as losses are concerned, the two obvious and expected factors are atmospheric pollution and agricultural land use, including improved drainage. The alterations they induce in a habitat are reflected in changes in species composition and numbers. That is perhaps the best argument for basing conservation on habitat restoration rather than the narrower and, in some ways, more appealing notion of species restoration. Atmospheric pollution is slowly but surely ameliorating, so urban lichens are beginning to return to the trees and to roofs in towns although acid rain, principally blown across from the continent, is declining far more slowly. Its direct and, more insidiously, indirect effects through soil and water acidification are still a cause for concern even as far as the lochans in the north of Scotland. The loss of lowland heath sites either through tree planting or building, although not entirely irreparable in the former situation, affects a wide diversity of both animals and plants. The situation has been recognized, however, and English Nature's 'Heathland Heritage' programme should do much both to restore that habitat and its flora and fauna. Of much greater concern, but fortunately also recognized, is the drying out of Britain's wetlands and changes in the water regimes through industrial and domestic extraction and discharge. Especially in the case of wetlands, the decline has affected almost every kind of organism found there – mammals, birds, amphibians, fishes, insects, aquatic invertebrates, vascular plants and algae. Here too there is hope. The deliberately engineered increases in wetlands and reedy places, pioneered by the RSPB and others or, on an individually smaller scale although collectively significant, the millennial restoration of village ponds and the increase in domestic ornamental ponds are helping to restore this sector of Britain's biodiversity.

A continuous increase in new species is caused by the constant influx of aliens – or perhaps better, non-native British species, since the definition of 'alien' is debatable! Many do not persist but in recent times the ecological consequences of some introductions, whether inadvertently or by human choice, have proved quite devastating. The new species that causes Dutch elm disease (*Ophiostoma novi-ulmi*), imported either in infected timber or on fungus-carrying beetles on Canadian rock elm (*Ulmus thomasii*), has brought about as great a decline in native elm as that 5000 years ago at the end of the Atlantic period. The importation of this timber reflected the increased demand for hard chine used in dinghy construction and this in its turn reflected the increased wealth and leisure of modern Britain. Dutch elm disease, therefore, is a salutary reminder that in environmental matters biological cause and effect are not always immediately obvious and predictable unless social conditions are also considered. Deliberate introductions have also caused problems: the zebra mussel and signal crayfish – the latter bringing a new crayfish disease – are classic examples of commercial 'development' while that of the American flatworm is an equally classic example of the consequences of importing ornamental plants with alien soil.

Inadvertent increase in the number of existing and new species probably also reflects an increase in range of response to human-induced, climatic warming. Such increased ranges of resident species range from freshwater algae such as *Hydrodictyon*, to flies and dragonflies while, not surprisingly, of new species, it is the most mobile such as species of butterflies that are best represented. If the climatological predictions are right, a far greater influx of new species of all kinds can obviously be expected this century. Even the most apparently immobile organisms get around eventually!

The one cause for real joy in species increase, however, must be for the many newly discovered or recognized species. Of course such cases tend to be in the taxonomic groups that are least known or where the paucity of taxonomic expertise is a truly limiting factor. Such discoveries are, in part, a spin-off from taxonomic and ecological activity in the second half of this century which has been notable as a time of immense importance for biological recording of every kind. Not only have the numbers of species recognized increased as a result of the immediate post-Second World War boom in taxonomic activity, now alas in decline, but the numbers of new and known species recorded have exceeded all those recorded in previous centuries. Moreover, the publication of several distribution atlases has provided a more exact understanding of where wildlife can be found in these islands. These atlases have generally resulted from the co-ordinated voluntary recording activities of tens of thousands of naturalists – a new phenomenon that will become of increasing importance in the future. Moreover, in the last quarter of the twentieth century, the need to monitor change accurately and to correlate it with environmental and biological changes was recognized as of increasing importance to an understanding of change and persistence of wildlife of all kinds. Indeed, the awarenesss that change of all kinds is a normal part of day-to-day living is now receiving public recognition.

That naturally raises a variety of questions concerning the future. This book provides two pointers. The statutory framework established to promote conservation of the wildlife of these islands is described in the first chapter and the two concluding chapters assess prospects for the forseeable future and outline how future changes should be tracked.

Statutory protection for wildlife in Britain is still in its infancy although the last 50 years have seen real progress. Good though it is in many ways, the legal protection afforded wildlife and the penalties for damage are not yet adequate. Even if the forthcoming Countryside Bill were to remedy this, it is still far too easy to damage Sites of Special Scientific Interest (SSSIs) irreparably. In any event, the manner in which they are statutorily defined makes no provision for an essential 'buffer' zone around the site. Hence ecological or agricultural erosion around the margins of the features of scientific importance cannot be guarded against adequately. Much the same applies to National

Nature Reserves (NNRs). Are they really big enough? Do not they need a surrounding 'buffer' zone and more or better defined channels for wildlife movement into, and exchange with, the wider countryside? Are NNRs exploited as they should be? Virtually no material provision for scientific work relevant to conservation is provided and their monitoring, although improved, still leaves much to be desired. Are they, therefore, really providing the maximum of information that they should to aid in conservation countrywide? This is debatable territory but many believe that far more should be done.

Change will inevitably occur, but how is it best detected and monitored – on statutorily protected sites or in the countryside as a whole? This is a question to which the recording community needs to give more attention. Dramatic changes have occurred both in the knowledge and tools available to them. Not only is far more known about the kinds, numbers, distribution and ecology of all wildlife but the methods for recording, storing, transmitting and publishing such information have been revolutionized by computers, while precise distributions can be achieved by handheld global location devices. Although the tools for future recording are of a degree of precision never before available, there are limiting factors on their effective deployment: the human factor and training in the making, compilation, organization and interpretaion of observations.

The British Isles has a proud record in its observers, unequalled in the rest of the world. For centuries, informed and dedicated generations of individual, volunteer naturalists have kept wonderfully informed eyes on our wildlife – a fact rightly acknowledged by several of the book's authors. It is these volunteers – one can hardly call them amateurs since they have included and continue to include many of Britain's best field specialists and taxonomists – who have not only kept the records but, often, noticed the changes and alerted the nation to them. The post-war trend towards accurate mass recording of species' distributions needs to be improved by using modern tools and extended to a variety of monitoring programmes. In this way the basic data needed for a deeper insight into the activities of organisms and how they persist, decline, or even become extinct could become available. Good conservation needs such information, preferably as contemporaneously as possible. I have no doubt that individual naturalists will adapt to these needs. Moreover, in the recently established National Biodiversity Network, a tool is being forged for recorders that will increase the accuracy and extent of their basic data, make them more widely available and enable them to be better and more effective for a variety of conservation needs.

Promising though this is, there is a major cause for concern. Although more of the population is probably interested in wildlife in a general way than ever before, the numbers capable of identifying species, or even interested in acquiring the ability to identify species, are in danger of decline. The reasons are twofold. First and most serious is the decline in taxonomic teaching of professional biologists: the nation's taxonomists are a declining and ageing group. This exacerbates the existing imbalance between the numbers of taxonomists concerned and the taxonomic needs of particular groups of organisms. Indeed, the numbers of taxonomists concerned is often in almost inverse proportion to the size and needs of different groups of organisms. Consider only the woefully small numbers of taxonomists concerned with the needs of the insects and fungi, the two largest taxonomic groups in Britain – indeed in the world. Consequently, the support provided to volunteer naturalists from professional taxonomists is declining or non-existent. Second, this decline is paralleled amongst the voluntary sector. Teaching in schools has reflected the decline in taxonomic teaching in colleges and universities and increasingly, and understandably, interest in wildlife stresses behaviour and ecology rather than classification. So despite an increase in naturalists who may well know much more about the natural biology of a single species or small group of species than ever before, there are probably far fewer who can identify the full range of species within any group of organisms. This decline in general taxonomic expertise is something that must be addressed if knowledge of Britain's biodiversity is to be maintained, let alone increased. The teaching of identification in the field is probably the single weakest link in our armoury in fighting the potential loss of biodiversity. After all, if it is not known what we possess, it is hardly possible to promote its recovery!

But I am an optimist and I have no doubt that this excellent book will help to stimulate concern for 'knowing one's wildlife'. Above all, I hope that the pleasure of reading about Britain's wildlife will so intrigue and challenge readers that novices will take up and old hands return refreshed to the immensely rewarding and enjoyable study of wildlife of every kind 'in the field'.

Sir John Burnett
Chairman
National Biodiversity Network Trust
Oxford

December 2000

Chapter 1

Fifty years of statutory nature conservation in Great Britain

Earl of Cranbrook

ABSTRACT

In Great Britain, many strands have combined to forge public policy for nature conservation. Of these, three stand out: the influence of the voluntary sector, the role of Parliament and, latterly, a shift to European and international institutions as policy setters.

Government first recognised the need for national policies for nature conservation in the 1940s, in the context of planning for post-war reconstruction. The voluntary movement was involved from the beginning, setting out the case in 1941 with a conference on 'Nature Preservation', and contributing decisively to the proceedings of the Wild Life Conservation Special Committee and its report, Cmd 7122 (1947). The pre-legislative process of the 1940s also instituted the division between landscape preservation and public access, and the conservation of wildlife and natural features. As a consequence, the 1949 Act created two bodies: the National Parks Commission (later the Countryside Commission) and the Nature Conservancy (NC). This separation of responsibilities persisted until legislation of the 1990s brought together the two functions in Scotland and Wales.

Parliamentary interest has been continuous since the first Select Committee enquiry of 1957. Further measures were introduced, with the focus on species conservation and particular British preoccupations: the rolling tide of Protection of Birds Acts, 1954, 1964, 1967; the Deer Act, 1963; Conservation of Seals, 1970; Badgers, 1973; Conservation of Wild Creatures and Wild Plants, 1975. Government amended the status of the Nature Conservancy in 1973. None of these measures remedied the fundamental flaws in the 1949 Act: the lack of any legal obligation on owners or occupiers and absence of restriction on existing land use. The Wildlife and Countryside Act of 1981 was intended to remedy this deficiency and, for the first time, imposed obligations on owners and occupiers; a practical weakness was remedied by the 1985 amending Act. Through the 1980s, SSSI renotification consumed a major part of NCC's resources, and also engendered huge controversies. There was scepticism over the capability of available measures to conserve wildlife in the face of threats, natural or anthropogenic.

But, from the 1970s, the UK was also negotiating international measures that affected domestic policies. The source of political initiative began to move. Ratification of the Convention on International Trade in Endangered Species (CITES) in 1973 required new national law, duly enacted in 1976. The terms of the Council of Europe's

Berne and Bonn Conventions had to be met and, after joining the EEC, emerging Directives had to be implemented by domestic legislation.

Scottish law required separate enactment of nature conservation legislation, and the special standing of Scotland was always recognised administratively within the NC/NCC. The creation of three country agencies by the 1990 Act intensified that division. In Scotland and Wales, for the first time, responsibility for countryside and access issues was merged with nature conservation. New approaches have developed, and devolution may accentuate the divergence in application of the broad intention of the law. However, the basic content of UK conservation legislation is now largely determined by international or European obligations. The role of Parliament has diminished; for instance, the EC Habitats and Species Directive was translated into UK law in 1994 simply by a Regulation, and its implementation is subject to the scrutiny of Brussels. Within the emerging mix of devolved administrations, in the application of policies for nature conservation a reliable mechanism will be needed to ensure that the UK nationally meets its EU and international obligations.

1 Introduction

It is now over 50 years since the Nature Conservancy was founded. The earliest of a cluster of demi-centennial dates to celebrate was the announcement in Parliament, on 21 April 1948, of the intention to form a national nature conservation service; Cyril Diver was appointed the first Director General on 1 November 1948; by an Order in Council on 4 March 1949, the objects of the Privy Council Committee for Agricultural Research were extended to include nature conservation, and the Nature Conservancy was constituted; the first meeting of Conservancy members (with the statutory Scottish Committee) was held on 11 March, and the Conservancy's Royal Charter was dated 23 March 1949.

Looking back after 40 years, Sir William Wilkinson recognised four phases: the Royal Charter period, 1949–65, the Research Council years, 1965–73, the pre-Wildlife and Countryside Act, 1973–81, and the post-1981 period (NCC 1990). To that framework, the phase of the country agencies must now be added. In this first half-century of the statutory nature conservation bodies, from the original, partially realised vision of a national nature conservancy and biological service to the present devolved structure, two external sources of initiative have also influenced public policy for the conservation of wildlife and natural features in Great Britain: voluntary conservation organisations (VCOs) and Parliamentarians. Over the same period, we have witnessed a progressive transition from national issues to international conventions and European legislation as the dominant target setters. These influences form recurrent themes in my short history.

This remains a personal account. Space is limited, so I have perforce been selective, and partial, openly taking advantage of my own involvement in parts of the story without, I hope, detriment to the rounded picture.

2 The inspiration

The crucial initiative represented a triumph for the persistence of VCOs. The chance was seized in the unlikely context of mid-World War national planning for reconstruction (Stamp 1969; Nicholson 1970; Sheail 1976; NCC 1984; Adams 1986; Evans 1997). In 1941, seeing an opportunity to implement its longstanding aspirations, the Society for the Promotion of Nature Reserves (SPNR) – founded in 1912 by Charles Rothschild (Rothschild and Marren 1997) – marshalled some 15 voluntary organisations in a conference on Nature Preservation. Thus prompted, the Government formed a Nature Reserves Investigation Committee whose conclusions were fed into the cross-ministerial (Scott) Committee on Land-use in Rural Areas. Scott led to John Dower's (1945) report on National Parks; in response, in July 1945, the Government set up the Committee on National Parks, chaired by Sir Arthur Hobhouse. The vision of the era was expressed by a participant:

> As, one after another, our cities were bombed, plans were put in hand for their rebuilding. They were to have a green ring of rural land, productively used but not urbanized. There were to be large tracts set aside for quiet enjoyment. But enjoyment of what? Clearly the natural or semi-natural vegetation of mountain, moorland and coast, and with it the wild life ... Gradually, the concept of nature conservation began to fit in as part of the picture of our land for the future. Naturalists met town and country planners; they did not clash.
>
> (Stamp 1969: xiv)

Dower had conceded that the objectives of national parks were not sufficient to deliver a national nature conservation policy. Hobhouse endorsed this divergence by creating two committees: one to cover access and related countryside issues; the other the Wild Life Conservation Special Committee (England and Wales) (WLCSC) chaired successively by Julian Huxley and Arthur Tansley. These produced separate reports, appearing in 1947 as Cmd 7121 and 7122 respectively. Parallel action in Scotland provided Cmd 6631 and 7235 in the same year, and Cmd 7814 in 1949. From this dichotomy sprang the institutional separation of the statutory bodies covering, respectively, conservation of wildlife and natural features and the wider countryside objectives of landscape preservation and access. The partition of responsibility for different, but related, aspects of land use survived more than 40 years before legislation of the 1990s reversed it in Wales and Scotland. National parks were not accepted in Scotland at the time but, 50 years later, this designation is now considered appropriate for a few large areas which are of national importance for their outstanding natural heritage and for the opportunities they provide for public enjoyment (SNH 1998b).

The WLCSC was dominated by scientists and academics who, not unnaturally, saw research and education as the foundations of an effective conservation policy. Members made field visits to 56 sites; the Committee was also represented on a tour of the Swiss National Park and nature reserves, and took evidence on the US national park system. The vision of Cmd 7122 reflects these influences, calling for a national biological service with five main purposes: conservation, biological survey and research, experiment, education and amenity, for 'the peaceful contemplation of nature' (Huxley 1947: 19). These purposes would require the selection of suitable sites, under variable degrees of control. Six categories were proposed: National Nature Reserves; Conservation Areas (biological, physiographical, geological or landscape); National Parks; Geological Monuments (paralleling Ancient Monuments, for which model legislation already existed); Local Nature Reserves; and Local Educational Reserves. The service should be staffed by professionals, 'working on a sound long-term programme of research into the fundamental factors affecting wild life' (ibid.: 37) and be supervised by a Board, 'within the scientific organisation for which the Lord President of the Council is the responsible Minister.' In order to gain support from an existing body, the Committee recommended attachment to the Agricultural Research Council (ibid.: 55).

Some of WLCSC's views were new at the time. The introduction of the term 'conservation' was in itself a break with the prevailing tradition of nature 'preservation'. The Committee undertook a review of sites proposed for National Nature Reserves (NNR) in previous compilations, and produced a list of 73 for England and Wales (total estimated area approximately 70 000 acres). For these NNRs, the necessary safe-keeping would be satisfied either through Government custody (by compulsory purchase in the last resort) or through ownership by conservation bodies – at that time, represented only by the National Trust, the Corporation of the City of London, a couple of county naturalist trusts and the SPNR (Huxley 1947: 89). The role of human intervention in the creation and perpetuation of semi-natural environments was recognised: 'a conservation policy directed to maintaining any particular biological equilibrium entails constant vigilance and a fine-scale "management" of a kind comparable to the most highly developed farming' (ibid.: 21); also accepted was the futility of policies 'which are not widely understood and backed by public opinion' (ibid.: 41). These notions have a firm contemporary resonance.

The task facing the proposed institution was evaluated. Some serious future perturbations, resulting from introduced organisms, were unpredictable at that time (for example myxomatosis); others were lurking, their potential for damage unanticipated (Dutch elm disease, grey squirrels and feral muntjac). Only dimly discerned were the scale and impact of new agricultural practices. On farmland, it was appreciated that risk 'springs from efforts to bring under temporary cultivation an increasing quantity of marginal land, with the result that sites of the greatest scientific and cultural value are irrevocably destroyed'; 'developments in modern agricultural techniques and machinery are capable of producing very drastic changes in the landscape within short periods'; dangers 'follow upon ill-informed or indiscriminate destruction (for instance … by the improper use of insecticides) of species whose interrelations with others have not yet been ascertained'. Consideration of potential conflicts with landowners or occupiers dwelt chiefly on pest-control and sporting rights. The value of grazing in the maintenance of conservation interest was recognised, but the slow creep of neglect and its malign effects were not anticipated. Conflict with the policies of the Forestry Commission was foreseen.

3 The Nature Conservancy: Royal Charter to Research Council

The aspirations of Cmd 7122 were more than could be accepted in full by the Government of that time, which was uncertain about the principles and worried about the cost of the proposals. Notwithstanding, eight years after the Conference on Nature Preservation, the Nature Conservancy (NC) was eventually constituted with the status of a research council reporting to the Lord President, under a Royal Charter dated 23 March 1949 requiring it:

to provide scientific advice on the conservation and control of the natural flora and fauna of Great Britain; to establish, maintain and manage nature reserves in Great Britain, including the maintenance of physical features of scientific interest, and to organise and develop the research and scientific services related thereto.

Soon after its formation, further important functions were given to NC by the National Parks and Access to the Countryside Act 1949. These included the duty to notify 'sites of special interest' (SSSIs) to the local planning authority, and two useful powers, still operative: (S. 21) to join with local authorities in designating Local Nature Reserves (LNR), and (S. 16) to enter into management agreements with owners or managers of National Nature Reserves. Sir Arthur Tansley (at the age of 78) was appointed first Chairman. With him, there were five FRSs on the 18-person Council.

Initial Government funding was cautious: grant in aid of £100 000 for the first year, 1949–50 (equivalent to £2 million in 1998 terms). But within four years, finance became the main limiting factor to the NC's ambition for growth. *Plus ça change*! Only different, in the context of prevailing constraints, was the tactfully worded protest in the Annual Report 1953–54:

> The Nature Conservancy fully recognise the stringency imposed on the national budget by rearmament and other factors and wish, if they may be allowed to do so, to place on record their appreciation of the consistently helpful and under-standing attitude which Her Majesty's Government have shown. The Conservancy also fully appreciate that the potential value of their scientific, educational and Reserve management activities is as yet understood and accepted only by an informed minority. At this early stage there is still a need to demonstrate this value by tangible results and to bring it home to wider circles by simple and effective publications and illustrations.

Science was an important policy driver of the nascent Nature Conservancy, but the function of giving advice on management and control was also taken very seriously (Poore 1987). A Scientific Policy Committee assumed responsibility for determining the relative priorities in the work programme, and other matters 'likely to have more than a local bearing upon the protection and control of species and upon conservation generally' (NC 1953). An immediate priority was the establishment of research stations, with Merlewood and Furzebrook being the first, along with field stations on Nature Reserves, of which Moor House and Anacaun (Ben Eighe) were both operational by 1953 (NC 1954). Procedures were also set up to allocate research grants and studentships. Looking back, Ratcliffe (1977) identified research themes which contributed to fundamental scientific knowledge while also providing answers to practical problems:

- Floristic surveys of grasslands, building on the pioneering work of Tansley himself in the 1920s, provided a baseline from which to assess changes following the introduction of myxomatosis among rabbits in 1954, and to develop appropriate strategies to counter alterations in grazing regime by domestic stock;
- Participation in the International Biological Programme (IBP), aimed at studying the biological basis of productivity and human welfare, provided knowledge of the

natural limits to organic production and potential management processes for nature conservation;

- Grouse-moor research, prompted by concern among landowners and shooting tenants about decline in stock of Red Grouse, demonstrated the complexity of population dynamics;
- Studies beginning in the 1960s on the effects of pesticides (synthetic insecticides, herbicides and fungicides) in terrestrial, freshwater and marine habitats demonstrated the pervasive effects of pollutants, the complexity of pathways and the potentially catastrophic effects on vulnerable species – notably the top carnivores, raptorial birds and otters.

Myxomatosis was revealed as a dramatic tool for rabbit control in Australia in 1950, affected wild populations in France in 1952 and reached Britain in 1953 (Fenner and Ratcliffe 1965). The NC introduced monitoring programmes locally at first, more intensively in 1955 when it became clear that the reduction of rabbit grazing greatly affected the growth of herbaceous and woody plants on heaths, downs and other semi-natural grasslands. Allied research investigated transmission of the disease, and the effects on a predator, the buzzard. The disappearance of rabbits starkly revealed their role in maintaining semi-natural heath and grassland swards. Release of the sycamore as a woodland weed received less attention, but had a permanent impact across lowland Britain.

Pesticide research started in 1952. Experiments on roadside verges in 1953–54 demonstrated the effects of 2–4D, which was then in widespread use. In 1955, the Ministry of Transport forced agreement from NC that spraying could continue to be used as a tool where traffic hazards were perceived to be greatest (MoT 1955). In the same year, Government commissioned an enquiry, headed by Solly (later Lord) Zuckerman into persistent chemicals used in agriculture. Concern became acute by 1960, with clear evidence of bird deaths from seed corn dressed with aldrin, dieldrin and heptachlor, and consequent deaths of scavenging mammals and raptorial birds. Paradoxically, in 1959, racing pigeon interests had sought to exempt the peregrine from the special status given by the Protection of Birds Act 1954. In response to NC's advice that there was not enough up-to-date ornithological evidence available on the status of the species, the Home Office ordered an enquiry. By 1961 proof of a massive decline had emerged: in 338 known territories visited, peregrines were absent from 152 where they had once bred regularly, and only 69 pairs managed to rear young in the year. Alarmed by the evidence, Lord Shackleton raised the issue in Parliament on 25 April 1961. Other voices followed, but the agricultural lobby was strong. With restrained despair, the Director of Monks Wood wrote: 'The main worry caused by the persistent organochlorine insecticides is that they have polluted not only all of Britain, but the whole of the world' (Mellanby 1967).

The first organisational reform arose in 1965 when, on the advice of the Advisory Council on Scientific Policy, the Science and Technology Act combined the statutory bodies for life and earth sciences in a new Natural Environment Research Council (NERC) into which NC was absorbed. The loss of independence (and its separate Charter) roused anxiety among NC staff, but was supported 'as an act of faith' (NC 1965: 3). This confidence was justified to the extent that the NC retained a separate identity under its Council, continued to manage its stations and to undertake research relevant to its objectives, but nature conservation had to compete with funds for broader research into geology, oceanography and the Antarctic (Poore 1987).

Initial policy for establishing the series of National Nature Reserves (NNRs) closely followed the proposals of WLCSC: the 73 for England and Wales, plus 24 for Scotland, became the list for acquisition. Later, the adequacy of this series began to be questioned and in 1966 the Nature Conservation Review (NCR) was founded, to identify areas of national importance for nature conservation on a systematic basis, against the background of natural variation in wildlife and habitats. The results were not finally published until 1977, but the methodology has driven the selection of key sites for designation into the 1990s.

Conservation-oriented legislation over this period progressively added responsibilities. Under the Protection of Birds Act 1954, S. 10 (2) (b), NC was required to license taking or killing scheduled birds (or eggs) for scientific or educational purposes, and also to propose members of the Scientific Advisory Committees (NC 1955: 28–29). The Deer (Scotland) Act 1959, charged NC with the appointment of two members of the Red Deer Commission; by the Deer Act 1963, NC became the licensing authority for the capture or use of stupefying drugs for scientific purposes; by the Animals (Restriction of Importation) Act 1964, NC became the scientific authority. The Water Resources Act 1962 added further responsibilities, to meet which the work of the Hydrological Research Committee under Dr E. B. Worthington was extended – NC was still a Research Council! The Seals Act 1970 established a standing advisory committee, reporting annually to the Home Secretary. Although measures for the protection of badgers, culminating in the Badgers Act 1973, were driven more by welfare concerns than by conservation, NC (soon to become the NCC) was given a role as consultant in establishing special areas of protection and as the licensing authority for, among other things, capture for scientific purposes. Thus, piecemeal, the statutory body accumulated licensing, consenting, advisory or appointing duties in related legislation. However, these selective additions did not materially extend its remit; nor did they deliver the original broad vision of an integrated national nature conservation service. Indeed, the Parliamentary debates during the passage of the Nature Conservancy Council Act 1973 showed Government's resistance to this wider role, which would require new powers and far more generous funding (NCC 1984).

4 The Nature Conservancy Council and the Wildlife and Countryside Act 1981

The hot topic of 1971–72 for Government-funded research was Cmd 4814 (the Rothschild Report) advocating the separation of customer and contractor. There were also tensions for NC created by the promotional role, which was increasingly inappropriate for a research council. In evidence to Parliament, NC itself sought independence and, by the eponymous Act of 1973, the Nature Conservancy Council was constituted. In this separation, the new body regretfully relinquished the research stations so painstakingly built up over a quarter of a century; these remained with NERC. On the other hand, now directly sponsored by the Department of the Environment, the statutory agency was required to develop a positive relationship with Government and Ministers; it was also more clearly visible to the VCOs, and to Parliament.

The 1975 Conservation of Wild Creatures and Wild Plants Act extended species conservation, but site protection still relied on the 1949 Act. This Act, as shown above, had emerged as an adjunct to post-war town and country planning law, with the aim of ensuring the inclusion of considerations of nature conservation. Only planning authorities were required to be notified of SSSIs. The Act laid no legal obligation on owners or occupiers, and involved no restrictions on agriculture or forestry. Although these land uses in their traditional form had permitted the co-existence, and sometimes expansion, of characteristic wildlife important for conservation, practices had changed very substantially.

Worries about deterioration or loss of SSSIs first appeared in NC's 8th Annual Report, for 1956–57. Thenceforth, perceptions of the frailty of the designations system grew apace. In 1976, planning application for development affected 7% of SSSIs and 4% were actually damaged. An analysis in 1980 indicated that 8% of SSSIs suffered damage from all causes, of which agricultural improvement and cessation of traditional agricultural practices together formed just over half of the relevant areas; other major causes were fires, roads and pipelines, recreation and afforestation (NCC 1981: 18–19).

Early in its existence, NC had appreciated that the protection of designated sites depends on the participation of the managers of the land (or water) and, in 1953, initiated a process of communication with owners and occupiers of SSSIs. The large number of these people, and the effort required to trace them all and to prepare the precise information applicable to each site, was resource demanding. By the following year, considerably less that half those notified had replied; of these, most were favourable but a significant minority took exception. The level of opposition among landowners prompted a Parliamentary Question (28 June 1954) in response to which the Lord President gave assurances that, in all cases where tenants and proprietors could be traced, they would be consulted before further areas were notified to the local planning authorities.

With this experience, key provisions of the Wildlife and Countryside Act of 1981 therefore included the notification of owners and occupiers, placing on them the obligation to consult the NCC before undertaking potentially damaging operations (including specified changes in agricultural practices). Reflecting the informal system in existence, the Act required preliminary consultations before notification. When it became apparent that some landowners were prepared, wilfully, to damage or destroy the nature conservation interest during the consultation period, the 1985 amending Act introduced the present system, by which consultation follows notification, leading within nine months to confirmation (or otherwise).

The 1981 Act provided the NCC with a range of optional responses, including the offer of management agreements based on the concept of profits foregone. The emphasis was on the development of a voluntary system, engaging the co-operation of land user and agency to promote conservation benefits. It also roused new expectations for resources. Although the source has never been confirmed, it was believed that the Secretary of State had given assurances that compensatory management agreements would be fully funded. Lord Chelwood told the House of Lords that NCC's financial memorandum had suggested that the extra cost 'might be of the order of £600 000 to £700 000 on average per year' (*Hansard* 30 March 1981, col. 102).

Although NCC's structure and statutory position were unaltered, the new remit had profound effects. In Wilkinson's words:

Whilst the new legislation made significant gains in the protection of habitats it also brought about a radical shift in the balance of NCC's work, tending to skew it towards the administrative aspects of notification and away from wider countryside and more outgoing initiatives. Greater interest in our work from Departments and politicians meant greater intervention and control. Greater recognition, however, has also brought larger budgets, allowing NCC to expand to meet its tasks. The Somerset Levels in England, the Berwyns in Wales and Duich Moss in Scotland have highlighted both the effectiveness of the Wildlife and Countryside Act and anxieties about its application. Despite its faults and the cumbersome processes leading to delays, I believe the Act stems from a commendable attempt to be fair. In my view the Act has largely been successful though considerable problems remain, not least those of resources. In Scotland, however, attitudes towards nature conservation have been less supportive. The voluntary movement is less numerous and it has been convenient for some to blame a Great Britain body, with its headquarters in Peterborough, for the impact of this national legislation. Critics have to realise that if this Act is applied properly and conscientiously as we have sought to do, no amount of devolution will alter its impacts and some will continue to complain.

(NCC 1990: 9)

After 1973, NCC had redeveloped its scientific capability for research and survey for nature conservation needs, managed through a large science directorate. Dr D.A. Ratcliffe, who originally joined NC in 1956, served continuously as NCC's Chief Scientist from 1973 to 1989, and made an influential personal contribution to national nature conservation policies. During his last two years at the helm, NCC allocated one-seventh of its grant-in-aid to scientific support (excluding staff costs, grants and other operating charges) (NCC 1990: 133).

The long-running battle with forestry anticipated in Cmd 7122 culminated in these years. Conflict stemmed from the obligations of the Forestry Commission (FC) to increase the supply of home-grown timber and enlarge the forested estate, and NC/NCC's weakness in defence of areas of importance for nature conservation. Lowland wildlife sites were vulnerable through the felling of ancient or semi-natural broadleaf woods and replacement with conifer or mixed species, and through new plantations of largely non-native conifers on heath and moorland. The NCC set out its priorities in 1986 (NCC 1986) and, in the same year, an initial concordat was negotiated with the FC. Controversy was not stilled on open moor and hill, reaching epic proportions when, heavily subsidised by personal tax allowances favouring the highest earners among society, plantation threatened the huge open landscapes of the peatlands of northern Scotland (Ratcliffe and Oswald 1988).

Then, suddenly, in his budget of March 1988, the Chancellor (Nigel Lawson) removed commercial woodlands entirely from the scope of income tax and corporation tax, so that the expenses of planting and maintaining trees for timber production were no longer deductible (FC 1988). A new Woodland Grant was developed, to include guidelines on nature conservation and freshwater protection worked out in conjunction with NCC (FC 1990). Over the next few years, the FC reviewed its role, increasing its focus on wildlife stewardship (FC 1994). As the regulatory and development arms of FC progressively separated, nature conservation has remained a significant concern for both sectors. In 1998, it was the declared objective of Forest Enterprise (FE) 'to enhance the nature conservation value of the national forests as a whole and to safe-guard special habitats.' All 388 SSSIs situated on the FE estate were being managed in accordance with plans endorsed under memoranda of agreement or statements of intent signed in 1997 with the Countryside Council for Wales (CCW), English Nature (EN) or Scottish Natural Heritage (SNH), and £1.7M from their own resources were allocated towards the £5M New Forest LIFE programme (FE 1998).

At the close of this phase, in search of the elusive grand vision and backed by col-leagues on the Royal Commission on Environmental Pollution, on 22 November 1989 I presented the Environment Protection Bill in the House of Lords. Modelled on health and safety legislation, this showed how an integrated scheme could be provided on the foundation of a general duty of environmental protection and conservation falling on employers, employees and persons engaged in potentially damaging activities. How-ever, events were moving fast and, after a good Second Reading debate in the new year, I withdrew the Bill in deference to its near homonym.

5 The Environmental Protection Act 1990: Country Agencies

Scottish law had required separate enactment of nature conservation legislation in 1949, and subsequently. From the start, within NC the special standing of Scotland was recognised administratively with the appointment of a Director Scotland, and a Scottish Committee with special status laid down by charter. The initial England and Wales Committee was soon divided into one for each country. The NCC was similarly organised, with a strong headquarters including a Chief Scientist Directorate, and three country administrations, each led by a director and guided by a governing committee. Operational delivery was through a regional structure: by 1989–90, there were eight regions in England, four in Scotland and three in Wales.

Devolution was taken a stage further by the announcement by the Secretaries of State for the Environment (Nicholas Ridley), Scotland (Malcolm Rifkind) and Wales (Peter Walker), in July 1989, that three independent country agencies would be cre-ated. This decision prompted the resignation of one NCC member (Lord Buxton of Alsa) and – by his own admission – angered the chairman who summarised his objec-tions, widely shared by VCOs and many environmental scientists, in a black-covered annual report (NCC 1990). Responding to such opinions, the House of Lords Select Committee on Science and Technology promoted an enquiry, chaired by FM the Lord Carver, whose report (House of Lords 1990) was influential in persuading Govern-ment to accept amendments to the Bill. In effect, the 1990 Act did not alter the princi-ples of existing nature conservation legislation (as contained in the 1981 Wildlife and Countryside Act) but did create three country Councils. Each was charged with the full range of functions formerly held by NCC within its own area, but constrained to exercise jointly those pertaining to Great Britain as a whole and to international nature conservation obligations through a Joint Nature Conservation Committee (JNCC). This was to be presided over by an independent chairman, and include among its members three other independents: the Chairman of the Countryside Com-mission (CC) (now reduced to England-only scope) and – for the first time bringing in opinion from the Province – two from the Northern Ireland Council for Nature Con-servation and the Countryside.

Despite his reservations, Sir William Wilkinson agreed to extend his term with the NCC for the final year, and was joined on the Council by the three chairmen designate: Michael Griffith, as chairman of the Committee for Wales, Magnus Magnusson KBE in that role for Scotland and myself for England. The new bodies formally came into existence in late 1990, when the chairmen took office and set about recruitment of staff, first of whom were the three Chief Executives: Ian Mercer (Wales), Roger Crofts (Scotland) and Derek Langslow (England). The next few months were hectic, forging new organisations while working with the Director General of NCC, Timothy Hornsby, to ensure that there were no gaps in the effectiveness of the old.

The JNCC staff were assigned from the three country agencies, among them senior scientists involved in continuing important conservation science and survey programmes of Great Britain-wide scope, such as the Marine Nature Conservation Review (Hill *et al.* 1996) and the Geological Conservation Review (Ellis *et al.* 1996). Externally, there were perceptions that JNCC somehow did, or should, function to perpetuate characteristics and traditions of NCC. Uncomfortable with his remit, the first Chairman, Sir Fred Holliday (a former Chairman of NCC) resigned after five months in September 1991. He was succeeded by the Earl of Selborne KBE FRS, under whom the Committee's strategy was reviewed and its role clarified. The third Chairman of JNCC, Sir Angus Stirling, was appointed in 1997.

The Countryside Council for Wales (CCW) was charged with a dual mandate, inheriting unaltered the functions of the Countryside Commission in Wales along with those of NCC. This combined role is expressed in the agency's statement in 1998: 'our long term mission is for a Welsh countryside and coast which is rich in wildlife and landscape and sustained, through care, for future generations to enjoy'. The Nature Conservancy Council for Scotland had a short life of one year; alone among the new agencies, it was required to establish a review committee to advise on appeals against SSSI notification. In 1992, following a further Scottish enactment, it was replaced by Scottish Natural Heritage (SNH), which inherited the review committee. SNH absorbed the Countryside Commission for Scotland. Its combined countryside and nature conservation remit includes obligations towards sustainable development:

> Our task is to secure the conservation and enhancement of Scotland's unique and precious natural heritage – the wildlife, the habitats and the landscapes which have evolved in Scotland through the long partnership between people and nature. We advise on policies and promote projects that aim to improve the natural heritage and support its sustainable use. Our aim is to help people enjoy Scotland's natural heritage responsibly, understand it more fully and use it wisely so that it can be sustained for future generations.
>
> (SNH 1998a)

In January 1994, as a sequel to the launch of the UK Sustainable Development Strategy (Secretaries of State 1994) and Biodiversity Action Plan (BAP), the Secretary of State for the Environment, John Gummer, initiated a six-month study to consider the implications of merger between English Nature and the Countryside Commission (DoE 1994a). The conclusions were adverse, and the proposal rejected (DoE 1994b; see also *English Nature and Countryside Commission*, Commons *Hansard*, 30 November 1994, col. 1202–3). With the amalgamation of the Countryside Commission (CC) and the Rural Development Commission on 1 April 1999 to form the Countryside Agency, EN has been confirmed as the sole single-function statutory nature conservation agency in UK.

At the start of 1999 Michael Griffith and Magnus Magnusson remained in office but, from 1 April 1998, the chair of EN was held by Barbara Baroness Young of Old Scone, until her resignation in November, 2000.

6 Voluntary movement

Much attention was given in NC's early years to fostering VCOs who were seen, then as now, as an essential complement to the statutory agency (Poore 1987). There are many instances of close personal and institutional relations between the statutory bodies and the voluntary conservation movement. It was, for instance, at a meeting chaired by Lord Hurcomb, NC's chairman, that the World Wildlife Fund (WWF) was set up in 1961.

Prominent among these parallel setters of national conservation policy has been SPNR, which in 1976 became the Society for the Promotion of Nature Conservation and later, in 1981, the Royal Society for Nature Conservation (RSNC), since 1991 joined wjth 46 local trusts as the Wildlife Trusts (S. Lyster, pers. comm.) The first local trust was the Norfolk Naturalists (1926). There followed Yorkshire (1946), Lincolnshire (1948), Cambridgeshire, Leicestershire and the West Midlands by the late 1950s and, after the first biennial conference in 1960, others with quickening pace; by 1962, 27 trusts existed in England and (south) Wales, with seven more under discussion. Fostered by Max Nicholson, NC's long-serving Director General, the relationship was always strong, emphasising the respect held by the statutory body for the conservation management capacity of the voluntary movement. In 1954 Woodwalton Fen, the first of SPNR's reserves (acquired through the generosity of Christopher Cadbury, another key figure behind the growth of the Trusts) was leased to NC; Hickling Broad and Weeting Heath were early NNRs by agreement with the Norfolk Trust. At the Trusts' second biennial Conference held in Norwich (May 1962) shared working arrangements for SSSI protection were agreed. Subsequently, at a critical stage in their expansion, NCC grants enabled Trusts to become more professional, to appoint conservation officers, as well as education and marketing staff, and thereby to build on the voluntary effort (R. Crane, pers. comm.). A new relationship began in 1991 with EN's Reserves Enhancement scheme which, in 1997–98 supported 30 Wildlife Trusts (and three other voluntary organisations) to enhance the management and promote public enjoyment of SSSI reserves in their care (EN 1998).

By 1998, there were 46 local Wildlife Trusts covering the UK and Isle of Man, claiming 3100 members, and managing some 2300 nature reserves covering almost 200 000 acres. Wildlife Trusts secured £35M from the Heritage Lottery Fund, partly to buy new reserves or extend existing ones, but mainly to fund a five-year programme of capital improvements to existing reserves and have benefited from some £6 million worth of projects under the Landfill Tax Credit Scheme (S. Lyster, pers. comm.). Local co-operative initiatives abound, linking a wildlife trust, a statutory body and third-party interests as joint funders, for instance in Cumbrian action plans for mountain massifs (Soane 1997).

With an even longer history, the Royal Society for the Protection of Birds (RSPB) has played a role no less significant. Birds have had an important role in the history of conservation, and specific legislative measures ran from the the the first Birds Protection Act of 1869 to the most recent of 1967. RSPB Council contributed to the first nature reserves list formulated by the NRIC, appearing in Cmd 7122. In the 1950s, RSPB significantly addressed the issues of egg collecting (at that time a popular hobby), falconry and persecution of birds of prey. NC and, later, NCC supported an influential Advisory Committee on Birds. Alongside this committee, RSPB was party to the deliberations culminating in the epoch-making 1954 Protection of Birds Act. RSPB membership grew phenomenally from the 1960s, topping 1 million (with Young Ornithologists' Club members) by 1998; the acquisition of reserves, many of them SSSI, shows a parallel steep rise. With this backing, RSPB has developed unmatched skills at presenting the conservation case, both to the public and to Parliamentarians.

Co-operation with NCC produced *Red Data Birds in Britain* and the collaborative *Birds of Conservation Concern*. For 30 years, RSPB has been involved in a 10-yearly census of UK breeding seabirds (Lloyd *et al.* 1991), latterly in co-operation with JNCC's Seabird 2000. In the 1990s, RSPB took the lead for 24 bird species (and the medicinal leech and stinking hawk's-beard) in the UK Biodiversity Action Plan. Among the voluntary bodies, RSPB has been a leader in monitoring UK responses to international conventions and European Directives (see below) and admonishing policy failure where this has been detected: on CITES (Gammell 1977; RSPB 1992); Ramsar (Williams 1980); the Birds Directive (Gammell 1987; Pritchard 1995); Habitats and Species Directive (Housden *et al.* 1991; O'Sullivan *et al.* 1993); Convention on Biological Diversity (Avery 1995; Wynne *et al.* 1995). The legal challenge mounted by RSPB over the development of Lappel Bank SPA established case law (Croner 1997).

7 Parliament and nature conservation

Although NC and NCC were obliged to submit annual reports to Government and Parliament, none of these important and fascinating documents has, to my surprise, been preserved in the archives of the present Department of Environment, Transport and the Regions. The Parliamentary libraries are almost as bare, with nothing available earlier than NCC's report for 1987 (G. R. Dymond, pers. comm.), although there is a complete collection for subsequent years. Only in the library of English Nature was I able to consult the full series.

Parliamentarians appointed to Council have provided a conduit for the expression of NC/NCC's concerns. A good example was Lord Hurcomb, a long-serving chairman of NC. As the result of amendments introduced by him, the first usage of the terms 'having regard to' and 'shall take into account' (in relation to environmental protection or nature conservation) are found in the Electricity Act 1957, setting a precedent for much future legislation.

Parliamentary action for nature conservation has not been lacking among MPs, who have over the years been assiduous in asking written and oral questions, and raising issues in debate. The first of many House of Commons Select Committee enquiries was held in 1957, but I can find only one full debate on an annual report – in the Lords, on 17 February 1988. With freer procedures for Private Members' Bills, Lords have also served to test the legislative will of Parliament and Government. Thus, after five years deliberation by two national Committees (in which NC was involved, together with the RSPB and other voluntary bodies), a first version of the great revising and consolidating Protection of Birds Bill was taken through the Lords in 1953 by Viscount Templewood. With the benefit of this trial, in the subsequent Session it was reintroduced with some modification to the Commons by Lady Tweedsmuir MP, passed, and felicitously moved in the Upper House by her husband (*Hansard*, 15 April, 1954); here it received 40 amendments, but ultimately passed into law, establishing the principle of 'reverse listing'.

Throughout the 1970s, conservation legislation continued to be brought forward by working alliances of voluntary organisations, responsive Parliamentarians and the statutory body. The focus remained largely national, however; for example, concerned by the decline in rare plants documented by its countrywide survey, the Botanical Society of the British Isles, with the Council for Nature and RSNC, briefed Lord Beaumont of Whitley who introduced a Wild Plants Protection Bill in January 1974, and again (the Bill having been lost by the election that brought in the previous Labour administration) in November of the same year. In parallel, informed by his personal connections with NCC, the Mammal Society and the British Herpetological Society among others, my late father introduced the Conservation of Wild Creatures Bill (the second time round, on 21 November 1974, Lord Cranbrook was absent through illness and the Bill was reintroduced on his behalf by Lord Wynne-Jones), affording protection to non-avian animals considered to be endangered in Britain as a result of human activity, especially collectors; the schedule (which NCC was given powers to amend) was short, comprising the greater horseshoe bat, mouse-eared bat, dormouse, sand lizard, smooth snake, natterjack toad and large blue butterfly, all hitherto unprotected. After passing in the Lords, these two Bills were amalgamated, adopted by Peter Hardy MP (now Lord Hardy of Wath) who had achieved a high place in the House of Commons ballot for Private Members' Bills, and passed through both Houses with Government support as the Conservation of Wild Creatures and Wild Plants Act 1975. In the last year of his life, prompted by concern for a wider variety of species, Lord Cranbrook returned with an amending Bill to the 1975 Act, which (after further amendment in the process) was passed by the Lords in May 1978. It was not taken up in the Commons, so – with customary persistence – he re-introduced it at the start of the next Session. He died on 22 November 1978, after the First Reading, when the date for the more significant Second Reading was already set. I was encouraged to apply for a Writ and take my seat promptly, and (uniquely I guess) thereby made my maiden speech to my parent's Bill – an excessively moving occasion, on which other Lords paid recognition to his lifetime's contribution. The Bill on that occasion was introduced by Lord Skelmersdale, who subsequently withdrew it in the face of assurances by the Government of more comprehensive legislation within a short time.

This responsibility fell on the incoming Conservative administration. In 1980, consultative papers were issued, and in December the Wildlife and Countryside Bill was introduced to the Lords by Lord Bellwin, who remarked:

> It is something of a break with tradition for the Government to introduce legislation of this kind on subjects which have been well cared for in the past by Private Member's Bills. The cause of nature conservation has been well served in the past, by the efforts of voluntary organisations and their parliamentary supporters.
>
> (*Hansard*, 16 December 1980, col. 983)

Government was unprepared for the level of controversy that the measures of the Bill would raise. The Second Reading debate highlighted the concerns of all interests. By the end of March 1981, when finally passed by the Lords, the Bill had broken all previous records in terms of numbers of amendments and hours of debate; it had yet to go through the Commons! The debates on marine nature reserves, spearheaded by Lord Craigton, were prolonged and hard fought. So, too, were those for the protection of bats, from which one of my amendments (to safeguard bats in roof spaces) was lampooned in the Australian press as the most ridiculous legislative proposal of the year. Fortunately, both measures were taken on board by Government amendments in the Other Place.

In 1989, as already noted, the proposal by Government to create three new country agencies roused great concern, focusing on the need to preserve the strong science base of NCC and the capacity to give advice on Great Britain-wide and international conservation issues. Parliamentarians responded promptly. At Westminster, the All Party Conservation Group of Both Houses held a special meeting on 9 November 1989, addressed by representatives of RSNC, RSPB, WWF, the Council for the Protection of Rural England, CC and the Countryside Commission for Scotland. The House of Lords Select Committee on Science and Technology (at that time there was no such body in the Commons) staged an enquiry and, with support from all quarters of the House, Lord Carver (chairman of the committee's enquiry) introduced amendments to the Bill requiring the establishment of the JNCC, ultimately accepted by Government.

8 International Conventions and European Directives

The impact of the plumage trade on wild populations had been a main concern in the pioneering British legislation for bird protection. Extended among the international community, to cover trade in living animals of all kinds or animal parts (skins, feathers and so on), these considerations led to the 1973 Convention on International Trade in Endangered Species (CITES). The Government initially believed that existing national law sufficed to allow UK to ratify, but was ultimately persuaded that this was not the case. The 1976 Act was thus first in a new series of conservation measures required for the implementation of international conventions.

Other international instruments had greater direct influence on domestic nature conservation policies. First among these was the 1971 Convention on Wetlands of International Importance especially as Waterfowl Habitat, commonly known (after the place where it was negotiated) as the Ramsar Convention and signed by the UK in 1973 (Williams 1980). The Convention lays down standards, but designation ('listing') is optional. In the UK this became the responsibility of the territorial Secretaries of State; the statutory body identifies sites that meet the criteria, consults affected owners, occupiers and local authorities and proposes to Departments (DETR for England). The UK Government has required that Ramsar sites must also be SSSI; the additional protection afforded to listed sites is detailed in Planning Policy Guidance Note 9 (Nature Conservation) 1994. After an initial burst of enthusiasm in 1976 (including, in East Anglia, Hickling Broad and Bure Marshes, and the Ouse Washes), listing proceeded slowly until the 1990s; but by the close of the decade there were 59 Ramsar sites in England, some of them composite, several very large: Morecambe Bay 39 759 ha, New Forest 28 001 ha, Upper Solway Flats and Marshes 29 950 ha (EN 1998).

The Bonn Convention on the Conservation of Migratory Species of Wild Animals (1979), which requires concerted action by states within whose borders there are threatened populations of migratory species, had limited impact on UK policy. The Berne Convention on the Conservation of European Wildlife and Natural Habitats, drawn up by the Council of Europe in 1979, had more significance. Although the UK became an adherent to both Conventions, the obligations thereby assumed could not be backed by enforceable legal sanctions, as are instruments of the European Union.

The UK accession to the Treaty of Rome in 1973 coincided with the emergence of the EC's environmental policy. Since there were no original environmental clauses in the Treaty, the means of developing this policy lay through items of Community legislation which member states were then obliged to implement. The principal EC instruments are the Regulation, which is directly applicable law in member states, and the Directive, which is binding as to the results to be achieved, but leaves member states discretion in the form by which it is incorporated into national law. The Directive on the Conservation of Wild Birds was proposed in 1976, notified in 1979 and required formal compliance by 6 April 1981 (Haigh 1987). Thus it was that, in 1981, while the Wildlife and Countryside Bill was intended, in part, to tackle the glaring weaknesses of site protection under the 1949 domestic legislation, the stimulus for bringing measures forward at that particular time was the requirement to implement the EC Directive on Wild Birds. The passage of the Act also permitted Government to ratify the Berne and Bonn Conventions.

Implementation of the Birds Directive requires the preservation of a sufficient diversity of habitats for all naturally occuring species of wild birds, coupled with the designation of Special Protection Areas (SPAS) to conserve the habitats of certain rare species and migratory species. The EC subsequently agreed the more rigorous 1992 Directive on the conservation of natural habitats and of wild fauna and flora (Habitats and Species Directive). This aims at establishing 'favourable conservation status' for habitat types and species of European interest, listed in detailed annexes, by the designation of Special Areas of Conservation (SACs). The UK transposed the Directive into national law by the Conservation (Natural Habitats etc.) Regulations 1994 (Hughes 1994), without the careful consideration of primary legislation. As a Great Britain-wide obligation, the JNCC has been responsible for transferring the Directive's objectives to the circumstances of Great Britain. The country agencies share responsibility for selection, carry the duty of consultation and have advised the Secretaries of State on designation. Together, these two Directives will result in the establishment of a coherent European ecological network of sites, to be known as Natura 2000. The first implementation report by member states was due by 5 June 2000; thereafter, the EU Commission will review the series and report by 5 June 2002.

In parallel, UK has ratified the Convention on Biological Diversity, agreed at Rio in 1992. Implementation is through the domestic UK Biodiversity Action Plan, which has statutory force. Targeted are nationally endangered species and vulnerable habitats (cf. Tither 1998). Derek Langslow was the first chairman of the targets group, responsible for the preparation of Action Plans. By the end of the process (anticipated in 1999), some 400 Species Action Plans and 38 Habitat Action Plans will have been published.

9 The new millennium

Since 1991, the different remits and separate geographical responsibilities of the country agencies have encouraged each to develop its own character and distinctive approach. In many programme areas, new ventures have grown from the sound foundation of NCC undertakings; examples are EN's Species Recovery Programme, initiated in April 1991 but presaged by the work of Whitten (1990), the Wildlife Enhancement Scheme initiated in 1992, and Natural Areas (to show the importance of wildlife everywhere, and emphasise its local character), which joined forces with CC in the development of the Countryside Character Areas. Fulfilling the percipience of Cmd 7122, increasing use has been made of the provisions allowing declaration as NNR of land held and managed by 'approved bodies' – now including the enlarged army of Wildlife Trusts, National Trust, RSPB, Forestry Commission, local authorities and, in England, one PLC. In the seven years to 31 March 1998, the number of English NNRs rose from 128 to 191 (of which 43 are wholly or partially owned and managed by approved bodies), and the area from 42 270 ha to 73 374 ha (Marten 1994; EN 1998). Section 21 of the 1949 Act has also continued to prove its value, with the number of declared LNRs in England rising from 222 (11 011 ha) in 1991 to 598 (29 032 ha) in 1998 (EN 1998).

Each country agency was faced with the task of completion of the notification and renotification process required by the 1981 Act, under inherited Guidelines (safeguarded by the JNCC). The Council of EN has reserved to itself the duties of notification and of confirmation when unresolved objections are outstanding, while CCW and SNH adopted delegated procedures allowed under the 1990 Act. In England, by 1998 there was a total of 3987 SSSIs notified, covering 967 365 ha and involving 32 000 owners and occupiers. The programme was approaching the end of this active phase, with the full series of sites selected through the Geological Conservation Review and all but two of 27 rivers notified (EN 1998).

For SSSIs existing in 1991, the agencies inherited from NCC a portfolio of 'profit foregone' agreements exercised under the 1981 Act. While it is a travesty to represent all of these as payment for doing nothing, the high costs and relatively poor value for money of some agreements attracted media attention. Determined moves have shifted the balance. In 1997–98, 84% of management agreements issued by English Nature were 'positive' (including the Wildlife Enhancement Scheme); yet 'compensatory' agreements, 16% by number, still absorbed 30% of total costs (EN 1998). Moreover, regrettably, despite improved relations with owners and occupiers achieved by all country agencies and the successful new positive management schemes, closer monitoring programmes now in place have shown continued decline in nature conservation features of SSSIs. In England, statistics for overall 'special' feature condition in 1997–98 showed that 57% were favourable, 15% unfavourable but improving, 16% unfavourable with no change, and 12% unfavourable and declining; 0.09% of the resource had been destroyed, and 0.51% partly destroyed. Of the damaging activities, development (overwhelmingly, the continued extraction of peat under planning permission) accounted for 8% of the area damaged; the remainder suffered chiefly from neglect (some lowland heaths, fens and mires) or agricultural activities (predominantly overgrazing of upland sites). In both cases, the condition remains reversible with appropriate land use changes (EN 1998).

Statistics such as these have been brought to public attention by VCOs, taken up by the media and aired in Parliament. In 1995, having won a place in the ballot and obtained cross-party support, James Couchman MP introduced a Wildlife Bill which, among other things, would have strengthened the provisions of Nature Conservation Orders, prohibited purely compensatory management agreements, introduced provisions for Restoration Orders and required the notification of statutory undertakers. After amendment in Committee, the Bill was passed, received a first Reading in the Lords but failed thereafter for lack of time. A new version of a Wildlife Bill, again with cross-party support, received a first Reading in the Commons on 3 November 1998, but was lost with the close of the session. Further Private Member's aspirations for nationwide legislation have now been largely met by The Countryside and Rights of Way Act 2000.

The 1997 Manifesto of the incoming Labour Administration contained a commitment to strengthen the protection of Britain's wildlife. In September 1998, DETR (acting for England and Wales) issued a consultation document in which the Environment Ministers for both territories in their joint foreword expressed the 'need to ensure that the statutory framework both encourages positive action, and properly empowers the conservation agencies in identifying *and protecting* areas of national conservation importance' (DETR 1998; emphasis in original). Anticipating devolution, the Scottish Office set up a Land Reform Policy Group in October 1997, and in 1998 declared its intention to reform the SSSI system, as outlined in a separate consultative document (Scottish Office, undated but 1998); the intention was elaborated by the Secretary of State for Scotland, speaking beside Loch Lomond on 2 February 1999, in the form of proposals since put to the Scottish Parliament.

Both consultative documents acknowledged the requirement of European law that the boundaries of Natura 2000 sites (that is, SPAs under the Birds Directive and SACs under the Habitats and Species Directive) should be based on scientific grounds only. Both also reassert existing UK policy that terrestrial and coastal sites should be underpinned by SSSI designation, although it now appears uncertain that SSSI criteria will remain common among the components of a devolved UK.

It will evidently be important that the scheme of inter-administration concordats envisaged in devolution legislation should work to ensure that standards for the selection of EU sites, and for the subsequent maintenance of good ecological condition, will be applied uniformly across the UK, including Northern Ireland. The test will come early in the twenty-first century, when the European Commission begins its validation exercise. Meanwhile, the House of Lords Select Committee on the European Communities, Subcommittee C (Environment, Public Health and Consumer Affairs) has conducted an enquiry into the implementation of the Habitats and Species Directive across the EU (House of Lords 1999, 2000).

The UK Biodiversity Action Plan involves local and central government, the statutory agencies, VCOs and industry and commerce in a concerted attempt to halt the decline in biodiversity in the UK, and to recoup those past losses that are still recoverable. The Plan is a binding national objective, which has fully engaged the VCOs (Wynne *et al.* 1995) and will shape and drive the work of the statutory agencies in Britain and Northern Ireland. Co-ordinated Local BAPs are emerging throughout the country. In England, the regional development agencies and emerging regional administrations will need to take account of these national objectives for nature conservation, implemented through Local BAPs (RTSD 1999).

There has hardly been a year when some friend of nature conservation has not pointed to the inadequacy of Government grant-in-aid. Additional resources, and relaxation of controls on running costs, formed the basis of the first recommendation of the Environment, Transport and Regional Affairs Committee report on English Nature (House of Commons 1998), to which the Government has responded with supplemetary funds for 1999–2001. Fortunately, other sources have proliferated and, after initial hesitation, the statutory bodies have formed appropriate partnerships and been successful applicants: for example, in 1997–98 EN established the £18 million Tomorrow's Heathland Heritage programme with Heritage Lottery Fund support, and led a £4 million joint project for work on marine candidate Special Areas of Conservation (cSACs), half funded by the EU LIFE programme (EN 1998).

The legislation in 1949 recognised nature conservation as a public good, appropriate for government funding, for the benefit of people and not exclusively for wildlife. Implicit in this approach is belief that the natural world (including biodiversity and natural features) is important functionally and aesthetically to current and future generations. In this sense, nature conservation as an activity, and as a set of personal and organisational beliefs, is now more widely endorsed by the public and by Government than ever before. But the pace and scale of environmental change in Britain have accelerated.

The statutory agencies need the ultimate back-up of legal sanction, but the operation of law has shown itself to be an insufficient tool. Legislation did not prevent the ravages of Dutch elm disease, nor the spread of grey squirrels, feral mink or muntjac; it has not halted the deterioration of habitats through legitimate activities such as land drainage and water abstraction; cannot compel the reoccupation of abandoned or under-used land, or the destocking of over-exploited pastures for the benefit of nature conservation; and has not controlled pervasive pollution, such as atmospheric acidification.

The nation continues to need the parallel commitment of confident VCOs and persistent Parliamentarians. European and international intitiatives have taken the lead, and citizens must find routes to influence these remoter institutions while also engaging with the closer devolved and regional administrations that will dictate local solutions. Contemporary policies place biodiversity firmly in the context of sustainable development. The present generation must strive to safeguard this precious nature conservation resource for the benefit of all to come.

The succeeding contributions in this volume delineate the challenge and evaluate the means of attaining this objective.

ACKNOWLEDGEMENTS

I am grateful to librarians at English Nature, Peterborough, and the House of Lords, who have skilfully found key literature. Derek Langslow kindly read a draft text and commented from the depth of his experience; Jennifer Beamish, Chris Fancy, Simon Lyster, Robin Crane and Tim Sands have provided helpful notes.

REFERENCES

Adams, W.M. (1986) *Nature's Place: Conservation Sites and Countryside Change*, London: Allen & Unwin.

Avery, M., (ed.) (1995) *Developing Species and Habitat Action Plans*, Biodiversity Challenge Group.

Beamish. J. (1998) *A history of SPNR-SPNC-RSNC-The Wildlife Trusts: The last fifty years*, unpublished research brief.

Croner (1997) *Croner's Environmental Management Case Law: Special Report*, Issue 5.

DETR (1998) *Sites of Special Scientific Interest: Better Protection and Management. A consultation document for England and Wales*, London: Department of the Environment, Transport and the Regions.

DoE (1994a) *Bringing together the Countryside Commission and English Nature*, Department of the Environment, news release, 27 January 1994.

DoE (1994b) *Countryside Commission and English Nature: New working arrangements announced*, Department of the Environment, news release, 7 October 1994.

Dower, J. (1945) *National Parks in England and Wales*, Cmd 6628, London: HMSO.

Ellis, N.V., Boffin, D.Q., Campbell, S., Knill, J.L., McKirdy, A.P., Prosser, C.D., Vincent, M.A. and Wilson, R.C.L. (1996) *An Introduction to the Geological Conservation Review*, Peterborough: Joint Nature Conservation Committee.

EN (1998) *English Nature 7th Report*, Peterborough: English Nature.

Evans, D. (1997) *A History of Nature Conservation in Britain*, 2nd edn, London and New York: Routledge.

Fenner, R. and Ratcliffe, F. N. (1965) *Myxomatosis*, Cambridge: Cambridge University Press.

FC (1988) *Forestry Commission 68th Annual Report and Accounts 1987–1988*, Edinburgh: HMSO.

FC (1990) *Forestry Commission 70th Annual Report and Accounts for the year ended 31 March 1990*, London: HMSO.

FC (1994) *Forestry Commission 73rd Annual Report and Accounts for the year ended 31 March 1993*, London: HMSO.

FE (1998) *Forest Enterprise Annual Report and Accounts 1997–98*, Edinburgh: The Stationery Office.

Gammell, A. (1977) *Certain Aspects of the Application of the CITES of Wild Fauna and Flora*, Sandy: Royal Society for the Protection of Birds.

Gammell, A. (1987) *Manual on European Council Directives on the Conservation of Wild Birds*, Brussels: European Environmental Bureau.

Haigh, N. (1987) *EEC Environmental Policy and Britain*, 2nd edn, London: Longman.

Hill, T.O., Emblow, C.S. and Northen, K.O. (1996) *Marine Nature Conservation Review: Sector 6. Inlets in eastern England: area summaries*, Peterborough, Joint Nature Conservation Committee (Coasts and Seas of the United Kingdom – MNCR series).

Hobhouse, A. (1947) *National Parks Committee*, Cmd 7121, London: HMSO.

Housden, S., Thomas, G., Bibby, C. and Porter, R. (1991). Towards a habitat conservation strategy for bird habitats in Britain, *RSPB Conservation Review* 5, 9–16.

House of Commons (1998) *English Nature*. Environment, Transport and Regional Affairs Committee, Session 1997–98, 9th report. HC 790, London: The Stationery Office.

House of Lords (1990) *Nature Conservancy Council*. Select Committee on Science and Technology. Session 1989–90, 2nd Report. HL Paper 33, London: HMSO.

House of Lords (1999) *Biodiversity in the European Union: interim report: United Kingdom measures*. Select Committee on European Communities, Session 1998–9, 18th Report. HL Paper 100. London, The Stationery Office.

House of Lords (2000) *Biodiversity in the European Union. Final Report: International Issues.* Select Committee on European Communities, Session 1998–9, 22nd Report. HL Paper 119. London, The Stationery Office.

Hughes, P. (1994) *The Habitats Directive and the UK Conservation Framework and SSSI System*, House of Commons Library, Research Paper 94/90.

Huxley, J.S. (1947) *Conservation of Nature in England and Wales*, Ministry of Town and Country Planning. Cmd 7122, London: HMSO.

Lloyd, C., Tasker, M. and Partridge, K. (1991) *The Status of Seabirds in Britian and Ireland*, London: Poyser.

Lyster, S. (1998) Unpublished notes on the Wildlife Trusts.

Marren, P. (1994) *England's National Nature Reserves*, London: T. and D. A. Poyser.

Mellanby, K. (1967) *Pesticides and Pollution*, London: Collins.

MoT (1955) Ministry of Transport and Civil Aviation, Circular 718, 31 August 1955.

NC (1953) *Reports of the Nature Conservancy for the period up to 30 September 1952*, London: HMSO.

NC (1954) *Report of the Nature Conservancy for 1953*, London: HMSO.

NC (1955) *Report of the Nature Conrervancy for 1954*, London: HMSO.

NC (1957) *Report of the Nature Conservancy for 1956*, London: HMSO.

NC (1958) *Report of the Nature Conservancy for 1957*, London: HMSO.

NC (1965) *Report of the Nature Conservancy for 1964*, London: HMSO.

NERC (1970) *Report of the Natural Environment Research Council 1969–70*, London: Natural Environment Research Council.

NCC (1981) *Nature Conservancy Council Report for 1980–81*, Peterborough: Nature Conservancy Council.

NCC (1984) *Nature Conservation in Great Britain*, Shrewsbury: Nature Conservancy Council Interpretive Branch.

NCC (1986) *Nature Conservation and Afforestation in Britain*, Peterborough: Nature Conservancy Council.

NCC (1990) *16th Report*, Nature Conservancy Council.

Nicholson, M. (1970) *The Environmental Solution: A Guide for the New Masters of the World*, London: Hodder & Stoughton.

O'Sullivan, J., Pritchard, D. and Gammell, A. (1993) Saving Europe's wildlife? The EC Habitat Directive, *RSPB Conservation Review, 1993.*

Poore, M.E.D. (1987) Changing attitudes in nature conservation: the Nature Conservancy and Nature Conservancy Council, *Biological Journal of the Linnean Society*, 32, 179–187.

Pritchard, D.E. (1985) *Britain's Implementation of the EC Birds Directive*, Sandy: Royal Society for the Protection of Birds.

Ratcliffe, D.A. (1977) Nature conservation: aims, methods and achievements, *Proceedings of the Royal Society of London B 197*, 11–29.

Ratcliffe, D.A. and Oswald, P.H. (eds) (1988) *Birds, Bogs and Forestry. The peatlands of Caithness and Sutherland*, Peterborough: Nature Conservancy Council.

Rothschild, M. and Marren, P. (1997) *Rothschild's Reserves: time and fragile nature*, Israel: Balaban Publishers.

RSPB (1992) *Proposed New EC CITES Regulation: a preliminary RSPB view*, Sandy: Royal Society for the Protection of Birds.

RTSD (1999) *Sustainable Development – Devolved and Regional Dimensions*, London: Round Table on Sustainable Development.

Scottish National Parks Survey Committee (1947) *National Parks: a Scottish survey*, Cmd. 6631, Edinburgh: HMSO.

Scottish National Parks Committee and the Scottish Wild Life Conservation Committee (1947) *National Parks and the Conservation of Nature in Scotland*, Cmd 7235, Edinburgh: HMSO.

Scottish National Parks Committee and the Scottish Wild Life Conservation Committee (1949) *Final Report on Nature Reserves in Scotland*, Cmd 7814, Edinburgh: HMSO.

Scottish Office (1998) *People and Nature. A new approach to SSSI designations in Scotland. Consultation Paper*, Edinburgh: SOAEFD, Scottish Office.

Secretaries of State (1994) *Sustainable Development: the UK Strategy*. [Presented to Parliament by the Secretaries of State for the Environment and for Foreign and Commonwealth Affairs, the Chancellor of the Exchequer, the President of the Board of Trade, the Secretaries of State for Transport, Defence, National Heritage and Employment, the Chancellor of the Duchy of Lancaster, the Secretaries of State for Scotland, Northern Ireland, Education and Health, the Minister for Agriculture, Fisheries and Food, the Secretary of State for Wales and the Minister for Overseas Development by Command and of Her Majesty] Cm 2426, London: HMSO.

Sheail, J. (1976) *Nature in Trust: the history of nature conservation in Britain*, Glasgow and London: Blackie & Son.

Soane, I. (1997) Action plans for mountain massifs, *Cumbrian Nature*, 3 (1).

SNH (1998a) *Progress and Plans 1998*, Battleby: Scottish Natural Heritage.

SNH (1998b) *National Parks for Scotland. A consultation paper*, Battleby: Scottish Natural Heritage.

Stamp, D. (1969) *Nature Conservation in Britain*, London: Collins.

Tither, M. (ed.) (1998) Talking action, *English Nature Magazine*, 38, 6–7.

Whitten, A.J. (1990) *Recovery: a proposed programme for Britain's endangered species*, [CSD Report No. 1089] Peterborough: Nature Conservancy Council.

Williams, G. (1980) *The Ramsar Convention*, Sandy: Royal Society for the Protection of Birds.

Wynne, G., Avery, M., Campbell, L., Gubbay, S., Hawkswell, S., King, T., Newbery, M., Smart, P., Steel, J., Stones, S., Stubbs, A., Taylor, J., Tydeman, C. and Wynde, R. (1995) *Biodiversity Challenge: an agenda for conservation in the UK*, 2nd edn, Sandy: Royal Society for the Protection of Birds.

Chapter 2

Flowering plants

Timothy C. G. Rich

ABSTRACT

The flowering plants of Great Britain and Ireland are among the best documented in the world. There are currently about 1390 native species and over 1100 well-established aliens.

The post-war agricultural revolution had a huge impact on the floras. Between 1930–60 and 1987–88, the Botanical Society of the British Isles (BSBI) Monitoring Scheme found significant changes in frequency of 24% of the species in England (13% decline, 11% increase), 12% in Scotland (4.4% decline, 7.4% increase), and 19% in Ireland (11% decline, 8% increase). The main declines were in plants of grassland, heathland, aquatic and swamp habitats and arable weeds, while introduced species had spread. The Countryside Survey 1990 found an overall reduction in botanical diversity since 1978 in Britain.

About 15 new plants have been recorded since 1973, mostly previously overlooked or newly described taxa. New sites have been discovered for some rare species. A few native species have spread naturally (especially halophytes on roadsides), but more have spread from gardens. Many alien species have increased. Twenty-one flowering plants in Great Britain and 11 in Ireland (excluding critical species) have been lost since detailed records began. Six species have probably become extinct since 1973 in Great Britain and one in Ireland, but there is worryingly little information on the current status of critical species (many of which are endemic). Many other species are known to be declining. A few critically rare species are benefiting from intensive conservation work.

Direct habitat loss and changes in management due to agriculture are probably still the major factors causing change. Legislation has been tolerably successful in protecting some sites and species, but has failed in the wider countryside and the Common Agricultural Policy has been disastrous. Recent agri-environment schemes such as Countryside Stewardship are generating some optimism. Further losses are inevitable.

1 State of knowledge

The flora of Great Britain and Ireland is among the best documented in the world. Although the general taxonomy and distribution of the flora has been known since the 1960s, over the last 25 years huge progress has been made on the detail.

A wealth of literature, millions of botanical records and magnificent if under-utilised herbarium collections are held by institutions and individuals. A bibliographic database is currently being compiled by Botanical Society of the British Isles (BSBI), augmenting Simpson (1960).

1.1 Taxonomy

Recent floras of the British Isles by Stace (1997) and for Ireland alone by Webb, Parnell and Doogue (1996) have succeeded the respected Clapham, Tutin and Warburg (1952) era. Another more detailed flora of the British Isles is in the process of being published (Sell and Murrell 1996). There are recent monographs of critical genera including *Euphrasia, Limonium, Rosa, Rubus* and *Taraxacum* (another on *Hieracium* by Sell and Murrell is nearly complete) and guidance is available on identifying other critical groups (see for example Rich and Jermy 1998). The completion of *Flora Europaea* (Tutin *et al.* 1964–80) has established the European context (Webb 1983; Walters 1984). A cytological catalogue is in preparation by R. J. Gornall and J. P. Bailey.

1.2 Distribution

Distributional data based on 10-km squares of the national grid have been available for nearly 40 years in the *Atlas of the British Flora* (Perring and Walters 1962). This standard reference work, once a milestone in phytogeography, is now being updated for the millennium as *Atlas 2000*. There is a more recent update for Wales (Ellis 1983). In England and Wales there are numerous county floras published or in preparation, though relatively few are available for Scotland and Ireland (the account of progress in local floras by McCosh (1988) is now somewhat dated). Some small areas have even more detailed floras (for example the islands of Roaring Water Bay, Akeroyd *et al.* 1996). A recent list of vice-counties for which plants have been recorded is available for Ireland (Scannell and Synnott 1987), and another is in preparation for Great Britain to update Druce (1932).

A major achievement has been the compilation of reference sources on the huge number of alien plants in Great Britain (Clement and Foster 1994; Ryves, Clement and Foster 1996). In Ireland aliens have received less attention, but recently work has begun on a checklist (S. C. P. Reynolds, pers. comm.).

Distribution maps of British and Irish species in Europe are slowly being published in *Atlas Flora Europaea*. The wider European geographical relationships have been revised by Preston and Hill (1997), replacing the classic work of Matthews (1955).

The development of standardised recording methods has been a major advance to improve the quality of the records (Rich and Smith 1996; Rich *et al.* 1996b).

1.3 Ecology

A huge amount of ecological data is available. Autecological accounts of over 200 plants have been published as biological floras in the *Journal of Ecology*. Detailed studies on many common species have been compiled by the Unit of Comparative Plant Ecology, which has also developed a unifying plant strategy theory which classifies plants according to their responses to stress and disturbance (Grime, Hodgson and Hunt 1988). Publication of the national vegetation classification (NVC), which replaces the classic work of Tansley (1949), is complete (Rodwell *et al.* 1991–2000).

An ecological database has been set up at York (Fitter and Peat 1994). A comprehensive ecological flora integrating all the data would be valuable.

1.4 Conservation

In general the conservation status of plants is well established, with the exception of difficult plant groups such as *Utricularia* and *Rubus*, and Palmer (1995, 1996) has drawn up a strategic approach to conserving plants in the UK. The conservation status of the rarest species has recently been revised for the third edition of the *Red Data Book* for vascular plants (Wigginton 1999). A review of the nationally scarce species (Stewart *et al.* 1994) was carried out between 1990 and 1993, though there are some limitations (Rich, FitzGerald and Kay 1996; Rich 1997). The BSBI Monitoring Scheme has also indicated which of the commoner species may have declined since 1960 (Rich and Woodruff 1990, 1996).

2 Data capture

2.1 Mapping schemes

The tradition of collecting data on plant distributions dates from the seventeenth century. The use to which the data have been put has changed from medicine and curiosity to phytogeography and recently to applied conservation and environmental audit. Recent and current national/international mapping schemes, largely carried out by the voluntary sector, are as follows:

- BSBI Monitoring Scheme: a sample survey of Great Britain and Ireland in 1987–1988 to assess changes in the flora since 1960 and establish a baseline for future monitoring (Rich and Woodruff 1990, 1996).
- Scarce Plants Project: creation of a computer database of British Nationally Scarce species, which was used to revise the list of Nationally Scarce species, update distribution maps and assess changes in their frequency (Stewart *et al.* 1994); 238 flowering plants were included.
- Aquatic plants: a compilation of records of aquatic species with ecological accounts (Preston and Croft 1997).
- *Red Data Books*: reviews of the rarest species in Ireland (Curtis and McGough 1988) and Great Britain (Wigginton 1999).
- *BSBI Atlas 2000*: a resurvey of the *Atlas of the British Flora* for the year 2000.
- Individual species surveys, e.g. *Populus nigra* (Milne-Redhead 1990) and *Viscum album* (Briggs 1999).

There are also numerous county floras and more local studies. Since the 1980s a large amount of species and habitat data has been collected during river corridor surveys, Phase 1 and 2 surveys, environmental audits and so on, for conservation and environmental organisations (e.g. wildlife trusts, statutory agencies, local authorities) and developers. However, relatively few of these data get into the mainstream biological recording systems, mainly because of confidentiality or copyright clauses or simply because the records are not passed on (an exception is Northern Ireland; McKee 1999).

2.2 *Data quantity and quality*

There is a wealth of floristic data collected by volunteers; some estimates are of 6 million botanical records alone. Most data are held in card files by the BSBI vice-county (VC) Recorders and other individuals, although a substantial proportion are computerised at least in summary form by the Biological Records Centre, Monks Wood (BRC). Some, of rather mixed quality, are also held by local record centres. Recently the BSBI have stimulated many VC Recorders to compile records on computer databases as part of the *Atlas 2000* project, and have started to compile a database of plants under threat. This will allow electronic retrieval and transfer of data more easily in the future.

The distribution data have significant and unrecognised weaknesses for four main reasons. First, the records are usually collected on an *ad hoc* basis by amateurs of differing expertise who spend varying amounts of time and effort in each recording unit, resulting in taxonomically and geographically biased data. Data collected by professionals are also variable in quality, and are rarely validated. Second, as coverage is not comprehensive, the records are samples but are rarely treated as such. For instance, Rich and Woodruff (1990) estimated coverage for the *1962 Atlas of the British Flora* at 49%. Third, information about how the data were recorded is not collected or presented, allowing no assessment or correction of bias (Rich and Woodruff 1992). Fourth, there is a marked reluctance to audit databases and their derived maps, with most being accepted at face value (Rich, FitzGerald and Kay 1996; Rich 1998). A degree of standardisation in sampling to minimise the taxonomic and geographic bias is long overdue at a national level, as is beginning to happen in Europe (Rich *et al.* 1996b; Bremer 1997).

3 Workforce

In contrast with many European countries where floristic work is still highly valued and professionally funded, in Great Britain and Ireland much of the inventory work is carried out by amateurs. The main organisations involved in the floristic work on flowering plants are as follows:

- *The Botanical Society of the British Isles*. This charity is the main botanical society in Great Britain and Ireland, primarily concerned with taxonomy and distribution (membership *c.* 2760). It has a well-organised system of VC recorders, though most information is disseminated nationally by BRC.
- *The Botanical Society of Scotland* (formerly the Botanical Society of Edinburgh). A moderate-sized society involved with all aspects of botany in Scotland and elsewhere (membership *c.* 350).
- *The Wild Flower Society*. A smaller and less scientifically based charity than the BSBI, but whose importance is increasingly educating and nurturing beginners (membership *c.* 730).
- *Plantlife*. A charity in Great Britain and Ireland set up to conserve British and Irish wild plants (membership *c.* 11 000). Some data are collected on selected rare taxa.

Statutory agencies, environmental consultancies, NGOs, etc., also contribute indirectly.

The combined active workforce contributing to floristic work is probably about 1000 botanists, with another 1000–3000 contributing in a minor way. Only 10% of BSBI members are under 35 and 47% are over 55 (Roper 1994), with most graduates now learning molecular biology rather than their plants. There is often a high degree of overlap between the societies (for example, *c.* 50% of the BSBI membership belong to Plantlife). The requirement for professional botanists/surveyors by the environmental organisations and companies has resulted in better employment prospects than there was 25 years ago.

There is much to do to educate the workforce, a key feature being to improve the quality of the records (Rich and Smith 1996). The introduction of taxonomic workshops (such as those run by the BSBI or the Institute of Ecology and Environmental Management) and identification qualifications (IDQs) run by the Natural History Museum or the Certificate in Biological Recording and Species Identification from the University of Birmingham are welcome first steps in raising standards.

There are only three professional taxonomists working full-time on the flora in Great Britain and none in Ireland but probably another fifty contribute significantly part-time supplemented by another 100 or so amateurs (A. C. Jermy, pers. comm.).

4 Synopsis of change to 1973

Many of the factors affecting the British flora up to 1973 were reviewed by specialists at a conference in 1969 (Perring 1970), but an excellent overview is given by Ratcliffe (1984). Similar impacts probably also occurred in Ireland, but on a much less intense scale.

Widespread systematic changes in management of the countryside after the Second World War undoubtedly affected the flora significantly. The post-war agricultural revolution had a huge impact; widespread drainage, the application of fertilisers and herbicides, improved crops, deep ploughing, increased stocking densities and so on all markedly affected grassland species. Rapidly changing arable practices affected arable weeds; Wilson (1992) gives an excellent summary. Hedges were removed from fields. Stock were withdrawn from lowland heaths and remote grasslands which became rank or reverted to woodland resulting in loss of species of open habitats, while other parts of heaths were reclaimed for agriculture or buildings. Subsidies were paid to overgraze the uplands resulting in impoverished turf. Broad-leaved woodlands were replanted with conifers or coppice cycles were neglected, and conifers were planted on bogs, heaths and mires, and upland pastures. Aquatic habitats everywhere were becoming eutrophicated with sewage and agricultural runoff, and the effects of uncontrolled pollution were locally catastrophic. Rivers were canalised and deepened, often in misguided efforts to prevent floods with consequent loss of aquatic plants and flood-plain meadows. After *Brucellosis* in cattle was linked to stagnant water, many ponds were fenced or filled in resulting in small plants being ousted by robust swamp species. Fens and mires were drained, and peat cutting intensified or stopped completely. The coasts also came under pressure from development, reclamation, recreation and sea defence works. Transport infrastructure destroyed many habitats but created new networks along which plants spread. Many aliens spread dramatically resulting in some of the most obvious changes in the flora.

Cumulatively, these effects were having a profound impact on the flora. Although only 12 species were extinct in Great Britain by 1973 (Wigginton 1999), about one third of the known localities of rare plants had been lost with agriculture as the main cause (Perring 1970, 1974). The general pattern of mechanisation, intensification and increased scale of operations was set to continue for the next 25 years.

5 Species in Red List categories, scheduled species and Species Recovery Programmes

5.1 Numbers of taxa

In the British Isles, there are currently about 1390 native species (*c.* 2200 including *c.* 230 *Taraxacum*, *c.* 325 *Rubus* and over 260 *Hieracium* microspecies; *Ulmus* and *Ranunculus auricomus* microspecies remain to be described), and over 1100 reasonably well-established aliens (Kent 1992 *et seq.*). Of the native species, *c.* 1375 occur in Great Britain and about 820 in Ireland. There are *c.* 450 endemic species in the British Isles (*c.* 20% of the flora) contained mostly in the critical genera *Alchemilla* (1 species), *Euphrasia* (9), *Hieracium* (150+), *Limonium* (7), *Rubus* (*c.* 210), *Sorbus* (15) and *Taraxacum* (39), and 10 non-critical species and 29 endemic subspecies (Rich *et al.* 1999). Given the continued taxonomic work and doubt as to the native/introduced status of some taxa, these figures are likely to be continuously revised.

5.2 Red List categories

The British Red List species have recently been revised using the current International Union for Conservation of Nature and Natural Resources (IUCN) criteria (Wigginton 1999). 267 species (*c.* 18 % of native non-critical flora) qualify for inclusion as Red List species. The following numbers of species and subspecies are included in each IUCN (1994) category:

- Critically endangered: 25 (1.8%) taxa
- Endangered: 44 (3.2%) taxa
- Vulnerable: 136 (9.9%) taxa
- Lower risk (near threatened): 40 (2.9%) taxa
- Data deficient: 2 (0.1%) taxa
- Extinct: 21 (1.5%) of which two have been reintroduced (see also below)

The Irish Red Data Book (Curtis and McGough 1988) followed the IUCN criteria available at that time. It is currently being updated, with surveys commissioned for rare and threatened plants across the Republic of Ireland, and information being drawn from *Atlas 2000* recording with a few selected species surveys for Northern Ireland. The numbers (and % of native non-critical flora) of taxa for those categories are:

- Extinct: 11 (1.3%) taxa
- Endangered: 6 (0.7%) taxa
- Vulnerable: 44 (5.4%) taxa
- Rare: 78 (9.5%) taxa
- Non-rare or threatened: 16 (1.9%) taxa
- Indeterminate: 6 taxa

5.3 Statutorily protected species

In Great Britain, 111 (8% of the non-critical native flora) species are listed on Schedule 8 of the Wildlife and Countryside Act 1981 (as amended 1998), which prohibits the picking, uprooting, destruction or sale of any listed species. The Act also prohibits the unauthorised intentional uprooting of any plant, though most species may still be picked. Of the European legislation, 11 species are protected under the Bern Convention and nine under the EC Habitats and Species Directive (excluding species regulated by trade or licenses).

In the Irish Republic, 61 (7.4%) species were listed on the Flora (Protection) Order 1999 under Section 21 of the Wildlife Act 1976. In Northern Ireland, 55 (6.7%) species are listed on Schedule 8 of the Wildlife (NI) Order 1985 (an amended list is in preparation).

5.4 Species recovery programmes

While attempts have been made in Great Britain for years to conserve and recover some rare plants, species recovery programmes really took off, at least in England, following publication of the proposed programme for Great Britain's protected species (Whitten 1990). English Nature's Species Recovery Programme, which now funds Plantlife's 'Back from the Brink' project and other work, is now a major force in driving plant conservation efforts. It has realistic funding for species on the programme, and gives significant hope for the future of some rare species.

In Wales, Scotland and Northern Ireland there are no specific species recovery programmes, though individual species which require attention are being addressed through the Biodiversity Action Plans (e.g. the two remaining plants of *Saxifraga hirculus* in Northern Ireland). There are no specific recovery programmes in the Republic of Ireland.

5.5 Number of priority species and species of conservation concern

In the UK, there are 82 priority flowering plant species (UK Biodiversity Group 1998) and 211 species of conservation concern (UK Biodiversity Steering Group 1994). Of these plants, 71 (3.3% of the total flora) now have Species Action Plans. No comparable figures are available for the Republic of Ireland.

6 Changes 1973–98

6.1 BSBI Monitoring Scheme

The main study to address changes relevant to this period is the BSBI Monitoring Scheme, which was a sample hectad survey of Great Britain and Ireland to assess changes in the vascular plant flora since the *Atlas of the British Flora* (Perring and Walters 1962). Presence/absence records from an 11% sample of 10-km squares made before 1960 were compared with those collected in 1987–88 (Rich and Woodruff 1990, 1996). It was only possible to make a relatively crude comparison due to variation in recording effort (16% more records were collected for the 1987–88 period due to a more intensive survey with significant regional variations) but the main trends were clear.

Significant changes were found in the number of 10-km squares recorded for at least 24% of the species in England (13% decline, 11% increase), 12% in Scotland (4.4% decline, 7.4% increase), and 19% in Ireland (11% decline, 8% increase) (Table 2.1). The patterns of change differed between countries. In England the main trends were declines in plants of grasslands, heathland, aquatic and swamp habitats and arable weeds, while introduced species had spread. In Scotland there were losses of plants of unimproved grassland and arable weeds, contrasting with an increase in aliens and aquatic and swamp species (the latter may be due to more intensive sampling). In Ireland the declines occurred in plants of open, unimproved and calcareous grassland, coasts and arable weeds, and introduced species, and possibly woodland/scrub/hedge species had spread (the latter may be due to more consistent recording of woody species). Data for Wales were too variable to analyse because of the small sample size and marked increase in recording effort. Changes were least in the uplands and most marked in the English Midlands. Species of high quality habitats such as calcareous grassland were declining most and were clearly a priority for conservation. Although aliens were generally increasing there was little evidence that they were doing so at the expense of the native flora.

The Monitoring Scheme was a very crude, insensitive means of monitoring the flora as it was based on presence/absence at a 10-km square level; many localities could be

Table 2.1 Number of species increasing or decreasing in frequency by country and major habitat between pre-1960 and 1987–88 (from the BSBI Monitoring Scheme, Rich and Woodruff 1990).

Habitat	England		Scotland		Ireland	
	Increase	*Decrease*	*Increase*	*Decrease*	*Increase*	*Decrease*
Woodland, scrub, hedges etc.	12	15	0	1	18	13
General grassland	10	1	1	3	1	1
Calcareous grassland	0	22	0	1	0	16
Wet grassland	4	17	3	1	8	6
Unimproved grassland	5	22	0	12	1	8
Open grassland	3	12	2	3	3	12
Moors, heaths, acid grassland, etc.	0	24	0	2	2	6
Uplands	0	2	1	1	0	4
Aquatics and swamps	6	24	11	2	3	8
Coast	1	8	2	2	1	15
Arable weeds	13	31	1	10	5	18
Introductions	110	17	65	10	44	20
Others	10	1	2	4	2	6
Total	174	196	88	52	88	133

lost within a square without change being detected. Many additional changes will no doubt be shown in the maps in the forthcoming *Atlas 2000*.

6.2 The Countryside Survey

The Countryside Survey provides information on the stock and change in land cover, landscape features and habitats of Great Britain (Barr *et al.* 1993), together with national statistics on the rural countryside. A similar project is being run in Northern Ireland.

Surveys were carried out in 1978, 1984, 1990 and in 1998–99 (data from the latter not yet available) using satellite coverage of the country combined with field survey of a stratified random sample of 508 squares of 1 km^2, including detailed plots. The 1-km squares were classified into arable, pastural, marginal upland and upland landscape types, and field data were analysed by broad habitat type.

The 1990 survey found significant changes in the composition and ecological quality of the vegetation since 1978 with an overall reduction in botanical diversity. In arable landscapes, much of the change was due to rotation between grassland and tilled land, with reduction of the already limited diversity. Hedge bottoms became dominated by species associated with intensive land use. Verges and especially streamsides showed declines in meadow plants and increase in competitive species.

Pastural landscapes showed the highest degree of change overall. Semi-improved grasslands showed loss of diversity, especially of species of unimproved mesotrophic meadows, the latter also declining in verges, hedge bottoms and streamsides. Verges became more overgrown, and streamsides lost aquatic margins and wet meadow species. The ground flora of woodlands became more grassy and less rich, with evidence of disturbance.

In marginal upland landscapes, the changes were variable. Some semi-improved grasslands and upland grassland mosaics showed a small increase in diversity of plants of unimproved and infertile soils. Verges became more overgrown as in the lowlands, and streamsides lost wet meadow species. Woodlands declined in diversity. Hedge bottoms increased with species associated with intensive grassland increased but lost woodland species. Moorlands showed an increase in diversity, especially in species of grasslands at the expense of heathland species. The uplands were relatively stable, though woodland and upland grassland mosaic species richness decreased, while moorland species increased, especially those species associated with flushes.

6.3 Floristic change and habitat change

Habitat change is the predominant, but not the only, cause of change in species distribution and abundance. Habitat and floristic surveys have generally been run separately but their combination can aid interpretation and improve estimated of rates of change dramatically. The similarities in findings of the BSBI Monitoring Scheme and the Countryside Survey described above give confidence that both are reflecting the same trends despite the differences in detail and timescale.

Floristic changes assessed from presence/absence data in grid squares seriously under-estimate the extent of change compared with detailed habitat or site-based studies. As yet there are few studies to do this due to the limited habitat data available compared with species data, but no doubt more will become available when Phase 1 and Phase 2 habitat surveys are repeated.

For example, Byfield and Pearman (1996) studied decline in Dorset heathland species and their habitats by repeating Good's (1948) survey of 1931–37. Only *c.* 25% of the populations of 41 rare heathland species surveyed in 390 stands had survived; 35% of the sites had been destroyed completely, while others had reverted to woodland or closed vegetation. Similarly in the Llyn peninsular, Blackstock *et al.* (1995) found a 51% loss of dry heathland and a 95% loss of wet heathland between 1920–22 and 1987–88, and one species, *Genista anglica*, was noted as only still present in 20 out of 70 former sites. Set in the context of loss of 75% of British heaths since the 1830s (Farrell 1989), the overall loss of heathland species is catastrophic.

Recent estimates of rates of change for all habitats would be valuable, and provide powerful arguments for conservation (cf. Ratcliffe 1984). For instance, the oft-quoted loss of 95% of wildflower meadows since the war should be reflected in a 95% decline in meadow species, which would put most grassland plants into the IUCN Critically Endangered category.

6.4 Extinctions

The 21 species in Great Britain and 11 in Ireland that have become extinct since detailed records began are listed in Table 2.2. *Bupleurum rotundifolium, Campanula perscicifolia* and *Euphorbia villosa* were included as extinct by Perring and Farrell (1983) but are now regarded as introductions. In the British Isles as a whole only 15 species have become extinct (*c.* 1% of the native flora), the other 17 still surviving in either Great Britain or Ireland. Seven species have probably gone extinct since 1973, six in Great Britain, one in Ireland.

Of these species, only *Bromus interruptus* appears to be extinct in the wild on a world scale, ironically the year after Perring (1974) discussed its distribution. Attempts to conserve it in its last site in Cambridgeshire by rotovation were unsuccessful, though miraculously it survived in cultivation. It is reputed to be endemic, but given its association with clover and sainfoin fields and a decline matching that of other casuals, it may have originated elsewhere in the world.

The nativeness of several extinct species is open to question. *Galeopsis segetum* was regarded as native in arable fields in Caernarvonshire, and an infrequent casual elsewhere (including Anglesey); such an odd distribution pattern is not good evidence for native status and consequently its demise since 1975 may be of no consequence. *Agrostemma githago*, if it ever was native, is now confined to a few maintained sites but it is still sporadically recorded as a casual (e.g. it was recorded in seven hectads during 1987–88 for the BSBI Monitoring Scheme). While some records undoubtedly originate from sown seed (it is frequent in so-called 'wildflower' seed mixtures), others are difficult to distinguish from germination of long-buried seed which may be viable for at least 20 years. Conversely, *Ajuga genevensis* in Berkshire has recently been regarded as introduced, but I suspect it is native.

Crepis foetida, an annual/biennial recorded from about 15 sites mostly on the coast in south-east England, was last seen on Dungeness in about 1980. Fortunately it has been reintroduced from native seed held at Cambridge Botanic Garden, and natural regeneration is now occurring with over 100 plants in 1997 (UK Biodiversity Group 1998). Similarly, the rare annual *Filago gallica*, extinct in mainland Great Britain since 1955, was reintroduced under Plantlife's 'Back from the Brink' project (Rich *et al.* 1999).

Table 2.2 Plants extinct in Great Britain and Ireland since detailed records began; — absent.

Species	Great Britain	Ireland	Notes
Agrostemma githago	Extinct	Extinct	Still occurs sporadically, either as an alien or from seed bank
Ajuga genevensis	Extinct	—	Native status requires clarification
Anthemis arvensis	Locally frequent	Extinct	
Arnoseris minima	Extinct	—	
Atriplex pedunculata	Recently rediscovered	Extinct	The Irish record requires confirmation
Bromus interruptus	Extinct	—	Doubtfully native
Carex buxbaumii	Four sites in Scotland	Extinct	Native material retained in cultivation
C. davalliana	Extinct	—	
Centaurium latifolium	Extinct	—	
Crepis foetida	Reintroduced	—	
Euphorbia peplis	Extinct	Extinct	Still present in Channel Islands
Filago gallica	Reintroduced	—	
Galeopsis segetum	Extinct	—	Very doubtfully native
Holosteum umbellatum	Extinct	—	
Luzula pallidula	Not seen since 1993	Extinct	
Matthiola sinuata	Rare in south-west	Extinct	Has declined everywhere
Neotinea maculata	Extinct	Locally abundant in west	
Otanthus maritimus	Extinct	Rare in south-east	
Pinguicula alpina	Extinct	—	
Polygonum maritimum	Spreading in south	Extinct	Possibly not an established member of the flora in Ireland
Rubus arcticus	Extinct	—	
Sagina boydii	Extinct	—	Of doubtful origin
Saxifraga rosacea	Extinct	Locally abundant in west	
Scandix pecten-veneris	Scattered	Extinct	
Scheuzeria palustris	Rare in Scotland, extinct in England	Extinct	
Serratula tinctoria	Locally common	Extinct	
Spiranthes aestivalis	Extinct	—	
Tephroseris palustris	Extinct	—	
Trichophorum alpinum	Extinct	—	

Native material had been maintained in cultivation by D. McClintock since 1948, and it was possible to put it back into the original site in Essex in 1994 using historical records to determine the ecological requirements. It has slowly increased.

Neotinea maculata appeared briefly on sand dunes at the north end of the Isle of Man in 1966–80, presumably from wind-blown seeds from western Ireland where it is still locally abundant. The plants were fenced to 'protect' them but disappeared after suppressed by rank *Festuca* (L. S. Garrad, pers. comm.).

There is very little information on how many critical species may be extinct, which is particularly worrying as many are endemic. The only known site of *Hieracium hethlandiae* was quarried away in 1976 (it survives in cultivation). *Taraxacum sarniense* would be extinct in the Channel Islands were it not for its recent inclusion in the more widespread *T. ciliare*. There are now no known sites for *Rubus dobuniensis*.

6.5 Declining species

A number of species are on the verge of extinction. *Luzula pallidula*, an annual to short-lived perennial of open peaty or sandy ground, has not been seen in Great Britain for five years, or in Ireland for nearly 30 years (Fig. 2.1). In Great Britain, *Schoenoplectus triqueter*, once present in the Thames, the Medway and the Arun, was by 1995 reduced to one clump on the Tamar. Material collected in 1994 has been propagated and planted out, and in 1997 several clumps were growing well. In Ireland it is still locally frequent in the upper parts of the Shannon Estuary but has also declined there since it was first discovered.

Another precarious rarity is *Damasonium alisma* (Birkinshaw 1994; Rich *et al.* 1996a) which used to occur on the margins of ponds in over 100 localities in England.

Figure 2.1 Distribution of *Luzula pallidula* in the British Isles. o Native pre-1990. ● Native post-1990.
 × Casual, pre-1990. ∗ Casual post-1990.

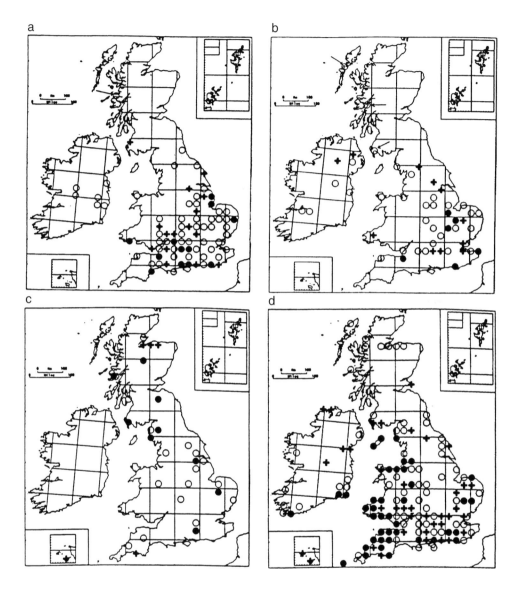

Figure 2.2 Maps of selected species showing decreases between pre-1960 and 1987–88. The symbols on the maps are enlarged so that they are clear to read and do not imply comprehensive coverage. o Recorded only for the *Atlas* between 1930 and 1960 (or before 1960 in Ireland); • Recorded for both the *Atlas* and the Monitoring Scheme; + Recorded only for the Monitoring Scheme (1987–1988). A predominance of open circles suggests a decline, and a predominance of pluses, an increase. (a) *Orchis morio* (b) *Ranunculus circinatus* (c) *Teesdalia nudicaulis* (d) *Stachys arvensis*.

Reproduced from Rich and Woodruff 1990 with permission from JNCC.

By the early 1980s it had decreased to one site in Surrey, but following pond restoration work, it amazingly reappeared from long-dormant seed banks at two ponds in Buckinghamshire in 1989 and 1990. Although further restoration work has brought it back temporarily to six ponds, it has not been possible to maintain the populations in the wild for more than a few years. Accumulation of silt on the bottom of the ponds does not hinder germination, but could affect seedling establishment. In cultivation plants grow prolifically but are sensitive to herbivory by water snails; in the wild molluscs may take time to recolonise newly restored ponds restricting *Damasonium* to the period immediately after restoration. Climate also plays a role, as might be expected for an annual at the northern edge of its range, with plants reappearing in ponds in 1996 after the hot summer of 1995.

Numerous other examples of decline of rare and Nationally Scarce species are given in Wigginton (1999) and Stewart *et al.* (1994), though few species have been studied in detail (*Orchis ustulata* is an excellent example of a species which has; Foley 1992). Fig. 2.2 shows selected examples of decline in more widespread species from the BSBI Monitoring Scheme. *Orchis morio*, a widespread plant of unimproved grassland, has declined significantly in Great Britain, as have many other grassland species. *Ranunculus circinatus* has declined overall in aquatic habitats. *Teesdalia nudicaulis* has declined from heaths in the lowlands, though persists elsewhere. *Stachys arvensis*, an arable weed, has declined in the north, east and centre of Great Britain, presumably because of the use of herbicides.

6.6 New species

Additions to the flora may be new arrivals, or overlooked long-established species. Of the genuine new species since 1973, three new European *Serapias* tongue orchids have been recorded recently. With tiny wind-blown seeds and global warming, the evidence available to date indicates they have good credentials to be native. In 1989, *S. parviflora* was found in a remote spot on the Cornish coast. Plants reappeared in the two successive years, and again in 1998. In 1992, *S. lingua* was found in a patch of wild ground in Guernsey, and in 1998 three plants were discovered in an improved ley in South Devon. Both of these are generally southern European species of dry and wet grassland, dune slacks and olive groves, but are currently spreading in Brittany (Bargain *et al.* 1991). In 1996, one plant of *S. cordigera* was found in an old quarry in Kent, with two plants present in 1997. This is a widespread species from around the Mediterranean north to Finistere which grows in grassland, heaths, open woodland and scrub. *Ophrys bertolonii* was found in Dorset in 1976 but is now thought to have been planted (Stace 1997).

6.7 Overlooked species

The prize for an overlooked species must go to *Sorbus domestica*, a medium-sized tree now known in six sites around the Severn Estuary (Fig. 2.3). It was first discovered in two sites in Glamorgan by M. Hampton, and subsequently single trees were found in four sites in Gloucestershire by M. Kitchen and colleagues. Although a planted tree was known in the Wyre Forest, the existence of native populations was unsuspected, and it must be lurking elsewhere.

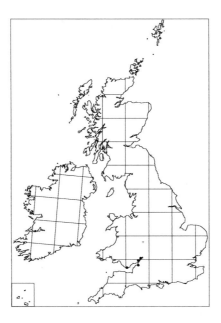

Figure 2.3 Distribution of *Sorbus domestica* in Great Britain ● Native post-1990. + Planted?, pre-1990.

Figure 2.4 Distribution of *Gentianella uliginosa* in Great Britain (plants recently reported from Scotland are *G. amarella*). o Native pre-1990. ● Native post-1990.

Other stunning discoveries are of the small yellow composite *Crepis praemorsa* new to western Europe in a very well-botanised unimproved meadow in Cumbria in 1988, *Gagea bohemica* on rock ledges in Wales in 1964 although not identified until 1974, *Gentianella ciliata* in chalk grassland in Buckinghamshire, and *Pulmonaria obscura* in woodland in Suffolk. Other newly discovered, identified or described taxa include *Alchemilla gracilis* in Northumberland, *Atriplex praecox* and *A. longipes* around the coasts, *Dactylorhiza lapponica* in Scotland, *Epipactis youngiana* in Northumberland and around Glasgow, numerous new segregates of the *Limonium binervosum* group, and *Petrorhagia prolifera* in East Anglia.

Some native species are now known to be more common than in 1973 because they have been previously under-recorded. For example, *Schoenus ferrugineus*, whose sole known site in Scotland in 1952 was destroyed by flooding for a reservoir, has been discovered in six new sites since 1979 with a total population of *c.* 12 000 plants (Cowie and Sydes 1995). Other taxa which had been overlooked include *Alopecurus aequalis* in Ireland, *Carex chordorrhiza* in Scotland, *Carex divisa* in Ireland, *Cerastium brachypetalum* in Kent, *Ceratophyllum submersum* in Ireland, *Gentianella uliginosa* in England (Fig. 2.4; now rejected for Scotland), *Hydrilla verticillata* in Scotland in 1999, *Luronium natans* native in Ireland, *Oenanthe pimpinelloides* in Ireland and much more commonly in England, *Pedicularis sylvatica* subsp. *hibernica* in Wales, *Populus nigra* in Ireland, *Poa infirma* eastwards to Sussex, *Rorippa islandica* scattered in Ireland and locally common in south Wales, *Scorzonera humilis* in Wales, and *Vaccinium uliginosum* in Somerset.

6.8 Increasing species

Fig. 2.5 shows examples of species which are spreading, selected from the BSBI Monitoring Scheme. *Cerastium glomeratum* has increased markedly throughout the British Isles, possibly due to resistance to herbicides. *Chamerion angustifolium* spread widely in Great Britain during the latter half of the nineteenth century, but in Ireland the spread began much later and is continuing today. *Meconopsis cambrica* has increased in Scotland and England probably resulting from garden escapes (it is native in Wales); other native species which have also increased because of garden escapes include *Ranunculus lingua*, *Taxus baccata* and *Polemonium caeruleum*. A number of halophytes are spreading inland on roads treated with de-icing salt (Scott and Davison 1982); *Puccinellia distans* and *Cochlearia danica* have increased dramatically on roadsides in the last 20 years.

Some of the rarer species are also spreading. *Scrophularia umbrosa* appears to have been expanding for some time. *Polygonum maritimum* has been spreading recently along the south coast of England with a sudden spate of records from Hampshire, Isle of Wight and both halves of Sussex well to the east of previous records. Another Red Data Book species, *Erica ciliaris*, seems to be spreading outwards into Dorset and Hampshire from its stronghold in the moist heathlands around Poole Harbour, and there are other newly discovered populations in Cornwall and Somerset. *Atriplex peduculata* has been recorded from about 16 saltmarshes scattered around the southeastern coasts of England and an unconfirmed one in Connemara (Webb and Akeroyd 1991). It was thought extinct in the 1930s but was discovered in a new site in Essex in 1987, perhaps brought by migrating wildfowl (Leach 1988). The neo-endemic

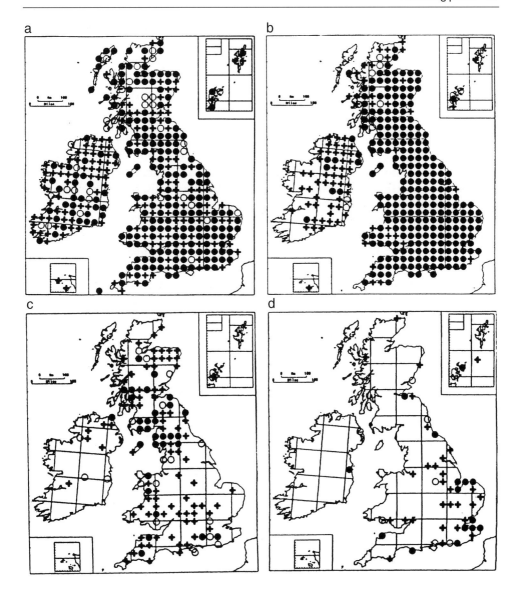

Figure 2.5 Maps of selected species showing increases between pre-1960 and 1987–88. For details of the symbols see Fig. 2.2. (a) *Cerastium glomeratum* (b) *Chamerion angustifolium* (c) *Meconopsis cambrica* (d) *Puccinellia distans.*

Reproduced from Rich and Woodruff 1990 with permission from JNCC.

allotetraploid *Senecio cambrensis* appears to have arisen anew in Edinburgh, as well as spreading in Wales.

The range of *Himantoglossum hircinum*, an orchid of calcareous grassland with wind-dispersed seeds, has gone through several phases of expansion and contraction. An increase between the 1900s and 1930s was attributed by Good (1936) to climate warming, but may be due to the increase in the amount of unmanaged grassland suitable for colonisation during the agricultural depression as the plant declined after the Second World War despite further amelioration of the climate (J. S. Rodwell, pers. comm.). Since the mid-1980s, *Himantoglossum* has again increased, some plants even being found pushing their way up through tarmac near Bristol; this time climate may be having a significant effect (Carey and Brown 1994).

6.9 Species benefiting from recovery programmes

Other rare species are recovering as a result of intensive conservation work. Perhaps the best-known example is *Cypripedium calceolus*, a famous rare plant which has survived in one site in Yorkshire as a solitary individual since the 1930s. Since 1983 seedlings have been raised for re-establishment in the wild under the Sainsbury Orchid Conservation Project at the Royal Botanic Gardens, Kew (Ramsey and Stewart 1998). Over 1000 young plants have now been transplanted back into 16 former locations, and although mortality is high and variable between sites, the early indications for recovery in the long term are good.

Carex depauperata is rare across much of its European range. It has a scattered, if generally southern, distribution in Great Britain and Ireland, being recorded from 14 sites (Fig. 2.6). By the early 1970s only one plant was known near Cheddar, the second-to-last site in Surrey being lost under a landslip in 1972. In 1973, however, it was surprisingly discovered new to Ireland in County Cork; the Cheddar populations were reinforced with cultivated plants in the late 1980s (Birkinshaw 1990), and in 1992 it was rediscovered in the old Surrey site in a gap in the tree canopy created during the Great Storm of 1987. Further opening of the vegetation and canopy clearance has now

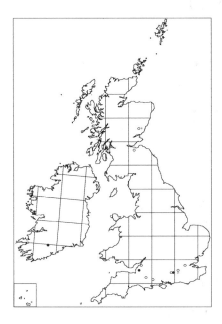

Figure 2.6 Distribution of *Carex depauperata* in the British Isles. o Native pre-1990. • Native post-1990.

increased the populations to over 50 with many seedlings at Cheddar and four at Godalming. The recent discovery of old specimens from Scotland and England and details of the lost Welsh site suggest it may occur elsewhere.

Other species in recovery programmes include *Alisma gramineum*, *Althaea hirsuta*, *Gentianella anglica*, *Liparis loeselii*, *Pyrus cordata*, *Rumex rupestris*, *Salvia pratensis*, *Thlaspi perfoliatum*, and *Viola perscicifolia*.

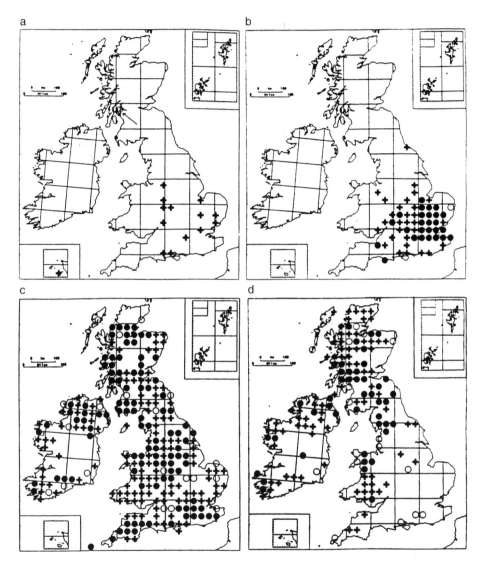

Figure 2.7 Maps of selected alien species showing increases between pre-1960 and 1987–88. For details of the symbols see Fig. 2.2. (a) *Crassula helmsii* (b) *Lactuca serriola* (c) *Rhododendron ponticum* (d) *Epilobium brunnescens*.

Reproduced from Rich and Woodruff 1990 with permission from JNCC.

6.10 Aliens and introductions

There have been major spreads in alien species, though some of the apparent increase is due to more consistent recording by botanists. Examples of aliens which the BSBI Monitoring Scheme found to have spread are given in Fig. 2.7. *Crassula helmsii* is an aquatic regularly thrown out from aquaria and ornamental ponds and has spread markedly, as have other aquarium plants such as *Lagarosiphon major* and *Elodea nuttallii*. Casuals such as *Lactuca serriola* are spreading on roadsides and waste ground, and a few garden escapes such as *Rhododendron ponticum* are now having significant impacts on the local flora. *Epilobium brunnescens* is unusual for an alien in that it is spreading in damp places in the mountains; in some localities such as Ben Bulben in Ireland it now forms a significant proportion of the turf. Two other species which have spread dramatically are *Galeobdolon argentatum* and *Fallopia japonica*.

A comparison of the ecological and habitat characteristics of increasing alien and native plants showed that they are functionally indistinguishable, though aliens are more likely to be clonal, polycarpic perennials with erect leafy stems and transient seed banks (Thompson, Hodgson and Rich 1995).

Significant changes in casuals associated with ballast, grain and wool shoddy (e.g. *Lepidium perfoliatum*, *Bupleurum rotundifolium*) have taken place as their modes of introduction have changed. The major current sources are from gardens (including bird seed) and possibly so-called 'wildflower' seed mixtures.

7 Factors affecting change

7.1 Agriculture

Direct habitat loss and changes in management due to agriculture are probably still the major factors causing decline in flowering plants throughout the British Isles, as they were 25 years ago. The pace of change has rapidly accelerated in Ireland. The BSBI Monitoring Scheme attributed losses of grassland taxa to ploughing, drainage, herbicides, fertilisers, etc., declines in arable weeds to herbicides and seed screening, declines in heathland species to reclamation and neglect, and declines in aquatics and swamp species due to eutrophication, drainage and reclamation (Rich and Woodruff 1996). Secondary impacts such as scrub development on neglected land, or from intensive use of land adjacent to ponds (cf. Williams *et al.* 1998), have also affected some species.

7.2 Woodland management

Many broad-leaved woodlands are no longer being coppiced, leading to suppression of species which benefit from the temporary increase in light (e.g. *Euphorbia amygdaloides*). Many such species have long-dormant seeds, and will probably recur after resumption of management. Grazing by stock or browsing by deer is now a significant factor hindering tree regeneration in many upland woodlands.

Forestry remains a continuing threat, though the incidence of planting semi-natural habitats is declining, possibly following policy changes arising from the public outcry over the disastrous development of the Flow Country in the 1980s. Conifer plantations tend to cast dense shade, restricting plants to woodland rides or streamsides. Plantation clearances can result in temporary increases in a few species from the seed bank (e.g.

Deschampsia flexuosa), but in general they offer a poor habitat and a poor substitute for the previous land use. The forestry industry is now attempting to restore habitat diversity and quality in many of its woodlands.

7.3 Pollution

Other slow and insidious although significant effects are occurring through air and water pollution, though these are difficult to quantify in isolation and occur mainly within the context of vegetation. Air pollution mainly affects plants through deposition to soils, with the severity of impact depending on abiotic factors such as soil nutrient status and buffering capacity. Long-term nitrogen deposition has led to nutrient enrichment in several ecosystems, resulting in competitive exclusion of characteristic species by nitrophilous plants, especially in oligotrophic–mesotrophic soil conditions (Bobbink, Hornung and Roelofs 1998). Deposition of the various pollutants can lead to soil acidification in weakly buffered environments (Woodin and Farmer 1993), resulting in changes in community structure, and plants may also be less resistant to secondary stresses such as pathogens or frost damage. Ozone pollution may also cause direct stress injury to plants.

The differentiation of changes in the flora driven by climate warming have yet to be clearly separated from natural variations. The sensitivity of vegetation to seasonal weather types demonstrated at Bibury, Gloucestershire (Dunnett *et al.* 1998) suggests there could be profound changes in communities as the climate warms.

Water pollution is now primarily eutrophication from atmospheric deposition (Stevens, Ormerod and Reynolds 1997), sewage and agricultural runoff. These result in shifts to plants of eutrophic waters and often impoverishment of the flora. As many aquatic plants are readily dispersed, damaged sites can recover quite rapidly once the sources of pollution have been removed and nutrient levels lowered.

7.4 Anthropogenic pressure

Housing and industrial development on green-field sites is a continuing threat with ever increasing expansion of urban areas. Objections to new quarries, open-cast mines, roads and railways often receive widespread publicity, but their actual impacts compared with those of agriculture are relatively minor. Development for recreation (for example marinas) also affects some sites. Abstraction of peat for horticulture or for Irish power stations is devastating large areas of bog.

Collecting wild plants for horticulture can result in the loss of some populations such as rare orchids, but collection of herbarium specimens now has a negligible impact, though perhaps again except for some very rare plants.

In contrast, escapes from gardens and cultivation now form about 8% of the flora, some of which may be causing significant impacts on some species. For instance, the spread of *Rhododendron ponticum* is threatening a number of sites including sea cliffs on Lundy (the sole habitat for the rare endemic *Coincya wrightii*), and the woodlands at Killarney. Similarly, bird-sown *Cotoneaster* species are also invading some calcareous grasslands, rocks and screes, and are very difficult to eradicate.

Accidentally introduced pests and pathogens have already affected some species. Dutch elm disease caused the loss of many trees in the 1970s and 1980s, though many survived as suckers and have since regrown. Alder disease, die-back in oak, diasy rust and smut and hollyhock rust for instance are all currently spreading.

7.5 National and international policies

British legislation such as the Wildlife and Countryside Act 1981 has been tolerably successful in protecting some sites and species. While the dedication of many of the statutory nature conservation staff is unquestionable, the lack of political will and consequently resources places serious constraints on their abilities to protect nature. Placing plants on the schedules of protected species has helped prevent extinction as attention tends to be focused on them, but many rare species occur outside protected sites and are still vulnerable. Experience has shown that rare plants have to be listed on SSSI schedules and not simply be present on SSSIs to have resources allocated to their conservation.

The Wildlife and Countryside Act 1981 has failed in the wider countryside where agriculture and mismanagement have probably been equally to blame for losses. There have been too many contradictions between the conservation and agricultural policies which should have been integrated; few, if any, 'environmental protection' measures in the agricultural support system (e.g. set-aside, Environmentally Sensitive Areas) bring any significant benefits to plants in general (Sotherton 1998). However, two recent schemes, Countryside Stewardship in England and Tir Cymen in Wales, have been excellent in promoting appropriate management of selected habitats and changing attitudes. Wildlife Enhancement Schemes and Reserve Enhancement Schemes are also giving positive benefits. Limestone Pavement Protection Orders have stopped damage to this rare habitat in Great Britain, but in Ireland large areas of scrubby pavement are being cleared in the Burren. Another exception has also been the conservation of ancient woodlands, the ancient woodland inventories of England and Wales (Spencer and Kirby 1992) being of particular value in raising awareness. Conservation of the flora in the wider countryside will be helped by implementation of the UK Biodiversity Action Plan, and the provision of adequate funding for whole-farm conservation initiatives such as Tir Gofal in Wales.

International policies such as the Common Agricultural Policy have been disastrous for wildlife overall, with their heavy subsidies again often in direct conflict with nature conservation requirements. The introduction of effective environmental protection incentives as part of the grants is a step back in the right direction. The EC Habitats and Species Directive is also a major piece of positive legislation in refocusing policy on internationally rare habitats and species. Other international legislation such as the Ramsar convention has probably had little impact on plant conservation.

7.6 Other factors

Numerous other factors are also affecting the flora of Great Britain and Ireland. Over-abstraction of water from lowland rivers during drought and consequent water-table drawdown of wetlands has had damaging impacts on aquatic plants. The impacts of isolation on dispersal and gene flow remain to be studied, but are likely to become significant in an increasingly fragmented countryside.

Failure to manage the countryside has probably had almost as large an impact as loss of habitat. The lessons of over-protection (for example, fencing a nature reserve and letting nature take its course) of sites are slowly being learnt, and disturbance is now accepted as a requirement for some habitats such as heathlands and dune slacks. For instance, *Corrigiola litoralis* nearly became extinct in South Devon when its gravelly pond margin was fenced from trampling by cattle, and it was rapidly overgrown by taller plants. Similarly, neglect of management has probably had a profound impact; the catalogue of serious damage to 'Rothschild's Reserves', regarded as top sites in Great Britain for nature conservation before the First World War (Rothschild and Marren 1997), is another stark pointer to the increasing importance of the habitat management handbooks published by the statutory agencies.

8 Prognosis for the next 25 years

There will undoubtedly be further loss of floral biodiversity in Great Britain and Ireland over the next 25 years. The continuous, patchwork loss of sites will continue, and plants will slowly disappear. This will increasingly be reflected in the national distribution maps as the last sites for each species in each 10-km square are snuffed out. Rates of loss from site and habitat-based data would give a much faster rate of change.

The main areas of research required to conserve flowering plants are more detailed taxonomic and distribution work on endemic, critical and infraspecific taxa, autecology and management requirements of species for conservation, population biology and genetic variation. I am concerned that the current rush towards elucidating genetic variation will divert attention from the more urgent requirement of investigating the optimum management requirements of rare species and implementing them. I am also concerned about the current fad of reintroducing rare plants to lots of sites; the importance of the extant sites is being diminished and in some cases reintroductions are already being carried out at the expense of conserving the existing sites.

Improvements in the quality of the record base on which conservation priorities and changes in the flora are assessed are also required, and a long-term plan for future botanical recording and monitoring should be drawn up (a recommendation of the BSBI Monitoring Scheme). This is now too late for *Atlas 2000*, but it should be possible to compile a comprehensive *Atlas of the British and Irish Floras* for 2025, based on full abstraction of data from major herbaria, literature, and validated data from other environmental databases, as well as systematically collected field data from the period 2015–225. The problems encountered when comparing floristic data from the BSBI Monitoring Scheme with that from the *Atlas of the British Flora* (see Rich and Woodruff 1990) suggest that more systematic, detailed approaches will be required to monitor the flora in the future, and considerable thought will have to be given before the BSBI Monitoring Scheme is run again (a practical technique for standardised repeatable floristic survey has been demonstrated by Rich *et al.* 1996b). Floristic work needs to be coupled to habitat surveys such as the ITE Countryside Survey to help explain the changes that are taking place. Once tetrad atlases have been completed for each county, work may move towards Red Data Books, detailed studies of particular species or habitats, or the reworking of older historical floras.

Another cause for concern is the impact of climate warming, and especially its indirect effects, though the effects are difficult to predict with certainty. Dryer summers may result in loss of pasture as the balance shifts further from stock to arable, and the need for water could result in further damage due to over-abstraction of water and construction of new reservoirs. If the summers become wetter, species characteristic of droughted habitats or those with continental distributions may be at risk. Land will be required to relocate people from low-lying coastal areas. The influx of alien species will also change as a consequence of changes in world trade. However, climate-driven changes in the flora are likely to be insignificant compared with the secondary consequences of a reorganisation of society at national and global scales.

More proactive plant conservation work is required. One can only hope that the increased sense of public ownership of the countryside will begin to make the politicians take notice and direct countryside policy and protection accordingly. Agri-environment schemes have a large potential role in conservation (Ovenden, Swash and Smallshire 1998) and should be strengthened and funded. I hope we are not reduced to the same state as The Netherlands where every wild plant is valuable because there are so few left (R. van der Meijden, pers. comm.).

ACKNOWLEDGEMENTS

I would like to thank A. Bowley, A. O. Chater, N. Hodgetts, R. A. Jones, N. Lockhart, R. Mitchell, D. Pearman, M. Ramsey, T. C. E. Wells, R. Weyl, M. Wigginton, R. Woods, and M. Wyse-Jackson. The distribution maps were plotted using DMAPW by Alan Morton. The BSBI Monitoring Scheme maps are reproduced with permission from JNCC.

REFERENCES

Akeroyd, J. R. (ed.) (1996) *The Wild Plants of Sherkin, Cape Clear and adjacent Islands of West Cork*. Sherkin Island: Sherkin Island Marine Station.

Bargain, B., Bioret, F. and Monnat, J.-Y. (1991) *Orchidées de Bretagne*. Bannalec: Imprimerie Régionale.

Barr, C. J., Bunce, R. H., Clarke, R. T., Fuller, R. M., Furse, M. T., Gillespie, M. K., Groom, G. B., Hallam, C. J., Hornung, M., Howard, D. C. and Ness, M. J. (1993) *Countryside Survey 1990*. London: Department of the Environment.

Birkinshaw, C. R. (1990) *The ecology of Carex depauperata and its reinforcement at Cheddar Wood, Somerset*. [CSD Report No. 1152]. Peterborough: Nature Conservancy Council.

Birkinshaw, C. R. (1994) Aspects of the ecology and conservation of *Damasonium alisma* Miller in western Europe. *Watsonia*, 20, 33–39.

Blackstock, T. H., Stevens, J. P., Howe, E. A. and Stevens, D. P. (1995) Changes in the extent and fragmentation of heathland and other semi-natural habitat between 1922 and 1987–1988 in the Llyn Peninsular, Wales, UK. *Biological Conservation*, 72, 33–44.

Bobbink, R., Hornung, M. and Roelofs, J. G. M. (1998) The effects of air-borne nitrogen pollutants on species diversity in natural and semi-natural European vegetation. *Journal of Ecology*, 86, 717–738.

Bremer, P. (1997) Over de volledigheid van de inventarisatie van een kilometerhok. *Gorteria*, 23, 144–154.

Briggs, J. (1999) *Kissing Goodbye to Mistletoe? The results of a national survey aimed at discovering whether mistletoe in Britain is in decline*. London: Plantlife.

Byfield, A. and Pearman, D. A. (1996) *Dorset's Disappearing Heathland Flora*. Sandy: RSPB.

Carey, P. D. and Brown, N. J. (1994) The use of GIS to identify sites that will become suitable for a rare orchid, *Himantoglossum hircinum*, in a future changed climate. *Biodiversity Letters*, 2, 117–123.

Clapham, A. R., Tutin, T. G. and Warburg, E. F. (1952) *Flora of the British Isles*. Cambridge: Cambridge University Press.

Clement, E. J. and Foster, M. C. (1994) *Alien Plants of the British Isles,* London: BSBI.

Cowie, N. R. and Sydes, C. (1995) *Status, distribution, ecology and management of Brown bog-rush, Schoenus ferrugineus.* [Scottish Natural Heritage Review No. 43]. Edinburgh: Scottish Natural Heritage.

Curtis, T. G. F. and McGough, H. N. (1988) *The Irish Red Data Book*. Vol. 1. *Vascular Plants,* Dublin: Stationery Office.

Druce, G. C. (1932) *The Comital Flora of the British Isles*. Arbroath: T. Buncle.

Dunnett, N. P., Willis, A. J., Hunt, R. and Grime, J. P. (1998) A 38 year study of relations between weather and vegetation dynamics in road verges near Bibury, Gloucestershire. *Journal of Ecology*, 86, 610–623.

Ellis, R. G. (1983) *Flowering Plants of Wales*. Cardiff: National Museum of Wales.

Farrell, L. (1989) The different types and importance of British heaths. *Botanical Journal of the Linnean Society,* 101, 291–299.

Fitter, A. H. and Peat, H. J. (1994) The ecological flora database. *Journal of Ecology*, 82, 415–425.

Foley, M. J. Y. (1992) The current distribution and abundance of *Orchis ustulata* L. (Orchidaceae) in the British Isles – an updated summary. *Watsonia*, 19, 121–126.

Good, R. (1936) On the distribution of the Lizard orchid (*Himantoglossum hircinum*) Koch. *New Phytologist*, 35, 142–170.

Good, R. (1948) *A Geographical Handbook of the Dorset Flora*. Dorchester: Dorset Natural History and Archaeological Society.

Grime, J. P., Hodgson, J. G. and Hunt, R. (1988) *Comparative Plant Ecology*. London: Unwin Hyman.

IUCN (1994) *IUCN Red List Categories*. Gland: The World Conservation Union.

Kent, D. H. (1992) *List of Vascular Plants of the British Isles*. London: BSBI.

Leach, S. J. (1988) Rediscovery of *Halimione pedunculata* (L.) Aellen in Great Britain. *Watsonia*, 17, 170–171.

Matthews, J. R. (1955) *Origin and Distribution of the British Flora*. London: Hutchinson.

McCosh, D. J. (1988) Local floras – a progress report. *Watsonia*, 17, 81–89.

McKee, F. (1999) The vascular plant database for Northern Ireland (NI) and NI Atlas 2000 project. *BSBI News*, 80, 10–12.

Milne-Redhead, E. (1990) The BSBI Black poplar survey, 1973–1988. *Watsonia*, 18, 1–5.

Ovenden, G. N., Swash, A. R. H. and Smallshire D. (1998) Agri-environment schemes and their contribution to the conservation of biodiversity in England. *Journal of Ecology,* 35, 955–960.

Palmer, M. A. (1995) *A UK Plant Conservation Strategy: A strategic framework for the conservation of the native flora of Great Britain and Ireland*. Peterborough: Joint Nature Conservation Committee.

Palmer, M. A. (1996). A strategic approach to the conservation of plants in the United Kingdom. *Journal of Applied Ecology*, 33, 1231–1240.

Perring, F. H. (ed.) (1970) *The Flora of a Changing Britain*. [BSBI Conference Report no. 11], Hampton: E. W. Classey.

Perring, F. H. (1974) Changes in our native vascular plant flora, in Hawksworth, D. L. (ed.) *The Changing Flora and Fauna of Britain*. London: Academic Press, pp. 7–25.

Perring, F. H. and Farrell, L. (1983) *British Red Data Books*. Vol. 1. *Vascular plants,* 2nd edn. Lincoln: Society for the Promotion of Nature Conservation.

Perring, F. H. and Walters, S. M. (1962) *Atlas of the British Flora*. London: Thomas Nelson and Sons.

Preston, C. D. and Croft, J. M. (1997) *Aquatic Plants in Britain and Ireland*. Colchester: Harley Books.

Preston, C. D. and Hill, M. O. (1997) The geographical relationships of British and Irish vascular plants. *Botanical Journal of the Linnean Society*, 124, 1–120.

Ramsey, M. M. and Stewart, J. (1998) Re-establishment of the lady's slipper orchid (*Cypripedium calceolus* L.) in Great Britain. *Botanical Journal of the Linnean Society*, 126, 173–181.

Ratcliffe, D. A. (1984) Post-medieval and recent changes in British vegetation: the culmination of human influence. *New Phytologist*, 98, 73–100.

Rich, T. C. G. (1997) Scarce plants in Great Britain: have some been over-looked and are others really scarce? *Watsonia*, 21, 327–333.

Rich, T. C. G. (1998) Squaring the circles – bias in distribution maps. *British Wildlife*, 9, 213–219.

Rich, T. C. G., Alder, J., McVeigh, A. and Showler, A. J. (1996a) *Stars in our eyes* [Back from the Brink Project Report No. 73]. London: Plantlife.

Rich, T. C. G., Donovan, P., Harmes, P., Knapp, A., McFarlane, M., Marrable, C., Muggeridge, N., Nicholson, R., Reader, M., Reader, P., Rich, E. and White, P. (1996b) *Flora of Ashdown Forest*. East Grinstead: Sussex Botanical Recording Society.

Rich, T. C. G., Gibson, C. and Marsden, M. (1999) Conservation of Great Britain's Biodiversity VI: Re-introduction of the extinct native species *Filago gallica* L. (Asteraceae), Narrow-leaved Cudweed. *Biological Conservation*, 91, 1–8.

Rich, T. C. G., FitzGerald, R. and Kay, G. M. (1996) *Review and Survey of Scarce Vascular Plants*, Unpublished contract report to Scottish Natural Heritage.

Rich, T. C. G., Hutchinson, G., Randall, R. D. and R. G. Ellis (1999) List of plants endemic to the British Isles. *BSBI News*, 80, 23–27.

Rich, T. C. G. and Jermy, A. C. (1998) *Plant Crib 1998*. London, BSBI.

Rich, T. C. G. and Smith, P. A. (1996) Botanical recording, distribution maps and species frequency. *Watsonia*, 21, 161–173.

Rich, T. C. G. and Woodruff, E. R. (1990) *BSBI Monitoring Scheme 1987–1988*. [Chief Scientist's Directorate Report No. 1265]. Peterborough: Nature Conservancy Council.

Rich, T. C. G. and Woodruff, E. R. (1992) The influence of recording bias in botanical surveys: Examples from the BSBI Monitoring Scheme 1987–1988. *Watsonia*, 19, 73–95.

Rich, T. C. G. and Woodruff, E. R. (1996) Changes in the floras of England and Scotland between 1930–1960 and 1987–1988: The BSBI Monitoring Scheme. *Biological Conservation*, 75, 217–229.

Rodwell, J. R., *et al.* (eds) (1991–2000) *British Plant Communities*. Cambridge: Cambridge University Press.

Roper, P. (1994) The BSBI membership survey. *BSBI News*, 67, 7–10.

Rothschild, M. and Marren, P. (1997) *Rothschild's Reserves*. Rehovot: Balaban.

Ryves, T. B., Clement, E. J. and Foster, M. C. (1996) *Alien Grasses of the British Isles*. London, BSBI.

Scannell, M. J. P. and Synnott, D. M. (1987) *Census Catalogue of the Flora of Ireland*. London: The Stationery Office.

Scott, N. E. and Davison, A. W. (1982) De-icing salt and the invasion of road verges by maritime plants. *Watsonia*, 14, 41–52.

Sell, P. D. and Murrell, G. (1996) *Flora of Great Britain and Ireland*. Vol. 5. *Butomaceae – Orchidaceae*. Cambridge: Cambridge University Press.

Simpson, N. D. (1960) *A Bibliographical Index of the British Flora*. Privately published.

Sotherton, N. W. (1998) Land use changes and the decline of farmland wildlife: an appraisal of the set-aside approach. *Biological Conservation*, 83, 259–268.

Spencer, J. W. and Kirby, K. J. (1992) An inventory of ancient woodland for England and Wales. *Biological Conservation*, 62, 77–93.

Stace, C. A. (1997) *New flora of the British Isles*. 2nd edn., Cambridge: Cambridge University Press.

Stevens, P. A., Ormerod, S. J. and Reynolds, B. (1997) *Final Report on the Acid Waters Survey for Wales*. Bangor: ITE.

Stewart, A., Pearman, D. A. and Preston, C. D. (1994) *Scarce Plants in Great Britain*. Peterborough: Joint Nature Conservation Committee.

Tansley, A. G. (1949) *The British Islands and their Vegetation*. Cambridge: Cambridge University Press.

Thompson, K., Hodgson, J. G. and Rich, T. C. G. (1995) Native and alien invasive plants: more of the same? *Ecography*, 18, 390–402.

Tutin, T. G. *et al.* (eds) (1964–1980) *Flora Europaea*. Cambridge: Cambridge University Press.

Walters, S. M. (1984) The relation between the British and the European floras. *New Phytologist*, 98, 3–13.

Webb, D. A. (1983) The flora of Ireland in its European context. *Journal of Life Sciences of the Royal Dublin Society*, 4, 143–160.

Webb, D. A. and Akeroyd, J. R. (1991) Inconstancy of sea-shore plants. *Irish Naturalists' Journal*, 23, 384–385.

Webb, D. A., Parnell, J. and Doogue, D. (1996) *An Irish Flora*. 7th edn., Dundalk: Dundalgan Press.

Whitten, A. J. (1990) *Recovery: A Proposed Programme for Great Britain's protected species* [CSD Report No. 1089]. Peterborough: Nature Conservancy Council.

Wigginton, M. J. (ed.) (1999) *British Red Data Books*. Vol. 1. *Vascular Plants*, 3rd edn., Peterborough: JNCC.

Williams, P. J., Biggs, J., Barr, C. J., Cummins, C. P., Gillespie, M. K., Rich, T. C. G., Baker, A., Baker, J., Beesley, J., Corfield, A., Dobson, D., Culling, A. S., Fox, G., Howard, D. C., Luursema, K., Rich, M. D. B., Samson, D., Scott, W. A., White, R. and Whitfield, M. (1998) *Lowland Ponds Survey 1996*. London: Department of the Environment, Transport and the Regions.

Wilson, P. J. (1992) Great Britain's arable weeds. *British Wildlife*, 3, 149–161.

Woodin, S. J. and Farmer, A. M. (1993) Impacts of sulphur and nitrogen deposition on sites and species of nature conservation importance in Great Britain. *Biological Conservation*, 63, 23–30.

UK Biodiversity Steering Group (1994) *Biodiversity: the UK Steering Group report*. Vol. 1. *Meeting the Rio Challenge*. Vol. 2. *Action Plans*. London: HMSO.

UK Biodiversity Group (1998) *Action Plans*. Vol. 2. *Vertebrates, Vascular Plants*. Peterborough: English Nature.

Chapter 3

Ferns and allied plants

Christopher N. Page

ABSTRACT

Pteridophytes (ferns and allied plants) represent a collective group of vascular plants native to Great Britain and Ireland which have been particularly critically studied from the point of view of detailed ranges, species abundance and habitats. Their study has a history spanning well over 150 years. Patterns of change are characterized by complex dynamics of both losses and gains, and these can differ substantially between the different included groups.

This account presents a brief overview of this important loss–gain equation set in an overall habitat perspective. It uses fern and fern ally examples to illustrate contrasting situations on the basis of biological evidence. It is concluded that although many of the changes have been, and continue to be, influenced in either their direction, extent or speed of progression by humans, nevertheless these influences should be seen as a dressing, however substantial, to an underlying process of natural change which is an integral part of the progression of post-glacial evolutionary bio-rediversification in our unusually species-depauperated insular pteridophyte flora. It is thus at least as important to allow new opportunities for new taxa to continue to appear, and especially for new hybrids to form, as it is to preserve just the array of species that we already have.

1 Introduction

1.1 Scope

The ferns and allied plants (ferns, clubmosses and quillworts, and horsetails – collectively referred to here as the native pteridophytes) are in a more or less constant state of change. This account presents a brief overview of dynamics and significance of this change within the insular flora of Britain and Ireland. Previous reports of vascular plant change (e.g. Perring 1970, 1974; Harley and Lewis 1984; Rich and Woodruff 1996) have incorporated pteridophytes along with the much larger group of flowering plants. Rich (Chapter 2, this volume) deals with flowering plants separately, and covers many of the more political issues which also relate to changes in pteridophytes. This report, which should be read in conjunction to that of Rich, presents an overall loss–gain equation for pteridophytes set in a species, hybrid and habitat perspective, and emphasizes biological aspects pertinent to an ecological and evolutionary understanding of these changes. The importance of the need for comprehensive onward recording, supplemented by detailed 'case studies', is particularly emphasized.

1.2 Biological background

The ferns and allied plants of Britain and Ireland are few in number for their latitude, with no more than 53 species of ferns, nine of clubmosses and quillworts and eight of horsetails known (71 native pteridophyte species; Page 1997). In a wider perspective, however, this small contingent is nevertheless particularly rich in Atlantic species (Page 1987), and forms a unique element on a European scale (Birks 1976; Webb 1983). In this it parallels the similarly significant Atlantic element among native bryophytes (Ratcliffe 1968; Hodgetts 1997). Further, in the pteridophytes, an exceptional number of hybrids are now known: at a conservative estimate, there are at least an additional 25 native hybrid ferns, one or more possible clubmosses and at least nine of horsetails – with, in the latter genus, the known native hybrids clearly outnumbering the species. Thus with over 100 species and hybrid taxa combined, and with additional subspecies of several of the taxa becoming increasingly recognized, what the pteridophyte flora may appear to lack in initial diversity is compensated for both in regional interest and in biotic complexity in an insular setting.

This pteridophyte flora has been studied both extensively and intensively for well over 150 years, with a current field-based biological flora (Page 1997a) which develops directly from the mould established by Newman (1844). Pteridophyta have always been included in most national and county floras of Britain and Ireland, so that despite the always small numbers of pteridologists, Britain and Ireland has certainly now the most intensively taxonomically and distributionally known pteridophyte flora in the world. This is not to say that there are not still new discoveries to be made. For it is largely since the affirmation of the importance of cytological chromosome analysis (largely post-Manton 1950) in confirming the status of species complexes and especially of hybrid taxa, that the very existence of, and distributional data in relation to, the large number of such taxa has so substantially advanced.

For pteridophytes, these islands thus form an important natural laboratory for the study of insular biogeography (Page 1987), in which the hybrid numbers in relation to species appear particularly exceptional. In the longer term, such hybrids are themselves the building blocks of potential new species in evolution through the route of allopolyploidy (see for example Manton 1950; Walker 1973, 1979; Haufler 1989, 1996). Hence, although pteridophytes are collectively the oldest native vascular land-plants, evidence from accumulated experimental sources (Dyer 1979; Dyer and Page 1985; Camus *et al.* 1996; Page 1997a) shows that many still retain the continued ability to respond to fresh opportunities and challenges as environmental conditions promote.

Pteridophytes thus remain good indicators of the very processes involved in such change. The theory has been advanced (Page 1997b) that the hybrids, as well as the formation of unusually edaphically and climatically adapted infraspecific taxa, are all part of a natural evolutionary process of progressive natural bio-rediversification in our post-glacially depauperated pteridophyte flora in an environmentally conducive insular setting. In this process, humans are playing an unwitting, but significant, additional role through the opening of successions of new habitats of considerable diversity. This account presents a brief overview of evidence of this continuing change.

2 Information bases

In Britain and Ireland, pteridophytes have a long history of critical taxonomic study. Among British pteridologists, there have always been strong links between interest in the fossil as well as the living species (Kidston and Gwynne-Vaughan 1907; Bower 1908, 1923–28, 1935; Seward 1931; Sporne 1962; Harris 1973; Thomas 1991; Colinson 1996; Poole and Page 2000) with a consequent stimulus to evolutionary approaches to the study of taxa. From the standpoint of extant species, this adds a strongly experimental approach to the taxonomic one. This route, closely integrating the potentials of laboratory and glasshouse in close combination, was particularly championed by Manton (1950), and following the path which she set through several decades such studies have helped to greatly expand our knowledge of the complexities of native genera such as *Asplenium* (e.g. Lovis 1955; Sleep 1985), *Dryopteris* (Gibby and Walker 1977), *Polystichum* (Sleep 1971), *Polypodium* (Shivas 1961a, b) and *Equisetum* (Page 1972a, b); see also Walker (1973, 1979), Lovis (1977) and Gibby (1991) for important overviews. As a basis for such studies, monographic surveys of pteridophytes always have had, and will continue to have, important pioneering roles to play (Holttum 1973, 1982), providing the essential fundamental foundations on which to base subsequent more exacting experimental analyses as new techniques come to the fore. More recently there have been many such developing chemotaxonomic and molecular approaches (Cooper-Driver 1976; Wolfe *et al.* 1994; Dubuisson 1996; Pahnke *et al.* 1996; Rumsey *et al.* 1996; Schneller and Holderegger 1996; Bridges *et al.* 1998). These are helping to indicate genetic variation levels in populations, to further define and delimit critical taxa, and to provide additional evidence of phyletic interrelationships at a variety of levels. Results from application of these techniques are of highest value when adequately considered in relation to evidence from all other sources. A recent extensive study of *Athyrium flexile* in Scotland (McHaffie 1998, 1999) has achieved highly worthwhile and surprising results by combining careful greenhouse study with diligent field observation backed by molecular analysis.

Of extant species ranges, species abundance and habitats, there is also an extensive knowledge base, much of it beginning with Victorian naturalists, especially Newman (1844), although its origins stretch back to Bolton (1785) – the first book in the world devoted entirely to ferns. In the last 25 years, there has been a burgeoning of knowledge of the field biology and ecology of both ferns and fern-allies (e.g. Page 1967, 1972, 1988, 1997a; Lovis 1977; Lawton 1982; Petersen 1985), as well as many aspects of their experimental biology (Page 1978; Dyer 1979; Dyer and Page 1985; Camus *et al.* 1996). There has also been an important progression in understanding their role as experimental tools (e.g. Manton 1950; Page 1978; Bell 1985; Raven 1985; Haufler 1996), as fundamental indicators in landscape change (Page and McHaffie 1991), in the process of natural biodiversity restoration (Page 1997b), and as surviving indicators of the locations of former glacial refugia (Vogel *et al.* 1999). Much of this wide overall information base relates not only to species, but also to hybrids and to some of the circumstances conducive to the occurrence of the free-living gametophyte generation, which is essential for successful sexual reproduction. The beginnings of more exacting studies of population dynamics of native ferns have also been undertaken in the last 25 years (Edwards 1982; Grime 1985; Willmot 1985; McHaffie 1997b), as well as the beginnings of different scientifically structured measures towards fern conservation (Lindsay *et al.* 1992; Page *et al.* 1992).

There remains much that can be further elucidated in relation to ecology of ferns and fern allies, including – of their gametophyte stages – the roles which the dynamics of novel taxa are playing in relation to processes of natural bio-rediversification, and the careful application of this knowledge to specific conservation issues. There is also an enormous amount of work to be done, scarcely yet started even on the international scene, applying the knowledge gained of ecological and evolutionary processes and their environmental interactions on the living pteridophytes with the enormous scope and growing interest and wealth of data in the fossil field.

From our uniquely strong and diverse existing information base, we have solid foundations for our knowledge not only of detailed species distributions and often of species abundance of the extant taxa, but also at least a working knowledge of the complex dynamics of the ecological and evolutionary processes that are constantly active in building and modifying the field picture that we see and record. Despite these potentials, the last 25 years have seen a substantial reduction in the number of pteridologists in academic institutions, with former specialists being retired and few replacement posts made, in a field in which even as recently as the 1970s Britain led the world. Future pteridologists are needed with the ability to continue to make the vital connections between taxonomy, pteridophyte potentials and the principles and processes involved and field dynamics, if we are to continue to build on what has already been achieved in understanding evolutionary processes and their interpretation, as well as maintaining a knowledge base to apply to sound conservation management.

3 Records and recording knowledge

Pteridophytes are, in many ways, ideal subjects for field recording. Many are quite habitat-specific, and most are large enough to be easily spotted. Fronds of ferns (shoots of clubmosses and horsetails) are readily amenable to being collected (always with secateurs) and easily pressed (if done immediately) with minimal harm to the perennial plant, while preserving the necessary taxonomic details necessary when further confirmation is required. Their photographs can also be additionally informative for recording purposes. They also have a high collective diversity of form and habit, sufficient to enable the ready scoring of many taxa with a high degree of confidence in overall field surveys, and for more unusual occurrences of species of non-conforming appearance to signal themselves readily for further attention. [As a result of biological recording, in the 1970s I coined the term 'diversity' to represent the morphological and ecological differences found among ferns. This usage has become broadened to 'biological diversity' as used in the Convention on Biological Diversity where it covers genetic, organismal and ecological diversity; that usage goes back to 1980, and the contraction 'biodiversity' was first made in 1985 (Harper and Hawksworth 1994).] Furthermore, in virtually all vegetation types, it is possible to undertake some (often complementary) aspects of pteridophyte recording at all times of year. Many of the species of pteridophytes which are wintergreen can be more conspicuous when other rank vegetation has died down, such as *Asplenium* in Cornish hedges,

Polypodium in high epiphytic sites, and *Equisetum* subgenus *Hippochaete* fringing the eroding outer margins of coarse streambank vegetation.

The inclusion of pteridophytes in vascular plant atlases, in the work of the Botanical Society of the British Isles and in the roles of botanical county recorders and many amateur naturalists in Britain and Ireland, has stimulated data capture to a level and on a scale which is at least comparable with those of most flowering plants. The inclusion of pteridophytes within overall vascular plant surveys has further helped to update the general knowledge base with respect to overall species change (Rich and Woodruff 1996). Furthermore, the production in the 1970s of specific pteridophyte atlases for both Europe (Jalas and Suominen 1972) and, in greater detail (10-km square) for Britain and Ireland (Jermy *et al.* 1978), plus the availability of subsequent supporting taxonomic pteridophyte literature (Synnott [n.d.]; Page 1982; Jermy and Camus 1991; Merryweather 1991; Hutchinson and Thomas 1996; Page 1997a; Rich and Jermy 1998), importantly at a range of different levels, have stimulated onward data capture from a wide range of sources, so that the pteridophyte flora of Britain and Ireland is now one of the most extensively known in the world.

Additionally, the British Pteridological Society has formed an important forum for fern and fern-ally interest, and has spawned a number of local regional groups active over long periods. It has also accumulated a critical knowledge base, often at a detailed scale. A national pteridophyte atlas is currently planned, while many tetrad atlases at a vice-county or county scale valuably include ferns and allied plants. There is thus now detailed knowledge on this scale for a growing number of counties (e.g. Cumbria, Kent and Northumberland) as well as for the whole south-west peninsula (Devon in Ivimey-Cook 1984, plus Cornwall in French *et al.* 1999 and 2000). Especially for local ranges and for critical groups, there is also a growing awareness of the importance of countering recording bias (cf. Rich and Smith 1996; Rich and Woodruff 1992; Rich, Chapter 2, this volume).

Currently 14 species of native pteridophytes are listed as nationally scarce and a further 12 taxa as Red Data Book taxa (38%) (Cooke 1996). A valuable table summarizing pteridophyte species protected by national laws in Britain, Ireland and Europe has already been presented elsewhere by Jermy (1998), to which the reader is referred for these aspects, while maps of pteridophyte ranges will be included in detail in the forthcoming specific pteridophyte atlas, and are thus not repeated within this account. Further, because many of the more political aspects of native pteridophyte knowledge follow closely in features and trends those already discussed with respect to flowering plants (see Chapter 2, this volume), this chapter focuses more particularly on establishing the more biological aspects that are significant in understanding the changes in range which pteridophytes specifically demonstrate, and which are thus ones of ultimate importance in understanding as a basis for structuring sound strategies for their conservation.

4 Susceptibility to range change

High sensitivity of pteridophyte ranges to influence of externally imposed environmental factors exists largely because of the tendency of ferns, horsetails, quillworts and clubmosses to mainly occupy ecologically marginal habitats, which are often too limiting for similar success of most flowering plants (Page 1987). Within these already narrowly prescribed conditions there may be little flexibility for tolerance of additional variation by sudden imposed exposure to new fluctuations and extremes, which are consequently most likely to be beyond the sphere of tolerance for many. For pterido-phytes on a world scale (Page 1979a, b), most species thus thrive best where conditions remain most constant.

Habitat changes which are especially inimical to many include both losses in humid-ity regimes and over-saturation by new light levels associated with disturbance of forest canopy cover, as well as high sensitivity to artificially high levels of grazing exposure. Additionally, for lithophytic and epiphytic species their high dependency directly on incidental rainwater supplies exposes them particularly to atmospherically derived pol-lutants in the same way as bryophytes and lichens of these habitats. In addition, excep-tional tolerance of low nutrient regimes is a valuable enabling strategy in many pteridophytes. Consequently sudden enhanced nutrient exposure, such as that from high nitrification levels, may rapidly swing the competitive balance in high light habi-tats totally away from pteridophytes in favour of the success of rank flowering plant vegetation, especially grasses.

Although many of these factors have been operative in causing change to pteridophyte ranges over long time spans, additional degrees of exposure to many of them in concert (including grazing, pollutants and probably nitrification levels) seem to have been especially significant in the last 25 years.

5 Biological potentials for area and range extensions

Two potentials of pteridophytes are of importance in promoting change in either area or range (and sometimes both): those which can be achieved through vegetative growth and those which can be achieved via dispersal of their minute airborne spores.

Potential increase through vegetative growth results from the great longevity of many pteridophytes sporophytes, and is particularly characteristic of taxa possessing long-creeping subterranean vegetative rhizomes (in Britain and Ireland notably *Equisetum* and *Pteridium*, for example). In *Equisetum*, formation of additional colo-nies through water-borne rhizome fragmentation occurs among several species and hybrids, achieving colonial replication of rare taxa along river banks, lake shores and even marine shorelines (Page and Barker 1985). In *Pteridium*, the history and strategies of spread of the plant continue to be the subject of detailed study (Page 1976, 1986; Birnie and Miller 1986; Weaver 1986; Pakeman and Marrs 1993; Kendall *et al.* 1995; Pakeman *et al.* 1995; Whitehead and Digby 1995; Robinson and Page 2000). Colonial expansion alone has been estimated to add an overall increase of 1–3% in overall area per annum (Taylor 1986, 1988), while rhizome growth can also enable sparse bracken to turn rapidly to dense bracken with change in management regimes (Marrs *et al.* 1997; Smith 2000).

Potential increase especially in range through spore dispersal opens opportunities for both more distant colonisation of new sites by pteridophytes of all types and, espe-cially, rapid invasion of new virgin habitats almost wherever these may arise. Stimulat-ing discussion on an appreciation of fern edaphic niche requirements is presented by Petersen (1985) and of overall ecological factors limiting the contribution of pterido-phytes to a local flora are presented by Grime (1979, 1984, 1985). There are rather few direct studies on airborne pteridophyte spore diversity, but recent studies which have centred on the monitoring of airborne spores of bracken have tended to confirm their ubiquitous nature (e.g. Lacey and McCartney 1994; Caulton *et al.* 1995). The high mobility of pteridophyte spores enables pteridophytes to invade new territory, especially under sudden post-disaster scenarios at a variety of scales. Furthermore, dispersed spores which have already formed soil spore banks (Dyer and Lindsay 1992) may germinate on subsequent exhumation when natural soil erosion events take place.

6 External factors affecting change in range

Constant changes in either range or abundance are probably always happening on shorter or longer time scales to virtually all of our native species. Those changes which are likely to cause greatest conservation concern are, however, those which appear to be mainly additional to the balance of purely natural processes, especially when the trends are ones of steady diminishment and when similar trends are reflected by unrelated species. Factors causing such diminishment in pteridophyte ranges are often difficult to prove and exactly quantify, and probably remain numerous and diverse. The major ones are presented in summary form in Table 3.1.

A notable outcome of presenting these data in assembled form is that it illustrates not only how many taxonomically and habitat differentiated pteridophytes are threatened, but how many and diverse are the factors influencing them. It also displays how many of these factors are manifestly human induced, and thus by definition could be controlled.

For some species, while they are decreasing in some habitats, expansion may be occurring in others. A particular example might be *Asplenium trichomanes* ssp. *quadrivalens*. This can be suffering from pollution effects on mural habitats in urban areas, but can be simultaneously spreading on walls in rural west-coast areas of high humidity as old wall mortars gradually decay to an advanced state of suitable maturity for colonization. Such changes in balance are, however, of chief occurrence amongst the naturally more widespread taxa, and taxa of more restricted range seldom have similar opportunity for populations to counterbalance elsewhere.

However, in contrast to this negative side of change in range and abundance, there are also two positive sides to be seen. These are sites in which old taxa could potentially recover or in which new taxa *per se* are actually appearing.

New taxa can, of course, appear in an area in which they have been previously unrecorded through immigration by airborne spores, and this is almost certainly their most frequent method of origin. But an additional route of recolonisation comes from the important discovery in the last decade of the occurrence of natural banks of viable fern spores surviving in soil, at least of the large number of brown-spored species, although spores of bracken and green-spored species (for example, *Osmunda regalis* and *Hymenophyllum tunbrigense* probably do not survive – Dyer and Lindsay 1992; Lindsay *et al.* 1992). Although further work is required yet to prove several aspects of this, including the time for which spores actually survive in soil under field conditions such spores could potentially provide sources from which, when re-exhumed by natural or by deliberate management processes, new fern gametophytes and then sporophytes of the local genotypes of former species could arise.

The second of these positive aspects is in the formation of natural hybrids which may be new for sites or new to science. In practice, it is difficult to separate the number of

Table 3.1 A selection of native pteridophyte species known or suspected of showing continuing over-all decline set against likely causative agents of range and/or abundance decline.

Taxon	Suspected factors (in order of probable importance)
Diphasiastrum × issleri	hfb, clsh, nrs
Huperzia selago	hfb, wt
Lycopodiella innundata	hfb, wt
Lycopodium annotinum	hfb, clsh, aff, nrs, shbs
L. clavatum	hfb, clsh, nrs, shbs
Equisetum hyemale	grz, clsh, nrs, nie
E. variegatum	grz, clsh, nrs, nie
E. pratense	clsh, nrs, shbs
×Asplenophyllitis confluens	poll, lfbh, nie
×A. microdon	poll, lfbh, nie
Asplenium billotii	poll, lfbh, nie, drow
A. marinum	poll, shbs, nie
A. septentrionale	poll, clsh, san, shbs, aff
A. ruta-muraria	poll, dhd, drow
A. viride	poll, shbs, dhd
A. ×alternifolium	poll, clsh, san, shbs, aff
A. ×clermontiae	poll?
A. ×contre	poll?
A. ×murbeckii	poll, shbs
Blechnum spicant	poll, lfbh, law, nie
Botrychium lunaria	poll, lop, bren, shbs, nie
Ceterach officinarum	poll, shbs
Cystopteris fragilis	poll, nrs, drow
C. montana	clsh, nrs, shbs
Cryptogramma crispa	clsh
Dryopteris aemula	lfbh, law
D. expansa	clsh, ugr, shbs, lauw
D. carthusiana	wdr, lauw
D. cristata	wdr, sthb, nie
D. submontana	dhd
Gymnocarpium dryopteris	clsh, ugr, lauw

continued on next page

Table 3.1 A selection of native pteridophyte species (cont.).

Taxon	Suspected factors (in order of probable importance)
Gymnocarpium robertianum	clsh, dhd
Ophioglossum vulgatum	poll, lop, bren, shbs, nie
Osmunda regalis	wdr, ugr, shbs, lauw
Phegopteris connectilis	clsh, ugr, lauw
Pilularia globulifera	poll, wdr, ldh, nie
Polystichum lonchitis	clsh, ugr, shbs
P. aculeatum	law, san, shbs
Pteridium pinetorum	law, nie
Thelypteris palustris	wdr, nie, sthb
Trichomanes speciosum	wrfc
Woodsia alpina	wrfc, clsh, nrs
W. ilvensis	wrfc, clsh, nrs

Key to likely causative agents
aff   afforestation programmes
bren   bracken encroachment
clsh   climatic shifts (especially affecting populations near the limit of their range)
dhd   direct original wild habitat destruction
drow   destruction and/or repointing of old wall with new Portland cement mortars
hfb   heathland fireburn
law   loss of ancient coastal, lowland or upland woodland
ldh   lack of disturbance necessary for some habitats
lfbh   losses of field boundary hedges
lop   loss of old pasture
nie   nitrate enrichment
nrs   natural range shifts
poll   habitat pollution (aerial and/or groundwater, the aerial especially because of dependence on incoming rainwater)
san   sanitisation of former habitats (especially of former minesite tailings)
shbs   susceptibility (mainly through location) to herbicide spraying
sthb   stabilisation of habitats
ugr   upland (and occasional lowland) grazing regimes and related factors
wdr   wetland drainage (wetlands of all types)
wrfc   weak recovery from former collection, where species are already near to the margin of their range

Key to table 3.3 opposite
adw   artificially drained woodland, including coniferous plantation woodland
hes   hedges of earth and stone (chiefly in south-western Britain, southern Ireland and the Channel Islands)
oqu   old quarry workings
owm   old wall mortar (of either brick or stone-built walls)
mmw   metalliferous mine tailings
rsb   road-side banks and/or other roadside earth and/grassed verge areas
rsd   road-side ditches
rwl   railway line sides, including cutting banks, embankment surfaces and lineside ballast arrays
tsb   trackside banks, including tracks beside streams

For further details of ecology and parentage of hybrid taxa see Page (1997)

Table 3.2 Number of actual known hybrid taxa of pteridophytes compared to the number of known parents in Great Britain and Ireland.

Genus	Number of native species	Number of known hybrid combinations	Ratio of known hybrids to species
Diphasiastrum	1 (+1 extinct)	1	1:2*
Equisetum	8 (+1 extinct)	7	1:2.8
Asplenium	8	8	1:1
Dryopteris	9	8	1:1.2
Polypodium	3	3	1:1*
Polystichum	3	3	1:1*

* all wild hybrid combinations possible are actually known.

Table 3.3 Comparison of occurrence of hybrid pteridophytes recorded from sites created by humans in the short (25–30 year) and longer (100+ year) term in Great Britain and Ireland.

Hybrid taxon	Known in last 25–30 years	Known in last 100+ years	Habitat type
Asplenium ×alternifolium	x	x	mmt
A. ×clermontiae		x	owm
A. ×lusaticum	x		hes
A. ×sarniens	x		hes
A. ×ticinense	x		hes
×Asplenophyllitis confluens	x	x	owm
×A. jacksonii		x	hes
×Asplenophyllitis microdon	x	x	hes
Dryopteris ×deweveri	x	x	adw
Polypodium ×font-queri	x		owm
P. ×mantoniae	x	x	hes
P. ×shivasiae	x		hes
Polystichum ×bicknellii	x	x	hes, rsb
Equisetum ×bowmanii	x		rsb
E. ×dycei	x		rsd, thus
E. ×font-queri	x		rsd, rsb, rwl
E. ×litorale	x	x	rsd, rwl, oqw
E. ×willmotii	x		tsb

key on previous page

such hybrid taxa actually known from the number of pteridologists who are available to recognize them. But even bearing this in mind, Britain and Ireland appear to have an exceptionally high proportion of known wild hybrids in the native pteridophyte flora (Table 3.2). Some hybrids may acquire a permanent niche and achieve vegetative longevity, and can become the basis of future species should some of their own spores achieve eventual genetic stabilization through chromosome doubling resulting in *de novo* allopolyploid taxa. Other hybrids (probably at least equal in amount) are less successful than their parents and, using the example of *Equisetum*, it has proved possible to predict which hybrid combinations are likely to produce ecologically successful progeny and which are least likely (Page and Barker 1985).

But for hybrids and for new colonizations there is a noteworthy trend (Table 3.3): for fresh colonists to appear in manifestly man-made situations and, in the case of hybrids, for the known number of taxa as well as the number of their stations to be increasing with time especially in man-made habitats. Records suggest too that multiple origins of similar hybrids are possible. Notable records for this occur in *Polypodium ×mantoniae*, *Equisetum ×dycei*, *E. ×font-queri*, *E. ×litorale*, *E. ×rothmaleri*, arising widely *de novo* on a polytopic and polychronic basis.

7 Synoptic examples of native abundance and range changes by habitat

7.1 Outcropping rocks

Many rock habitats, especially those in upland areas, contain rich assemblages of ferns, and although often remote, changes in these can result from both general and specific factors. In upland areas particularly, the most widespread factors causing change are likely to be those due to progressive climatic amelioration or increasing aridity influencing both the vegetative and perhaps reproductive success of species, especially those of northern and arctic-alpine range that are already at or near to geographic and presumably ecological limits. Decline in *Woodsia* species in Scotland has been well documented and remains under current study by Dyer and Lindsay. Reproductive loss has been recorded in *Equisetum pratense* (Page 1988), with a steady northward migration in the occurrence of fertile shoots, which were frequent as far south as Teesdale in the mid-nineteenth century and today are rare in occasional seasons mainly in Perthshire and unknown further south. Where the outcropping rocks are of unusual mineral content, as is the case with serpentines, unusual fern assemblages often also flourish. These, however, can be very local, and present their own problems for assured conservation (Vogel 1997). Elsewhere, direct destruction of specific natural limestone pavement habitats, formerly considered relatively 'safe' from most destructive forces, now threatens at least locally several species of smaller ferns, among which *Asplenium viride*, *Dryopteris submontana* and *Gymnocarpium robertianum* are particularly characteristic (Webb and Glading 1998).

7.2 Heathlands

Heathland areas at a great range of altitudes, from the uplands to coastal fringes, are the main habitats *par excellence* for the many British species of clubmoss

(*Diphasiastrum, Huperzia, Lycopodiella, Lycopodium* and *Selaginella* species), and several of these can (and formerly almost certainly were) extensive and frequent in patches within the mosaic of living and decaying old woody heath vegetation – a vegetation type which develops its full mosaic of 'pattern and process' cycles (Watt 1947) only over a long time period (for example Gimmingham 1964; Barclay-Estrup and Gimmingham 1969; Gimmingham *et al.* 1979). However, clubmoss regeneration through spore establishment, mycotrophically supported gametophyte growth and subsequent sporophyte establishment is a particularly slow process progressing over many years. Sporophytes of probably all species are destroyed by the passage of fire, which removes almost every individual as well as their means to sexually and vegetatively reproduce and perranate. The recolonization of fireburn areas by clubmosses may thus be longer than the fire cycles ('muirburn' in Scotland) of human origin which are used as a management tool in promoting growth of young heather for game. At the other geographic extreme of Britain, heathland fires have been also identified as the factor responsible for the loss of extreme southern outlying sites for *Huperzia* clubmosses in Cornish heaths (Murphy and Bennallick in press).

Such fires, especially when combined with high and sustained grazing pressures, themselves stimulate the continued encroachment of upland bracken (*Pteridium*) both directly into heathland and into the opened canopy of surviving woodland enclaves (Watt 1955; Taylor 1985, 1986; Weaver 1986). Such bracken itself depresses further the biological diversity of the habitat, including the elimination of other ferns through its strong allelopathic effects (Gliessman and Muller 1972; Cooper-Driver 1976; Gliessman 1976). Furthermore, spraying to control bracken through aerial application of broad-spectrum pteridocidal herbicides (such as Asulox) is a new fern survival hazard. The short-term and long-term effects of such herbicides, especially on narrow ravine enclaves of surviving woodland fringes and their remaining fern fragments, without adequate pre- and post-treatment pteridological survey, have aroused conservation awareness, mainly in relation to subsequent vegetation restoration (Lawton 1982; Burn 1988; Marrs and Lowday 1992; Pakeman and Marrs 1993; Marrs and Pakeman 1995). Specific pteridological effects thus remain to be widely assessed.

7.3 Upland community and upland woodland cover

The significance of upland pteridophyte-rich vegetation communities either in fern gullies of long winter snow-lie or in small fragments of upland woodland and typically rich in ferns, horsetails and clubmosses have long been specifically recognized by ecologists (Pigott 1956; McVean 1958a; McVean and Ratcliffe 1962; McVean 1964; Ratcliffe 1960, 1977), and have more recently received more focused attention from pteridologists (Page 1988; McHaffie 1998, 1999). Grazing exclosure experiments demonstrate very clearly how constant grazing by stock (especially sheep) and deer, in addition to mountain hares and rabbits, continues to have a profound causative effect on steady diminishment to fragments of such upland woodland cover and associated fern species. For species such as *Dryopteris expansa*, its frequent confinement to only the most grazing-inaccessible ledges, and the progression to this extreme through human use of the uplands (e.g. O'Sullivan 1977; Dimbleby 1984) provides a particularly vivid example of the effect that constant heavy upland grazing continues to maintain artificially. Most upland ferns are probably similarly diminished, with the possible exception of

Cryptogramma crispa, which is very unpalatable and thus can survive in grazed areas when other species do not (McHaffie 1997). In valley bottoms, *Osmunda regalis* is also grazing-sensitive, and in the Scottish west Highlands, for example, has continued a trend noted earlier by McVean (1958b) in often becoming restricted to small grazing-inaccessible islands within steep-sided peaty lochans or to islets of difficult grazing access offshore within larger Scottish lochs.

7.4 Ancient woodlands

Ancient woodlands or 'wildwood' have a long and continuous history which has fascinated ecologists in both Britain and Ireland (Tansley 1939, 1953; McCracken 1971; Mitchell 1976; Rackham 1980, 1986; Peterken 1981; Woodell 1985; Bullard 1987). Such ancient woodlands are an extremely important habitat for pteridophytes, not only for the species numbers and individual fern abundance often contained, but also for a number of range-restricted species which may be well represented (Richards and Evans 1972; Roberts 1979; Page 1988; Ratcliffe *et al.* 1993) and the fern genetic diversity which can be potentially harboured (e.g. Rumsey *et al.* 1996).

The sensitivity of most woodland ferns to sudden environmental change, however, and especially to episodes of large-scale canopy removal, parallels that of many bryophytes of these habitats (e.g. Edwards 1986; Hodgetts 1997) and makes the presence of several pteridophytes particularly valuable indicators of ancient woodland sites in which deciduous canopy-cover has a long history of continuity. For sporophytes of *Hymenophyllum tunbrigense* and the gametophytes of the rare *Trichomanes speciosum* for example, occupied sites are ones which show remarkably little variation of diurnal and annual temperature and humidity regimes (Rich *et al.* 1995; Rumsey *et al.* 1998). Indeed, it is usually only in the wetter climates, where frequent cloud cover to a large extent compensates for the presence of a woodland canopy, that many of our woodland ferns are able to survive in the open, especially in the shelter of ravines (e.g. Page 1988; Church 1990). Within ancient woodland areas, I number among good indicators of long-term canopy cover the presence of substantial populations of large individuals of clump-forming ferns such as *Dryopteris expansa*, *D. aemula* and *Blechnum spicant*, large continuous areas of *Gymnocarpium dryopteris* and *Phegopteris connectilis*, the woodland horsetails *Equisetum sylvaticum* and *E. pratense*, large-sized patches of *Hymenophyllum tunbrigense* and *H. wilsonii* on rocks and extensive bank-top patches of all *Polypodium* species. Many of these acquire large size only by slow growth over extended periods of time under particular constancy of conditions. Similarly, Rackham (1975:126) notes that within long-established woodland, ground vegetation zones tend to be relatively stable and determined by long-term factors rather than by the year-to-year fluctuations of weather or changes in the structure of the tree canopy. Old-established woodland canopies clearly provide vital and unique short- and long-term substantial environmental buffers to the vegetative and species diversity within, of which the pteridophytes are integral components of high indicator value.

Additionally, as in upland woodland, animal grazing, especially by mammalian herbivores, continues to provide a further insidious way in which many of our woodland areas, including ancient ones, have already floristically degenerated, in which tree regeneration itself is suppressed (e.g. Durno and McVean 1959; Gimmingham 1977; Proctor *et al.* 1980) and in which the fern component becomes widely replaced by grasses (Page 1988). Grazing exclosures examined by the author after runs of 10–25+ years in relation to Scots pine in Scotland, to upland woodland in north Wales, and to Killarney oakwood in south-west Ireland all show that the exclusion of grazing alone can be sufficient for former natural woodland multi-aged tree-stand structure to slowly regenerate and for internal woodland habitats to begin to reform. Into this, native ferns and other pteridophytes can successfully return, often from small enclaves to which they have become restricted such as along the sides of occasional surviving grazing-inaccessible ravines (Page 1988). The re-establishment of the continuity of the forest canopy additionally provides a simple and effective natural way of suppressing the vigour of bracken, even when its spread and density may have already been promoted by the former grazing regime.

7.5 Old pasture

Old pasture grasslands are an important habitat particularly of the slow-growing colonial *Ophioglossum vulgatum* and *Botrychium lunaria*, the presence of which are good indicators of long-established turf with lack of disturbance (Page 1988). Both also depend upon obligate mycotrophic relationships with soil fungi, and are slow growing both at the gametophyte and sporophyte stage. Both also have a reputation for having a sporadic and perhaps somewhat cyclical occurrence in sites in which they have been formerly known (Page 1987) and which appear to be easily disturbed. In contrast to the relatively slow rates of change characteristic of the long historic period of settlement of such sites (e.g. Jones 1988; McDonald 1988), the modern cumulative effects of agriculture through mechanization, intensification and increased scale of operations have been commented upon also by Perring (1970, 1974) and by Rich (Chapter 2; this volume). Although once widespread, both of these ferns are now scarce and local and their very wide demise probably results from a combination of many factors, of which the ploughing of old turf and the loss of old water-meadow pasture are events which are subsequently difficult to reverse.

7.6 Wetlands

Areas of bog, marsh, fen and carr are important wetland habitats for a number of specially adapted pteridophytes belonging to diverse and often ancient groups (Page 1988). In bog areas, careful monitoring is yet required of populations of *Lycopodiella innundata* to establish clearer patterns underlying the dynamics of their long-term appearance and disappearance. A similar need also applies to *Pilularia globulifera*. Both are intolerant of vegetative competition, and probably require habitats constantly reopening. For *Pilularia*, these islands have probably the most extensive surviving stands in Europe: a great deal of survey work has been undertaken and for some sites active remediation attention is focusing (e.g. Scott *et al.* 1999). In other areas, especially fen and woodland carr, drainage programmes of even small enclaves continue to reduce many former habitats for such species as *Dryopteris carthusiana* (Edwards 1982). In many areas, this species continues a century-long history of steady and extensive disappearance. *Thelypteris palustris*, which once characteristically pioneered many of the wettest sites within fenland habitats, continues a widespread decline and has been presumed extinct in the last 25 years in many former sites from Scotland to

Cornwall. In addition to direct drainage threats, total exclusion of natural sources of former disturbance regimes may also have locally harmful effects (Page 1987; Page and McHaffie 1991), and is probably the cause of wide losses of *Dryopteris cristata* and its hybrids from some remaining fenland habitats. In changing conditions in the Norfolk Broads fens, however, acidification appears to be increasing sphagnum areas and forming a new substrate for the prothalli of *Dryopteris cristata* (A.C. Jermy pers. comm.). There is also some evidence that this species showed a temporary increase following periods of extensive flooding of the reedbeds and consequent widespread silty depositions after the widespread Norfolk floods of 1952, and it has recently been found successfully colonizing an irrigated area of old metal mine tailings in Cornwall, in the absence of any known other colony in the county. Episodes of local disturbance, involving the creation of temporarily bared moist mineral surfaces (e.g. Wheeler 1980; Moss 1983), are clearly valuable in maintaining the full diversity of wetland habitats for such pteridophytes.

7.7 Lake margins, streamsides, ponds and ditches

The shorelines of lake margins, streamsides and ditches can be important pteridophyte habitats. Such sites offer appropriately scoured virgin surface habitats presenting opportunity for the occurrence of spore colonization as well as for spread by vegetative stem and rhizome fragments. Such sites have, fortunately, suffered less from despoilation than many others. For example, acidic lake shores continue to provide strandline habitats particularly suited to colonization by *Osmunda*, muddy ones for colonization by *Pilularia*. for the latter species, further factors causing a degree of disturbance of the mud surface (such as waterfowl and sometimes cattle), as well as a certain amount of natural scouring and fluctuations in water table levels may be desirable in maintaining the openness of habitats for annual recolonization. Where an aquatic marginal habitat is sufficiently both base-rich and silica-rich (such as over many deep clay substrates), several species of *Equisetum* are usually characteristic colonists: the species further differentiate both by water depth and by its degree of aeration. Where additionally opened by people, some of these habitats, including those of canal banks and ditches, have provided important arenas for prothallial growth and outcrossing, and in such disturbed wetsites many new hybrids in *Equisetum* have been discovered in the last 25–30 years (Page 1982, 1997; Page and Barker 1985).

7.8 Coastal habitats

Because of the preponderance of exposed bare mineral surfaces free of most other vegetation competition, both soft and hard rock coasts can contain a wide range of pteridophyte species. The diversity and number of these is not always appreciated because of the difficulty of visual access to many such sites. However, evidence on mainland areas compared with similar habitats on nearby off-shore islands, and from old dated mainland specimens preserved in herbaria, suggests that on many mainland rocks there has been as steady demise in numbers of individuals and in the sizes achieved by individual plants compared with those still persisting on nearby islands (Page 1988 for the Bass Rock and Channel Islands; Dyer 1996 for Iona; and Gastrang 1998 for Rathlin Island as examples). This is a little appreciated trend in insidious diminishment of ferns on many mainland coasts which probably began a century or more ago, but which appears to have continued apace over the last 25 years. Pollution (as well as drying up) of seeping groundwater supplies and their nitrification in many areas remains a particularly significant factor in a long overall process of pteridophyte decline.

7.9 Field boundary hedges

Field boundary hedges, especially 'hedges' of earth and stone of the Cornish type, are important and often rich habitats for pteridophytes, especially along many field boundaries and along those of the deep water-lanes of southern Ireland, south Devon, Cornwall and the Channel Islands. Many such hedges have a long and stable history, some dating from Iron Age times (e.g. Mitchell 1976; Balchin 1983; Dimbleby 1984; Woodell 1985). These habitats, although manifestly of human construction, contain niches which combine the characteristics of those of natural woodlands and rocky slopes, and have in consequence plant communities of their own, many of the components of which can be zoned on specific hedge aspects. As many such hedges were constructed before much of the surrounding woodland cover was itself destroyed, they have formed refuges potentially harbouring ancient genetic diversity, and their evolutionary and conservational importance should be viewed in this perspective. Numerous ancient hedges, especially those between small fields, have been removed in the last 50 and especially 25 years, to establish larger fields for modern machinery (e.g. Ratcliffe 1984). Those adjacent to roads and lanes have probably suffered less direct depletion, although are most directly influenced by changes in hedge-trimming regimes as well as the effects of passing traffic. In relation to the latter factors, a current survey of those in west Cornwall by the author (Page, in prep.) shows a hitherto unrecognized trend of progressive floristic impoverishment of ones along roads carrying the heaviest of modern vehicular traffic, resulting in diminishment of the greater part of their fern diversity.

7.10 Rocks and walls

These equally manifestly human-made habitats have had both positive and negative effects on the ranges of fern species. Praeger (1934) pointed out long ago, for example, that the great majority of habitats for *Ceterach officinarum* in Ireland were walls of human construction, and that the species must have spread widely geographically as a result of their construction. Across the whole of Britain and Ireland, this might also apply also to *Asplenium ruta-muraria* and *A. trichomanes* ssp. *quadrivalens*. Old churchyard walls, and some of those associated with old railway and canal constructions (e.g. Busby 1976), remain pteridologically important habitats. In recent years, probably over the last 50 but especially through the last 25, there has been a great diminishment of these species from many walls in urban areas. Causes of this are probably a combination of background airborne pollutants, and the repointing of many walls with new Portland cement mortars (rather than the lime mortars with which many were originally made – see Page 1987). The slow and insidious effects of pollutants are further commented upon by Rich (this volume).

7.11 Quarries, mine spoil and abandoned industrial areas

Disused quarries and mine-spoil areas (including flooded areas, clay pits, mineral tips, tailings and associated buildings), and other areas of former industrial abandonment (including canals and railway line trackbed, stations and cuttings) have provided pteridologically valuable and evolutionarily informative sites. Through the process of disturbance and exposure of some unusual mineral surfaces to colonisation, initially many of these provide unusually stressful environments (e.g. Parsons 1983), but into many of which an array of fern and fern-allies have spread during their period of post-industrial abandonment, either on a seral or often on a potentially more permanent basis (Page 1987). The potential value of such sites as wildlife refuges has become increasingly recognized in the last 25 years (e.g. Davis 1979; Page 1987), and mineral working in clayey areas still gives potential sites for colonisation by *Pilularia globulifera* (e.g. Whisby in Lincolnshire – A.C. Jermy, pers. comm. 1999). However, mainly during this same period many quarry areas have found alternative use as landfill sites for dumping urban waste. Furthermore, with remarkable ecological insensitivity, many mine-tailing sites have been infilled and grassed over in the name of relandscaping. The resultant sanitized sites of little floristic value now replace pteridologically exciting mine sites, as far apart as Strontian in Ardnamurchan, western Scotland, in several former metal mineworkings in North Wales, in coal-workings of the Forest of Dean in Gloucestershire and around some Cornish tin mines. All have been subject to costly land-restoration schemes, resulting in extensive unnecessary loss of their geological and industrial archaeological interest of which the colonizing fern diversity was a unique and integral component.

8 Aliens and introductions

From other parts of the world, several pteridophytes (especially ferns) have been introduced for horticultural ornamental use into both Britain and especially Ireland. Many have long persisted vegetatively within the confines of the original gardens in which they were planted. Most have achieved spore fertility in our climate, but despite this, few seem to have successfully established themselves subsequently from spores and then usually only as scattered or solitary individuals (e.g. Rickard 1994). Additionally a very few have spread by accessory vegetative means. The most widespread examples are given in Table 3.4.

Most of the taxa occurring in or near gardens and establishing by local spore spread do so, as do many native species, mostly on freshly eroded earth banks by footpaths, tracksides and so on. Such colonizations are more characteristic of western (and especially south-western) gardens in Britain and Ireland than elsewhere, largely under the influence of mild and moist Atlantic climates.

However, there appear to be severe limits to what can be achieved by most aliens in terms of successful establishment within existing native vegetation when this is undisturbed. Competition levels which allow the continued occupancy of already vegetated habitats by the existing taxa and the exclusion of most new invaders (which must surely frequently arrive by spores) appears high. Remarkably few introduced aliens thus seem (without human intervention) to travel far, and over time, surprisingly little change in occurrence patterns of dispersal achievement amongst introductions appears to arise beyond small shifts on a largely local basis.

Table 3.4 Some non-native adventive pteridophytes in Great Britain and Ireland persisting or establishing locally (the list is on a sample basis only and is not intended to be all-inclusive as there are many additional taxa for which there are scattered and occasional records).

Species	Occurrence pattern	Main means of persistence	Main means of dispersal
Azolla filiculoides	flu	lrg	vfwf
Onoclea sensibilis	lbw	lrg	vfwb
Matteuccia struthiopteris	lbw	lrg	vfwb
Equisetum ramosissimum	lbw	lrg	vfwb
Selaginella kraussiana	ghe	lrg	lrf
Cyrtomium falcatum	lbs	lrg	?
Dicksonia antarctica	gvs	lrg	lss
Pteris cretica	gvs	eph	lss
Polystichum polyblepharum	gvs	lrg	lss
Pyrrosia confluens	gve	lrg	lrf

Key:
eph ephemeral
flu fluctuating population on an annual basis
ghe glasshouse escape, with an ability to persist below staging through advanced states of glasshouse decay (and as an escape in adjacent stonework in at least one Cornish garden)
gve gardens vicinity as epiphyte
gvs gardens vicinity on eroding slopes
lbs local by sea (in a habitat similar to that of its overseas native occurrence)
lbw local in occurrence, mainly or exclusively by water
lrf local rhizome fragmentation
lrg local rhizome growth
lss limited spore spread on a chiefly local basis
vfwb vegetative fragments mainly water-borne
vfwf vegetative fragmentation dispersal by waterfowl

Such biotic competition levels involved in the intimate pattern and process of the dynamics of the micro-environments in which the lifecycles take place operate, it must be concluded, at a level and with an intensity of rigour that we have, as yet, fully to appreciate, let alone to explain.

9 Future research and observations required

The making of fundamental field observations is the very basis of virtually all laboratory research relating to species, as well as to all conservation. In pteridophytes, as in many groups, continuing field observations remain much needed. Yet today, the making of field observations is increasingly becoming regarded as a non-essential part of top-flight fundable academic research activity. How often do we hear modern academic seminars, lectures and the like, which are strong in technical detail, founder on a deficiency of fundamental field data to which that research ultimately applies?

There are continuing needs for the following:

- There is a fundamental need to identify species correctly if records made are to form a reliable basis for both current environmental assessments and from which future comparisons can be made. It is also vital to have adequate herbarium specimens, particularly of critical material, non-destructively collected and permanently deposited in recognized herbaria as a basis for future taxonomic verification and comparison. There is a need here for sustained support of records centres and herbaria at a local as well as national level.

- New data are continually needed in refining our picture of pteridophyte ranges, including recording of continuing change. Various important aspects of the forward work of pteridophyte recording and monitoring by the British Pteridological Society have been outlined by Jermy (1998). Relating to this is the need for careful 'case studies' monitoring, with sustained observations at population level, the details of the 'pattern and process' of change in specific populations. Many such observations are as open to the amateur as to the professional to make.

- There is a need to develop a better understanding and, where possible, quantification of many fundamental aspects of pteridophyte lifecycles in the field, coupling laboratory techniques with sound field observation. Valuable recent studies such as with the field-recognizable gametophytes of *Trichomanes* (e.g. Rumsey and Sheffield 1990; Rumsey *et al.* 1998) as well as our developing knowledge of the role of soil spore banks point to important ways forward.

- For most species, little information is available yet of the distribution of, and variations in, underlying genetic diversity with habitat. We need to know whether 'hotspot' enclaves of high diversity exist (both morphological and molecular, if different), of the conditions which have promoted the special formation and/or survival of such sites, and how such knowledge can best be incorporated into sensitive conservation strategies. Linked to this, the use of experimental cultivation of critical taxa under greenhouse conditions through each stage of the lifecycle, are techniques which still remain of highest value in understanding the significance of genetic traits identified.

- The use of good field photography is an irreplaceable tool in basic recording. Photographic records are invaluable not only in potentially setting population change against a backdrop of change in habitat and vegetation, but also the relationships between vegetation and the physical environment, such as erosional activity. Proctor *et al.* (1980) show clearly the value of the application of repetitive photography for the monitoring of change in plant communities and provide a especially valuable model, though there remain remarkably few published examples of this.

Places for permanent publication of many of these observations exist through the various publications of BSBI and the British Pteridological Society. Collectively such observations provide evidence of value in the focussing of priorities in future conservation management schemes.

10 Strategies for the conservation of native species

Strategies important for native pteridophyte conservation fall under two heads: those actions important in maintaining the species which we have, and those actions important in allowing new opportunities for taxa to continue to appear.

10.1 Actions important to maintain species

- Control of grazing regimes. Many pteridophytes are very much more affected (and usually destroyed) by high artificially maintained grazing regimes than was appreciated even only a few years ago. Ferns, clubmosses and horsetails are all special grazing targets of a variety of animals, and high and sustained artificial grazing regimes are responsible directly for the widespread loss of many pteridophyte species from habitats as diverse as mountain rocks to lowland Atlantic woodland. There is clearly great potential yet to be conservationally gained by simple exclusion of artificially high grazing regimes from very many potentially fern-rich habitats.
- Control of fire regimes. Fire is especially inimical to survival of most clubmosses species communities of which build up only very slowly over only long periods of time (see 'heathlands' above). In all heathland areas, but in the uplands especially in areas of the most broken topography, such as along upland stream ravines, long-term continuity of sufficient unburnt areas are needed if many clubmoss species are to return adequately.
- Control of airborne and water-borne pollutants and of drainage. Widespread multiple effects of pollution of both of air and water appear to extensively affect many pteridophytes in ways which are insidious and thus often difficult to perceive or quantify. Additionally, draw-down of water tables and direct drainage of other habitats is similarly causing diminishment of pteridophytes from many habitats in which they formerly were present, from fenlands to coastal flushes and valley-side seepage lines. Adequate control of pollution sources, nitrification of water supplies, and water table levels are issues which have yet to be totally addressed.
- Sensitive conservation management of abandoned mineral workings, quarries and similar areas of former industrial exploitation. It is important that already existing abandoned areas of exposed bare mineral surfaces are not unnecessarily grassed-over and sanitized and important quarries not used arbitrarily as landfill sites in the name of progress, to create artificially grassed areas of little floristic or other wildlife potential. The pteridological value of quarries and similar areas needs to be fully recognized as important elements in conservation strategies for such sites.
- Adoption of specific pteridological surveys. For conservation assessments, specific pteridological surveys are not usually built-in to more general surveys of sites of species threat, high floristic diversity or into ones which are the special focus of unusual habitat interest. One aspect of this is the need for specific before-and-after surveys relating to bracken control spraying operation (Robinson and Page 2000). A second aspect is the need for specific pteridological surveys in sites such as rocky upland habitats, and especially upland screes. For in these sites ferns often most thrive as pioneers, and are typically eventually replaced by alpine flowering plants if stabilization occurs. Richness in one is thus often the reciprocal of the other, and the significance of this needs to be addressed in the structuring of appropriate management assessments.

10.2 Actions that allow new opportunities for taxa to continue to appear

Opportunities exist for dynamic habitat management. In many protected areas (and especially, but not exclusively, in wetlands), exclusion of natural sources of disturbance may itself have locally harmful effects in diminishing pioneering elements of pteridophyte diversity which can be present. Dynamic but sensitive management within these areas to continually recreate a variety of newly open areas from which seral successions begin can be pteridologically important. Examples include the creation of new bare mud surfaces in wetland areas mimicking flood deposition surfaces, the exposure of bare ditch sides and banks by streams mimicking natural erosion slopes, and the recreation of fresh exposure of bared damp mineral surfaces in quarries, such as happened formerly around old metalliferous mine sites and within many old quarry workings. Such active aspects of management follow and augment often irregularly episodal processes of natural disturbance regimes which usually benefit pteridophyte diversity. A certain degree of experimentation is probably always necessary and desirable, based on sensible local experience and comparison with other habitats and natural events, and the results of such experimentation require sensitive incorporation into onward management strategies.

11 Conclusions and prognosis

Few future changes are totally predictable, although several general principles can be learned. Change can be rapid where there are rapidly changing environments, and there are several types of environmental change to which pteridophytes are collectively or specifically highly sensitive. Different species have different levels of tolerance to environmental shifts and to different suites of underlying causes. Further, responses may vary within a single species in different parts of its range, where differentially influenced by factors operating either sympatrically or more locally. Additionally, enclaves of higher genetic diversity may exist in different parts of ranges, with certain genotypes enjoying different levels of tolerance to particular variables. The loss or partial loss of such formerly established species continues to occur widely, usually directly or indirectly through human activity, and the degree to which this is likely to be modified in future depends on choices within the human sphere.

Change, however, is also a fundamental part of the evolutionary process, and pteridophytes are no exception. Although many fern, horsetail and clubmoss changes have been, and continue to be, influenced in their direction, extent or speed of progression by humans, nevertheless these influences should be seen as a dressing, however substantial, to an underlying process of natural change which is an integral part of the progression of post-glacial, evolutionary bio-rediversification and floristic recovery in our unusually species-impoverished insular pteridophyte flora (Page 1988, 1997b). Change is a vital element of the dynamics of such biodiversity recovery. It is thus at least as important to allow opportunities for new taxa to appear, and especially for new hybrids to form actively, as it is to preserve the array of species that we already have.

By placing emphasis on gaining an ecological and evolutionary understanding of the reasons for change, there is considerable potential in many pteridophytes as relatively sensitive bio-indicators of multiple aspects of change. We need to learn to read what factors are indicated by each species and the thresholds involved. Such indicator values of pteridophytes are important aspects yet to be quantified. We are likely to find a far greater environmental application of them as sensitive bio-indicators in the forthcoming century than we have in the last.

ACKNOWLEDGEMENTS

I am particularly grateful to Dr R. M. Bateman, Dr Adrian Dyer, Clive Jermy, Dr Heather McHaffie, Rose Murphy, Dr Michael Proctor and Dr Tim Rich for many valuable and constructive comments on the manuscript of this article.

REFERENCES

Barclay-Estrup, P. and Gimmingham, C. H. (1969) The description and interpretation of cyclical processes in a heath community. I. Vegetational change in relation to the *Calluna* cycle. *Journal of Ecology*, **57**, 737–758.

Bell, P. R. (1985) The essential role of Pteridophyta in the study of land plants. *Proceedings of the Royal Society of Edinburgh*, B **83**, 1–4.

Birks, H. J. B. (1976) The distribution of European pteridophytes: a numerical analysis. *New Phytologist*, **77**, 257–287.

Birnie, R. V. and Miller, D. R. (1986) The bracken problem in Scotland: a new assessment using remotely sensed data, in Smith, R. T. and Taylor, J. A. (eds) Bracken. *Ecology, Land Use and Control Technology*. Carnforth: Parthenon Publishing, pp. 43–64.

Bolton, J. (1785) *Filices Britannicae*. Vol. 1. Leeds: Binns.

Bower, F. O. (1908) *The Origin of a Land Flora*. London: Macmillan.

Bower, F. O. (1923–28) *The Ferns (Filicales)*. 3 vols., Cambridge: Cambridge University Press.

Bower, F. O. (1935) *Primitive Land Plants*, London: Macmillan.

Bridges, K. M., Ashcroft, C. J. and Sheffield, E. (1998) Population analysis of the type localities of some recently recognised taxa of British *Pteridium (Dennstaedtiaceae: Pteridophyta)*. *Fern Gazette*, **15**, 205–213.

Bullard, P. (1987) *A Revised Inventory of Gloucestershire's Ancient Woodlands*. Gloucester: Gloucestershire Trust for Nature Conservation.

Burn, A. (1988) Bracken and nature conservation, in Senior Technical Officer's Group Wales (ed.) *Bracken in Wales*. Bangor: Nature Conservancy Council, pp. 47–55.

Busby, A. R. (1976) Ferns in canal navigations in Birmingham. *Fern Gazette*, **11**, 269.

Camus, J. M., Gibby, M. and Johns, R. J. (eds) (1996) *Pteridology in Perspective*. London: Royal Botanic Gardens, Kew.

Caulton, E., Keddie, S. and Dyer, A. F. (1995) The incidence of airborne spores of bracken Pteridium aquilinum (L.) Kuhn. in the rooftop airstream over Edinburgh, Scotland, UK, in Smith, R. T. and Taylor, J. A. (eds) *Bracken: an environmental issue*. Aberystwyth: International Bracken Group, pp. 82–89.

Church, A. R. (1990) Recent finds of Killarney fern (*Trichomanes speciosum Willd.*) in Arran. *Glasgow Naturalist*, **21**, 608–614.

Colinson, M. E. (1996) 'What use are fossil ferns?' – 20 years on: with a review of the fossil history of extant pteridophyte families and genera, in Camus, J., Gibby, M. and Johns, R. J. (eds) *Pteridology in Perspective*. London; Royal Botanic Gardens, Kew pp. 187–394.

Cooper-Driver, G. (1976) Chemotaxonomy and phytochemical ecology of bracken, *Botanical Journal of the Linnean Society*, **73**, 35–46.

Davis, B. N. K. (1979) Chalk and limestone quarries as wildlife habitats. *Minerals and the Environment*, **1**, 48–56.

Dimbleby, G. W. (1984) Anthropogenic changes from Neolithic through medieval times. *New Phytologist*, **98**, 57–72.

Dubuisson, J.-Y. (1996) Evolutionary relationships within the genus *Trichomanes sensu lato (Hymenophyllaceae)* based on anatomical and morphological characters and a comparison with RBCL nucleotide sequences: comparative results, in Camus, J. M., Gibby, M. and Johns, R. J. (eds) *Pteridology in Perspective*. London: Royal Botanic Gardens, Kew, pp. 285–287.

Durno, S. E. and McVean, D. N. (1959) Forest history of the Beinn Eighe Nature Reserve. *New Phytologist*, 58, 228–236.

Dyer, A. F. (ed.) (1979) *The Experimental Biology of Ferns*. London: Academic Press.

Dyer, A. F. (1996) *Asplenium marinum* in Iona Abbey. *Pteridologist*, 3, 7–12.

Dyer, A. F. and Lindsay, S. (1992) Soil spore banks of temperate ferns, *American Fern Journal*, 82, 89–122.

Dyer, A. F. and Page, C. N. (eds) (1985) *Biology of Pteridophytes*. Edinburgh: Royal Society of Edinburgh.

Edwards, M. E. (1986) Disturbance histories of four Snowdonia woodlands and their relation to Atlantic bryophyte distributions. *Biological Conservation*, 37, 301–320.

Edwards, P. (1982) The appearance and disappearance of a *Dryopteris carthusiana* colony. *Fern Gazette*, 12, 224.

French, C., Murphy, R., and Atkinson, M. (1999) *Flora of Cornwall*. Camborne: Wheal Seton Press.

French, C., Murphy, R. and Page, C. N. (2000) *Atlas of the Ferns and Fern Allies of Cornwall* (in press).

Gibby, M. (1991) The development of laboratory-based studies in fern variation, in Camus, J. M. (ed.) *The History of British Pteridology*. London: British Pteridological Society, pp. 59–63.

Gibby, M. and Walker, S. (1977) Further cytogenetic studies and a reappraisal of the diploid ancestry of the *Dryopteris carthusiana* complex. *Fern Gazette*, 11, 315–324.

Gimmingham, C. H. (1964) Dwarf-shrub heaths, in Burnett, J.H. (ed.) *The Vegetation of Scotland*. Edinburgh: Oliver and Boyd, pp. 232–288.

Gimmingham, C. H. (1977) The status of pinewoods in British ecosystems, in Bunce, R. G. H. and Jeffers, J. N. R. (eds) *Native Pinewoods of Scotland*. Cambridge: Institute of Terrestrial Ecology, pp. 1–4.

Gimmingham, C. H., Chapman, S. B. and Webb, N. R. (1979) European heathlands, in Specht, R.L. (ed.) *Ecosystems of the World*. Vol. 9A. *Heathlands and Related Shrublands*, Amsterdam: Elsevier, pp. 365–413.

Gliessman, S. P. (1976) Allelopathy in a broad spectrum of environments as illustrated by bracken. *Botanical Journal of the Linnean Society*, 73, 95–105.

Gliessman, S. P. and Muller, C. H. (1972) The phytotoxic potential of bracken, *Pteridium aquilinum* (L.) Kuhn. *Madrono*, 21, 299–304.

Grime, J. P. (1979) *Plant Strategies and Vegetation Processes*. Chichester: J. Wiley and Sons.

Grime, J. P. (1984) The ecology of species, families and communities of the contemporary British flora. *New Phytologist*, 98, 15–33.

Grime, J. P. (1985) Factors limiting the contribution of pteridophytes to a local flora. *Proceedings of the Royal Society of Edinburgh*, B, 86, 403–421.

Harley, J. L. and Lewis, D. H. (eds) (1984) *The Flora and Vegetation of Britain: origins and changes – the facts and their interpretation*. London: Academic Press.

Harper, J. L. and Hawksworth, D. L. (1994) Biodiversity: measurement and estimation. *Philosophical Transactions of the Royal Society of London*, B, 345, 5–12.

Harris, T. M. (1973) 'What use are fossil ferns ?', in Jermy, A. C., Crabbe, J. A. and Thomas, B. A. (eds) *The Phylogeny and Classification of the Ferns*. London: Academic Press, pp. 41–44.

Haufler, C. (1989) Towards a synthesis of evolutionary modes and mechanisms in homosporous pteridophytes. *Biochemical Systematics and Ecology*, 17, 109–115.

Haufler, C. (1996) Species concepts and speciation in pteridophytes, in Camus, J., Gibby, M. and Johns, R. J. (eds) *Pteridology in Perspective*. London: Royal Botanic Gardens, Kew, pp. 291–305.

Hodgetts, N. G. (1997) Atlantic bryophytes in Scotland. *Botanical Journal of Scotland*, 49, 375–385.

Holttum, R. E. (1973) Posing the problem, in Jermy, A. C., Crabbe, J. A. and Thomas, B. A. (eds) *The Phylogeny and Classification of the Ferns*. London: Academic Press, pp. 1–10.

Holttum, R. E. (1982) The continuing need for more monographic studies of ferns. *Fern Gazette*, 12, 185–190.

Hutchinson, G. and Thomas, B. A. (1996) *Welsh Ferns*. 7th edn, Cardiff: National Museums and Galleries of Wales.

Ivimey-Cook, R. B. (1984) *Atlas of the Devon Flora*. Exeter: Devonshire Association.

Jalas, J. and Suominen, J. (eds) (1972) *Atlas Florae Europaeae*. Vol 1, Pteridophyta, Helsinki: Committee for Mapping the Flora of Europe and Societas Biologica Fennica Vanamo.

Jermy, A. C. (1998) Fern recording and monitoring: a programme to take the BPS into the next century. *Pteridologist*, 3, 5–12.

Jermy, A. C., Arnold, H. R., Farrell, L., and Perring, F. H. (1978) *Atlas of Ferns of the British Isles*. London: Botanical Society of the British Isles and The British Pteridological Society.

Jermy, A. C. and Camus, J. (1991) *The Illustrated Field Guide to Ferns and Allied Plants of the British Isles*. London: Natural History Museum Publications.

Jones, M. (1988) The arable field: a botanical battleground, in Jones, M. (ed.) *Archaeology and the Flora of the British Isles*, Oxford: Oxford University Committee for Archaeology and BSBI, pp. 86–92.

Kendall, A., Page, C. N. and Taylor, J. A. (1995) Linkages between bracken sporulation rates and weather and climate in Britain, in Smith, R. T. and Taylor, J. A. (eds) *Bracken: an environmental issue*. Aberystwyth: International Bracken Group, pp. 77–81.

Kidston, R. and Gwynne-Vaughan, D.T.(1907) On the fossil *Osmundaceae*. Parts I–IV. *Transactions of the Royal Society of Edinburgh*, 45, 759–780.

Lacey, M. E. and McCartney, H. A. (1994) Measurement of airborne concentrations of spores of bracken (*Pteridium aquilinum*). *Grana*, 33, 91–93.

Lawton, J. H. (1982) Vacant niches and unsaturated communities: a comparison of bracken herbivores at sites on two continents. *Journal of Animal Ecology*, 51, 573–595.

Lindsay, S. Sheffield, E., and Dyer, A. F. (1992) Soil spore banks, fern conservation and isozyme analysis, in Ide, J. M., Jermy, A. C. and Paul, A. M. (eds) *Fern Horticulture, Past, Present and Future Perspectives*. Andover: Intercept Publishing, pp. 289–283.

Lovis, J. D. (1955) The problem of *Asplenium trichomanes*, in Lousley, J. E. (ed.) *Species Studies in the British Flora*. London: Botanical Society of the British Isles, pp. 99–103.

Lovis, J. D. (1977) Evolutionary patterns and processes in ferns. *Advances in Botanical Research*, 4, 229–415.

Manton, I. (1950) *Problems of Cytology and Evolution in the Pteridophyta*. Cambridge: Cambridge University Press.

Marrs, R. H. and Lowday J. E. (1992) Control of bracken and the restoration of heathland. II. Regeneration of the heathland community. *Journal of Applied Ecology*, 29, 204–211.

Marrs, R. H. and Pakeman, R. J. (1994) Bracken control and heathland restoration in Breckland, in Smith, R. T. and Taylor, J. A. (eds) *Bracken: an environmental issue*. Aberystwyth: International Bracken Group, pp. 160–172.

Marrs, R. H., Pakeman, R. J., Le Duc, M. G. and Paterson, S. (1997) Bracken invasion in Scotland. *Botanical Journal of Scotland*, 49, 347–356.

McCracken, E. (1971) *The Irish Woods since Tudor Times*. Newton Abbott: David and Charles.

McDonald, A. (1988) Changes in the flora of Port Meadow and Picksey Mead, Oxford, in Jones, M. (ed.) *Archaeology and the Flora of the British Isles*. Oxford: Oxford University Committee for Archaeology and Botanical Society of the British Isles, pp. 76–84.

McHaffie, H. S. (1997) Acquiring a taste for ferns. *Pteridologist*, 3, 98.

McHaffie, H. S. (1998) The Biology of *Athyrium distentifolium* and *A. flexile*. PhD thesis, University of Edinburgh.

McHaffie, H. S. (1999) *Athyrium distentifolium* var. *flexile*: an endemic variety. *Botanical Journal of Scotland*, **51**, 227–236.

McVean, D. N. (1958a) Snow cover and vegetation in the Scottish Highlands. *Weather*, **13**, 197–200.

McVean, D.N. (1958b) Island vegetation of some West Highland freshwater lochs. *Transactions of the Botanical Society of Edinburgh*, **37**, 200–208.

McVean, D. N. (1964) Moss and fern meadows, in Burnett, J. H. (ed.) *The Vegetation of Scotland*. Edinburgh: Oliver and Boyd, pp. 514–521.

McVean, D. N. and Ratcliffe D. A. (1962) *Plant Communities of the Scottish Highlands* [Nature Conservancy Monographs No. 1.]. London: HMSO.

Merryweather, J. (1991) *The Fern Guide*. Preston Montford: Field Studies Council.

Mitchell, F. (1976) *The Irish Landscape*. London: Collins.

Moss, B. (1983) Norfolk Broadland; experiments in the restoration of a complex wetland. *Biological Reviews*, **58**, 521–567.

Newman, E. (1844) *A History of British Ferns and Allied Plants*. London: Van Voorst.

O'Sullivan, P. E. (1977) Vegetation history and the native pinewoods, in Bunce, R. G .H. and Jeffers, J. N. R. (eds) *Native Pinewoods of Scotland*. Cambridge: Institute of Terrestrial Ecology, pp. 60–69.

Page, C. N. (1967) Sporelings of *Equisetum arvense* in the wild. *British Fern Gazette*, **9**, 335–338.

Page, C. N. (1972a) An assessment of inter-specific relationships in *Equisetum* subgenus *Equisetum*, *New Phytologist*, **71**, 355–369.

Page, C. N. (1972b) An interpretation of the morphology and evolution of the cone and shoot of Equisetum. *Botanical Journal of the Linnean Society*, **65**, 359–397.

Page, C. N. (1976) The history and spread of bracken in Britain, *Proceedings of the Royal Society of Edinburgh*, B, **81**, 3–10.

Page, C. N. (1978) Ferns as taxonomic tools and the future of pteridology. *Transactions and Proceedings of the Botanical Society of Edinburgh*, **42** (Supplement), 37–41.

Page, C. N. (1979a) The diversity of ferns: an ecological perspective in Dyer, A. F. (ed.) *The Experimental Biology of Pteridophytes*. London: Academic Press, pp. 9–56.

Page, C. N. (1979b) Experimental aspects of fern ecology, in Dyer, A. F. (ed.) *The Experimental Biology of Pteridophytes*. London: Academic Press, pp. 551–589.

Page, C. N. (1982) *The Ferns of Britain and Ireland*. Cambridge: Cambridge University Press.

Page, C. N. (1986) The strategies of bracken as a permanent ecological opportunist, in Smith, R. T. and Taylor, J. A. (eds) *Bracken: ecology, land use and control technology*. Carnforth: Parthenon Publishing, pp. 173–181.

Page, C. N. (1988) *Ferns: their habitats in the landscape of Britain and Ireland*. Glasgow: Collins New Naturalist.

Page, C. N. (1990) Taxonomic evaluation of the fern genus *Pteridium* and its active evolutionary state, in Thomson, J. A. and Smith, R. T. (eds) *Bracken Biology and Management*. Sydney: Australian Institute of Agricultural Science, pp. 23–34.

Page, C. N. (1997a) *The Ferns of Britain and Ireland*. 2nd edn, Cambridge: Cambridge University Press.

Page, C. N. (1997b) Pteridophytes as field indicators of natural biodiversity restoration in the Scottish flora. *Botanical Journal of Scotland*, **49**, 405–414 .

Page, C. N. and Barker, M. A. (1985) Ecology and geography of hybridisation in British and Irish horsetails. *Proceedings of the Royal Society of Edinburgh*, B, **86**, 265–272.

Page, C. N., Dyer, A. F., Lindsay, S. and Mann, D. G. (1992) Conservation of pteridophytes: the ex-situ approach, in Ide, J. M., Jermy, A. C. and Paul, A. M. (eds) *Fern Horticulture, Past, Present and Future Perspectives*. Andover: Intercept Publishing, pp. 269–278.

Page, C. N. and McHaffie, H. S.(1991) Pteridophytes as indicators of landscape changes in the British Isles in the last hundred years, in Camus, J. M. (ed.) *The History of British Pteridology*. London: British Pteridological Society, pp. 25–40.

Pakeman, R. J. and Marrs, R. H. (1993) Vegetation development on moorland after control of *Pteridium aquilinum* with asulam. *Journal of Vegetation Science*, **3**, 707–710.

Pakeman, R. J., Marrs, R. H., Howard, D. C. and Barr, C. J. (1995) Predicting the effects of climate and land-use change on the spread of bracken, in Smith, R. T. and Taylor, J. A. (eds) *Bracken: an environmental issue*. Aberystwyth: International Bracken Group, pp. 69–76.

Parsons, P. A. (1983) *The Evolutionary Biology of Colonizing Species*. Cambridge: Cambridge University Press.

Perring, F. H. (ed.) (1970) *The Flora of a Changing Britain* [Botanical Society of the British Isles Conference Report No. 11.]. Hampton: E. Classey.

Perring, F. H. (1974) Changes in our native vascular plant flora, in Hawksworth, D. L. (ed.) *The Changing Flora and Fauna of Britain*. London: Academic Press, pp. 7–25.

Peterken, G. F. (1981) *Woodland Conservation and Management*. London: Chapman and Hall.

Petersen, R. L. (1985) Towards an appreciation of fern edaphic niche requirements. *Proceedings of the Royal Society of Edinburgh*, B, **86**, 93–103.

Pigott, C. D. (1956) The vegetation of Upper Teesdale in the north Pennines. *Journal of Ecology*, **44**, 545–586.

Poole, I. and Page, C. N. (2000). A fossil fern indicator of epiphytism in a Tertiary flora. *New Phytolgist*, 148, 117-125.

Praeger, R. L. (1934) *The Botanist in Ireland*. Dublin: Dublin University Press.

Proctor, M. C. F., Spooner, G. M. and Spooner, M. F. (1980) Changes in Wistman's Wood, Dartmoor: photographic and other evidence. *Report and Transactions of the Devonshire Association for the Advancement of Science*, **112**, 43–79.

Rackham, O. (ed.) (1975) *Hayley Wood: its history and ecology*. Cambridge: Cambridgeshire and Isle of Ely Naturalists' Trust.

Rackham, O. (1980) *Ancient Woodland*. London: Edward Arnold.

Rackham, O. (1986) *The History of the Countryside*. London: J. M. Dent.

Ratcliffe, D. A. (1960) The mountain flora of Lakeland. *Proceedings of the Botanical Society of the British Isles*, **4**, 1–25.

Ratcliffe, D. A. (1968) An ecological account of Atlantic bryophytes in the British Isles. *New Phytologist*, **67**, 365–439.

Ratcliffe, D. A. (1977) *Highland Flora*. Inverness: Highlands and Islands Development Board.

Ratcliffe, D. A. (1984) Post-medieval and recent changes in British vegetation: the culmination of human influence. *New Phytologist*, **98**, 73–100.

Ratcliffe, D. A., Birks, H. J. B. and Birks, H. H. (1993) The ecology and conservation of the Killarney fern *Trichomanes speciosum* Willd. in Britain and Ireland. *Biological Conservation*, **66**, 231–247.

Raven, J. A. (1985) Physiology and biochemistry of pteridophytes. *Proceedings of the Royal Society of Edinburgh*, B, **86**, 37–44.

Rich, T. G. C. and Jermy, A. C. (1998) *A Crib for Ferns and Allied Plants*. London: Botanical Society of the British Isles.

Rich, T. G. C. and Smith, P. A. (1996) Botanical recording, distribution maps and species frequency. *Watsonia*, 21, 161–173.

Rich, T. G. C., Richardson, S. J. and Rose, F. (1995) Tunbridge filmy fern *Hymenophyllum tunbrigense* (Hymenophyllaceae: Pteridophyta) in south east England in 1994/1995. *Fern Gazette*, **15**, 51–63.

Rich, T. G. C. and Woodruff, E. R. (1992) The influence of recording bias in botanical surveys: examples from the BSBI Monitoring Scheme 1987–1988. *Watsonia*, 19, 73–95.

Rich, T. G. C. and Woodruff, E. R. (1996) Changes in the vascular plant floras of England and Scotland between 1930–1960 and 1987–1988: the BSBI monitoring scheme. *Biological Conservation*, 75, 217–229.

Richards, P. W. and Evans, G. B. (1972) Biological Flora of the British Isles: *Hymenophyllum*. *Journal of Ecology*, **60**, 245–268.

Rickard, M. (1994) Ferns – a case history, in Perry, A. R. and Gwynne Ellis, R. (eds) *The Common Ground of Wild and Cultivated Plants*. Cardiff: National Museum of Wales, pp. 51–57.

Roberts, R. H. (1979) The Killarney fern, *Trichomanes speciosum*, in Wales. *Fern Gazette*, **12**, 1–4.

Robinson, R. C. & Page, C. N. (2000) Protection of non-target ferns during extensive spraying of bracken, in Taylor, J. A. & Smith, R. T. (eds) *Bracken Fern: Toxicity, Biology and Control*. Aberystwyth: International Bracken Group, pp. 163–174.

Rumsey, F. J., Jermy, A. C. and Sheffield, E. (1998) The independent gametophytic stage of *Trichomanes speciosum* Willd. (*Hymenophyllaceae*), the Killarney Fern and its distribution in the British Isles. *Watsonia*, **22**, 1–19.

Rumsey, F. J., Russell, S. J., Barrett, J. A. and Gibby, M. (1996) Genetic variation in the endangered filmy fern *Trichomanes speciosum* Willd. in Camus, J., Gibby, M. and Johns, R. J. (eds) *Pteridology in Perspective*. London: Royal Botanic Gardens, Kew, pp. 161–165.

Rumsey, F. J. and Sheffield, E. (1996) Intergenerational ecological niche separation and the 'independent gametophyte' phenomenon in Camus, J., Gibby, M. and Johns, R. J. (eds) *Pteridology in Perspective*. London: Royal Botanic Gardens, Kew, pp. 563–570.

Schneller, J. J. and Holderegger, R. (1996) Colonising events and genetic variability within populations of *Asplenium ruta-muraria*. L., in Camus, J., Gibby, M. and Johns, R. J. (eds) *Pteridology in Perspective*. London: Royal Botanic Gardens, Kew, pp. 571–580.

Scott, M., Scott, S. and Sydes, C. (1999) A Scottish perspective on the conservation of Pillwort. *British Wildlife*, **10**, 297–302.

Seward, A. C. (1931) *Plant Life Through the Ages*. Cambridge: Cambridge University Press.

Shivas, M. G. (1961a) Contributions to the cytology and taxonomy of species of *Polypodium* in Europe and America. I. Cytology. *Journal of the Linnean Society (Botany)*, **58**, 27–38.

Shivas, M. G. (1961b) Contributions to the cytology and taxonomy of species of *Polypodium* in Europe and America. II. Taxonomy. *Journal of the Linnean Society (Botany)*, **58**, 39–50.

Sleep, A. (1971) *Polystichum* hybrids in Britain. *British Fern Gazette*, **10**, 208–209.

Sleep, A. (1985) Speciation in relation to edaphic factors in the *Asplenium adiantum-nigrum* group. *Proceedings of the Royal Society of Edinburgh*, B, **86**, 325–334.

Smith, R. T. (2000). The spread of bracken: an end-of-century assessment of factors, risks and land use realities, in Taylor, J. A. and Smith, R. T. (eds) *Bracken Fern: Toxicity, Biology and Control*. Aberystwyth: International Bracken Group pp 2-8.

Sporne, K. R. (1962) *The Morphology of Pteridophytes*. London: Hutchinson.

Synnott, D. [undated] *Ferns of Ireland*. Dublin: Folens.

Tansley, A. G. (1939) *The British Islands and their Vegetation*. Cambridge: Cambridge University Press.

Tansley, A.G. (1949) *Britain's Green Mantle*. London: Allen & Unwin.

Taylor, J. A. (1985) The relationship between land-use change and variations in bracken encroachment rates in Britain, in Smith, R. T. (ed.) *The Biogeographical Impact of Land-use Changes*. Norwich: Geo Books, pp. 19–28.

Taylor, J. A. (1986) The bracken problem: a local hazard and a global issue, in Smith, R. T. and Taylor, J. A. (eds) *Bracken: ecology, land use and control technology*. Carnforth: Parthenon Publishing, pp. 21–42.

Taylor, J. A. (1988) Bracken areas and encroachment rates in Wales, in Senior Technical Officer's Group, Wales (ed.) *Bracken in Wales*. Bangor: Nature Conservancy Council, pp. 32–35.

Thomas, A. S. (1963) Changes in vegetation since the advent of myxomatosis. *Journal of Ecology*, **48**, 356; **51**, 151–186.

Thomas, B. (1991) The study of fossil ferns, in Camus, J. M. (ed.) *The History of British Pteridology*. London: British Pteridological Society, pp. 7–15.

Vogel, J. C. (1997) Conservation status and distribution of two serpentine restricted *Asplenium* species in central Europe, in Camus, J., Gibby, M. and Johns, R. J. (eds) *Pteridology in Perspective*. London: Royal Botanic Gardens, Kew, pp. 187–188.

Vogel, J. C., Rumsey, F. J., Schneller, J. J., Barrett, J. A. and Gibby, M. (1999) Where are the glacial refugia in Europe? Evidence from pteridophytes. *Biological Journal of the Linnean Society*, **66**, 23–37.

Walker, T. G. (1973) Evidence from cytology in the clasification of ferns, in Jermy, A. C., Crabbe, J. A. and Thomas, B. A. (eds) *The Phylogeny and Classification of the Ferns*. London: Academic Press, pp. 91–108.

Walker, T. G. (1979) The cytogenetics of ferns, in Dyer, A. F. (ed.) *The Experimental Biology of Pteridophytes*. London: Academic Press, pp. 87–132.

Watt, A. S. (1947) Pattern and process in the plant community. *Journal of Ecology*, **43**, 490–506.

Watt, A. S. (1955) Bracken versus heather, a study in plant sociology. *Journal of Ecology*, **43**, 490–506.

Weaver, R. E. (1986) Use of remote sensing to monitor bracken encroachment in the North Yorks Moors, in Smith, R. T. and Taylor, J. A. (eds) *Bracken: Ecology, Land Use and Control Technology*. Carnforth: Parthenon Publishing, pp. 65–76.

Webb, D. A. (1983) The flora of Ireland in its European context. *Journal of Life Sciences of the Royal Dublin Society*, **4**, 143–160.

Webb, S. and Glading, P. (1998) The ecology and conservation of limestone pavement in Britain. *British Wildlife*, **10**, 103–113.

Wheeler, B. D. (1980) Plant communities of rich-fen systems in England and Wales. III. Fen meadows, fen grassland and fen woodland communities and contact communities. *Journal of Ecology*, **68**, 761–788.

Whitehead, S. J. and Digby, J. (1995) Morphological and physiological behavior of bracken at advancing fronts: implications for control, in Smith, R. T. and Taylor, J. A. (eds) *Bracken: an environmental issue*. Aberystwyth: International Bracken Group, pp. 155–159.

Willmot, A. (1985) Population dynamics of woodland *Dryopteris* in Britain. *Proceedings of the Royal Society of Edinburgh*, B, **86**, 307–313.

Willmot, A. (1989) The phenology of leaf life spans in woodland populations of the ferns *Dryopteris filix-mas* (L.) Schott and *D. dilatata* (Hoffm.) A. Gray in Derbyshire. *Biological Journal of the Linnean Society*, **39**, 387–395.

Wolf, P. G., Sheffield, E. and Haufler, C. H. (1991) Estimates of gene flow, genetic substructure and population heterogeneity in bracken (*Pteridium aquilinum*). *Biological Journal of the Linnean Society*, **42**, 407–423.

Wolf, P. G., Sheffield, E., Thomson, J. A. and Sinclair, R. B. (1994) Bracken taxa in Britain: a molecular analysis, in Smith, R.T. and Taylor, J.A. (eds) *Bracken: an environmental issue*. Aberystwyth: International Bracken Group, pp. 16–20.

Woodell, S. R. J. (1985) *The English Landscape: past, present and future*. Oxford: Oxford University Press.

Chapter 4

Mosses, liverworts and hornworts

Anthony J. E. Smith

ABSTRACT

Since 1973 Red Data Lists of 33 liverworts and 107 mosses have been prepared and nine liverworts and 28 mosses are now on the list of protected plants. Fifteen species are now thought to be extinct compared with 20 in 1973 and 26 species have been removed either because of taxonomic revisions or previous incorrect determinations. Six liverworts and 11 mosses have been described as new to science from the British Isles and 56 species previously not known in Britain have been added to the British list plus 39 resulting from taxonomic revisions. As of December 1998 four hornworts (four in Britain and three in Ireland), 296 liverworts (287 in Britain and 232 in Ireland) and 746 mosses (743 in Britain and 513 in Ireland) are recorded from the British Isles. Species Action Plans have been or are in the process of being drawn up and sites of bryological interest are being assessed. Possible changes likely to take place over the next 25 years are indicated.

1 Introduction

The study of mosses (phylum *Musci*), liverworts (phylum *Hepaticae*) and hornworts (phylum *Anthocerotae*) in the British Isles began seriously with the publication of John Jacob Dillenius's *Historia Muscorum* in 1742. During the nineteenth century there was active exploration by botanists to the extent that, as Preston (1991) points out, 'as early as 1827 Scottish botanists regarded the flora of England as tediously well worked.' He quotes Greville (1822–1828) as saying 'Every path in England … has been so assiduously explored that when a new plant so high in the scale of vegetation as a Moss is discovered, considerable interest is excited.' In 1896 the Moss Exchange Club, later to become the British Bryological Society (BBS) in 1924, was founded. A series of census catalogues listing the vice-county occurrence of mosses and liverworts was published culminating, at the time of Rose and Wallace's *Changes in the Bryophyte Flora of Britain* (1974), in the fourth edition of the liverwort catalogue (Paton 1965a) and the third edition for mosses (Warburg 1963). However, the first 40 years of the twentieth century was a period of relative inactivity when bryologizing consisted mainly of visiting sites of well-known bryological interest. Following the end of the Second World War, however, there was a revival of interest, particularly involving younger people, and this has been expanding ever since.

2 Bryophyte recording

The recording of bryophytes on a vice-county basis (the counties referred to in this chapter are Watsonian vice-counties) is an activity that is still vigorously pursued – an annual list of new records is published in the *Bulletin of the British Bryological Society* listing the discoveries of between 40 and 50 BBS members who are normally active in the field. Over the past 10–15 years there has been an overall diminution in the annual number of records, a reflection of the increasing difficulty of finding species new to mostly well-worked vice-counties. A new *Census Catalogue of British and Irish Bryophytes*, the nomenclature of which is followed here, has just been published (Blockeel and Long 1998). In 1960 the BBS launched its Bryophyte Distribution Maps Scheme which ran for 30 years and in which bryophytes were recorded on the basis of their presence in the 10-km grid squares of the British and Irish national grids. Altogether about 165 bryologists, of whom about 40 were active over the whole or part of the 30 years of the programme, participated in the scheme. Their efforts resulted in the publication of the *Atlas of the Bryophytes of Britain and Ireland* (Hill *et al.* 1991, 1992, 1994). Thus over the past 25 years there has been very active field bryology in Great Britain although, unfortunately, much less in Ireland. This, together with taxonomic revisions of critical groups, has resulted in the addition of many new species to the British and Irish floras (see Tables 4.4–4.6) and a greatly improved knowledge of frequency and distribution patterns, increases and decreases. However, there is now only one professional bryophyte taxonomist who does any work on British bryophytes, and that only sporadically, at the Royal Botanic Garden, Edinburgh.

During the 25 years covered by this chapter, there were only two other professional bryophyte taxonomists, one of whom retired in 1981. Additions to the British flora were mainly the result of field activities of members of the BBS or to taxonomic work by foreign bryologists.

3 Recording bryophytes

3.1 *Distribution maps*

Fig. 4.1 and Fig. 4.2 show the numbers of liverwort (including hornwort) and moss species recorded from each 10-km grid square. The mapping scheme has produced a more informative picture with regard to the frequency and change in status of individual species than the much longer standing vice-county recording scheme. Thus *Plagiomnium undulatum* and *P. elatum* are recorded from 112 and 101 British vice-counties, respectively, suggesting a not very great difference in frequency (with only four of the records of the latter species and none of the former being treated as pre-1950). By contrast the distribution maps (Hill *et al.* 1994) show 2043 grid squares (of which 3.5% are pre-1950) for *P. undulatum* and 352 (of which 13.6% are pre-1950) for *P. elatum*, giving a very different picture for the two species.

It has been argued that distribution maps reflect the areas of bryological activity rather than the real distributions of species. However, field experience leads to the belief that this is not so (see e.g. Crundwell 1992) although it is clear that evenness of coverage leaves a lot to be desired. Some areas such as Cheshire, Nottinghamshire, Shropshire, Staffordshire, parts of eastern Yorkshire, south-east and north-east Scotland and much of Ireland are under-worked while others, such as Cornwall, the Home

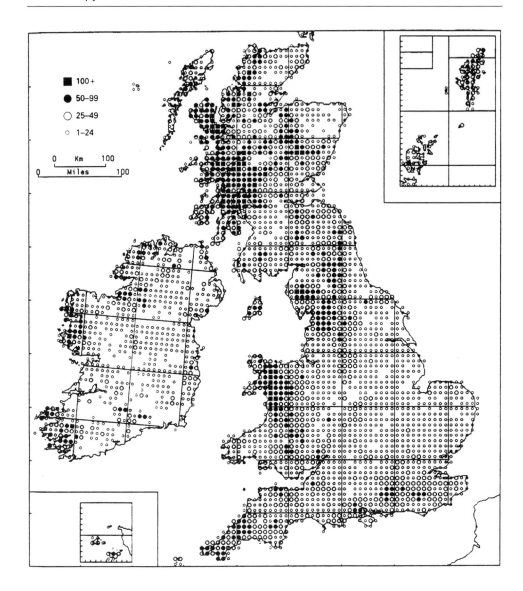

Figure 4.1 Map showing the number of liverwort species recorded per 10-km square of the national and Irish grids.

From Hill *et al.* 1991.

Counties, north-west Wales and Skye are extremely well worked. Even with these pro-visos in mind the distribution maps in Hill *et al.* (1991, 1992, 1994) give a reasonable picture for most species. Less satisfactory are the maps of critical or recently described or discovered species. Using Canonical Correspondence Analysis of distribution records of 64 liverwort species and environmental data Hill and Dominguezlozano (1994) were able to show that there was a close approximation between the expected numbers of species per 10-km square and the actual number per square. Fig. 4.3 shows

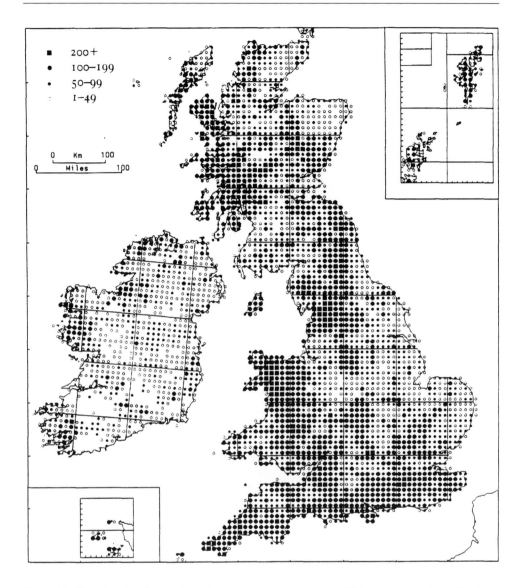

Figure 4.2 Map showing the number of moss species recorded per 10-km square of the national and Irish grids.

From Hill *et al.* 1991.

the ratio of actual to predicted number of liverworts per 10-km grid square. This map suggests that distribution maps of liverworts and, by implication of mosses, give a reasonable picture of the distribution pattern of each species.

How complete the data are for each square is debatable. Crundwell (1992) considers, or rather guesses, that square records represent a little over two-thirds of the species present in those squares and that perhaps one in 30 records would have to be deleted as erroneous.

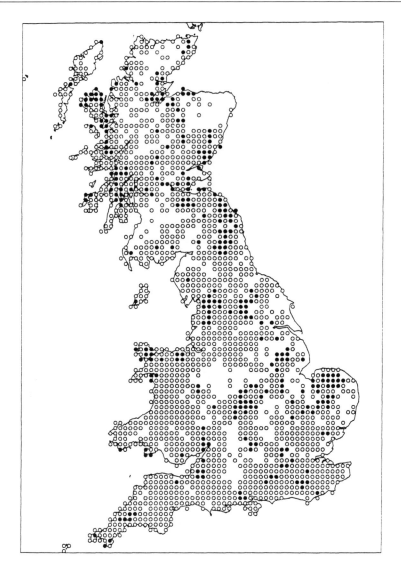

Figure 4.3 Map showing ratio of actual to predicted number of liverwort species per 10-km square of the national grid. Blank squares, with less than half the expected number, are likely to be under-recorded. Squares with open circles have an actual value within a factor of 2 of the predicted number of species. Squares with closed circles contain twice as many species as predicted.

The above comments apply only to Great Britain. Ireland is very under-recorded except in the extreme west and there is a far higher proportion of pre-1950 records. This is in part due to the fact that there are very few Irish bryologists – perhaps only five or six at present – and British bryologists going to Ireland tend only to visit interesting areas in the far west. Meetings of the BBS organized in less interesting parts are poorly attended, somewhat different from the situation with regard to meetings arranged in Britain.

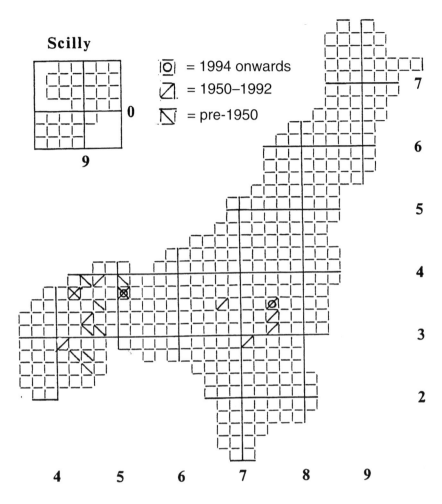

Scilly

|O| = 1994 onwards

|Z| = 1950–1992

|N| = pre-1950

Figure 4.4 Distribution map of *Andreaea rothii* showing changes in frequency in 2 × 2 km squares in West Cornwall.

By courtesy of Dr J. Holyoak and Mrs J. A. Paton.

With the data bank that has been built up it will be possible, with future recording schemes, to show any changes that have taken place in frequency on a national basis and relate these to changes in climate and in levels of various types of atmospheric and aquatic pollutants. The value of this may be seen in the change in frequency of species, such as *Campulopus introflexus* and *Orthodontium lineare* (see Table 4.10), that have already been monitored over many years.

In addition to recording on a national scale there has also been intensive local recording and the publication of bryophyte floras for a number of vice-counties including Berkshire (Bates 1995), Berwickshire (Long 1990), Durham (Graham 1988), Essex (Adams 1974), Hampshire (Crundwell and Rose, 1996), North Wales (Hill 1988), Radnorshire (Woods 1993), Surrey (Gardiner 1981), Sussex (Rose *et al.* 1991) and others still in

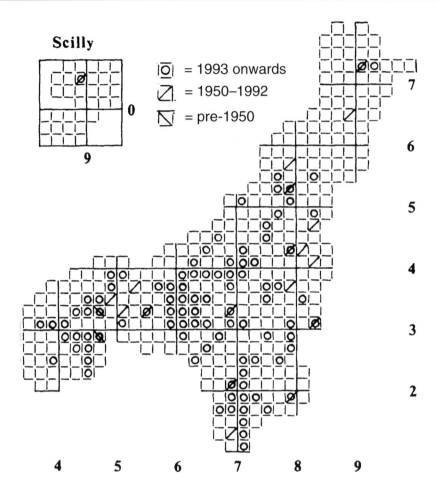

Figure 4.5 Distribution map of *Zygodon conoideus* showing changes in frequency in 2 × 2 km squares in West Cornwall.

By courtesy of Dr J. Holyoak and Mrs J. A. Paton.

progress. These provide greater detail of changes that have occurred as much of the recording is more intensive and sometimes carried out over a long period of time.

There are two examples that exemplify the value of such detailed studies in revealing changes taking place in bryophyte flora on a local scale. Of particular interest is a paper on the changing bryophyte flora of Oxfordshire (Jones 1991) where recording first began at the end of the seventeenth century, and there is a reasonably detailed picture of what has happened since Sibthorpe's *Flora Oxonniensis* of 1794. Table 4.1 shows the percentage frequency with which certain species were recorded in habitat lists made in 1940–50 and 1960–90 revealing increases and decreases of certain species in that period.

Since the late 1950s there has been intensive recording of bryophytes in Cornwall, firstly by J. A. Paton for the production of a bryophyte flora (Paton 1969) and later, from 1993 onwards, in west Cornwall by D. Holyoak. It has become clear that marked changes in frequency of certain species have taken place between the two periods of recording. Figs 4.4 and 4.5 show the distributions of *Andreaea rothii* and *Zygodon conoideus* in 2-km squares revealing the changes in frequency of these two species that have taken place. That these changes are not just due to vagaries in recording is illustrated by the map of *Zygodon viridissimus* which shows few changes (Fig. 4.6). It is of interest to note that Jones (1991) reported a marked increase in the frequency of *Z. conoideus* between 1940–50 and 1960–90 (see Table 4.1).

It seems likely that the change in frequency of the two species can be explained in terms of increased traffic pollution, and particularly of traffic generated NO_x, a consequence of the booming tourist industry in south-west England. *Andreaea rothii is* intolerant of even very low levels of added nutrients and has become almost extinct. On the other hand, it seems very likely that *Zygodon conoideus* has benefited from the consequent eutrophication and acidification of bark. These two species illustrate very well how intensive recording, even over a relatively short period of time, can reveal real changes in frequency and distribution not apparent in 10-km square distribution maps.

4 Red List species

The growing concern about threats to bryophytes, both in the British Isles and abroad, has led to a number of steps being taken. In Britain the BBS discontinued their specimen exchange scheme in the early 1960s and appointed a Conservation Officer in 1977. Twenty-eight mosses and nine liverworts have been added to Schedule 8 (Plants) of the Wildlife and Countryside Act, and to Schedule 8 (Part 1) of the Wildlife (Northern Ireland) Order, 1985 (see Table 4.2). A Red List of British liverworts and hornworts and a Red List of British mosses were drawn up by the Joint Nature Conservation Committee in 1998 (Table 4.2). A Red List is currently being prepared for the National Parks and Wildlife Service of the Irish Republic. Two species, the liverwort *Petalophyllum ralfssi* and the moss *Hamatocaulis vernicosus*, are currently protected in the Republic, and six liverworts and 16 mosses will appear on the next Flora Protection Order. Species Action Plans have been published for 16 species and are in preparation for a further 33. Biodiversity Action Plans for a range of habitats (Porley 1998) and a European bryophyte site register (Hodgetts 1995) are being drawn up.

5 Developments since 1973

5.1 State of knowledge in 1973

Rose and Wallace (1974) point out that the majority of British bryophytes were first recorded during the nineteenth century but that during the 30 years up to 1973 some 96 species were added and 3 deleted. This was due in part at least to the much more vigorous field activity commencing in the late 1940s. They gave three reasons for the additions:

1 Probable introductions.
2 The discovery of species (new to science in Britain or the discovery of species known elsewhere but not previously seen in the British Isles).
3 Taxonomic revisions leading to the segregation of new species.

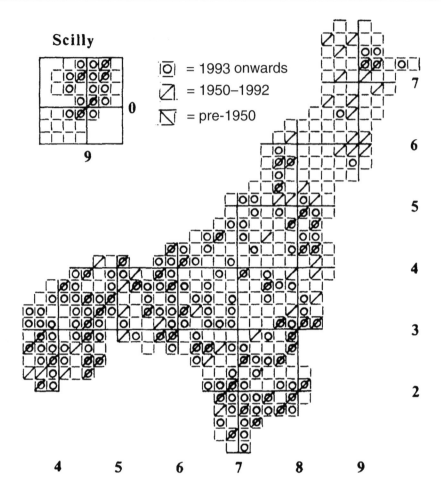

Figure 4.6 Distribution map of *Zygodon viridissimus* showing changes in frequency in 2 × 2 km squares in West Cornwall.

By courtesy of Dr J. Holyoak and Mrs J. A. Paton.

These three reasons apply equally to species added in the past 25 years. It should be noted that taxonomic reasons, in addition to altering the status of taxa, may also result in the description of new species and the resurrection of old ones, described from elsewhere, buried in the older literature.

Rose and Wallace (1974) also point out that it is difficult to ascertain that a particular species is really extinct, and Table 4.3, which lists the species thought to be extinct in 1973, illustrates this.

5.2 Losses 1973–98

In the list of 20 extinctions given by Rose and Wallace (1974) eight species can be removed (see Table 4.3). However, in the Red Data List of British Bryophytes (Table

4.2) there are an additional four species, *Bryum mamillatum*, *Encalypta brevicollis*, *Grimmia anodon* and *Pterygoneurum lamellatum*. There is also *Meesia triquetra*, found new to the British Isles in West Mayo, Ireland, in 1957 (Warburg 1958) that has since been lost as the result of peat cutting – although while a search was being made for it in 1998, *Paludella squarrosa*, last seen in Britain in 1916, was found new to Ireland. An Irish plant that is possibly extinct is *Hypnum uncinulatum*, not seen for more than 90 years, although no systematic search has been made for it.

Rose and Wallace (1974) also noted that in the preceding 30 years 11 liverworts and 14 mosses had been removed from the list because they were erroneously recorded or as a consequence of taxonomic revisions resulting in reduction to intraspecific rank or to synonymy. Since then another 26 species have been removed for the same reasons (see Table 4.4).

5.3 Decreases

It is also clear from the bryophyte atlas (Hill *et al.* 1991, 1992, 1994) that there has been a marked decrease in the frequency of many species, especially some epiphytes and those from wetland habitats. Rose and Wallace (1974), in discussing the reasons for such decreases, give a table listing diminishing species with possible reasons for their decline. There is very little that can be added to their remarks and it would be repetitive to do so except to provide data on selected species (Table 4.1) confirming that what they said in 1974 is still very true today.

However, data for the British Isles as a whole do not tell the full story and regional floras may reflect a different situation in particular areas. Thus 10-km square maps of *Hylocomium splendens* and *Rhytidiadelphus triquetrus* – common and widespread mosses with 9% and 7% of records pre-1950, respectively (Hill *et al.* 1994) – give no indication of decrease yet in lowland England they are much less frequent now than formerly. Jones (1991) pointed out that in Oxfordshire between 1940–50 and 1960–90 in habitat lists the percentage frequencies of *Hylocomium splendes* had decreased from 47% to 7.5% and of *Rhytidiadelphus triquetrus* from 37% to 17.5%. As these are large conspicuous species it is hardly likely that they would have been overlooked and the decrease is genuine. The declines of these two species in Oxfordshire are mirrored by similar declines in Berkshire (Bates 1995), Essex (Adams 1974), Hampshire (Crundwell *et al.* 1996), Surrey (Gardiner 1981), Sussex (Rose *et al.* 1991) and Warwickshire (Laflin 1971). Jones (1991) considers that it is reduced humidity resulting from improved drainage and better runoff that is mainly responsible rather than pollution. This is doubtful as there is abundant observational (e.g. Winner and Bewley 1978; Klein and Bliss 1984) and experimental (e.g. Raemakers 1987) evidence that indicates that acid deposition is a predominant factor.

In another example, between 1951–52 and 1989 there has been a marked decline in *Racomitrium lanuginosum*, the dominant species of upland *Racomitrium* heath in Britain. On the Carneddau in Snowdonia the cover of *Racomitrium* has decreased by 75% (D. A. Ratcliffe in Garvie 1990). This is not apparent in 10-km square distribution maps; pre-1950 records only constitute 9.6% of total 10-km square data. The reasons for the decrease are not clear but it may be due to grazing pressure. In the mountains of eastern Ireland there has been a decrease in species richness and there have been losses in hepatic mar communities in the west. This is due to over-grazing by sheep, a development stimulated by generous European Community farming subsidies.

There have been marked reductions in frequencies of some species of stubble fields and grass leys such as *Funaria fascicularis, Pleuridium subulatum, Tortula lanceola, T. modica, T. protobryoides* (Table 4.5) and *Wiessia squarrosa* (42% pre-1950 records). The reasons for this are not clear unless it is a result of changes in farming practices affecting more demanding species, as there does not seem to have been any lessening of commoner weedy species such as *Ditirchum cylindricum, Pleuridium acunminatum* and *Tortula acaulon,* nor can it be explained in terms of the plants being overlooked as, for many years, there has been systematic search of suitable habitats for various relatively recently described species such *Dicranella staphylina* and members of the *Bryum erythrocarpum* aggregate.

Another species which has almost vanished from lowland England, with 72% of these records pre-1950, is *Splachnum ampullaceum*, a coprophilous plant of wet heaths, moorland and bog, the spores of which are dispersed by flies. Crundwell (1994) suggests that this decrease is due to habitat destruction by drainage and agricultural improvement and the drying out of areas of wet heath. However, Stevenson (1996) draws attention to the use of chemicals to control cattle parasites rendering the cowpats sterile and affecting the number of flies – perhaps this is a further factor in the decline of *Splachnum ampullaceum*.

There has also been an alarming decrease in rare *Bryum* species (71% of records of seven species are pre-1950). The reasons for this are unknown as these species mostly occur on disturbed or sandy soil of which there is no lack. Another group of species that have also decreased are nine rarer *Grimmia* species (54% of records are pre-1950), all saxicolous and often well away from sources of pollution. The cause is most likely to be sulphur dioxide pollution as, unlike *Andreaea rothii* (Fig. 4.3), they have been decreasing for most of this century. However, collecting has also played a part – identification in the field is difficult or impossible and as plants are rare they are attractive to collectors. They are also very slow growing; many have no known means of propagation in the British Isles and so have suffered accordingly.

Rare *Orthotrichum* species, all of which are corticolous, have been decreasing since the beginning of the twentieth century (71% of records of four species are pre-1950) – this is mainly the result of atmospheric pollution but felling of mature trees is likely to have played a part as well. It should be borne in mind, however, that some of these saxicolous and corticolous species are common in parts of Europe and that their decrease began with the Industrial Revolution; they were already rare when systematic recording began. Interestingly the decrease in southern England of some corticolous species seems to have ceased, at least in some areas (see section 5.4 below).

Members of the *Amblystegiaceae* from various types of wetland have suffered very badly as a consequence of habitat destruction or modification. Rare or uncommon species such as *Campylium elodes* (35%), *Drepanocladus lycopodioides* (55%), *D. polygamus* (33%), *D. sendtneri* (51%), *Hamatocaulis vernicosus* (22%) and *Tomentypnum nitens* (38%) have decreased (figures in brackets are the percentage of pre-1950 10-km square records) throughout their ranges whilst the relatively common species *Calliergon giganteum, C. stramineum, Campylium stellatum, Drepanocladus aduncus, D. revolvens, Palustriella commutata, Warnstorfia exannulatus* and *W. fluitans* have decreased very markedly in lowland England although nowhere else to any extent.

5.4 Additions and increases

Over the past 25 years, partly as a consequence of more intensive field activity, taxonomic revisions and the preparation of floras (Smith 1978, and in prep; J. A. Paton 1999) there have been some 112 additions to the bryophyte flora of the British Isles, more than balancing the 26 deletions. Seventeen species new to science have been discovered in Britain (Table 4.6). Fifty-six species known elsewhere have been newly recorded in the British Isles (Table 4.7) and 39 have been added following taxonomic revisions resulting in the raising of intraspecific taxa to specific rank or to the description of new species (Table 4.8).

With regard to the latter, taxonomic revisions of critical groups such as the *Bryum bicolor* aggr. (Wilczek and Demaret 1976; Smith and Whitehouse 1978), the *Bryum capillare* aggr. (Syed 1973), the *Hypnum cupressiforme* aggr. (Smith 1998), *Pohlia* sect. *Pohliella* (Wilczek and Demaret 1970; Lewis and Smith 1977, 1978), *Racomitrium* (Frisvoll 1983, 1988) and *Schistidium* (Blom 1996) have resulted in a considerable number of additions, as well as a few reductions.

Most of the species described as new to science from Britain or previously unknown in the British Isles are either very small plants, often occurring in what are regarded as uninteresting habitats, or are likely to be confused with known species. It is likely that more new species remain to be found (see e.g. Table 4.9). Once discovered and the habitats known some of these new species have been found to be relatively common and widespread. Examples (with number of 10 km grid square records in brackets) include: *Dicranella staphylina* (623) first described in 1969 (Whitehouse 1969), *Fissidens celticus* (180) described in 1965a (Paton 1965), *Pohlia lutescens* (299) found in 1961 (Warburg 1965), *Bryum subelegans* described in 1973 (Syed 1973) and members of the *Bryum erythrocarpum* aggregate (50–1075) described in 1966 (Crundwell and Nyholm 1964). Only a few, such as *Pohlia crudoides*, *Didymodon mammilosus* and *Hygrohypnum polare*, are still only known from a single locality (and whether the latter two, which can only be identified microscopically, are likely to remain so as collection is prohibited under the provisions of the Wildlife and Countryside Act), suggesting that they are genuinely very rare in Britain. Seven of the species described as new to science from Britain since 1973 have since been found outside the British Isles (Table 4.6). *Dicranella staphylina*, described in 1969 (Whitehouse 1969) has also been found elsewhere since then.

Corticolous species are of particular interest as, while many have decreased as a consequence of atmospheric pollution, others such as *Alacomnium androgynum*, *Orthotrichum affine*, *Ulota crispa* aggr., and *Zygodon conoideus* have reappeared or have increased over the past 40 years in lowland England, particularly in the southeast, perhaps reflecting decreased levels in sulphur dioxide pollution.

A number of species have been increasing during the twentieth century, particularly certain epiphytes. It is possible that they are tolerant of higher levels of atmospheric pollution than other species and have been able to colonize new habitats in the absence of competition from other epiphytic bryophytes, lichens and algae eliminated by sulphur dioxide and acidification of bark. Again, they are easily recognized so it is unlikely that the increase in number of records is simply the result of more assiduous recording. Rose and Wallace (1974) give examples of species that were spreading at the time of writing their paper and further information about these, together with some additions, both native and introduced, are given in Table 4.10.

One species mentioned by Rose and Wallace (1974) but not in Table 4.10 because it has not increased on a vice-comital basis, is *Dicranoweisia cirrata*. Its increase on a local basis can be seen from Table 4.1; in Oxfordshire, its frequency in habitat lists increased form 35% to 92% between 1940–50 and 1960–90.

According to Rose and Wallace (1974), there were 681 moss species and 289 liverwort species (or more accurately three hornworts and 286 liverworts) known to the British and Irish flora. The numbers now stand at four hornworts (4 in Britain and 3 in Ireland), 296 liverworts (287 in Britain and 232 in Ireland) and 746 mosses (743 in Britain and 513 in Ireland (Blockeel and Long, 1998)).

6 Introductions and their spread

There have been three definite introductions reported since 1973 – the liverworts *Lophocolea bispinosa* and *Lophozia herzogeana* and the moss *Syntrichia amplexa* – and a remarkable spread of two mosses – *Campylopus introflexus* and *Orthodontium lineare* – and to a lesser extent of *Lophocolea bispinosa*, all species known before 1973. *Lophocolea bispinosa*, an Australasian species probably introduced with horticultural plants, was first discovered in the Abbey Gardens of the Scilly Island of Tresco in 1962 (Paton 1974) and has since spread to all the other islands. It has also been discovered in a garden and a nearby site on the Scottish island of Colonsay where it is thought to be an independent introduction (Wallace 1979). It has also recently been found in Dorset, West Sussex and Surrey, suggesting that it is spreading at least in southern England *Lophocolea semiteres*, another southern hemisphere species, was also found on Tresco in 1961 (Paton 1965b) and has spread to neighbouring islands. It was later found in Benmore Gardens, Argyll in 1972 (Long 1982) and, like the Scottish record of *L. bispinosa*, appears to have been an independent introduction. This species has also been found, sometimes abundantly, in 17 sites in The Netherlands and more than 38 sites in Belgium where it is thought to have been introduced independently on more than one occasion and to be spreading in both countries (Stieperaere 1994). It will be interesting to see if *L. semiteres* will show a similar spread in parts of Britain.

Syntrichia amplexa, a western North American plant, was first found mixed with modelling clay at a pottery class in Tunbridge Wells. The clay was traced back to a clay quarry in Leicestershire owned by an American company and hence its probable source (Side and Whitehouse 1974). The plant has recently been found in several sites in a single 10-km square in Leicestershire (H. L. K. Whitehouse pers. comm).

Much more spectacular have been the spreads of *Orthodontium lineare*, a Southern Hemisphere species, first found in Cheshire in 1910 (Burrill 1940) and *Campylopus introflexus*, another southern hemisphere species, discovered in Sussex in 1941 and Dublin in 1942 (Richards and Smith 1975). The spread of both species, in terms of number of 10 km squares in which they occur, is shown in Table 4.11. The *Campylopus is* now spreading in western continental Europe. In view of its competitive effect against less vigorous native bryophyte species, *C. introflexus* might be likened to the bryological equivalent of *Rhododendron ponticum*. *Orthodontium lineare* is now common in England and eastern Wales and is now rapidly spreading eastwards in continental Europe. It is out competing the less vigorous native *O. gracile*, which is now only known from nine 10-km grid squares although it was known from an additional 25 squares prior to 1950.

7 Proposals and prognosis for the next 25 years

Three conferences on the conservation of European bryophytes have been held and more are likely to be arranged. A workshop was also held in 1996 and the proceedings were published in *Lindbergia* 23(1) in 1998. The most important contribution of the second was of guidelines for the application of revised IUCN threat categories to bryophytes (Hallingbäck *et al.* 1998).

Although bryology is a very esoteric subject, attempts at interesting members of the public are being made by a travelling museum exhibition prepared by the BBS. In another similar attempt an account of what is in all probability one of the most, if not the most, threatened mosses in the world, *Thamnobryum angustifolium* (leading a tenuous existence at a single easily accessible site in Derbyshire), has raised such concern that an article about it appeared in the national press (Wainwright 1998). It is to be hoped that such approaches to increasing a public awareness of bryophytes will continue.

While there is relatively little threat – other than the effects of climate change and acid rain – to bryophytes in montane parts of Britain, it is likely that further losses will occur in lowland England and central Scotland as a result of continued urbanization, industrialization, pollution and 'improving' agricultural practices. While local and National Nature Reserves will help survival in limited areas, habitat destruction elsewhere will certainly continue. Under present legislation, sites of Special Scientific Interest (SSSIs) count for little. As an example of this, an SSSI on the Lleyn Peninsula in Gwynedd, containing abundant *Scopelophila cataractae*, a heavy-metal-tolerant species that is very rare on a world scale and two rare *Caphaloziella* species, was destroyed in the course of extending a golf course. The owners were fined £2000 – hardly a disincentive where profits are concerned.

It is likely that changes outlined under sections 5.3 and 5.4 will continue. Hopefully some of the species on the extinct list will be refound but others are likely to be added, mainly very rare species which have not been found for 50 years or more, either because they could not be refound at their original sites (e.g. *Seligeria carniolica)* or have not been looked for since their discovery (e.g. *Hygrohypnum polare*). Although species that, might occur in the British Isles are listed in Table 4.9 how many of these, if any, will turn up in the next 25 years remains to be seen. It is equally likely that totally unexpected species from far further afield than continental Europe may be found (similar to *Dicranum subporodictyuon* from Asia and Western Canada, or *Syntrichia amplexa* from eastern North America). Species added or removed as a result of taxonomic revisions are likely to be fewer as most of the larger critical groups occurring in the British isles have now been revised.

ACKNOWLEDGEMENTS

I am most grateful to Mr N. G. Hodgetts (Joint Nature Conservation Committee) for providing up-to-date information on protected and Red List species) and to Mr H. L. Fox (National Botanic Gardens, Dublin), and Dr N. Lockhart (National Parks and

Wildlife Service, Dublin). Dr D. Holyoak very generously made available three West Cornwall distribution maps and Mrs J. A. Paton kindly gave permission to use her data in these maps. Figs. 4.4–4.6 are reproduced by permission of Harley Books, Colchester.

Table 4.1 Percentage frequency with which certain species in Oxfordshire were recorded in habitat lists made in 1940–50 and 1960–90 (modified from Jones 1991).

Species	1940–50	1960–90
Epiphytes		
Frullania dilatata	60	31
Loohocolea heterohylla	42	90
Metzgeria furcata	47	58
Porella platyphylla	29	27
Radula complanata	16	12
Aulacomnium androgynum	5	56
Cryphaea heteromalla	12.5	0
Dicranoweissia cirrata	35	92
Hypnum cupressiforme agg.	40	85
Leucodon sciuroides	25	4
Orthotrichum affine	40	62
O. lyellii	33	8
Tortula laevipila	28	5
Zygodon conoideus	0	43
Terricolous species		
Hylocomium splendens	47	7.5
Isopterygium elegans	0	20
Pseudoscieropodium purum	24	25
Rhytidiadeiphus squarrosus	22	62
R. triquetrus	37	27.5

Table 4.2 Red List of British bryophytes compiled from the lists for liverworts and hornworts, and for mosses prepared by the Joint Nature Conservation Committee in 1998. Species in bold italic type are protected under the provisions of the Wildlife and Countryside Act, Schedule 8 (Plants), 1981.

Extinct mosses

Bryum lawersianum
Cynodontium fallax
Encalypta brevicollis
Grimmia anodon
G. elatior
Gyroweisia reflexa
Helodium blandowii
Lescuraea saxicola

Neckera pennata
Paludella squarrosa[3]
Pterygoneurum lamellatum
Sphagnum obtusum
Tortella limosella
Trematodon ambiguus
Weissia mittenii

Critically endangered liverworts

Fossombronia crozalsii
Lophozia rutheana

Marsupella profunda

Critically endangered mosses

Bryum mamillaturn
B. schleicheri
B. turbinatum
B. uliginosum
Buxbaumia viridis
Didymodon cordatus
D. glaucus
Ditrichum cornubicum
Ephemerum cohaerens

Orthotrichum gymnostomum[1]
O. obtusifolium
O. pallens
Philonotis marchica
Rhynchostegium rotundlfolium
Seligeria carniolica
Tetrodontium repandurn
Thamnobryum angustifolium

Endangered liverworts and hornworts

Cephaloziella dentata
C. integerrima
Jamesoniella undulifolia
Jungerrnannia leiantha

Lejeunea rnandonii
Phaeoceros carolinianus
Scapania paludicola
Southbya nigrella

Endangered mosses

Acaulon triquetrum
Anornodon attenuatus
A. longifolius
Atrichum angustatum
Bantramia stricta
Brachythecium erythrorrhizon
Bryum cyclophyllum
B. gemmiparum
B. marratii
B. neodamense
Ceratodon conicus
Daltonia splachnoides
Dicranum elongatum

Homomallium incurvatum
Hygrohypnurn styriacum
Leptodontiurn gemmascens
Micromitrium tenerum
Orthodontiurn gracile
Philonotis cernua
Plagiothecium piliferum
Scorpidium turgescens
Sematophyllum demissum
Sphagnum balticum
Taybria lingulata
T. tenuis
Tortula cernua

continued on next page

Table 4.2 Red List of British bryophytes (cont.)

Ephemerum stellatum	*Weissia squarrosa*
Eurhynchium pulchellum	**Zygodon forsteri**
Habrodon perpusillus	**Z. gracilis**

Vulnerable liverworts

Adelanthus lindenbergrianus	*Marsupella sparsifolia*
Cephaloziella baurngartneri	*Pallavicinia lyellii*
C. rnassalongi	**Petalophyllum ralfsii**
C. turneri	*Radula carringtonii*
Durnortiera hirsuta	**Riccia bifurca**
Geocalyx graveolens	*R. canaliculata*
Gyrnnocolea acutiloba	*R. crystallina*
Gymnomitrion apiculatum	*R. huebeneriana*
Herbertus borealis	*Scapania praetervisa*
Lejeunea holtii	*Sphaerocarpus texanus*
Marsupella arctica	*Telaranea nernatodes*

Vulnerable mosses

Arnblystegium saxatile	**H. vaucheri**
Andreaea frigida	**Mielichhoferia mielichhoferiana**
Aplodon wormskjoldii	*Myurella tenerrima*
Blindia caespiticia	*Otenidium procerrimum*
Brachythecium starkei	*Physcomitrium eurystomum*
Bryum calophyllum	*Pohila crudoides*
B. knowitoni[1]	*Pseudoleskeella nervosa*
B. stirtonii	**Saelania glaucescens**
B. warneum	*Schistidium atrofuscum*
Campylium halleri	*S. pruinosum*
Cryphaea lamyana	*Seligeria brevifolia*
Cyclodictyon laetevirens	*S. diversifolia*
Desmatodon leucostoma	*Splachnum vasculosum*
Dicranum bergeri	*Syntrichia norvegica*
Eurhynchium meridionale	*Thamnobryum cataractarum*
Fissidens serrulatus	*Timmia austriaca*
Grimmia ovalis	*Tortula cuneifolia*
G. sessitana	*Weissia condensa*
G. unicolor	*W. levieri*
Hygrohypnum molle	*W. multicapsularis*
H. polare	*W. wimmeriana*[2]
Hypnum revolutum	

[1] Almost certainly extinct.
[2] Now reduced to a variety of *Weissia controversa*.
[3] Found in West Mayo, Ireland in 1998 (unpublished).

Note:

Hamatocaulis vernisocus is on the list of protected species but is not on the Red List. The reasons for its inclusion on the former list are obscure.

Table 4.3 List of bryophytes not seen in Britain for many years and apparently extinct (modified from Rose and Wallace, 1974). Species in bold type have since been refound.

Anomobryum juliforme[1]	Neckera pennata
Anomodon attenuatus	Paludella squarrosa[4]
Cynodontium fallax	**Paraleucobryum longifolium**
Fontinalis dolosa[2]	**Pseudoleskeella nervcosa**
F. hypnoides[2]	Scapania crassiretis[2]
Grimmia crinita[3]	**S. parvifolia**
G. elatior	Sphagnum obtusum
Gyroweisia reflexa	Tortella limosella
Helodium blandowii	Trematodon ambiguus
Lescuraea saxicola	Weissia mittenii

[1] Since reduced to synonymy with A. julaceum.
[2] Now considered to be erroneously recorded from Britain.
[3] Almost certainly introduced.
[4] Found in West Mayo, Ireland in 1998 (unpublished).

Table 4.4 Species removed from the British and Irish lists since 1973.

Liverworts

Lophocolea cuspidata[1]	Scapania crassiretis[3]

Mosses

Amblystegium juratzkanum[1]	G. stirtonii[1]
Anomobryum concinnatum[2]	G. subsquarosa[1]
A. juliforme[1]	Pohila acuminata[2]
Bryum obconicum[1]	P. polymorpha[2]
B. purpurascens[3]	Pottia asperula[1]
B. rufifolium[2]	P. commutata
Campylium protensum[2]	Sphagnum imbricatum[3]
Ephemerum intermedium[2]	S. recurvum[3]
Fontinalis dixonii[2]	S. rubellum[2]
F. dolosa[3]	S. subfulvum[2]
F. hypnoides[3]	Thuidium hystricasum[2]
Grimmia meuhlenbeckii[3]	Weissia wimmeriana[2]

[1] Reduced to synonymy with another species.
[2] Reduced to an infraspecific taxon of another species.
[3] Removed because of incorrect determination or specimens of doubtful provenance.

Table 4.5 Percentage of pre-1950 10-km grid square records of widespread species that are decreasing (from Hill *et al.* 1992, 1994).

Aloina arnbigua	53	*Pogonarurn nanurn*	39
Antitrichia curtipendula	43	*Pterygoneurum ovatum*	63
Bryum intermedium	65	*Splachnum ampullaceum*	33
Cryphaea heteromalla	24	*Tortula lanceola*	38
Dicranella crispa	59	*T. modica*	32
Funaria fascicularis	36	Eight rare *Bryum* species	71
Grimmia donniana	23	Nine rare *Grimmia* species	54
Orthotrichum striatum	32	Five rare epiphytic *Orthotrichum* species	63

Table 4.6 Species described as new to science from the British Isles since 1973.

Liverworts

Fossombronia fimbriata	*Plagiochila atlantica*[1]
F. maritima	*P. britannica*[1]
Leiocolea fitzgeraldii	*Telaranea longii*[3]

Mosses

Brachythecium appleyardii[2]	*D. plurnbicola*[1]
Bryoerythrophyllum caledonicum	*Molendoa warburgii*[1]
Bryum dunense[1]	*Pictus scoticus*
Didyrnodon mamillosus[1]	*Sphagnum skyense*
D. tomaculosus	*Tharnnobryum cataractacum*
Ditrichum cornubicum	

[1] Species since found outside the British Isles.
[2] Recently found in Germany (unpublished).
[3] Introduced with horticultural plants from an unknown source (Paton 1992).

Table 4.7 Species previously known elsewhere recorded as new to the flora of the British Isles since 1973. Species in bold type have since proved to be widespread.

Liverworts

Lophocolea bispinosa[1]	*Nardia insecta*
Lophozia herzogiana[1]	*Plagiochila norvegica*
Marsupella arctica	*Scapania lingulata*
M. profunda	*S. mucronata*
Metzgeria temperata	*S. paludicola*

Mosses

Andreaea blyttii	*Pohlia flexuosa*
A. mutabilis	*P. proligera*[5]
Bryum archangelicum	*P. scotica*
B. stirtonii	*Racomitrium affine*[6]
B. subelegans	**R. ericoides**
Cinclidotus riparius[2]	*R. nacounii*
Dicranum subporodictyon	*Rhizomnium magnifolium*
Ditrichum gracile	*Sanionia lycopodioides*
Encalypta brevicollis[3]	*Schistidium crassipilum*[6]
Fissidens limbatus	*S. elegantulum*
F. pusillus[1]	*S. flaccidum*
Funaria pulchella	*S. frigidum*
Grimmia tergestina	*Scopelophila cataractae*
Gymnostomum viridulum	*Seligeria brevifolia*
Hedwigia stellata	*S. campylopoda*
Hennediella macrophylla	*S. diversifolia*
Heterocladium wulfsbergii	*Sematophyllum substrumulosum*
Hygrohypnum styriacum	*Sphagnum affine*
Hypnum uncinulatum[4]	*S. angustifolium*
Leptobarbula berica	*S. fallax*
Philonotis marchica	*Tortula amplexa*[1]
Pohlia camptotrachela	*T. solmsii*

[1] Thought to be introduced.
[2] Deleted from the British and Irish lists prior to 1973 but since reinstated.
[3] Based upon a single Scottish specimen dated 1871 (Horton 1980).
[4] Deleted from the Irish list prior to 1973 but since reinstated (Ando andTowensend 1980).
[5] The plant described in Smith (1978) as *Pohila proligera* is *P. andalusica* (= *P. annotina*) but *P. proligera* s.s. has since been shown to occur in the British Isles (Shaw 1981).
[6] *Racomitrium affine* sensu (Smith 1978) is *R. sudeticum*. *R. affine* s.s. is a different species (Frisvoll 1988).
[7] Likely to prove widespread and common.

Table 4.8 Additions to the British flora resulting from taxonomic revisions since 1973.

Liverworts and Hornworts

Cephaloziella nicholsonii[1]	*P. porelloides*[1]
Plagiochila killarniensis[1]	*Phaeoceros carolineanus*[1]

Mosses

Andreaea alpestris[1]	*Hymnostylium insigne*[1]
A. frigida[1]	*Hypnum andoi*[1]
A. megstospora[2]	*H. jutlandicum*[1]
A. sinuosa[2]	*H. lacunosurn*[1]
Bryum elegans[1]	*H. resupinatum*[1]
B. funkii[1]	*Pseudoleskeella rupestris*[1]
B. torquescens[1]	*Racomitriurn elongaturn*[2]
Callialaria curvicaulis[1]	*Rhynchostegiella litorea*[1]
Ceratodon conicus[1]	*Rhytidiadeiphus subpinnatus*[1]
Dichodontium flavescens[1]	*Schistidiurn conferturn*[1]
Dicranurn flexicaule[1]	*S. rivulare*[1]
Drepanocladus cossonii[1]	*S. robustum*[1]
Ephemerurn minutissimum[1]	*Sphagnum angustifolium*[1]
Fissidens gracilifolius[1]	*S. flexuosum*[1]
Grimmia arenaria[1]	*S. inundatum*[1]
G. austrofunalis[1]	*Tetrodontium repandum*[1]

[1] Previously treated as an infraspecific taxon of another species.
[2] Species described new to science.

Table 4.9 Continental European bryophytes not known from the British Isles but which are likely to occur.

Liverworts

Anastrophyllum assimile	*Lophocolea minor*
A. michauxii	*Lophozia ascendens*
Cephalozziella arctica	*L. laxa*
C. elegans	*Metzgeria simplex*
Frullania jackii	

continued on next page

Table 4.9 Continental European bryophytes (cont.).

Mosses

Amblystegium subtile	Hypnum pratense
Brachythecium curtum	H. recurvatum
B. fendleri	Lescuraea radicosa
Bryhnia novae-angliae	Mnium blyttii
Dicranella howei	Pohlia sphagnicola
Ditrichum pallidum	Pseudoleskeella tectorum
Fissidens arnoldii	Rhodobryum spathulatum

Note:
The original list also included *Narida insecta* and *Cinclidotus riparius* but these have since been reported from the British Isles (Blackstock 1995; Blockeel 1998).

From Crundwell 1992.

Table 4.10 Number of vice-county records of species that have increased in the twentieth century at the time of publication of successive Census Catalogues. The dates represent the time of publication of the following catalogues: Ingham (1907), Ingham (1913), Duncan (1926), Wilson (1930), Warburg (1963), Paton (1965a), Corley and Hill (1981), and Blockeel and Long (1998).

	1913	1930	1965	1981	1998
Nowellia curvifolia	42	46	71	87	92
Ptilidium pulcherrimum	12	22	56	68	74
	1907	1926	1963	1981	1998
Atrichum crispum	13	16	16	24	27
Campylopus introflexus[1]	—	—	26	101	110
Dicranum flagellare	5	4	15	16	19
D. montanum	13	17	34	51	66
D. tauricum	3	7	25	48	64
Herzogiella seligeri	3	5	14	18	20
Orthodontiurn lineare[2]	—	3	64	94	101
Platygyrium repens	—	—	1	9	23
Tortella inflexa[3]	—	—	15	19	22

1 First recorded in England in 1940 (Richards and Smith 1975).

2 First specimen collected in 1910 (Burrill 1940).

3 The earliest herbarium specimen is dated 1904 but the plant was not recognised in Britain until 1957 (Wallace 1957; Hill *et al.* 1992).

Table 4.11 Data from the BBS mapping scheme showing the spread of *Orthodontium lineare* and *Campylopus introflexus* since their discoveries in England in 1910 and 1941, respectively. The numbers represent the total of 10-km grid square records.

Year	Orthodontium lineare		Campylopus introflexus	
	Britain	Ireland	Britain	Ireland
1940	45	1	—	—
1950	—	—	3	2
1960	136	2	37	22
1975	556	3	369	97
1991	919	7	980	225

REFERENCES

Adams, K.J. (1974) The flora: bryophytes, in Jermyn S.J., *Flora of Essex*. Colchester: Harley Books, pp. 227–271.

Ando, H. and Townsend, C.C. (1980) *Hypnum uncinulatum* Jur. reinstated as an Irish species. *Journal of Bryology*, **11**, 185–189.

Bates, J.W. (1995) A bryophyte flora of Berkshire. *Journal of Bryology*, **18**, 503–620.

Blackstock, T. (1995) *Nardia insecta* Linndb., an addition to the liverwort flora of Britain, with cytological observations and a comparison with *Nardia geoscyphus* (De Not.) Lindb. *Journal of Bryology*, **18**, 485–492.

Blockeel, T.L. (1998) *Cinclidotus riparius* reinstated as a British and Irish moss. *Journal of Bryology*, **20**, 109–120.

Blom, H.H. (1996) A revision of the *Schistidium apocarpum* complex in Norway and Sweden. *Bryophytorum Bibliotheca*, **49**, 1–333.

Burrill, W.H. (1940) A field study of *Orthodontium gracile* (Wilson) Schwaegrichen and its variety *heteropterum* Watson. *The Naturalist*, **785**, 295–302.

Crundwell, A.C. (1992) The bryophytes of Britain and Ireland in a European context, in Hill, M.O., Preston, C.D. and Smith, A.J.E. (eds) *Atlas of the Bryophytes of Britain and Ireland*. Vol. 3. *Mosses (except Diplolepideae)*. Colchester: Harley Books, pp. 9–16.

Crundwell, A.C. (1994) 80/2. *Splachnum ampullaceum*, in Hill, M.O., Preston, C.D. and Smith, A.J.E. (eds) *Atlas of the Bryophytes of Britian and Ireland*. Vol. 3. *Mosses (Diplolepideae)*. Colchester: Harley Books, p. 48.

Crundwell, A.C., Bowman, P. and Rose, F. (eds) *The Flora of Hampshire*. Colchester: Harley Books, pp. 325–341.

Crundwell, A.C. and Nyholm, E. (1964), The European species of the *Bryum erythrocarpum* complex. *Transactions of the British Bryological Society*, **4**, 597–637.

Duncan, J.B. (1926) *A Census Catalogue of British Mosses*. 2nd edn, Berwick-upon-Tweed: British Bryological Society.

Frisvoll, A.A. (1983) A taxonomic revision of the *Racomitrium canescens* group (Bryophyta, Grimmiales). *Gunneria*, **41**, 1–180.

Frisvoll, A.A. (1988) A taxonomic revision of the *Racomitrium heterostichum* group (Bryophyta, Grimmiales) in N. and C. America, N. Africa, Europe and Asia. *Gunneria*, **59**, 1–289.

Gardiner J.C. (1981) A bryophyte flora of Surrey. *Journal of Bryology*, **11**, 747–841.

Garvie, M.R. (1990) *Studies on the effect of atmospheric deposition on the woolly hair moss* (*Racomitrium lanuginosum*). MSc thesis, University of Wales.

Graham, G.G. (1988) *The flora and vegetation of County Durham*. Durham: The Durham Flora Committee and the Durham County Conservation Trust.

Greville, R.K. (1822–28), *Scottish Cryptogamic Flora*. 6 vols. Edinburgh: MacLachlan & Stewart.

Hallingbäck, T., Hodgetts, N., Raymaekers, G., Schumacher, R., Sérgio, C., Söderstrom, L., Stewart, N. and Vána, J. (1998) Guidelines for the application of the revised IUCN threat categories to bryophytes. *Lindbergia*, **23**, 6–12.

Hill, M.O. (1988) A bryophyte flora of North Wales. *Journal of Bryology*, **15**, 377–481.

Hill, M.O. and Dominguezlozano, F. (1994) A numerical analysis of the distribution of liverworts in Great Britain, in Hill, M.O., Preston, C.D. and Smith, A.J.E. (eds) *Atlas of the Bryophytes of Britain and Ireland*. Vol. 3. *Mosses (Diplolepideae)*. Colchester, Harley Books, pp. 11–20.

Hill, M.O., Preston, C.D. and Smith, A.J.E. (eds) (1991) *Atlas of the Bryophytes of Britain and Ireland*. Vol. 1. *Liverworts (Hepaticae and Anthocerotea)*. Colchester: Harley Books.

Hill, M.O., Preston, C.D. and Smith, A.J.E. (eds) (1992) *Atlas of the Bryophytes of Britain and Ireland*. Vol. 2. *Mosses (except Diplolepideae)*. Colchester: Harley Books.

Hill, M.O., Preston, C.D. and Smith, A.J.E. (eds) (1994) *Atlas of the Bryophytes of Britain and Ireland*. Vo1. 3. *Mosses (Diplolepideae)*. Colchester: Harley Books.

Hodgetts, N.G. (1995) Bryophyte site register for Europe including Macaronesia, in European Committee for the Conservation of Bryophytes (ed.) *Red Data Book of European Bryophytes*. Trondheim: European Committee for the Conservation of Bryophytes, pp. 195–291.

Horton, D.G. (1980) *Encalypta brevipes* and *E. bricolla*: new records from North America, Iceland, Great Britain and Europe. *Journal of the Hattori Botanical Laboratory*, **11**, 209–212.

Ingham, W. (1907) *A Census Catalogue of British Hepatics*. Darwen: Moss Exchange Club.

Ingham, W. (1913) *A Census Catalogue of British Hepatics*. 2nd edn, Darwen: Moss Exchange Club.

Jones, E.W. (1991) The changing bryophyte flora of Oxfordshire. *Journal of Bryology*, **16**, 513–549.

Klein, R.M. and Bliss, M. (1984) Decline in surface coverage by mosses on Camels Hump Mountain, Vermont, possible relationship to acidic deposition. *Bryologist*, **87**, 128–131.

Laflin T. (1971), 'Bryophytes', in Cadbury, D.A., Hawkes, J.G. and Reader, R.C. (eds) *A Computer Mapped Flora of Warwickshire*. London: Academic Press.

Lewis, K. and Smith, A.J.E. (1977) Studies on some bulbiliferous species of *Pohlia* section *Pohliella* I. *Journal of Bryology*, **9**, 539–556.

Lewis, K. and Smith, A.J.E. (1978) Studies on some bulbiliferous species of *Pohlia* section *Pohliella* II. Taxonomy. *Journal of Bryology*, **10**, 9–27.

Long, D.G. (1982) *Lophocolea smeiteres* (Lehm.) Mitt. established in Argyll, Scotland. *Journal of Bryology*, **12**, 113–115.

Long, D.L. (1990) Bryophytes of Berwickshire VC 80, in Braithwaite, M.E. and Long, D.G. (eds) *The Botanist in Berwickshire*. Berwick: Berwickshire Naturalists Club, pp. 69–98.

Paton, J.A. (1965a) *Census Catalogue of British and Irish Hepatics*. 4th edn, British Bryological Society.

Paton, J.A. (1965b) *Lophoclea semiteres* (Lehm.) Mitt. and *Telarabea murphyae sp. nov.* established on Tresco. *Transactions of the British Bryological Society*, **4**, 775–779.

Paton, J.A. (1965c) A new British moss, *Fissidens celticus* sp. nov. *Transactions of the British Bryological Society*, **4**, 780–784.

Paton, J.A. (1969) A bryophyte flora of Cornwall. *Transactions of the British Bryological Society*, **51**, 669–756.

Paton, J.A. (1974) *Lophocolea bisponosa* (Hook. & Tayl.) Gottsche, Lindenb. & Nees established on the Isles of Scilly. *Journal of Bryology*, **8**, 191–196.

Paton, J.A. (1992) *Telaranea longii sp. nov.* in Britain and a comparision with *T. murphyae* Paton. *Journal of Bryology*, **17**, 289–295.

Paton, J.A. (1999) *The Liverwort Flora of Britain and Ireland*. Colchester: Harley Books.

Preston, C.D. (1991) History of bryophyte recording in the British Isles, in Hill, M.O., Preston, C.D. and Smith, A.J.E. (eds) *Atlas of the Bryophytes of Britain and Ireland*. Vol. 1. *Liverworts (Hepaticae and Anthocerotea)*. Colchester: Harley Books, pp. 13–20.

Porley, R. (1998) Recording matters. *Bulletin of the British Bryological Society*, **71**, 21–22.

Raemakers, G. (1987) The effects of simulated acid rain and lead on the biomass, nutrient status and heavy metal content of *Pleurozium schreberi* (Brid.) Mitt., *Journal of the Hattori Botanical Laboratory*, **63**, 219–230.

Richards, P.W. and Smith, A.J.E. (1975) A progress report on *Campylopus introflexus* (Hedw.) Brid. And *C. polytrichoides* De Not. in Britain and Ireland. *Journal of Bryology*, **8**, 293–298.

Rose, F., Stern, R.C., Matcham, I.V. and Coppins, B.J. (1991) *Atlas of Sussex Mosses and Liverworts*. Brighton: Booth Museum of Natural History.

Rose, F. and Wallace, E.C. (1974) Changes in the bryophyte flora of Britain, in Hawksworth, D.L. (ed.) *The Changing Flora and Fauna of Britain*. London: Academic Press, pp. 27–46.

Shaw, A.J. (1981) A taxonomic revision of the propaguliferous species of *Pohlia* (Musci) in North America. *Journal of the Hattori Botanical Laboratory*, **50**, 1–81.

Side, A.G. and Whitehouse, H.L.K. (1974) *Tortula amplexa* (Lesq.) Steere in Britain. *Journal of Bryology*, **8**, 15–18.

Smith, A.J.E. (1978) *The Moss Flora of Britain and Ireland*. Cambridge: Cambridge University Press.

Smith, A.J.E. (1997) The *Hypnum cupressiforme* complex in the British Isles. *Journal of Bryology*, **19**, 751–774.

Smith, A.J.E. and Whitehouse, H.L.K. (1978) An account of the British species of the *Bryum bicolor* complex including *B. dunense* sp. nov. *Journal of Bryology*, **10**, 29–47.

Stevenson, R. (1996) The disappearance of *Splachnum ampullaceum* Hedw. *Bulletin of the British Bryological Society*, **67**, 43.

Stieperaere, H. (1994) *Lophocolea semiteres* (Lehm.) Mitt. In Belgium and The Netherlands, another antipodal bryophyte spreading on the continent. *Lindbergia*, **19**, 29–36.

Syed, H. (1973) A taxonomic study of *Bryum capillare* Hedw. and related species. *Journal of Bryology*, **7**, 265–326.

Wainright, M. (1998) Rare moss clings on to a low profile. *The Guardian*, 12 May, 2.

Wallace, E.C. (1957) *Tortella inflexa* (Bruch) Broth. in Britian. *Transactions of the British Bryological Society*, **3**, 303.

Wallace, E.C. (1979), *Lophocolea bisoinosa* (Hook. f. & Tayl.) Gottsche, Lindenb. & Nees established on the Isle of Colonsay, Scotland. *Journal of Bryology*, **10**, 675–577.

Warburg, E.F. (1958) *Meesia tristicha* Bruch & Schimp. in the British Isles. *Transactions of the British Bryological Society*, **3**, 160–162 .

Warburg, E.F. (1963) *Census Catalogue of British Mosses*. 3rd edn, Cardiff: British Bryological Society.

Warburg, E.F. (1965) *Pohlia pulchella* in Britain. *Transactions of the British Bryological Society*, **4**, 160–162.

Whitehouse, H.L.K. (1969) *Dicranella staphylina*, a new European species. *Transactions of the British Bryological Society*, **5**, 757–764.

Wilczek, R. and Demaret, E. (1970) Les *Pohlia* propagulifères de Belgique (Bryaceae). *Bulletin de Jardin Botanique National de Belgique*, **40**, 405–422.

Wilson, A. (1930) *A Census Catalogue of British Hepatics*. 3rd edn, Berwick-upon-Tweed: British Bryological Society.

Winner, W.F. and Bewley, J.D. (1978) Terrestrial mosses as bioindicators of SO_2 pollution. Synecological analysis and the index of atmospheric purity. *Oecalogia*, **35**, 221–230.

Woods, R.C. (1993) *The Flora of Radnorshire*. Cardiff: National Museum of Wales.

Chapter 5

Larger fungi

Roy Watling

ABSTRACT

Five reasons are explored with examples for the impressive changes in the larger fungi (macromycetes) of Britain over the last 25 years. The description of new species, the splitting of aggregate taxa, the greater availablity of new identificatory manuals allowing better recording, and the expansion of the distribution of known species all play their part. Recording from previously poorly studied areas of the British Isles, extinctions and commercial collecting are all addressed.

1 Introduction

There are five reasons for the changes that have been experienced in this group of organisms over the last 25 years, two of which might only be considered cosmetic while the others are very fundamental. Together all five have brought about a spectacular change and recent examples will be given to illustrate each of these facets. In fact the two so-called secondary parameters are also very real as they reflect an important phase in the understanding of larger fungi and their distribution in Britain and Ireland. What do we mean by a changing mycota anyway? At this stage in our knowledge we can only use as a baseline the information available in species lists and herbarium records; sadly no long-term critical monitoring has ever been undertaken in this country.

2 Description of new species

It appeared to many that the description of 120 new species of agarics 40 years ago (Orton 1960) was rather excessive, surely marking the end of a chapter, but since then the same trend has been maintained.

The early mycologists apparently had a rather wide concept of their species, accepting great variation – a tradition which held until well into the second half of the twentieth century. The significance of this variation, if it existed, has now been addressed and it can be conclusively shown that the presence of many new species in British mycota is real and they are not just 'splits' of familiar, already identified taxa, namely species complexes. These new taxa have probably been members of the British mycota for a very long time, albeit unrecognized even though many are quite conspicuous and easy to spot. There has also been the inevitable splitting of existing taxa. For instance when I introduced two new members each to the *Leccinum scabrum* and *L. versipelle* complexes over 30 years ago

(Watling 1969) the move was frowned upon; but gradually, if in some cases reluctantly, this solution was accepted even though it was suggested that further additions would undoubtedly come in the future. Now we must go one stage further and adopt the findings of Smith *et al.* (1966, 1967) and Lannoy and Estades (1995) adding at least a further nine taxa in the genus *Leccinum* to those occurring in Britain. That is 16 more than 50 years ago (Pearson and Dennis 1948), 15 more than in the checklist of Dennis *et al.* (1960) and nine more than in the first volume of the *British Fungus Flora* (Watling 1970). However, such drastic modifications pale into insignificance when the numbers of new species are taken into account. In less studied groups such as the jelly fungi and their allies it is probably less difficult to understand why this group is presently undergoing a drastic reassessment (Roberts 1992, 1993, 1994).

The description of new species and the splitting of existing species give to any present-day species list a very different appearance from its counterpart 25 years earlier. Unless voucher material and a fully experienced mycologist are on hand direct comparisons are rather difficult. This makes the work of staff untrained in mycology but wishing to analyse a site's biodiversity almost impossible. There is little doubt that such changes over and above the cosmetic, nomenclaturial changes, which in themselves can lead to confusion, will continue into the near future. A small number of species have been added to the British list after critical revision of herbarium material. Thus *Squamanita contortipes* was recognized in the herbarium of the Royal Botanic Garden, Edinburgh, among collections of *S. odorata* and separated by Bas (1965) as a provisionally new species to science, but later found to have been described previously from North America (Redhead *et al.* 1994). There are other examples.

3 Better recording

There has been a welcome shift in the last 25 years in continental Europe, particularly in the Mediterranean countries, from eating macromycetes to studying them for their own sake although mycophagy still is of high priority; many organizations have taken on a new life and new ones have been formed. This phenomenon has spread throughout Europe and is manifest in Britain also with a very much more discerning public demanding not just mushroom pickers' guides and coffee-table books, although many are aesthetically pleasing, but more complete and more accurate accounts of fungi. Indeed naturalists and conservation bodies have become more aware recently that fungi can add an additional subject of political interest to their repertoire, especially as biodiversity is a now a media-known byword. In Britain reliable texts are now in the hands of competent amateurs who are also equipped with excellent compound microscopes and computers. Thus many more fungal taxa are being recognized by many more people, correctly identified, and recorded, contrasting with only 25 years ago when Reid reported (Reid 1974).

The results of such familiarity show that fungi thought to be confined to a particular locality, because there are more people who can identitfy them with confidence, are now known from other sites. Thus *Boletus satanus,* because of its supposed rarity, was selected as a species in the Biodiversity Action Plan and in this capacity appeared on a British postage stamp; it also was selected as part of Plantlife's 'Back from the Brink' campaign (Marren 1998). *B. satanus* had appeared in at least 40 different sites before 1970, all south of a line joining the Wash and Pembroke, and rarely occurs northwards in the UK; it is still known in 22 of these sites since 1990 and even in 1998 new sightings were being made. These results are in marked contrast to the modest two records for 1949 in the British Mycological Society's database. The increase has been due to dedicated amateur collectors. There are therefore many even rarer species in the British macromycota than that bolete. *Battarrea phalloides* is a rather spectacular stalked puffball, which was once thought to have a very restricted distribution in Britain; although undoubtedly rare, it is much more widespread than thought before this new surge of active collecting (Pegler *et al*. 1995; Watling *et al*. 1995). It is apparently continuing to spread and has now been found for the first time in Oxfordshire (Fortey 1998). *Hygrocybe calyptriformis*, because of its infrequent appearance and beautiful colour, was selected for the poster advertising the European Mycological Congress held at Kew in 1992, the main theme of which was fungal conservation. With the introduction of the Wax Cap Survey by the Conservation Special Interests Commmittee of the British Mycological Society, this species too has been found to be not infrequent in many unimproved grasslands, even on garden lawns: indeed it occurs on a grass verge in the centre of Edinburgh. Similarly the distribution, certainly in Scotland, of *Clavaria zollingeri*, a Red Data List entry, is surely greater than first thought. As recording relies on the presence of fruiting bodies, sudden fruiting may go unnoticed unless an area is constantly monitored. The recording of the presence of a species relies on the collector being at the site at the right time as a few days later it may have disappeared, or it may have fructified just weeks before. Indeed species may appear suddenly and may disappear just as suddenly. The beauty of having a well-informed workforce is that collecting at a locality often continues all the year around so hitting the site at its peaks of production. Relying on foray records, as we have done in the past when the forays were traditionally so slavishly held at specific times, can give misleading information on distribution; this is now being addressed by the 'amateur' input.

Even during the preparation of this account two fungi long thought to have been lost from our mycota have been refound by amateur collectors, namely *Cytidia salicina* (previously only known from Kinrara, near Forres, in 1878), and *Stereopsis vitellina* (collected in Speyside at the turn of the twentieth century). The former has been refound in Kielder Forest in Northumberland, and the latter not far from its original site near Aviemore.

The recording of these fleshy, putrescent organisms is fraught with difficulties and demands that in any biodiversity programme fungi are treated as a special case (Watling 1995); the components may be small and inconspicuous or, if the fruiting bodies are conspicuous, short lived. Even after intensive recording in a single area the number of species may continue to increase, albeit slowly, for 20 years or more (Watling 1995). This often results from the fact that some larger fungi may only be found once in a given period (Tofts and Orton 1998) or may not even appear in the same locality again. Richardson (1970) showed that new species were being added to his intensively studied quadrants in a pinewood even in the fifth year.

4 New British records

Since 1960 in the agarics alone, as this date is the only recent benchmark we possess, roughly 17 new species have been added to the British List annually. Thus in the genus *Coprinus* there has been a 33% increase and in *Russula* 53.8%. Such spectacular figures have occurred not only in the agarics but also in other groups such as the jelly fungi (*Hymenomyctes* and *Heterobasidiae*).

An inspection of the New British Records series in *The Mycologist* in the four years from 1994 to 1997 indicates that 32 species new to the British Isles have been considered; this gives some measure of the rate of additions to the British List of species originally described from other parts of the world but now found in Britain. However, according to the editors the waiting list for publication in this item is extremely long, confirming that this figure is only the tip of the iceberg!

The species reported were known not just from Europe but some might also have been expected ultimately to occur in Britian. There are still some obvious absentees from Britain, for example *Amanita caesarea* and *Albatrellus ovinus*, both near by in continental Europe. It is doubtful whether such have been previously missed because of their eye-catching characters such as bright colour or size. This also applies to, for example, *Omphalotus olearius*, now known from the south of England (Pegler 1977); we must wait and see whether that fungus becomes established or is a purely transient member of the mycota as perhaps *Melanomphalia nigrescens* might be (Jefferies and Watling 1989). *Tulostoma niveum* may have been in Britain much longer and just missed because of its rather cryptic colours and specialized habitat (Pegler *et al.* 1995; Fleming *et al.* 1998) but even this species has very recently been found in a new locality far from the west-coast site; at the new locality there is a similiar vegetation type to its other sites but a very different climatic regime (Liz Holden pers. comm.). It might be suggested that *Omphalotus* has been introduced from a more tropical climate, perhaps the Mediterranean where it is common, and has established itself here because of changes in climate. Such examples might support the hypothesis for global warming but such conclusions are unfounded on the data available, especially as the vegetative state of the fungus may have existed for many years without fruiting. We do not know enough about the fruiting patterns of macromycetes to use these organisms as indicator species of climatic change with any certainty.

Undoubtedly, *Suillus placidus* recorded from Bedgbury Arboretum has been introduced with exotic conifers in much the same way as *S. grevillei* (now widely distributed in the UK) was with European larch on the Atholl Estate in 1830 (Watling 1965); in fact so widespread is the latter that many mycologists take this species for granted not realizing it is an alien in much the same way as some of the weed plants introduced with consignments of wool. *S. placidus* may have been happily growing on the introduced substrate for a long time without fruiting just as there may be many others waiting to fruit, but it has not appeared to have spread even though the characteristic ectomycorrhizal host is found elsewhere in the British Isles. Certainly mycorrhizas with a distinctive morphology have been found which cannot be linked to any known sporophore but which can be linked with certainty to a particular genus (A. Taylor pers. comm.). Among the stipitate hedgehog fungi, a group which has recently been selected for monitoring woodland quality (Gulden and Hanssen 1992), three species – *Bankera violascens*, *Hydnellum auratile*, *Sarcodon glaucopus* – have all recently been added to the British list. Has the quality of the Speyside woodland really improved that much? I doubt it, but again fruiting and the attendance of a observant mycologist are key factors in recognizing unusual fungi in our mycota. Undoubtedly the ability to identify a fungus in the absence of its fruiting body will take a step forward with the use of molecular techniques (Bruns *et al.* 1998) and place the use of macromycetyes as indicators of environmental change on a more substantiatable basis.

There will always be exotic fungi which may appear suddenly in Britain and then disappear. One group especially noted for this pattern of events is the phalloids (Ramsbottom 1953; Reid 1974; Spooner 1994; Pegler *et al.* 1995). Even when established in certain warmer areas of the UK, their appearance north of the border causes great interest, for example the occurrence of *Clathrus cibarius* at Musselburgh near Edinburgh in 1996 and 1997. *Chlorophyllum molybdites,* although since vanished, was found for the first time in Britain near Edinburgh in an amenity planting at a public swimming pool; it was undoubtedly introduced in compost (Watling 1991) and is one of a long line of fungi some of which establish themselves and increase in distribution because of human activity, for example *Leucocoprinus birnbaumii;* formerly found in botanic garden greenhouses, this is now quite common in supermarket plantings and propagation nurseries, and also with house plants. *L. lilacinogranulosus,* which has accompanied *L. birnbaumii* and *L. brebisonnii* in the Edinburgh greenhouses since the 1960s, although only formally recorded as British by Reid (1989), has not made this transition. *Amanita nauseosa* might be another example; described from Royal Botanic Gardens, Kew as *Lepiota nauseosa* in 1918, it occurred there again in 1963 (Reid 1966). It has not been seen since but it did occur among members of the *Zingberaceae* in the tropical greenhouses at the Royal Botanic Garden, Edinburgh for one season only (Watling 1980). All these species have probably been introduced from abroad and maintained in our mycota because of horticultural practices that favour them. *Gymnopilus dilepis* is the most recent recruit (Watling 1998). In contrast, *Mycena alphitophora* (syn. *M. osmundicola*) is probably less common than it was 25 years ago because *Osmunda* has been replaced by other substrates, notably coconut husk, in the cultivation of orchids. Except for *L. birnbaumii,* which has established itself precariously in a range of heated environments, the others are unlikely to expand their distribution.

In the outside environment *Clitocybe pruinosa* (syn. *C. radicellata*) has been brought to my attention growing in a rock garden in Edinburgh around a planting of *Veronica* that had been recently been purchased from a nursery. With the expansion of the nursery trade, the introduction of new techniques and an increase in the commodity trade with a range of new countries, we will see many such fungi appearing: *Melanotus textilis* on doormats manufactured in south-east Asia is an example. The truffle *Hydnangium carneum* has been found on several occasions in the British Isles since it was first recorded from the Glasgow Botanic Garden in 1875; it is associated with an introduced *Eucalyptus* species with which it is ectomycorrhizal. It was probably introduced from Australia very early in the history of translocation of myrtaceous plants to Europe. However, a newcomer is the similiarly associated and related agaric *Laccaria fraterna,* found in 1997 in a garden in East Lothian, Scotland (Last and Watling 1998). Although comparatively few in overall number these records constitute very important additions to our mycota.

5 Movement of fungi

Calocera pallidospathulata, although only described as new just before the period under review, was either in this country before and went unnoticed or had been introduced; with such a prominent fruit-body the former is less likely but it is now well established throughout the plantations of Great Britain growing on fallen conifer branches. The centre of expansion was probably the plantations of east Yorkshire but it was rapidly recognized in many places thereafter (Ing 1990). The speed of spread is in keeping with what we know of introductions in other organisms, and it is interesting to speculate from whence it came; was it the New World or from Indo-Malaysia from where another familiar fungus, *Serpula lacrymans*, probably came many generations ago? North America is probably more likely. Another jelly fungus, *Tremiscus helvelloides*, again a fungus with an expanding distribution, also first appeared in east Yorkshire but in 1914, and has extended to areas close to the original site at Sandsend, where it grows along the rides on sawdust/soil mixtures in plantations. Later it was found in Herefordshire, then in Warwickshire and several localities in and around Glamorgan; it appeared to be spreading west and in Wales it was apparently associated with quarry workings. More recently it has been recorded from Durham and from Stirlingshire at the edge of a willow invasion front on disturbed soil in old gravel workings (Munro and Watling 1993). It appears, therefore, to be slowly spreading both north and east. A similar expansion of the range of this species has been recorded in Scandanavia by Torkelsen (1998).

Also on disturbed, often nitrogenous, soil has been found *Paurocotylis pila*, a rather prominent, bright orange-red relative of the truffles. It was described from New Zealand and subsequently occurred in Nottinghamshire (Dennis 1975) in a quarry. It was 17 years later before it was found again, this time in a potato patch in Yorkshire (Barker and Watling 1993), 400 m from where it was again found in 1998. It is now known from several sites in a wood at Binscarth in Orkney and a site in West Lothian (Watling 1997) where it was found again in 1999. The most recent records are from a garden in Darlington (A. Legg pers. comm.) and a herb garden in Edinburgh; in the last two seasons it has been found in Sheffield and North Yorkshire. In January 2000 it even turned up in Edinburgh. The possible connection between the three Lothian sites is that the owner of the Edinburgh site regularly visits the other sites in the Lothians where it occurs; is this only a coincidence? The species has been subsequently refound in New Zealand (Pegler *et al.* 1992). Although it was considered introduced by Cannon *et al.* (1985), some doubt has been raised to this suggestion by Pegler and his colleagues (1992).

Paurocotylis pila is a rather unfamiliar species to the general mycologist, but *Schizophyllum commune*, the split-gill fungus, is not; it too has expanded its distribution in the British Isles in recent years. From a basically southern distribution (Watling 1978) it is now found as far north as the central belt and far west of Scotland. This normally lignicolous fungus has found a new habitat niche although it is true worldwide that it will grow on a wide range of substrates including human tissues (Watling and Sweeney 1971). Demonstrated by Webster (1995) as the fungus found emerging from the split sides of silage bags in Devon, it is known in post-1987 records for Scotland and Yorkshire under similar conditions. It apparently has switched from growing on logs etc. to straw substrates, although in 1884 Worthington G. Smith had already recorded it on silage in a silo. In the late 1960s it was found on straw near Edinburgh by a research team working on asthma-causing fungi and actinomycetes. Prior to these sightings it was only known in Scotland as isolated records, for example, on a seed box in a nursery, on driftwood, and on pitchy timber imported from West Africa. A similar, although much earlier expansion, not restricted to silage, occurred in Denmark after the Second World War (H. Knudsen pers. comm.). It is too early to speculate as to whether this kind of migration northwards is a result of global warming; some might use this as an example but the reasons for the fruiting of fungi are at present so poorly known it is dangerous to make unsupported conclusions. One only has to examine the records for the 1998 collecting season when, except for certain 'hot-spots', production of fruiting bodies has been at an all-time low. Yet some species have fructified in quantities never seen before, for example, hundreds of basidiomes of the rare *Rozites caperata* in a single wood in Deeside in autumn 1998, whereas at the same site the usual members of the macromycota were in unusually low number. *Suillus flavidus*, which is characteristic of the remnant Caledonian pine forests, has been collected several times now in pine plantations in Shetland where the seedlings had been germinated and grown in nurseries within the vicinity of the Speyside pinewoods (Watling 1992) before planting out, suggesting that mycorrhizas had been formed there. With the absence of competitors in its newly planted environment, the bolete continued to flourish. The felling of old conifer woodlands in the south of England has resulted in several species quite familiar to the mycologist at the turn of the twentieth century now being now confined to the northern outliers both in remnants of the Caledonian pine forest and rich plantations such as Culbin Sands, Morayshire. One such macromycete is the thelephoroid *Boletopsis leucomelanea*. Other more widespread agarics have in contrast recently taken up unfamiliar ecological niches. *Omphalina pyxidata*, commonly found in grasslands near the sea, has become a pathogen of winter wheat in England; a recent change in agricultural practice has offered a chance for this fungus to change its habitat.

6 Uncharted waters

The contents of the known mycobiota of the British Isles has changed dramatically with the addition of a whole series of fungi from habitats infrequently or never surveyed before. Thus montane (Watling 1987, 1988), coastal (Watling and Rotheroe 1989; Rotheroe 1993) and island communities have given a new perspective to the species known to occur in Britain. The fungi of the western islands of Scotland have come under particularly close scrutiny and their mycota has been summarized by Dennis (1986). The northernmost island archipelagos have been dealt with by Watling (1992) for Shetland, and Watling *et al.* (1999) and Dennis and Spooner (1992, 1993) for Orkney. Many arctic-alpine elements have been recognized and added to our mycota, for example *Gerronema marchanitiae* (Watling and Romero 1989) and *Multiclavula vernalis* (Watling and Fryday 1992). Even the mountainous areas of England and Wales have now been visited and the records made extended the known range of distribution of many familiar, once thought exclusive, Scottish species, for instance *Amanita nivalis* in the Lake District (Watling 1985) and *Phyconis luteovitellina* in the Snowdonia massif in Wales. Rotheroe (1993) has carried out a full survey of sand-dune systems around the coast of Britain and has added many taxa formerly not known from the British Isles.

7 Commercial collecting

There has been much recent discussion of the deleterious effects of picking fungi, especially edible forms, by commercial collectors. With so few scientifically conducted experiments, and none in Britain, we know little about the fruiting patterns of larger fungi. Apparently, climatic conditions play a greater role (Norvell 1995) in the lifecycle of some macromycetes than the effects of gathering, certainly in the case of *Cantharellus cibarius*, so it is hard to show unequivocally that this is significant in the changing British mycobiota. Certainly quantitatively at a single site in the border counties of Scotland, more colonies and fruit bodies were found in 1997 during the second of a three-year study. This had in fact been mirrored in other parts of Britain where colonies of this fungus fruited even in areas in which it had not previously been known; these fruitings were probably a result of climatic differences. The momentum has been upheld during 1998 fruiting even starting as early as mid-May and continuing into late November.

The above contrasts with *Boletus edulis*, which has had two catastrophically poor years, the 1998 season being the worst in my memory, including areas where no commercial collecting is or has been allowed. What factors this year spurred *Rozites*, mentioned above, to fruit and *C. tubiformis* to produce in 1997 a strip under *Picea* at Kindrogan, Perthshire, 2 m wide by over 100 m to produce many thousands of fruiting bodies? We may never know, or we are certainly a long way from knowing, because of lack of research. Interpreting macromycete data is fraught with danger.

8 Extinctions

The only example of a member of the British larger fungi becoming extinct is the puffball relative *Myriostoma coliforme*, which occurred near Kings Lynn until it was last seen in 1880. But there are numerous examples in the literature of fungi thought to be near extinction or which have disappeared but then suddenly reappeared. *Poronia punctata* is an excellent example of a fungus thought to have disappeared because of the loss of dray horses, but it is again a prominent member of the British mycota in the south of the country on pony droppings. The appearance of this fungus in adjacent areas may be linked to the movement of those ponies from the New Forest being selected as pets. However, sporadic fruitings show how dangerous it is to assume a fungus is extinct, even when the time interval of its non-appearance is measured in tens of years.

9 Conclusions

We are in an exciting period in the understanding of British macromycetes, the relationships between species and between the taxa and their ecosystems. It is therefore doubly disappointing that there is no teaching of systematic mycology in our British institutes and that in 1998 two experienced mycologists who specialized in the larger fungi retired with only one being replaced very recently. We cannot blame natural history wardens taking fungus lists at face value and monitoring the wrong species unless careful teaching is available. Alas, even botanists still think of larger fungi as being the contents of a mushroom picker's basket or purely objects found by those going on a natural history walk in the autumn. Although larger fungi have attracted much media coverage recently in connection with the exploitation of natural resources, global warming and a code of conduct for picking and conservation, they still do not command the attention they deserve. I can only emphasize that fungi really are a special case (Watling 1995). Larger fungi are directly important in soil-cycling activities while mycorrhizas, in addition to being food for invertebrates, are also, and perhaps less obviously, nourishment for mammals. In a single example during an in-depth study of the reasons for the decline of the red squirrel in Scotland, the stomach contents of 12 squirrels killed in nature in various ways were examined in Edinburgh and all contained large numbers of spores of *Elaphomyces*, *Melanogaster ambiguus* and *Leccinum* species (Turnbull 1995).

Undoubtedly the largest effect on the macromycota over the last 25 years is a result of the destruction of specific habitats (Ing 1996) and of changes in land management, namely forestry practices, horticultural and agricultural activities. These will no doubt change again in the future and more than once; we may even see changes quite soon now that set-aside is currently in operation.

REFERENCES

Barker, G. and Watling, R. (1993) Profiles of fungi 49: *Paurocotylis pila* Berk. *The Mycologist*, 7, 14.

Bas, C. (1965) The genus *Squamanita*. *Persoonia*, 3, 331–364.

Bruns, T.D., Szaro, T.M., Gardes, M., Chillings, K.W., Pan, J.J., Taylor, D.L., Horton, T.R., Kretzer, A., Garbelotto, M. and Li, Y. (1998) A sequence database for the identification of ectomycorrhizal basidiomycetes by phylogenetic analysis. *Molecular Ecology*, 7, 257–272.

Cannon, P., Hawksworth, D.L. and Sherwood-Pike, M.A. (1985) *The British Ascomycotina: an annotated checklist*. Slough: Commonwealth Agricultural Bureaux.

Dennis, R.W.G. (1975) New or interesting British microfungi. *Kew Bulletin*, 29, 157–179.

Dennis, R.W.G. (1986) *Fungi of the Hebrides*. Royal Botanic Gardens, Kew.

Dennis, R.W.G., Orton, P.D. and Hora, F.B. (1960) A new checklist of British agarics and boleti. *Transactions of the British Mycological Society*, 43 (Supplement), 1–225.

Dennis, R.W.G. and Spooner, B.J. (1992) The fungi of North Hoy, Orkney I. *Persoonia*,14, 493–507.

Dennis, R.W.G. and Spooner, B.J. (1993) The fungi of North Hoy, Orkney II. *Persoonia*, 15, 155–168.

Fleming, L.V., Ing, B. and Scouller, C.E.K. (1998) Current status and phenology of fruiting in Scotland of the endangered fungus *Tulostoma niveum*. *The Mycologist*, 3, 126–131.

Fortey, R.A. (1998) *Battarraea* in Oxfordshire. *The Mycologist*, 12, 159–160.

Gulden, G. and Hanssen, E. W. (1992) Distribution and ecology of stipitate hydnaceous fungi in Norway with special reference to the question of decline. *Sommerfeltia*, 13, 1–56.

Ing, B. (1990) Profiles of fungi 25: *Calocera pallidospathulata* Reid. *The Mycologist*, 4, 34.

Ing, B. (1996) Red data lists and decline of macromycetes in relation to pollution and loss of habitat, in Frankland, J., Magan, N. and Gadd, G.M. (eds) *Fungi and Environmental Change*, Cambridge: Cambridge University Press, pp. 61–69.

Jefferies, D.W. and Watling, R. (1989) Profiles of fungi 18. *Melanomphalia nigrescens* Christiansen. *The Mycologist*, 3, 43.

Lannoy, G. and Estades, A. (1995) *Monographie des Leccinum d'Europe*. La Roche-sur-Foron: Chevalier.

Last, F.T. and Watling, R. (1998) The first record of *Laccaria fraterna* in Britain. *The Mycologist*, 12, 152–153.

Marren, P. (1998) On the trail of the Devil's own toadstool. *Plantlife*, August, 6–7.

Munro, I. and Watling, R. (1993) Profiles of fungi 51: *Tremiscus helvelloides* (DC.: Pers.) Donk. *The Mycologist*, **7**, 86.

Norvell, L. (1995) Loving the chanterelle to death? *McIlvainea*,12, 6–25.

Orton, P.D. (1960) New check list of British agarics and boleti. Part III. Notes on genera and species in the list. *Transactions of the British Mycological Society*, 43, 159–439.

Pearson, A.A. and Dennis, R.W.G. (1948) Revised list of British agarics and boleti. *Transactions of the British Mycological Society*, 31, 145–190.

Pegler, D.N., (1977) *Omphalotus olearius (Agaricales) in Britain. Kew Bulletin*, 32, 5–7.

Pegler, D.N., Laessøe, T. and Spooner, B.M. (1995) *British Puffballs, Earthstars and Stinkhorns.* Kew: Royal Botanic Gardens.

Pegler, D.N., Spooner, B.M. and Young, T.W.K. (1992) *A Revision of the British Hypogeous Fungi.* Kew: Royal Botanic Gardens.

Ramsbottom, J. (1953) *Mushrooms and Toadstools*, London: Collins.

Redhead, S.A., Ammirati, J.F., Walker, G.R., Norvell, L.L. and Puccio, M.B. (1994) *Squamanita contortipes*, the Rosetta stone of a mycoparasitis agaric genus. *Canadian Journal of Botany*, 72, 1812–1824.

Reid, D.A. (1966) Coloured illustrations of rare and interesting fungi. Part l. *Nova Hedwigia* 11 (Suppl.), 1–32.

Reid, D.A. (1974) Changes in the British macromycetete flora, in Hawksworth, D.L. (ed.) *The Changing Flora and Fauna of Britain*. London: Academic Press, pp. 79–86.

Reid, D.A. (1989) Notes on some leucocoprinoid fungi from Britain. *Mycological Research*, **93**, 413–423.

Richardson, M.J. (1970) Studies on *Russula emetica* and other agarics in a Scots pine plantation. *Transactions of the British Mycological Society*, 55, 217–225.

Roberts, P. (1992) Spiral-spored species of *Tulasnella* from Devon and the New Forest. *Mycological Research*, **96**, 233–236.

Roberts, P. (1993) Allantoid-spored species of *Tulasnella* from Devon and the New Forest. *Mycological Research*, **93**, 212–220.

Roberts, P. (1994) Long-spored species of *Tulasnella* from Devon and the New Forest. *Mycological Research*, **98**, 1235–1244.

Rotheroe, M. (1993) The macrofungi of the sand dunes, in Pegler, D.N., Boddy, L., Ing, B. and Kirk, P.M. (eds) *Fungi of Europe: Investigation, Recording and Conservation*. Kew: Royal Botanic Gardens, pp. 121–138.

Smith, A.H., Theirs, H.D. and Watling, R. (1966) A preliminary account of the North American species of *Leccinum* section *Leccinum*. *Michigan Botanist*, 5(3A), 131–179.

Smith, A.H., Theirs, H.D. and Watling, R. (1967) A preliminary account of the North American species of *Leccinum* sections *Luteoscabra* and *Scabra*. *Michigan Botanist*, **6** (3A), 107–154.

Spooner, B.J. (1994) *Aseroe rubra* at Oxshott. *The Mycologist*, 8, 153.

Tofts, R.J. and Orton, D. (1998) The species accumulation curve for agarics and boleti from a Caledonian pinewood. *The Mycologist*, **12**, 98–101.

Torkelsen, A.-E. (1998) *Tremiscus helevelloides* – its distribution in the Nordic countries. *Agarica*, 15, 59–66.

Turnbull, E. (1995) Not only nuts in May. *The Mycologist*, 9, 82–83.

Watling, R. (1965) Notes on British boleti: IV. *Transactions of the Botanical Society of Edinburgh*, 40, 100–120.

Watling, R. (1969) Records of boleti and notes on their taxonomic position. *Notes from the Royal Botanic Garden, Edinburgh*, 29, 265–272.

Watling, R. (1970) *Boletaceae; Gomphidiaceae; Paxillaceae*, in Henderson, D.M., Orton, P.D. and Watling, R. (eds) *British Fungus Flora: Agarics and Boleti*, vol. 2. Edinburgh: Royal Botanic Garden, pp. 1–127.

Watling, R. (1978) The distribution of larger fungi in Yorkshire. *Naturalist London*, **103**, 39–57.

Watling, R. (1980) *Amanita nauseosa* – a foreign visitor? *Bulletin of the British Mycological Society*, **14**, 23.

Watling, R. (1985) Observations on *Amanita nivalis* Grev. *Agarica*, **6**, 327–335.

Watling, R. (1987) Larger arctic-alpine fungi in Scotland, in Laursen, G.A., Ammirati, J.F. and Redhead, S.A. (eds) *Arctic and Alpine Mycology II*, New York: Plenum Press, pp. 17–45.

Watling, R. (1988) A mycological kaleidoscope. *Transactions of the BritishMycological Society*, **90**, 1–28.

Watling, R. (1991) A striking addition to the British mycoflora. *The Mycologist*, **5**, 23.

Watling, R. (1992) *The Fungus Flora of Shetland*. Edinburgh: Royal Botanic Garden.

Watling, R. (1995) Assessment of fungal diversity; macromycetes, the problems. *Canadian Journal of Botany*, **73** (Supplement), 15–24.

Watling, R. (1997) Biodiversity of lichenised and non-lichenised fungi in Scotland, in Fleming, V.L., Newton, A.C., Vickery, J.A. and Usher, M.B. (eds) *Biodiversity in Scotland status, trends and initiatives*. Edinburgh: The Stationery Office, pp. 77–88.

Watling, R. (1998) Profiles of fungi 94: *Gymnopilus dilepis*. *The Mycologist*, **12**, 61.

Watling, R., Eggeling, T. and Turnbull, E. (1999) *The Fungus Flora of Orkney*. Edinburgh: Royal Botanic Garden.

Watling, R. and Fryday, A. (1992) Profiles of fungi 44: *Multiclavula vernalis* (Schwein.) Petersen. *The Mycologist*, **6**, 67.

Watling, R., Gucin, F. and Isiloglu, M. (1995) *Battarrea phalloides* – its history, biology and extension of its distribution. *Nova Hedwigia*, **60**, 13–18.

Watling, R. and Romero, A. (1989) Profiles of fungi 17: *Gerronema marchantiae* Singer & Clém. *The Mycologist*, **3**, 42.

Watling, R. and Rotheroe, M. (1989) Macrofungi of sanddunes. *Proceedings of the Royal Society of Edinburgh*, B96, pp. 111–126.

Watling, R. and Sweeney, J. (1971) Observations on *Schizophyllum commune* Fries. *Sabouraudia*, **12**, 214–226.

Webster, J. (1995) *Schizophyllum* in hay bales. *The Mycologist*, **5**, 118.

Chapter 6

Microscopic fungi

Paul F. Cannon, Paul M. Kirk, Jerry A. Cooper and David L. Hawksworth

ABSTRACT

Owing to the imbalance between species numbers and scientific inputs, changes in microscopic fungus diversity and population structure over the past 25 years can only be analysed in general terms. However, the provision of modern identification manuals and checklists, and the introduction of computerized recording using databases by the British Mycological Society, have massively increased the amount and quality of information available for analysing and monitoring microfungus populations in Great Britain and Ireland during this period. Around 12 000 species of fungi are currently known in these islands, and around 120 species are added to this total every year, including the discovery of many new to science. A total inventory of the mycobiota is far from complete, and could rise to 20 000 species, making the group second only to the insects in species numbers.

Changes in agriculture and forestry practice, pollution and global warming are all having identifiable effects, and it is likely that many native species have become less common. There is some concern that fungi associated with plants in the UK which are rare or have restricted distributions are threatened. Some of these are plant pathogens and have traditionally been considered deleterious to the plant populations, but are likely to be in equilibrium with their hosts unless affected by external influences. There have probably been significant numbers of microfungi introduced, associated with crop plants, ornamentals and tree species, and some (notably *Ophiostoma novo-ulmi*) have had a significant affect on the landscape in many parts of the country.

Biodiversity action plans have been prepared for only three species of British microfungi. A Red Data List is not yet formally published, but a preliminary contribution listed six species as probably extinct and a further nine as endangered. There is some optimism for the future as fungi are considered more seriously by conservation agencies, and the amateur mycological community increases in size and expertise under the banner of the British Mycological Society's recording scheme.

1 Definition

This chapter concerns non-lichenized members of the *Ascomycota* (including their anamorphs), *Chytridiomycota*, *Zygomycota* and the *Uredinales* and *Ustilaginales* (*Basidiomycota*).

2 State of knowledge

The classification and recording of microfungi in the British Isles has changed out of all recognition since 1973 (Booth 1974). At that stage, available checklists were all seriously out of date, and basic reference texts were few and far between. For British microfungi, the only recent identification manuals available were Wilson and Henderson's book on rust fungi (1966), Dennis' *British Ascomycetes* (1968) and Ellis' first hyphomycete book (1971), which while global in coverage included a large number of British and Irish species. The *Mucorales* were dealt with in Zycha *et al.* (1969), but coelomycete identification relied on Grove (1935–37), and many hyphomycete groups were inaccessible to the field mycologist. During the 1970s and 1980s, available texts started to increase in number. Highlights were publication of the volume of *The Fungi* devoted to *Ascomycota* (Ainsworth *et al.* 1973), a major contribution to identification of the bitunicate ascomycetes (von Arx and Müller 1975), a second volume on pigmented hyphomycetes (Ellis 1976), a general hyphomycete generic treatment (Carmichael *et al.* 1980), and the first major manual on coelomycetes for nearly 50 years (Sutton 1980). A much enlarged third edition of Dennis' ascomycete book appeared (1978) with a small supplement following three years later (Dennis 1981).

A new checklist of the British *Ascomycota* appeared in 1985 (Cannon *et al.* 1985) which included both lichenized and non-lichenized taxa, and basic information on ecology and distribution for 5100 species that proved to be known from Great Britain and Ireland. This greatly simplified the identification process for field mycologists by summarizing the available information and providing references to further literature. However, the most significant text for microfungus identification and recording in recent years is undoubtedly *Microfungi on Land Plants* (Ellis and Ellis 1985, 1997). This provided brief descriptions and drawings of an enormous range of British microfungi arranged on a host/substratum basis, making identification feasible for non-specialists with a basic knowledge of fungal structures and a reasonable microscope. It has contributed in no small way to the explosion in recording of microfungi in the British Isles in recent years, which is described below. This book was followed by a companion volume dealing with microfungi from soil, dung and other miscellaneous substrata (Ellis and Ellis 1988). Specialist texts have also begun to appear, including identification handbooks for smuts (Mordue and Ainsworth 1984) and hypogeous fungi (Pegler *et al.* 1993).

By the early 1970s, some of the basic assumptions of fungal classification were being questioned, leading to a series of changes in the way in which the taxonomy of microfungi was approached. This has had three major effects. First was the integration of lichenized and non-lichenized ascomycete classifications, primarily through a series of *Outlines of the Ascomycetes*, the first appearing in 1982 (Eriksson 1982) and the latest now available on the world wide web (Eriksson 1999). The seventh edition of *Ainsworth & Bisby's Dictionary of the Fungi* (Hawksworth *et al.* 1983) expanded Eriksson's work greatly, and was influential in the widespread acceptance of the new ideas, and further major changes were charted in the eighth edition (Hawksworth *et al.* 1995).

The second change began about the same time, but is still in the process of taking effect, despite a major boost in the early 1980s from the second Kananaskis conference

(Kendrick 1980). Although many fungi have been recognized as pleomorphic (i.e. with morphologically distinct stages in their lifecycles) since the middle of the nineteenth century, until recently, for practical reasons classification of asexual and sexual morphs has remained largely separate. This has resulted in widespread confusion, potential 'double-counting' of species in geographic or host lists, has seriously hindered research on fungal lifecycles, and led to complexities in the nomenclatural rules which are misunderstood by almost all (Cannon and Kirk 2000). Integration of the asexual and sexual morph classification will cause some short-term problems, but will result in immense long-term benefits with a one-name one-species system as enjoyed by almost all other systematic biologists.

The third change has only just begun, with nucleic acid sequences (especially of ribosomal DNA) being analysed as a routine procedure in fungal classification. This has already had major effects with an explosion of phylogenetic research, the provision of powerful new tools in integrating anamorph and teleomorph classifications, the ability to remove host effects from consideration in the classification of pathogenic fungi (as well as to study coevolution), and much new information on species concepts and infraspecific classification.

A further major influence on mycological recording and conservation in Britain over the last 25 years has been the biodiversity debates developed by and originating from the Earth Summit held in Rio de Janeiro in 1992. However, the inclusion of fungi in environmental monitoring programmes and conservation strategies is problematic because of the large number of species involved (fungi are probably the second most speciose organism group in the world; Hawksworth 1991), only a small proportion of which even have basic descriptions, and their small size and cryptic nature which makes detection difficult (Cannon 1997). The importance of fungi in ecological processes (Christensen 1989) is better realized than ever before, and fungi are now becoming integrated into biodiversity action plans in several parts of the world.

3 Data capture

The past 25 years have seen a massive increase in mycological recording throughout the United Kingdom. Part of this has been the result of years of dedicated study by relatively small groups of expert amateur mycologists. One of the most significant outputs was the publication of the *Fungus Flora of Warwickshire* (Clark 1980), which represented the first serious attempt (at least in modern times) at charting the mycological biota of a British county. Of the various groups of microfungi, the rusts and smuts had already been reasonably comprehensively surveyed, but the recording of ascomycetes and their anamorphs did not begin methodically until 1968. Even then, it was recognized by the survey participants that this group remains under-sampled. Over 2600 species were recorded, including about 1200 microfungi, of which 29 were new to science. This massive endeavour has been followed by a number of others, including Willis Bramley's survey of Yorkshire fungi (1985), compilations of the fungi of the western Scottish islands (Dennis 1986) and of south-east England (Dennis 1995), and most recently a major list of the fungi of the New Forest (Dickson and Leonard 1996).

On a smaller scale, there have been several long-term studies of individual sites in southern England which have added considerably to knowledge of fungal diversity, as almost certainly the most comprehensively surveyed localities in the world. Fungi from the National Nature Reserve at Slapton Ley in south Devon have been recorded since 1969 (Hawksworth 1976, 1986; Dobson and Hawksworth 1996), and now more than 2500 species are known from this 200 ha site of which over 30 have proved new to science (D. Hawksworth, pers. comm.); the total for this one site is expected to to be at least 3000 species (Dobson and Hawksworth 1996). Esher Common in Surrey (c. 400 ha) has been actively surveyed over 25 years, with over 2900 species now known (Kirk unpubl.), and a small (4 ha) area of chalk grassland nearby at Mickleham is known to harbour over 1300 species despite a relatively uniform vegetation type and no permanent open water (A. Henrici, pers. comm.). It is clear that none of these sites is comprehensively surveyed – in all cases species are being added to the lists steadily and many niches have not seen seriously studied. A further indication of incomplete survey is that there is only about a 40% overlap between the species lists of Slapton and Esher even though there are many similarities between the two sites (Cooper unpubl.).

The other major influence over fungal survey and recording during the last years of the twentieth century has been the placing of the British Mycological Society (BMS)'s fungal survey work on a proper scientific footing with the introduction in 1985 of a computerized records database by the Foray Secretary of the time, David W. Minter. Up to this point, many of the alliteratively titled 'Fungal Forays' of the BMS had been largely social occasions, with little serious attempt to produce anything more than bare lists of fungi recorded from often ill-defined localities. The introduction of the database had a massive impact on the scientific value of the records, with information on associated organisms (plant or animal), grid reference, collector's identity, herbarium records, and so on being systematically gathered for the first time. This change in attitude to fungal recording has resulted in an accumulation of an enormous amount of data, which is now massively valuable as a scientific resource in its own right. It has also spawned a whole series of local recording groups (currently over 30) affiliated to the BMS, which are contributing data to the central repository.

With the help of initiatives such as the Joint Nature Consultative Committee's grant to computerize fungal records published in the UK since 1955 (which added over 110 000 records), the BMS Fungal Records Database (BMSFRD) now contains almost half a million entries, representing an estimated 12 000 species. As the editorial work involved in maintaining the database is considerable, there is a permanent backlog of records originating from local groups waiting to be incorporated. About 42% of the database records refer to microscopic fungi as used in this chapter. Not surprisingly, almost all of the most-recorded species in the database are conspicuous macrofungi, but *Xylaria hypoxylon* is currently sixth in the league table with 2204 records as of March 2000, and *Daldinia concentrica*, *Rhytisma acerinum*, *Nectria cinnabarina* and *Lachnum virgineum* are included in a list of 41 species with more than 1000 records (Cooper 2000). Numerous pointers suggest that the database is by no means a comprehensive information source for records of British microfungi: nearly 40% of species are recorded only once, and nearly 1000 species were added to the database in the five years to 1997. Individual groups of microfungi are clearly under-represented; for example the *Ophiostomatales*, for which only 131 records exist despite the Dutch elm disease epidemics caused by *Ophiostoma novo-ulmi*. However, the current gaps in information should not detract from what is now a massively valuable resource for British mycology.

Just how many fungi are recorded from Great Britain and Ireland is uncertain in the absence of overall checklists. While the figure of 12 000 species (including larger fungi, lichens, and slime moulds) used by Hawksworth (1991) in estimating fungal diversity is in agreement with the number now included in the BMSFRD, it is probably low in not accounting for many species not recorded for decades but at the same time will include synonyms and duplicates due to separately named states of pleomorphic fungi. The figure of over 16 500 species (including lichens) used in *The UK Biodiversity Action Plan* (Department of the Environment 1994) is, however, almost certainly an overestimate of the number known.

New fungi are being added to the British and Irish lists at an astonishing rate. During the period 1981–90, 459 species were described as new to science from these islands (Hawksworth 1993), almost 46 species per year. The number already named but not previously reported each year is uncertain but likely to average around 80 species. This suggests that we are in no way close to a complete inventory of the mycobiota of Great Britain and Ireland. Grove (1937) pointed out that an ardent collector could find new fungi in almost any site, and this is still the case today, especially if neglected habitats are explored such as lichen thalli (lichenicolous fungi; see Chapter 7, this volume) and insect exoskeletons. The number of species of *Laboulbeniales* known from insects in the British isles was increased from 43 to 120 in five years by the work of a single enthusiast (A. Weir, pers. comm.). Even the gardens of Buckingham Palace in central London have yielded two fungi new to science. We hesitate to guess at the number of fungi actually present in Great Britain and Ireland, but would not be surprised if the total eventually rose from the current 12 000 to 20 000 – a figure approaching the 22 500 insect species currently recognized in the UK (Department of the Environment 1994).

The major advance that has been achieved in recording since 1973 has, regrettably, been against a background of declining support for systematic mycology in the UK: there are now probably only two professional systematists in Great Britain whose jobs are primarily devoted to UK mycology (excluding lichenologists), although a further two in universities do undertake systematic work on particular groups. The number of systematic mycologists in the intergovernmental CABI Bioscience (incorporating the International Mycological Institute), which has a world remit, had been 13 for many years but was reduced to seven following a restructuring at the end of 1997; the organization is also now required to charge for identifications undertaken, which has greatly reduced the throughput of material from non-commercial sources.

4 Synopsis of changes to 1973

In preparing the account of microfungi for the 1973 meeting, Booth (1974) had to work against a background of largely unassembled data. He focused on plant pathogenic fungi occurring on cultivated plants, as they are generally better recorded than most other fungi. Attention was drawn to the sudden occurrence and spread of disease-causing species, and to how climatic factors governed the ability of species to establish or spread. At that date Dutch elm disease was still to reach its peak.

Booth was wisely reluctant to make statements about the presence or absence of other microfungi, and stressed the problems in recording them and how little was known of those present. While the data now becoming available (see above) has changed the situation, it will remain difficult to analyse past changes as so little recording was carried out before the 1970s.

5 Changes 1973–98

There are insufficient data for anything except generalizations concerning changes in the British fungal biota over the last 25 years, and it is only during the last five years or so that individual species other than plant pathogens have been monitored on a systematic basis. The overall level of fungal diversity in the UK may well have declined significantly, as habitats have been destroyed, and agricultural practices have become more intensive. Balancing this, although a few fungal species may have extended their range naturally to the UK along with newly recorded plant species, a significant number of human-mediated introductions have probably taken place. Some of these are pathogens associated with crop plants or forest trees. The best documented example in recent years must be that of *Ophiostoma novo-ulmi*, which spread throughout the UK in the 1960s and 1970s, introduced probably from both North America and Eastern Europe (Brasier 1990, 1991). *Colletotrichum acutatum* now occurs regularly in Britain as a pathogen of strawberry and lupin (Simpson *et al.* 1994; Reed *et al.* 1996), almost certainly introduced with infected host tissues. *Puccinia distincta* is spreading rapidly through southern England on wild and cultivated *Bellis* species, and may well have been introduced from Australia via continental Europe (Preece *et al.* 2000). *Melampsora allii-populina* now occurs sporadically in southern England, introduced with fast-growing poplar clones (Lonsdale and Gibbs 1996). Many more species must now be established in the UK after initial introduction with ornamental plants. This process must have occurred for at least the last 200 years, since the days of the early plant hunters, but is probably still increasing due to unauthorized plant introductions and imperfect quarantine regulation. A number of microfungi associated with introduced plants such as *Laurus* and *Eucalyptus* have UK records but are otherwise known primarily from warm temperate and tropical regions. There are probably more records of the fungus *Readeriella mirabilis*, an apparently obligate inhabitant of *Eucalyptus* leaf litter, from the UK than there are from Australia. Species in the genera *Circinotrichum*, *Gyrothrix* and *Zygosporium* are not infrequently found on leaf litter of *Laurus nobilis* and a few other coriaceous leaf types in the UK although primarily considered to be genera of the tropics and subtropics. Many of these are likely to be introductions (though the date of their first appearance is unknown), and their presence as part of the UK fungal biota may not be permanent. Many common soil and dung saprobes must have had their natural ranges extended dramatically through human activity, with the spread of domesticated animals and crop plants as well as incidentally in soil attached to shoes. It is quite probable that the rate of this process is increasing as the exotic holiday market expands.

6 Endangered and protected species

Fungi have only recently begun to be included in formal biodiversity action plans, and only one single Site of Special Scientific Interest (SSI) has so far been designated only for its importance for non-lichenized fungi – a lawn rich in *Hygrocybe* and other rare grassland species. Of the Biodiversity Action Plans published so far (Biodiversity Steering Group 1998–99; Watling 1999), only three include microfungi. *Hypocreopsis rhododendri* is known from the north Devon and Cornwall border and a few sites in western Scotland, and is classified as vulnerable due to damage and destruction of its habitat, hazel thickets. *Microglossum olivaceum* is considered to be vulnerable, and is known from a scattering of sites throughout western England and Wales. It is threatened by agricultural improvement of its grassland habitat, and by scrub encroachment. *Poronia puncata* should probably be treated as endangered; it is known in the UK primarily from the New Forest on dung from horses which have fed on unimproved pasture, with a few recent records from Oxfordshire.

The UK does not yet have a formal Red List for fungi, although one is currently at an advanced state of preparation. A provisional list (Ing 1992) included 46 ascomycetes and 63 rusts and smuts, of which six were believed to be already extinct (*Geoglossum peckianum*, *G. simile*, *Microstoma protractum*, *Melanotaenium cingens*, *M. hypogeaum* and *Ustilago marina*) and a further nine considered as endangered. The inconspicuous nature of many microfungi, their often ephemeral appearance and the lack of qualified recorders make the compilation of such lists uncertain, but the threats to habitats are clear and mycologists must continue to work with the conservation movement to ensure that site protection and management plans are appropriate for survival of fungal species.

A number of fungi (especially rust species) are threatened indirectly through the rarity of plant species on which they are dependent for survival (Helfer 1993; Ingram 1999). Again, many of these may be under-recorded, but by definition rusts of plants such as *Adiantum capillus-veneris* and *Goodyera repens* must be restricted in their occurrence. It is important to impress on conservation decision-makers that many of these species have coevolved with their hosts and are unlikely to pose serious threats as pathogens. Instead, they should be considered as equivalent in importance to their host plants (Cannon *et al.* 2001). Not all biotrophic fungi associated with rare plants in the UK are endangered; while the plant *Dryas octopetala* is restricted to scattered limestone outcrops in northern England and Scotland, its ascomycete associate *Isothea rhytismoides* is common in northern Scotland and may well occur almost everywhere its host is found (Cannon 1996).

7 Factors affecting changes

7.1 Agriculture

Changes in agricultural practices must have had a dramatic effect on British fungi over the last 25 years, although there is little hard evidence. The massive increase in use of pesticides, more effective weed control, increases in chemical fertilizer application, the destruction of hedgerows and the use of antibiotics in livestock farming must all have had significant impact. The effects on fungi in Britain are difficult to measure, but species associated with unimproved cereal production such as *Gibellina cerealis*, which has not been recorded in Britain in recent times, are almost certainly no longer a part of the UK list. The 'improvement' of grassland has undoubtedly contributed significantly to the decline of *Geoglossum* species which are typical of non-intensively managed pastureland. *Poronia punctata* only occurs on dung of horses which have fed on unimproved pasture or hay, and may be threatened further through well-meaning activities to supplement the winter diets of the feral ponies on which it now depends.

Hedgerow eradication has been accompanied by the decimation of the British elm population through the introduction of *Ophiostoma novo-ulmi*. The disease is only prevalent in mature trees as the bark beetle vector does not attack young stems, though the long-term future of *Ulmus* species in Britain must be in doubt owing to reduction in its reproductive capability. Linked with this decline must be a considerable reduction in occurrence of other microfungi associated with *Ulmus* such as *Quaternaria dissepta* and *Platychora ulmi*.

7.2 Woodland management

Effects are again difficult to quantify, but the reduction over the last 25 years in the area of old natural woodland, its replacement by conifer monoculture, the decline in traditional practices such as coppicing, and the increasing use of woodlands for recreational purposes must have had significant effects on fungal populations. These may be more apparent for mycorrhizal macrofungi than for microscopic species, as microfungal mycelial nets must generally be less extensive and less prone to damage by trampling. Scrub clearance has been identified as a specific threat to *Hypocreopsis rhododendri* as a component of its biodiversity action plan (see above). Draining of marshland has probably contributed to the reduced presence of *Cryptomyces maximus*, an associate of willow considered to be vulnerable by Ing (1992), although it is not clear whether this species was ever common in the UK.

7.3 Climate change

The effects of global warming are only beginning to be felt, and while there is as yet no direct evidence of its effect on fungal populations, the documented spread of species from Europe such as the poplar rust mentioned above (Lonsdale and Gibbs 1996) is almost certainly linked to this phenomenon. It is tempting to speculate that species with warm temperate distributions such as *Biscogniauxia mediterranea* (Spooner 1986) and *B. anceps* (Rogers *et al.* 1996) have spread recently to the UK as a result of climate change, but they may well have been overlooked in earlier years. Increased levels of insolation, ultra-violet radiation and ozone have been shown experimentally to affect populations of fungal leaf associates (Ayres *et al.* 1996), and soil fungal community structure appears to change in response to raised atmospheric carbon dioxide levels (Jones *et al.* 1998).

The celebrated storms over southern Britain in 1987 and 1990 may or may not be a manifestation of climate change, but they have had a dramatic short-term effect on microfungus populations. Some species, for example *Hypoxylon fragiforme* and *Biscogniauxia nummularia* (this last considered rare by Dennis 1978), have increased in abundance (or at least in fruit-body production) dramatically with the increase in availability of recently killed *Fagus* trunks – although these effects are presumably not permanent unless further catastrophic weather conditions occur. Conversely, leaf-inhabiting fungi associated with beech such as *Apiognomonia errabunda* may well have declined, and species adapted to life under closed canopies in beechwoods must also have been seriously affected.

7.4 Pollution

The effects of atmospheric pollution, including acid rain, have been documented to some extent in recent years for microfungi, in addition to the much more detailed information available for lichenized fungi (Chapter 7, this volume). Magan *et al.* (1996) showed that phyllosphere and endophytic fungi were sensitive to increased sulphur dioxide and ozone concentration. *Rhytisma acerinum* was considered to be an indicator species for sulphur dioxide pollution (Bevan and Greenhalgh 1976), and the species is now more prevalent in urban areas than it appeared to be ten years ago. This may be the result of reduction in the use of high-sulphur coals, but could also be explained by other human-mediated practices such as changing policy on street and park cleaning. Increased prevalence of species such as Black Spot of roses caused by *Diplocarpon rosae* does, however, seem to be primarily correlated with falling sulphur dioxide levels as would be expected from the study of Saunders (1966).

Pollutants in streams have been shown to have significant effects on aquatic microfungi (Bermingham 1996), and the increase in nitrification of Britain's water systems must have had a dramatic effect on fungal population structures.

8 Prognosis for the next 25 years

Many of the trends identified in fungal population patterns we have discussed are likely to continue, and it is only a small comfort that we are now in a much better position to monitor change than we were in 1973. There is some cause for optimism, however, as environmental quality achieves a higher and higher profile in the public eye. Reduction in pesticide use is likely, changing policy in agriculture such as the increased adoption of set-aside may favour fungi, and management policies for natural habitats will be more enlightened. The development of genetically modified crops is likely to benefit fungal populations in general due to pesticide reduction (if commercial planting is ever allowed), although the incorporation of fungicide resistance or genes coding for proteins toxic to fungi might have a devastating long-term effect on saprobes as well as pathogens.

Other changes in factors affecting fungal occurrence and abundance are less likely to be beneficial. Climate change is likely to cause extinctions worldwide, and the effects on British and Irish fungi are likely to be significant. In particular, the lag between increasing temperatures causing extinction of cold-adapted species and their replacement by fungi adapted to warmer climates may be significant, leading to a large (but not necessarily permanent) reduction in fungal species numbers. Increasing summer temperatures and seasonal drought may well cause a great reduction in fungal diversity; there is much anecdotal evidence linking fungal diversity to availability of water and many species may not be able to survive for long periods of drought.

The most significant grounds for optimism over the next 25 years for the fungi of Great Britain are that they are now considered to be much more important components of the ecosystem than was the case in the 1970s, and while professional capacity for identification, characterization and monitoring has reduced and is unlikely to recover in the short term, the amateur interest shows every sign of increasing. While the professional capability to support this burgeoning interest will be an ongoing concern, the challenge will be to harness this enthusiasm, through increased support for the BMSFRD and other recording initiatives; provide education to increase the scientific value of recording; and enhance communication with biologists of other disciplines to further increase awareness of the significance of fungi. If this book contributes to the achievement of that aim, it will have served a vitally useful purpose.

REFERENCES

Ainsworth, G. C., Sparrow, F. K. and Sussman, A. S. (eds) (1973) *The Fungi. An advanced treatise*, vol. 4A. London: Academic Press.

Arx, J. A. von and Müller, E. (1975) A re-evaluation of the bitunicate ascomycetes with keys to families and genera. *Studies in Mycology*, 9, 1–159.

Ayres P. G., Gunasekera, T. S., Rasanayagam, M. S. and Paul, N. D. (1996) Effects of UV-B radiation (280–320 nm) on foliar saprotrophs and pathogens, in Frankland, J. C., Magan, N. and Gadd, G. M. (eds), *Fungi and Environmental Change*. Cambridge: Cambridge University Press, pp. 32–50.

Bermingham, S. (1996) Effects of pollutants on aquatic hyphomycetes colonizing leaf material in freshwaters, in Frankland, J. C., Magan, N. and Gadd, G. M. (eds), *Fungi and Environmental Change*. Cambridge: Cambridge University Press, pp. 201–216.

Bevan, R. J. and Greenhalgh, G. N. (1976) *Rhytisma acerinum* as a biological indicator of pollution. *Environmental Pollution*, 10, 271–285.

Biodiversity Steering Group (1998–99) *UK Biodiversity Group Tranche 2 Action Plans*. London: HMSO.

Booth, C. (1974) The changing flora of microfungi with emphasis on the plant pathogenic species, in Hawksworth, D. L. (ed.), *The Changing Flora and Fauna of Britain*. London: Acdemic Press, pp. 87–95.

Bramley, W. G. (1985) *A Fungus Flora of Yorkshire, 1985*, Leeds: Yorkshire Naturalists' Union.

Brasier, C. M. (1990) China and the origins of Dutch elm disease: an appraisal. *Plant Pathology*, 39, 5–16.

Brasier, C. M. (1991) *Ophiostoma novo-ulmi* sp. nov., causative agent of current Dutch elm disease pandemics. *Mycopathologia*, 115, 151–161.

Cannon, P. F. (1996) Systematics and diversity of the *Phyllachoraceae* associated with *Rosaceae*, with a monograph of the genus *Polystigma*. *Mycological Research*, 100, 1409–1427.

Cannon, P. F. (1997) Strategies for rapid assessment of fungal diversity. *Biodiversity and Conservation*, 6, 669–680.

Cannon, P. F., Hawksworth, D. L. and Sherwood-Pike, M. A. (1985) *The British Ascomycotina: an annotated checklist*. Slough: Commonwealth Agricultural Bureaux.

Cannon, P. F. and Kirk, P. M. (2001) The philosophy and practicalities of amalgamating anamorph and teleomorph concepts. *Studies in Mycology*, 45, 19–25.

Cannon, P. F., Mibey, R. K. and Siboe, G. M. (2000) Microfungus diversity and the conservation agenda in Kenya, in Moore, D., Nauta, M. M. and Rotheroe, M. (eds), *Fungal Conservation: Issues and Solutions*, in press.

Carmichael, J. W., Kendrick, W. B., Conners, I. L. and Sigler, L. (1980) *Genera of Hyphomycetes*. Edmonton: University of Alberta Press.

Christensen, M. (1989) A view of fungal ecology. *Mycologia*, 81, 1–19.

Clark, M. C. (1980) *A Fungus Flora of Warwickshire*. London: British Mycological Society.

Cooper, J. A. (2000) http://www.ukncc.co.uk/bmspages/bmsfrd/stats.htm

Dennis, R. W. G. (1968) *British Ascomycetes*. Lehre: J. Cramer.

Dennis, R. W. G. (1978) *British Ascomycetes*, 2nd edn. Vaduz: J. Cramer.

Dennis, R. W. G. (1981) *British Ascomycetes. Addenda and Corrigenda*. Vaduz: J. Cramer.

Dennis, R. W. G. (1986) *Fungi of the Hebrides*. Kew: Royal Botanic Gardens.

Dennis, R. W. G. (1995) *Fungi of South East England*. Kew: Royal Botanic Gardens.

Department of the Environment (1994) *Biodiversity. The UK Action Plan*. London: HMSO.

Dickson, G. and Leonard, A. (1996) *Fungi of the New Forest: a Mycota*. London: British Mycological Society.

Dobson, F. S. and Hawksworth, D. L. (1996) The Slapton fungal (including lichen) survey: inventorying and documenting changes in the mycobiota. *Field Studies*, 8, 677–684.

Ellis, M. B. (1971) *Dematiaceous Hyphomycetes,* Kew: Commonwealth Mycological Institute.

Ellis, M. B. (1976) *More Dematiaceous Hyphomycetes*. Kew: Commonwealth Mycological Institute.

Ellis, M. B. and Ellis, J. P. (1985) *Microfungi on Land Plants*. London: Croom Helm.

Ellis, M. B. and Ellis, J. P. (1988) *Microfungi on Miscellaneous Substrates*. London: Croom Helm.

Ellis, M. B. and Ellis, J. P. (1997) *Microfungi on Land Plants*, revised edn. Slough: Richmond Publishing.

Eriksson, O. E. (1982) Outline of the ascomycetes – 1982. *Mycotaxon*, 15, 203–248.

Eriksson, O. E. (1999) Outline of the ascomycetes – 1999. *Myconet*, 3, 1–88. [also http://www.umu.se/myconet/99.Outline/99.outline.html]

Grove, W. B. (1935–37) *British Stem- and Leaf-Fungi (Coelomycetes)*, 2 vols. Cambridge etc.: Cambridge University Press.

Hawksworth, D. L. (1976) The natural history of Slapton Ley Nature Reserve. X. Fungi. *Field Studies*, 4, 391–439.

Hawksworth, D. L. (1986) The natural history of Slapton Ley Nature Reserve. XVII. Additions to and changes in the fungi (including lichens). *Field Studies*, 6, 365–382.

Hawksworth, D. L. (1991) The fungal dimension of biodiversity: magnitude, significance, and conservation. *Mycological Research*, 95, 641–655.

Hawksworth, D. L. (1993) The tropical fungal biota: census, pertinence, prophyaxis, and prognosis, in Isaac, S., Frankland, J. C., Watling, R. and Whalley, A. J. S. (eds) *Aspects of Tropical Mycology*. Cambridge: Cambridge University Press, pp. 265–293.

Hawksworth, D. L., Kirk, P. M., Sutton, B. C. and Pegler, D. N. (1995) *Ainsworth & Bisby's Dictionary of the Fungi*, 8th edn. Wallingford: CAB International.

Hawksworth, D. L., Sutton, B. C. and Ainsworth, G. C. (1983) *Ainsworth & Bisby's Dictionary of the Fungi*, 7th edn. Slough: Commonwealth Agricultural Bureaux.

Helfer, S. (1993) Rust fungi – a conservationist's dilemma, in Pegler, D. N., Boddy, L., Ing, B. and Kirk, P. M (eds) *Fungi of Europe: investigation, recording and conservation*. Kew: Royal Botanic Gardens, pp. 287–294.

Ing, B. (1992) A provisional Red Data List of British fungi. *Mycologist*, 6, 124–128.

Ingram, D. S. (1999) Biodiversity, plant pathogens and conservation. *Plant Pathology*, 48, 433–442.

Jones, T. H., Thompson, L. J., Lawton, J. H., Bezemer, T. M., Bardgett, R. D., Blackburn, T .M., Bruce, K. D., Cannon, P. F., Hall, G. S., Howson, G., Jones, C. G., Kampichler, C., Kandeler, E. and Ritchie, D. A. (1998) Impacts of rising CO_2 on soil biota and processes in terrestrial ecosystems. *Science*, 280, 441–443.

Kendrick, [W.] B. (1980) *The Whole Fungus*, 2 vols. Ottawa: National Museums of Canada.

Lonsdale, D. and Gibbs, J. N. (1996) Effects of climate change on fungal diseases of trees, in Frankland, J. C., Magan, N. and Gadd, G. M. (eds), *Fungi and Environmental Change*. Cambridge: Cambridge University Press, pp. 1–19.

Magan, N., Smith, M. K. and Kirkwood, I. A. (1996) Effects of atmospheric pollutants on phyllosphere and endophytic fungi, in Frankland, J. C., Magan, N. and Gadd, G. M. (eds) *Fungi and Environmental Change*. Cambridge: Cambridge University Press, pp. 90–101.

Mordue, J. E. M. and Ainsworth, G. C. (1984) *Ustilaginales* of the British Isles. *Mycological Papers*, 154, 1–96.

Pegler, D. N., Spooner, B. M. and Young, T. W. K. (1993) *British Truffles. A revision of British hypogeous fungi*. Kew: Royal Botanic Gardens.

Preece, T. F., Weber, R. W. S. and Webster, J. (2000) Origin and spread of the daisy rust epidemic in Britain caused by *Puccinia distincta*. *Mycological Research*, 104, 576–580.

Reed, P. J., Dickens J. S. W. and O'Neill, T. M. (1996) Occurrence of anthracnose (*Colletotrichum acutatum*) on ornamental lupin in the United Kingdom. *Plant Pathology*, 45, 245–248.

Rogers, J. D., Ju, Y.-M. and Candoussau, F. (1996) *Biscogniauxia anceps* comb. nov. and *Vivantia guadalupensis* gen. et sp. nov. *Mycological Research*, 100, 669–674.

Saunders, P. J. W. (1966) The toxicity of sulphur dioxide to *Diplocarpon rosae* Wolf causing black spot of roses. *Annals of Applied Biology*, 58, 103–114.

Simpson, D. W., Winterbottom, C. Q., Bell, J. A. and Maltoni, M. L. (1994) Resistance to a single UK isolate of *Colletotrichum acutatum* in strawberry germplasm from northern Europe. *Euphytica*, 77, 161–164.

Spooner, B. M. (1986) New or rare British microfungi from Esher Common, Surrey. *Transactions of the British Mycological Society*, 86, 401–408.

Sutton, B. C. (1980) *The Coelomycetes. Fungi imperfecti with pycnidia, acervuli and stromata.* Kew: Commonwealth Mycological Institute.

Watling, R. (1999) Launch of UK Biodiversity Action Plans for lower plants. *Mycologist*, 13, 158.

Wilson, M. and Henderson, D. M. (1966) *British Rust Fungi*. Cambridge: Cambridge University Press.

Zycha, H., Siepmann, R. and Linnemann, G. (1969) *Mucorales*. Lehre: J. Cramer.

Chapter 7

Lichens

Brian J. Coppins, David L. Hawksworth and Francis Rose

ABSTRACT

The improvement in our knowledge of lichens in Great Britain and Ireland since 1973 has been dramatic. The number of species known since 1973 has increased by 295 to 1663, and while five species have been lost in this period, 16 lichens previously considered extinct have been refound. The effects of agriculture, especially hypertrophication of bark and ammonium emissions, are now a major concern. Reductions in sulphur dioxide levels have, however, resulted in dramatic improvements in urban and other lowland areas. Woodland and heathland management practices threaten other species. The importance of lichens as a habitat or food for other organisms is stressed. Twenty-nine species are now protected under the Wildlife and Countryside Act, and Biodiversity Action Plans are being drawn up and implemented for a wide range of endangered species. The loss of professional expertise in lichenology is now a major concern, and the prognosis for the future is of a decline in rarer species and spread of many commoner ones: an equilibration.

1 State of knowledge

In 1973, 1368 obligately lichenized fungi were known in Great Britain and Ireland; today that total stands at 1653. A new checklist was issued in 1980 (Hawksworth *et al.* 1980) with updates in 1985 (Cannon *et al.* 1985) and 1993–94 (Purvis *et al.* 1993, 1994a); separate census catalogues have also been prepared for Ireland (Seaward 1984, 1994a) and Wales (Woods and Orange 1999). A history of lichenological exploration from 1568 to 1975 has been prepared (Hawksworth and Seaward 1977), and Mitchell (1993, 1995, 1996, 1998) has made a major contribution to the documentation of early Irish lichenology. Distribution maps started to appear regularly in *The Lichenologist* from 1973, and the first volume of distribution maps appeared in 1982 (Seaward and Hitch 1982).

A comprehensive account of the species known, with descriptions and keys, has been published (Purvis *et al.* 1992); the first such treatment since those of Smith (1918, 1926). European or world monographs of numerous genera have been prepared which have greatly assisted the development of knowledge about British lichens. Those not listed in Purvis *et al.* (1992) include: Egea and Torrente (1994), Ekman (1997), Hafellner (1993, 1994), Hertel (1995), Obermayer (1994), Printzen (1995), Purvis *et al.* (1995), and Scheidegger (1993).

Major attention has focused on hitherto underworked habitats such as mountainous areas (Fryday 1997; Gilbert and Fox 1985, 1986; Gilbert and Giavarini 1993; Gilbert *et al.* 1982, 1988, 1992; Purvis *et al.* 1994b), rivers (Gilbert and Giavarini 1997), urban habitats (Gilbert 1990), seashores (Fletcher 1975a, b), lowland terricolous habitats (Fletcher 1984) and churchyards (Chester 1998). The importance of parklands and woodlands with ecological continuity has been reinforced by further work. Indices of ecological continuity have been refined (Rose 1976, 1992) and widely used in site assessments, such as those of Fletcher (1982, 1993). An overview and classification of lichen communities in the British Isles was prepared by James *et al.* (1977).

There has also been a revival in the publication of introductory identification manuals, notably Alvin (1977), Broad (1989), Dobson (1979a, b and later editions), Jahns (1980), Laundon (1986), Orange (1994a), and Phillips (1980). Also produced have been a valuable series of keys to species in the Burren Hills (McCarthy and Mitchell 1988), a field guide to *Cladonia* (Hodgetts 1992a), a new key to sterile crustose lichens on rocks (Fryday and Coppins 1997) and a CD-ROM on *Parmelia* (British Lichen Society 1997).

Although our knowledge of lichens in the British Isles has developed to a degree unforeseen in 1973, there is still much work needed on less explored habitats and taxonomic groups. About 10 species are discovered for the first time each year, including the occasional foliose or shrubby species, and we estimate the actual number present to be about 1800.

2 Data capture

The first mechanically produced maps from the British Lichen Society's Mapping Scheme were published in 1971; a series of maps started to be issued in *The Lichenologist* in 1973, and a first volume of an *Atlas* in 1982 included 176 species (Seaward and Hitch 1982). Preliminary additional parts were prepared and a new series of separately published fascicles was started in 1995 (Seaward 1995).

The coverage has increased considerably since 1973, both in terms of 10-km squares with records and the number of species recorded in the squares (Fig. 7.1). The Scheme was a leader when designed in the 1960s and focused on mapping at the 10-km square level. The degree to which data can be interrogated is limited, and so to complement the current scheme the Society decided in 1999 to also adopt the decentralized BioBase recording system; this will be able to output data into both the 10-km mapping scheme and to Recorder, the system now being adopted for many other groups of organisms in the UK.

Further information on the development of the Society's Mapping Scheme is provided by Hawksworth and Seaward (1990); and Seaward (1988a, 1995).

Additional regional and county studies to have appeared since our previous survey include ones on Berkshire, Buckinghamshire and Oxfordshire (Bowen 1980), Carlow (Seaward 1979), Colonsay (Rose and Coppins 1983), Dartmoor (Giavarini 1990), Dorset (Bowen 1976), Durham (Graham 1988), Glamorgan (Orange 1994b), Gwynedd (Pentecost 1987), Hampshire (Rose and James 1974; Sandell and Rose 1996), Isle of Wight (Pope 1985), Laois (Seaward 1983), Lundy (James *et al.* 1996, 1997), Mull (James 1978), Northumberland (Gilbert 1980), Radnor (Woods 1993), Sussex (Rose *et al.* 1991), Warwickshire (Lindsay 1980), the West Yorkshire

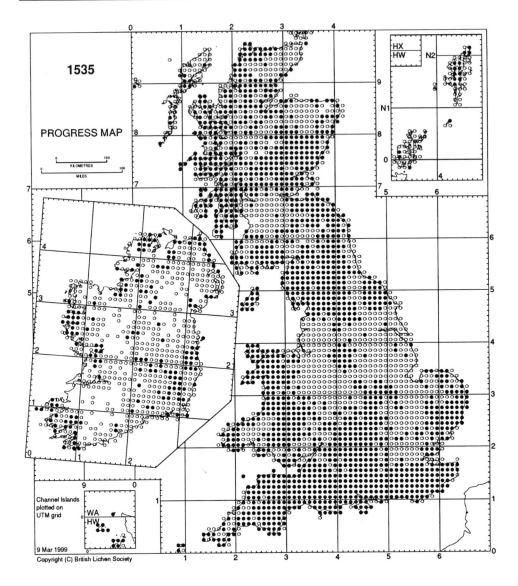

Figure 7.1 Progress in recording the lichens of Great Britain and Ireland. o less than 100 species;
        • over 100 species.

Data prepared by M. R. D. Seaward from the British Lichen Society's Mapping Scheme.

conurbation (Seaward 1975) and Yorkshire (Seaward 1994b). Detailed treatments of lichens in Devon, Somerset and Wiltshire are currently in advanced stages of preparation. Numerous other checklists, updates and studies of particular sites are cited in the *British Lichen Society Bulletin*.

Some regional record centres hold lichen data and the establishment of others is currently in progress. Various site surveys undertaken for the statutory conservation agencies and other bodies are held on file at various regional centres.

3 Workforce

Membership of the British Lichen Society has risen from 492 in 1976 to 604 in 1998 of whom 302 live in the British Isles. Of those resident in the British Isles, most are retired and amateur naturalists with a particular interest in lichens. No detailed information on the age spectrum is available, but we estimate the average age to be about 50. A study of the officers and members of the Council of the Society shows that many of those active in these roles are the same now as they were 25 years ago.

The number of systematic and ecological lichenologist professionals in permanent positions and actively working on British lichens in universities, museums, botanic gardens and other institutions is now seven, but these equate to two full-time equivalents in systematics, and two in ecology. These figures compare with 13 in 1976, which then corresponded to four full-time equivalents in systematics and five in ecology.

In 1976 there were seven British amateur lichenologists publishing on lichen systematics and distributions, and this figure has remained about the same. There are also currently about seven freelance lichenologists who undertake site survey consultancy work on an occasional basis.

We are extremely concerned at this trend, for the loss of available expertise in professional positions also means that support is dwindling for amateur and freelance lichenologists, conservation agencies and other bodies, ecologists, and experimental lichenologists. This is of particular importance now and in the near future, given the opportunities arising from implementation of national and European policies. (See below under Species protection.)

4 Synopsis of changes to 1973

The main changes which had taken place in the lichens of Great Britain and Ireland since the first records in 1568 to 1973 were of deterioration due to sulphur dioxide air pollution and habitat destruction, particularly the clearance or inappropriate management of woodlands. Maps were produced in our previous contributions showing pre- and post-1950 or 1960 distributions of representative species (Hawksworth *et al.* 1973, 1974) to illustrate the dramatic changes in species subjected to air pollution or the modification of woodland habitats. Species such as *Lobaria pulmonaria*, *Teloschistes flavicans* and *Usnea articulata* had been lost throughout their former ranges in central and eastern England. Others had become eliminated through much of their ranges in affected areas, for instance *Anaptychia ciliaris*, *Evernia prunastri*, *Parmelia caperata*, *P. perlata* and *Ramalina* and *Usnea* species.

In addition, some species able to tolerate moderate to high levels of sulphur dioxide or favouring substrata with a lower pH had extended their ranges into lowland areas (e.g. *Parmeliopsis ambigua*, *Lecanora muralis*). Most dramatic was the case of *Lecanora conizaeoides*, a species first discovered in Britain in Leicestershire around 1860, which had become one of the commonest lichens on trees through much of lowland England by the early decades of the twentieth century; it nevertheless remained absent, rare, or only on wood, in less polluted areas of western and northern Britain.

Dramatic changes in particular sites were also documented. Declines in Epping Forest had been particularly dramatic, declining from 86 species on trees in 1881–82 to 28 in 1968–70. Gopsall Park in Leicestershire went from 106 species on trees and wood in 1839–71 to 12 in 1968–69. More rapid declines occurred in sites near urban areas; the number of species on bark and wood at Bookham Common in Surrey fell from 48 in 1958 to 15 in 1969. In contrast, it was noted that many sites in areas not affected to a significant extent by air pollution and forest disturbance were much as documented in the mid-nineteenth century.

5 Changes 1973–98

In 1973 the number of obligately lichenized fungi known from Great Britain and Ireland was 1368, and rose to 1471 in 1980 (Hawksworth *et al.* 1980), 1561 in 1993 (Purvis *et al.* 1992, 1993), 1571 in 1994 (Purvis *et al.* 1994a) and 1663 by December 1999 (see Table 7.1 for post-1994 additions). This increase of 295 species is less than the actual number of new discoveries for two reasons. Preparations for the 1980 checklist, ascomycete checklist (Cannon *et al.* 1985) and lichen flora (Purvis *et al.* 1992) resulted in many names being reduced to synonymy, continuing the process begun by James (1965) in the preparation of the first modern checklist. Several additional species are now regarded as not correctly reported from the British Isles, while others thought to be extinct have had to be re-instated owing to recent finds (see below).

Many 'additions' have resulted from splits arising from critical taxonomic revisions, for example: '*Dermatocarpon hepaticum*' now includes *Catapyrenium boccanum*, *C. pilosellum* and *C. squamulosum*; *Lecania* '*erysibe*' – *L. erysibe* s. str., *L . hutchinsiae*, *L. inundata*, *L. turicensis* and *L. rabenhorstii*; '*Lecidea granulosa*' – *Trapeliopsis flexuosa*, *T. granulosa* and *T. pseudogranulosa*; '*Lecidea uliginosa*' – *Placynthiella dasaea*, *P. hyporhoda*, *P. icmalea* and *P. uliginosa* s. str.; *Peltigera* '*polydactyla*' – *P. lactucifolia*, *P. neckeri* and *P. polydactyla* s. str. This process is likely to continue well into the twenty-first century; indeed, at least 20 species are known to us that have yet to be formally described.

In 1973, 40 species were thought to be extinct in the British Isles (Hawksworth *et al.* 1974: Table 7.1), but of these 16 have been rediscovered (Table 7.2) and 11 are considered to have been incorrectly reported (Table 7.3). If a species is considered to be 'extinct' if not seen for 50 years (World Conservation Union 1994), then the current total for the British Isles stands at 46 (Table 7.4), although a few of these are inconspicuous species that may have been overlooked. Examples of rediscovered species are *Hypogymnia vittata* (found in east Sutherland in 1998), *Pseudephebe minuscula* (found in the northern Cairngorms in 1995), and *Ramalina capitata* (found in a Lincolnshire churchyard in 1998). Only five species are thought to have become extinct in the last 25 years; attempts to refind *Bryoria smithii*, *Gyalidea roseola*, *Hypogymnia intestiniformis*, *Pseudocyphellaria aurata* and *Umbilicaria spodochroa* in known locations have been unsuccessful.

No species are known to have become established in the British Isles as a result of introduction by humans, despite living lichens being imported on cork bark for horticultural use, and *Cladonia stellaris* for flower arrangements, architects and other models and wreaths. However, a few species may be relatively recent arrivals that are now established in small or very localized populations, e.g. *Parmelia submontana* on trees and gravestones in south-west Scotland, *Pleopsidium chlorophanum* (syn. *Acarospora chlorophana*) on a gravestone in an east Yorkshire churchyard, and *Ramalina capitata* on gravestones in a Lincolnshire churchyard.

Table 7.1 Lichen species added to the checklist of the British Isles since Purvis *et al.* (1994a). Only obligately lichenized species are included, and species marked with an asterisk (*) were added in 1999.

Arthopyrenia atractospora	Ionaspis rhodopsis	Ochrolechia arborea
Arthrorhaphis vacillans	Lecanora compallens *	Phylliscum demangeonii
Aspicilia moenium	L. hypoptella	Porina isidiata *
Bacidia caesiovirens	L. ochroidea	Porpidia ochrolemma
B. neosquamulosa*	L. xanthostoma	P. zeoroides
Baeomyces carneus	Lecidea ecrustacea	Pseudephebe minuscula
Biatora subduplex	L. haerjedalica	Pyrrhospora rubiginans
Buellia hyperbolica*	L. liljenstroemii	Ramalina capitata
B. papillata	L. luteoatra	Rhizocarpon anaperum
Caloplaca caesiorufella	L. mucosa	R. cinereonigrum
C. cerinelloides	L. porphyrospoda *	R. submodestum
C. maritima	L. promiscens	R. subpostumum
C. polycarpa *	L. promiscua	Rinodina colobinoides
Catillaria gilbertii	L. subspeirea	R. degeliana
Chrysothrix flavovirens	L. swartzioidea	R. fimbriata
Cladonia alpina	L. syncarpa	R. flavosoralifera
Coppinsia minutissima	Lecidella patavina	R. madeirensis
Cyphelium trachylioides *	L. subviridis *	Strigula tagananae
Dermatocarpon deminuens	Lepraria elobata	Thelocarpon opertum
Epigloea bactrospora	L. nylanderiana	T. sphaerosporum *
E. filifera	Leptogium subtorulosum	Thelopsis isiaca
Fellhanera viridisorediata *	Lithothelium phaeosporum	Verrucaria pachyderma
Halecania bryophila	Micarea deminuta	Xanthoria fulva
H. micacea	M. elachista	X. ucrainica
Hymenelia obtecta *	M. lapillicola	
Hypogymnia vittata	M. parva	

Table 7.2 Lichen species thought to be extinct in the British Isles in 1973, but since found or re-discovered. Numbers in parenthesis refer to the number of post-1960 10-km grid square records by November 1998.

Arctomia delicatula (4)	Lecanora achariana (4)
Arthonia arthonioides (51)	L. farinaria (36)
A. atlantica (14)	L. pruinosa (30)
A. elegans (189)	Pertusaria lactescens[b] (7)
Baeomyces carnosa (3)	P. bryontha (1)
Cladonia zopfii [syn. C. destricta] (27)	Ptychographa xylographoides (21)
Hypogymnia vittata[a] (1)	Ramalina capitatae (1)
Lecanactis amylacea (31)	

[a] Previous records probably erroneous.
[b] *Melanaria lactescens* and *M. urceolaria*.

Table 7.3 Species listed as extinct in 1973, but now considered to have been incorrectly reported from the British Isles.

Bryoria nitidula [= *Alectoria n.*]	*Phaeophyscia ciliata* [= *Physcia c.*]
Caloplaca furfuracea	*Physconia muscigena*
Cetraria cucullata	*Ramalina breviuscula*
Cladonia turgida	*R. dilacerata* [= *R. minuscula*]
Lecidea cinnabarina[a]	*Umbilicaria arctica*
Parmelia glabra	*U. rigida*

[a] All records apart from one, the provenance of which is uncertain, refer to either *Pertusaria pupillaris* or *Schismatomma quercicola*.

Table 7.4 Lichen species not reliably reported from the British Isles since 1900 or 1950.

Arthonia galactites	*L. olivacella*
A. myriocarpella[a]	*Lecanora congesta*
Arthothelium spectabile	*L. fuscescens*
Aspicilia tuberculosa	*L. populicola*
Bacidia auerswaldii[50]	*Lecidea advertens*
B. polychroa [= *B. fuscorubella*]	*L. confluentula* [= *L. matildae*][50]
B. subturgidula	*L. mucosa*
Biatora cuprea	*L. paraclitica*
Bryoria implexa [= *Alectoria trichodes auct. brit.*]	*Lecidella pulveracea*
Calicium quercinum	*Leptogium hildenbrandii*
C. trabinellum	*Nephroma helveticum*[50]
Caloplaca haematites	*N. resupinatum*
C. irrubescens	*Pertusaria hutchinsiae* [= *P. panyrga auct. brit.*]
C. pollinii	*Psorotichia pyrenopsoides*
Catillaria picila	*Pyrenula nitidella*
Cetraria juniperina	*Stereocaulon tomentosum*
Cetrelia olivetorum s. str.[b]	*Strigula stigmatella* s. str.[c]
Cladonia stellaris	*Thelotrema isidioides*
C. stygia[50]	*Toninia cumulata*
Collema conglomeratum	*T. physaroides*
Cryptothele rhodosticta	*T. opuntioides*
Lecania coeruleorubella	*T. tumidula*
L. fuscella	*Tornabea scutellifera* [= *Tornabenia atlantica*]

[50] Seen between 1900 and 1950, but not since.

[a] Taxonomic and biological status uncertain: it could be a lichenicolous fungus, or a saxicolous morph of *Arthonia mediella*.

[b] In the British Isles *C. olivetorum* s. lat. comprises four 'chemical species' – it is the olivetoric acid race (*C. olivetorum* s. str.) which has not been found since 1900.

[c] Reported from one site in the New Forest in the 1970s and could perhaps still be confirmed from there.

Attempts to refind the following in known locations have been unsuccessful, and they are feared extinct: *Bryoria smithii*, *Gyalidea roseola*, *Hypogymnia intestiniformis*, *Pseudocyphellaria aurata* and *Umbilicaria spodochroa*.

The following considered extinct by Purvis *et al.* (1994) have since been rediscovered: *Caloplaca crenulatella*, *Catillaria minuta*, *Endocarpon pusillum* s. lat., *Melaspilea amota*, *Pertusaria lactescens* and *Toninia rosulata*.

6 Factors affecting change

6.1 Agriculture

There is now evidence of local damage to lichens by herbicidal sprays. At Rotherfield Park, Hampshire, no herbicides are in use but few bark lichens occurred closer than 50 m from the boundary with adjoining arable land. However, damage to lichens in such cases may be caused by added wetting agents rather than the herbicidal chemicals themselves. Ammonia hypertrophication, arising from intensive animal husbandry and well documented in The Netherlands (van Herk 1999), has similar effects in Great Britain and Ireland in that it encourages the development of a species-poor but often luxuriant *Xanthorion* (Benfield 1994, 1998; Wolseley and Pryor 1999). It may also be inhibiting the re-establishment of many lichen communities that would be expected from the lowering of sulphur dioxide levels. There is a steady decline in the distribution and population sizes of several species that is giving cause for concern. For example, *Anaptychia ciliaris* has declined markedly in much of southern and eastern Britain, even on churchyard memorials. In north Hampshire it now survives in only four sites not obviously subjected to agricultural chemicals, and in Angus and Kincardineshire it is now known on only two trees. This phenomenon merits critical experimental research.

Apart from the release of excessive amounts of ammonia, high densities of livestock and the use of slurry can cause long-term physical damage to lichens. Cattle finding shelter under a large tree can rub the trunk to provide a polished, almost lichen-free surface, and one application of slurry can cake the surface of a tree trunk for up to eight years, thus killing all lichens below (Coppins 1998). The move from the use of granular fertilizers to slurry can be considered 'environmentally friendly', but for lichens the converse is apparent. Another example is a near ecological disaster, since averted, caused by a well-intentioned, organic farmer spraying slurry on a small area of pasture in which occurred several old ash pollards that hosted a rich lichen community with three Red-Listed species (Coppins and Coppins 1998).

6.2 Air pollution

Acid rain continues to be a problem, especially in the high rainfall areas of northern England and central Wales. Bark pH can be reduced to 3.5–4, and while the lichen communities appear healthy, the composition changes from *Lobarion* or *Parmelietum revolutae* to one of calcifuge species such as *Hypogymnia physodes*, *Ochrolechia androgyna* and *Pseudevernia furfuracea*, or to *Parmelietum laevigatae* communities in high rainfall areas. In the Cowal area west of Glasgow, to the east of Loch Long, the *Lobarion* and *Parmelietum revolutae* have now largely disappeared and been replaced with *Parmelietum laevigatae* down to loch level, except for a few sheltered *Fraxinus* trees. Historical records indicate that *Lobarion* was formerly widespread at loch level here, and rich *Lobarion* still occurs further to the west by Loch Fyne (e.g. around Inverary Park and the Loch Fyne coast) with the *P. laevigatae* confined to higher altitudes (100–150 m) as it is further to the north-west around Sunart. Studies on the effects of acid rain on lichens in western Britain, and elsewhere, are reviewed by Farmer *et al.* (1992). Acid rain has had less of an effect so far in the south of England; this may be due to a combination of the location of sources and the lower rainfall.

A general decrease in lichens with cyanobacterial partners is taking place and this may well be due to low amounts of acid rain, something indicated in the field by the replacement of the moss *Homalothecium sericeum* by *Hypnum cupressiforme*.

A spectacular collapse in sulphur dioxide levels in most of southern and central England has had a dramatic effect on the lichens present, with *Parmelia caperata*, *P. perlata* and many other leafy and shrubby species extending into suburban areas and even central London. Eight species last seen in London over 200 years ago are now back (Hawksworth and McManus 1989). Two lichens were recorded in Buckingham Palace Gardens in 1956, but 39 are now present there (Hawksworth 1999). This pattern is being repeated in many formerly polluted areas of lowland Britain; for example, Seaward (1993) has shown how the lichen diversity gradient in the West Yorkshire (Leeds and Bradford) conurbation has increased from 1972 to 1990. The species returning occur when mean levels of sulphur dioxide fall below the levels of tolerance suggested in the tables of Hawksworth and Rose (1970); no exceptions to that are known. The pattern of recolonization actually occurring provides additional support for the correlations these authors established between field occurrences and measured levels of this pollutant. As the Hawksworth and Rose correlations were logarithmic, the number of species able to grow rises rapidly with small falls of the pollutant (Seaward 1993). Interestingly, the pollution-tolerant *Lecanora conizaeoides* is declining in areas where it was dominant under higher levels of ambient sulphur dioxide and it can be quite difficult to find even in parts of the Midlands and Greater London. This rapid change has led to the phenomenon of 'zone skipping' where moderately tolerant lichens are not re-establishing but those generally considered to be more sensitive are (Gilbert 1992). Many crustose species and also old forest indicator species are not recolonizing, and ancient tree trunks tend to remain barren, presumably because their bark remains heavily impregnated with pollutants. Nevertheless, suburban areas are increasingly developing richer assemblages of lichens on bark than are found in many agricultural areas. The situation on churchyard memorials is similar, with suburban churchyards being richer than ones in intensively agricultural areas.

The extent of the reinvasion of *Parmelia caperata* is shown in Fig. 7. 2. Complementary maps for *P. perlata*, *Ramalina farinacea* and *Usnea* species in Great Britain and Ireland were published by Seaward (1998), who also categorized selected species by their ability to extend their ranges in areas with ameliorating environments in the period 1992–97. He found, for example, that *Evernia prunastri* had become extended by 7% or more, *P. caperata* by 4.5–6%, and *Lecidella elaeochroma* by only 2–4.5%; others had not shown any such expansion (e.g. *Graphis elegans* and *Ramalina fastigiata*).

Improvements in the lichens found in urban and other areas where sulphur dioxide levels have fallen can be expected to continue. Nevertheless, high-stack power stations remain of major concern; for example, the effects of the Pembroke Power Station are evident about 50 miles north-east in Radnor and Brecknock, while local effects near the station are minimal.

Furthermore, in the case of recolonizing species, the issue of whether the full range of genetic variation is represented by the invading populations has to be addressed. Crespo *et al.* (1999) found that only one of three rRNA genotypes of *Parmelia sulcata* occurred in areas of lowland Britain where the species had re-established. Conversely, deterioration in clean areas is expected to continue as high-level emissions and acid rain have their effects. The overall situation is therefore one of equilibration.

There is still no evidence of car exhausts or photochemical smog affecting lichen communities; species sensitive to sulphur dioxide thrive along roads which carry heavy tourist traffic in the Lake District, and increases in nitrogen oxide and ozone pollution in London have not inhibited lichen recolonization there. Lichens are well known as bioaccumulators of radionuclides and heavy metals, but, with the exception of copper runoff, these rarely seem to eliminate species (Richardson 1975; Nash 1996).

Limestone quarry dust has effects not dissimilar from those caused by ammonia emissions (see above), with heightened *Xanthorion* communities, and some normally rock-loving species can switch to trees with heavily impregnated bark (Gilbert 1976).

6.3 Woodland and tree management

Inappropriate woodland management is becoming more important than air pollution in the protection of epiphytic lichens. Of especial concern are the numerous species used in the Revised and New Indices of Ecological Continuity (Rose 1976, 1992), which require sites with a long history of woodland cover, as recolonization is exceptionally slow, or may never occur, once a species has been lost from a site. In the New Forest, species such as *Lobaria pulmonaria* have taken almost 300 years to migrate from the ancient Anses Wood to the adjacent and formerly clear-felled South Bentley Wood. The slowest species to re-establish are those of the *Lecanactidetum premneae*.

During the past 50 years there has been an encouraging, ever-increasing awareness of the importance of ancient woodlands. This arises both for reasons of national heritage and of biodiversity and nature conservation, with many ancient woodlands coming into the ownership or management of conservation organizations. Most of these woodlands have survived because they have been of use to humans as a source of timber or for sheltering or grazing livestock. Conservation action has in many instances been to exclude grazing by exclosure fencing and the cessation of all forestry practices. This policy of 'non-intervention' is something of a misnomer as it has intervened to break decades or even centuries of past practices of wood extraction, grazing, ivy cutting and the like. For some woodlands this has to date (at least) been beneficial to the woodland lichens, but in many others it is now having serious deleterious affects. Most 'old forest' lichens require sheltered yet well-lit situations, the idea that lichens like 'dank and shady places' being a total misconception. Depending on the characteristics and past history of the woodland, the results of 'non-intervention' can include excessive regeneration of trees and shrubs, and of climbers such as *Clematis*, *Hedera* and *Lonicera*, thus shading out the light-demanding lichens. *Hedera* is additionally important, as even if it occupies little space on the trunk, when it reaches the canopy it spreads so as to cause an 'umbrella affect', stemming much of the flow of water down the trunk as well as casting a dense shade.

The perniciousness of the invasive exotics *Rhododendron* and, to a lesser extent, *Prunus laurocerasus* are now well appreciated, with programmes of control or removal ongoing nationwide. In the New Forest, *Ilex* is a particular problem; once over a threshold level, the forest becomes impenetrable even to grazing animals and lichens are shaded out. In Horner Woods, Exmoor, areas of former wood pasture are becoming choked by an understorey of *Crataegus* and *Prunus spinosa*, coinciding with an alarming decline in lichens of the *Lobarion*, while in some other Exmoor woodlands

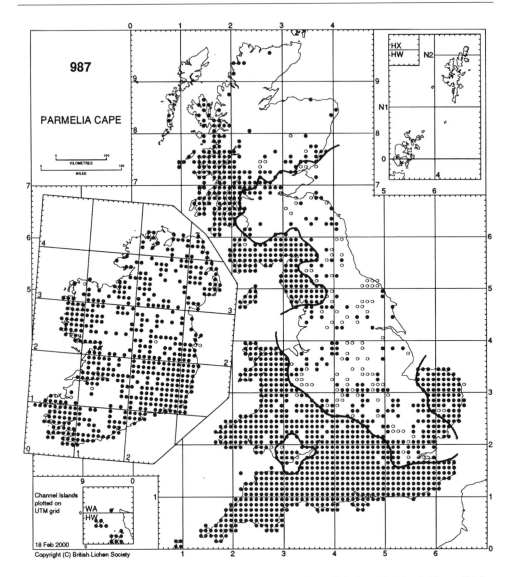

Figure 7.2 Expansion of the range of *Parmelia caperata* in Great Britain and Ireland. The solid line indicates the approximate limits of the species in 1973, although a few sites within the line were established before that date. o pre-1960 record (many from last century); • post-1960 record. Data prepared by M. R. D. Seaward from the British Lichen Society's Mapping Scheme.

excessive *Fagus* regeneration is causing similar problems. Such examples of scrub growth need active management, and at Eridge Park in West Sussex pigs are proving especially effective digging up *Pteridium* rhizomes as well as eating young *Betula*, *Rhododendron* and *Rubus*.

Idealistic moves to rid woodlands of non-native trees can also have a detrimental affect on lichen biodiversity. For example, sycamore (*Acer pseudoplatanus*) can host lichens requiring a high bark pH, and this may be highly significant where there has been a loss of *Ulmus* from Dutch elm disease or an absence or decline of *Fraxinus*. Conversely, mature *Larix* can provide niches for lichens requiring a very low pH.

At many woodland, parkland and wayside sites, rare and demanding lichens are often confined to just one or a few old trees alongside tracks or roads, and there have been many instances of local loss of notable (even scheduled) lichens due to the deliberate felling of such trees without prior consultation with regard to nature conservation. This problem has increased in recent years because of growing concerns over public safety and public liability and the fact that it is usually much less expensive to fell than to carry out tree surgery. Hopefully, the many veteran tree initiatives and local biodiversity programmes, now under way, will lead to a more considered approach in the future.

6.4 Heathland and grassland management

Ungrazed and uncut grasslands and heaths soon become lank and unsuitable for a rich lichen growth, and if left longer become invaded by scrub and new trees. The populations of *Buellia asterella*, *Fulgensia fulgens*, *Psora decipiens*, *Squamarina lentigera* and *Toninia sedifolia* in Breckland have collapsed dramatically. In order to maintain open ground and bring subsoil to the surface, topsoil stripping, ploughing and rotavating are being used on an experimental basis (Gilbert 1993).

Ground-dwelling lichens are also disappearing from lowland heaths. For example, Lavington Common in west Sussex has deteriorated from an optimum about 35 years ago, and by 1991 populations had seriously declined and *Cladonia sulphurina* had been lost. Remedial measures being taken by the National Trust include mowing the *Calluna* to form an open sward. Hot summer fires are catastrophic, but small patch or strip burning in March results in improvements in the lichen communities.

Lichens in chalk grasslands have declined catastrophically in many areas through reseeding, fertilizer applications and the loss of grazing, the last sometimes the result of over-protection of nature reserves (Gilbert 1993).

6.5 Marine and freshwater pollution

The specialized communities found on siliceous rocks in or along stream margins have become much better understood in recent years (Gilbert and Giavarini 1993), but water pollution has destroyed these assemblages in many streams where they used to occur. Lichens merit more attention in assessments of water quality, and this is now being appreciated via the biodiversity action plan for *Collema dichotomum* ('River jelly lichen') being led by the Environment Agency – Welsh Region.

Oil spills and detergents remain a matter of concern, but little new data on their effects have been presented since our last review.

6.6 Impact of other organisms

The loss of mature elms, mostly *Ulmus procera*, through Dutch elm disease caused by *Ophiostoma novo-ulmi*, has led to a collapse in the distribution of numerous species largely confined to this tree (Watson *et al.* 1988), most notably *Anaptychia ciliaris*, *Bacidia incompta* and *Caloplaca luteoalba*; even less restricted species such as *Parmelia acetabulum* have also suffered as a result. Indeed, the last species is now scarce in its former stronghold areas in eastern England.

A wide range of invertebrates are lichen feeders and can presumably affect the population dynamics of the lichens concerned, although this is a large subject of much-needed study. Grazers, especially tree slugs, have undoubtedly been partly responsible for the demise of some populations of larger lichens, such as *Lobaria pulmonaria*, especially where other factors such as acid rain or over-shading have reduced the vigour of the lichen thalli. With *L. pulmonaria*, slugs seem to have a predilection for grazing the raised ridges across the thallus, the site of production of the lichen's vegetative propagules (isidia and soralia). Thus, even though the slugs may not graze away entire thalli, they can effectively restrict their reproductive capacity and hence threaten the long-term survival of the population. It seems likely that factors affecting grazers such as tree slugs may have an indirect effect on lichens: it might seem fanciful to suggest that declining song-bird populations can adversely affect the success of lichen populations – but is it? Also, the number of slug-grazing hours per day is linked with both macroclimatic and microclimatic conditions, and the possibility of changes in lichen populations with regard to climate change need to take this into consideration. Likewise, the problems of over-shading in woodlands (see above) may also involve the 'slug factor'.

6.7 Over-collecting

Today, over-collecting is rarely a problem owing to the much increased awareness of naturalists to its dangers, but it remains a potential threat to very rare species which may be sought for private collections. A more real threat is inadvertent collecting of unfamiliar specimens for later identification; hence tutors of field courses are requested to avoid sites containing vulnerable species.

6.8 Public pressure

Trampling to a moderate degree may be beneficial to some lichens by simulating the effect of grazing (Gilbert 1993), but when intense will lead to the loss of specialized communities, such as those at high altitudes in the Lake District (Gilbert and Giavarini 1993) and the Cairngorms. Rock-climbing also remains a problem and some very rare species remain vulnerable to this, for example *Umbilicaria crustulosa* on the Langdale Pikes in Cumbria.

Public pressure has led to many 'environmental' clean-up programmes, which are not always beneficial to lichens. Most important among these is the use of old quarries, which can be spectacular lichen habits, as land-fill sites.

7 Effects on other wildlife

In addition to the often unappreciated extent of the ecological roles of lichens in nutrient cycling, as carbon sinks and in rock weathering, changing lichen communities can affect other forms of wildlife. Numerous invertebrates are dependent on lichens for food or camouflage (Seaward 1988b). This is especially true of moths (Richardson 1975), and correlations between the improvements of lichens following falls in sulphur dioxide and changes in the frequency of melanic morphs of *Biston betularia* have been investigated (Cook *et al.* 1990).

The extent to which fungi obligately occur on lichens has only come to be appreciated in the last 20 years. An annotated key to 218 lichenicolous fungi occurring in the British Isles was provided by Hawksworth (1983), but by the end of 1998 the number known had increased to 370, a staggering increase of 69%. This total is continuing to rise as more lichenologists and other mycologists become aware of these specialized fungi. The richest sites for such fungi are those with long histories of ecological continuity, and it may be that more note should be taken of them in site assessments; a woodland with *Plectocarpon lichenum* present on *Lobaria pulmonaria* could have a longer history of continuity than one where this fungus is not present. While a few of these fungi are pathogenic to their lichen hosts (e.g. *Athelia arachnoidea*), most appear to be commensals, form galls, or limited infections. Galls on lichens can also be produced by nematodes; these are scarcely studied but one gall on a *Cladonia glauca* specimen from Ireland yielded four species, of which two were new for science (Siddiqi and Hawksworth 1982).

Lichens on trees provide habitats sheltering a wide range of insects which are eaten by birds, and sites with luxuriant lichen communities therefore tend to be favoured by insectivorous birds. The long-tailed tit harvests *Evernia prunastri* for its nests and now that species is extending its range into formerly polluted areas it will be of interest to see whether this bird eventually becomes more frequent.

8 Species protection

The need for statutory protection for some lichens and lichen-dominated communities has steadily increased, and since the 1970s several Sites of Special Scientific Interest (SSSIs) have been designated largely or entirely on their lichen interest. Furthermore, detailed guidelines for the selection of sites of conservation importance for lichens, especially in woodlands, have been developed and agreed for use by the statutory conservation agencies in Great Britain (Hodgetts 1992b). This 'official' recognition of lichen interest is being carried forward into the designation of Special Areas of Conservation (SACs), as part of the implementation of the European Community's 1992 'Habitats Directive'. As yet, this directive does not list any individual lichens that must receive special protection, although the maintenance of a 'favourable conservation status' for reindeer lichens (*Cladonia* subgenus *Cladina*) is stipulated under Annex Vb. In the UK, the statutory protection of individual lichen species came into being in 1992, when 26 lichens were listed on Schedule 8 of the Wildlife and Countryside Act 1981, and since then a further three species have been added (Table 7.5).

The *Red Data Book* for the lichens of Britain (Church *et al.* 1997) lists 29 species as extinct, 27 as critically endangered, 30 as endangered and 91 as vulnerable, with a further 96 being regarded as data deficient and 91 as of lower risk (near threatened). A companion volume for Ireland is currently under review. Some regional or county 'Red Lists' have been published or are in preparation, e.g. Gainey (1997), Gilbert and Smith (1998) and Woods and Orange (1999). Further moves towards species protection have come from implementation of *Biodiversity: the UK Action Plan* (Department of the Environment 1994), and Biodiversity Action Plans (BAPs) have been prepared for 41 lichens (Department of the Environment 1995; UK Biodiversity Group 1999) (Table

Table 7.5 Priority lichen species for which there is legal protection, existing or proposed action plans, detailed dossiers or reports completed or in preparation, and recovery programmes underway.

	RDB	S8	BAP	Doss.	RP	Lit.
Alectoria ochroleuca	VU	+	+	SNH		
Arthothelium dictyosporum	Lr		+			
A. macounii	VU		+			
Aspicilia melanaspis	EN			SNH		
Bacidia incompta	VU		+			
Bellemerea alpina	CR		+			
Belonia calcicola	DD		+			
Biatoridium monasteriense	EN		+			
Bryoria furcellata	VU	+		SNH		
B. smithii	CR		+	EN		
Buellia asterella	CR	+	+		EN	
Calicium corynellum	CR		+			
Caloplaca aractina	CR		+			
C. flavorubescens	EN			SNH		
C. luteoalba	VU	+	+	EN, SNH		
C. nivalis	CR	+	+	SNH		
Catapyrenium psoromoides	CR	+	+	SNH	SNH	
Catillaria aphana	Lr		+			
C. laureri	VU	+		EN		2
C. subviridis	VU		+			
Catolechia wahlenbergii	VU	+		SNH		
Chaenotheca phaeocephala	CR		+			
Cladonia botrytes	CR		+	SNH		
C. convoluta	VU	+				
C. maxima	VU			SNH		
C. mediterranea	CR		+			
C. peziziformis	CR		+			
C. trassii [stricta]	VU	+		SNH		
Collema dichotomum	VU	+	+	SNH		
Enterographa elaborata	CR	+	+	?EN		
E. sorediata	Lr		+			
Graphina pauciloculata	VU		+			
Graphis alboscripta	Lr			SNH		
Gyalecta ulmi	EN	+	+	EN,SNH		
Gyalideopsis scotica	Lr		+	SNH		
Halecania rhypodiza	VU		+	SNH		
Heterodermia leucomelos	EN	+	+			
H. japonica [syn. H. propagulifera]	EN	+				
Hypogymnia intestiniformis	CR		+	SNH		
Lecanactis hemisphaerica	Lr	+	+			
Lecanora achariana	CR	+				
Lecidea inops	EN	+				

continued on next page

Table 7.5 Priority lichen species for which there is legal protection (cont.).

	RDB	S8	BAP	Doss.	RP	Lit.
Leptogium saturninum	VU			SNH		
Nephroma arcticum	EN	+		SNH	SNH	
Opegrapha fumosa	—		+			
O. paraxanthodes	Lr		+			
Pannaria ignobilis	VU	+		SNH		
Parmelia minarum	VU	+		EN		2
P. subargentifera	DD			SNH		
Parmentaria chilensis	VU	+		SNH		
Peltigera lepidophora	CR	+	+	SNH		
P. malacea	EN			SNH		
Pertusaria bryontha	CR	+	+	SNH		
Physcia tribacioides	EN	+				
Pseudocyphellaria aurata	CR		+			
P. lacerata	VU	+		SNH		
P. norvegica	—		+	SNH		
Psora rubiformis	VU	+		SNH		
Pyrenula dermatodes	CR			SNH		
Ramalina polymorpha	Lr			SNH		
Schismatomma graphidioides	VU		+	EN, SNH		
Solenopsora liparina	VU	+				
Squamarina lentigera	EN	+	+		EN	
Teloschistes chrysophthalmus	CR		+			
T. flavicans	VU			CCW, EN		1
Thelenella modesta	CR		+	SNH		
Toninia cumulata	VU			SNH		
Zamenhofia rosei	Lr		+			

[1] Gilbert and Purvis (1996).

[2] Sanderson (1994).

Notes:
RDB   Red Data Book category (Church et al. 1997).
S8   Listed on Schedule 8 of the Wildlife and Countryside Act 1981.
BAP   Biodiversity Action Plan published.
Doss.   Detailed species dossier or report prepared by or for CCW, EN or SNH. Use of italic indicates work in progress in November 1998.
RP   Species recovery plan in progress.
Lit.   Published reports on priority species.

7.5). For some species detailed species dossiers have, or are being, prepared for the governmental agencies and some species recovery plans are in progress (Table 7.5).

The Red Data Book and species BAP mechanisms are going a long way to ensure the protection of our rarer species, but we must not forget there are many species, although not considered 'rare' within the British Isles as a whole, that are either very rare regionally or on a European or global scale. For example, there are 38 lichens included in the EC Red List of macrolichens (Sérusiaux 1989) which are considered too common for inclusion in the British Red List (Church *et al.* 1997), and only one of these (*Pseudocyphellaria norvegica*) is a BAP species. Twenty-three (61%) of these macrolichens have their main centres of viable populations in the oceanic ancient woodlands of western Britain and Ireland. To ensure the 'favourable conservation status' of such species (there are many more among the microlichens!) lichenologists must increase their input into habitat BAPs, and continue to promote the awareness of habitats of international importance for lichens but overlooked by the EU's 'Habitat Directive', for example the 'climax' coastal hazelwoods of western Scotland and Ireland, which are as equivalent in their uniqueness to the British Isles as the 'recognised' machair (Coppins and Coppins 1997).

9 Prognosis for the next 25 years

Losses of rarer lichens will continue if current trends are maintained. Many are known from few sites and because of their slow regeneration and growth rates are extremely vulnerable. However, we are encouraged by the new awareness of lichens by the statutory agencies and NGOs, and also among naturalists generally. There now needs to be a commensurate commitment of resources to vigorously tackle management problems and reintroduce practices favourable to lichens in woodlands, heathlands and grasslands. We would welcome more attention being paid to habitat rather than species protection. While some species still plentiful in western Scotland, but almost extinct in England and Wales (e.g. *Lobaria amplissima*), will not be accorded a high conservation priority in a UK species context, the characteristic habitats containing them could.

There also needs to be more attention paid to old churchyards in England, which are of an extent not seen elsewhere in Europe. 'Better', 'tidier' management of churchyards may not always be beneficial to lichens, especially where attempts are made to clean memorials of lichen growth.

We are most concerned at the decline in the professional workforce concerned with British and Irish lichens to the equivalent of only two full-time positions (across four institutions). With the near-collapse of the teaching of organismal biology in universities, we wonder where the next generation of experts to replace the currently ageing one will come from. Retired professionals and skilled amateurs will become crucial, but without adequate funding for equipment, books and travel their potential will be limited. It is surely ironic that this decline in professional support coincides with a markedly increased awareness of the conservation importance of lichens and the explosion of international, national, regional and local initiatives (and hence funding opportunities) kindled by the 1992 Rio Earth Summit. Indeed, we are aware of several recent funding opportunities from statutory agencies and NGOs being lost because of the lack of available institutional and freelance lichenologists.

ACKNOWLEDGEMENT

We are indebted to Professor M. R. D. Seaward for preparing the maps presented from the British Lichen Society's Mapping Scheme.

REFERENCES

Alvin, K.L. (1977) *The Observer's Book of Lichens*. London: Frederick Warne.

Benfield, B. (1994) Impact of agriculture on epiphytic lichens at Plymtree, east Devon. *Lichenologist*, 26, 91–94, 317.

Benfield, B. (1998) Further observations at Plymtree, Devon, 1992–1997. *British Lichen Society Bulletin*, 83, 16–18.

Bowen, H.J.M. (1976) The lichen flora of Dorset. *Lichenologist*, 8, 1–33.

Bowen, H.J.M. (1980) The lichen flora of Berkshire, Buckinghamshire and Oxfordshire. *Lichenologist*, 12, 199–237.

British Lichen Society (1997) *Identification of Parmelia Ach*. [CD No. 1.] London: British Lichen Society.

Broad, K. (1989) *Lichens in Southern Woodlands*. [Forestry Commission Handbook No. 4.] London: HMSO.

Cannon, P.F., Hawksworth, D.L. and Sherwood-Pike, M.A. (1985) *The British Ascomycotina. An annotated checklist*. Slough: Commonwealth Agricultural Bureaux.

Chester, T. (1998) Churchyard project annual report 1998. *British Lichen Society Bulletin*, 83, 38–42.

Church, J.M., Coppins, B.J., Gilbert, O.L., James, P.W. and Stewart, N.F. (1997) ['1996'] *Red Data Books of Britain and Ireland: Lichens. Volume 1: Britain*. Peterborough: Joint Nature Conservation Committee.

Cook, L.M., Rigby, K.D. and Seaward, M.R.D. (1990) Melanic moths and changes in epiphytic vegetation in north-west England and north Wales. *Biological Journal of the Linnean Society*, 39, 343–354.

Coppins, A.M. (1998) *Lichen monitoring project – Crom and Florence Court 1993–1998*. Unpublished report for The National Trust.

Coppins, A.M. and Coppins, B.J. (1998) *Species Recovery Programme: Survey to assess the status of* Schismatomma graphidioides. Unpublished report for English Nature.

Coppins, B.J. and Coppins, A.M. (1997) Coastal hazelwoods and their lichens. *Native Woodland Discussion Group Newsletter*, 22(2), 27–29.

Crespo, A., Bridge, P.D., Hawksworth, D.L., Grube, M. and Cubero, O.F. (1999) Comparison of rRNA genotype frequencies of *Parmelia sulcata* from long established and recolonizing sites following sulphur dioxide amelioration. *Plant Systematics and Evolution*, 217, 177–183.

Department of the Environment (1994) *Biodiversity: the UK Action Plan*. London: HMSO.

Department of the Environment (1995) *Biodiversity: the UK steering group report. Volume 1: meeting the Rio challenge*. London: HMSO.

Dobson, F.S. (1979a) *Common British Lichens*. Norwich: Jarrold and Sons.

Dobson, F.S. (1979b) *Lichens. An illustrated guide to the British and Irish species*. Richmond: Richmond Publishing.

Egea, J.M. and Torrente, P. (1994) El género de hongos liquenizados *Lecanactis* (*Ascomycotina*). *Bibliotheca Lichenologica*, 54, 1–205.

Ekman, S. (1997) The genus *Cliostomum* revisited. *Symbolae Botanicae Upsaliensis*, 32(1), 17–28.

Farmer, A.M., Bates, J.W. and Bell, J.N.B. (1992) Ecophysiological effects of acid rain on bryophytes and lichens, in Bates, J.W. and Farmer, A.M. (eds) *Bryophytes and lichens in a changing environment*. Oxford: Clarendon Press, pp. 284–313.

Fletcher, A. (1975a) Key for the identification of British marine and maritime lichens I. Siliceous rocky shore species. *Lichenologist*, 7, 1–52.

Fletcher, A. (1975b) Key for the identification of British marine and maritime lichens II. Calcareous and terricolous species. *Lichenologist*, 7, 73–115.

Fletcher, A. (1982) *Survey and Assessment of Epiphytic Lichen Habitats*. Report for the Nature Conservancy Council.

Fletcher, A. (ed.) (1984) *Survey and Assessment of Lowland Heathland Lichen Habitats*. Report for the Nature Conservancy Council.

Fletcher, A. (1993) *Revised Assessment of Epiphytic Lichen Habitats*. [JNCC Report No. 170.] Peterborough: Joint Nature Conservation Committee.

Fryday, A.M. (1997) Montane lichens in Scotland. *Botanical Journal of Scotland*, 49, 367–374.

Fryday, A.M. and Coppins, B.J. (1997) Keys to sterile, crustose saxicolous and terricolous lichens occurring in the British Isles. *Lichenologist*, 29, 301–332.

Gainey, P.A. (1997) Lichens, in Spalding, A. (ed.) *Red Data Book for Cornwall and the Isles of Scilly*. Camborne, Cornwall: Croceago Press, pp. 18–36.

Giavarini, V.J. (1990) Lichens of the Dartmoor rocks. *Lichenologist*, 22, 367–396.

Gilbert, O.L. (1976) An alkaline dust effect on epiphytic lichens. *Lichenologist*, 8, 173–178.

Gilbert, O.L. (1980) A lichen flora of Northumberland. *Lichenologist*, 12, 325–395.

Gilbert, O.L. (1990) The lichen flora of urban wasteland. *Lichenologist*, 22, 87–101.

Gilbert, O.L. (1992) Lichen reinvasion with declining air pollution, in Bates, J.W. and Farmer, A.M. (eds) *Bryophytes and lichens in a changing environment*. Oxford: Clarendon Press, pp. 159–177.

Gilbert, O.L. (1993) The lichens of chalk grassland. *Lichenologist*, 25, 379–414.

Gilbert, O.L., Coppins, B.J. and Fox, B.W. (1988) The lichen flora of Ben Lawers. *Lichenologist*, 20, 201–243.

Gilbert, O.L. and Fox, B.W. (1985) Lichens of high ground in the Cairngorm mountains, Scotland. *Lichenologist*, 17, 51–66.

Gilbert, O.L. and Fox, B.W. (1986) A comparative study of the lichens occurring on the geologically distinctive mountains, Ben Loyal, Ben Hope and Foinaven. *Lichenologist*, 18, 79–93.

Gilbert, O.L., Fox, B.W. and Purvis, O.W. (1982) The lichen flora of a high-level limestone-epidiorite outcrop in the Ben Alder range, Scotland. *Lichenologist*, 14, 165–174.

Gilbert, O.L., Fryday, A.M., Giavarini, V.J. and Coppins, B.J. (1992) The lichen vegetation of high ground in the Ben Nevis range, Scotland. *Lichenologist*, 24, 43–61.

Gilbert, O.L. and Giavarini, V.J. (1993) The lichens of high ground in the English Lake District. *Lichenologist*. 25, 147–164.

Gilbert, O.L. and Giavarini, V.J. (1997) The lichen vegetation of acid watercourses in England. *Lichenologist*, 29, 347–367.

Gilbert, O.L. and Purvis, O.W. (1996) *Teloschistes flavicans* in Great Britain: distribution and ecology. *Lichenologist*, 28, 493–506.

Gilbert, O.L. and Smith, E.C. (1998) Lichens, in Kerslake, L. (ed.), Red Data Book for Northumberland. *Transactions of the Natural History Society of Northumbria*, 58(2), 273–288.

Graham, G.G. (1988) *The Flora and Vegetation of County Durham*. Durham: Durham Flora Committee and Durham County Conservation Trust.

Hafellner, J. (1993) Die Gattung *Pyrrhospora* in Europa. *Herzogia*, 9, 725–747.

Hafellner, J. (1994) On *Biatoridium*, a resurrected genus of lichenized fungi (*Ascomycotina, Lecanorales*). *Acta Botanica Fennica*, 150, 39–46.

Hawksworth, D.L. (1983) A key to the lichen-forming, parasitic, parasymbiotic and saprophytic fungi occurring on lichens in the British Isles. *Lichenologist*, 15, 1–44.

Hawksworth, D.L. (1999) Lichens (lichen-forming fungi) in Buckingham Palace gardens. *London Naturalist*, 78 (Supplement), 15–21.

Hawksworth, D.L., Coppins, B.J. and James, P.W. (1980) Checklist of British lichen-forming, lichenicolous and allied fungi. *Lichenologist*, 12, 1–115.

Hawksworth, D.L., Coppins, B.J. and Rose, F. (1974) Changes in the British lichen flora, in Hawksworth, D.L. (ed.) *The Changing Flora and Fauna of Britain*. London: Academic Press, pp. 47–78.

Hawksworth, D.L and McManus, P.M. (1989) Lichen recolonization in London under conditions of rapidly falling sulphur dioxide, and the concept of zone skipping. *Botanical Journal of the Linnean Society*, 100, 99–109.

Hawksworth, D.L. and McManus, P.M. (1992) Lichens: changes in the lichen flora on trees in Epping Forest through periods of increasing and ameliorating sulphur dioxide pollution. *Essex Naturalist*, 11, 92–101.

Hawksworth, D.L. and Rose, F. (1970) Qualitative scale for estimating sulphur dioxide air pollution in England and Wales using epiphytic lichens. *Nature*, 227, 145–148.

Hawksworth, D.L., Rose, F. and Coppins, B.J.(1973) Changes in the lichen flora of England and Wales attributable to pollution of the air by sulphur dioxide, in Ferry, B.W., Baddeley, M.S. and Hawksworth, D.L. (eds) *Air Pollution and Lichens*. London: Athlone Press of the University of London, pp. 330–367.

Hawksworth, D.L. and Seaward, M.R.D. (1977) *Lichenology in the British Isles 1568–1975*. Richmond: Richmond Publishing.

Hawksworth, D.L. and Seaward, M.R.D. (1990) Twenty-five years of lichen mapping in Great Britain and Ireland. *Stuttgarter Beiträge zur Naturkund*, 456, 5–10.

Hertel, H. (1995) Schlüssel für die Arten der Flechtenfamilie *Lecideaceae* in Europa. *Bibliotheca Lichenologica*, 58, 137–180.

Hodgetts, N.G. (1992a) *Cladonia: a field guide*. Peterborough: Joint Nature Conservation Committee.

Hodgetts, N.G. (1992b) *Guidelines for Selection of Bioilogical SSSIs: non-vascular plants*. Peterborough: Joint Nature Conservation Committee.

Jahns, H.M. (1980) *Collins Guide to the Ferns, Mosses and Lichens of Britain and Northern and Central Europe*. London: William Collins.

James, P.W. (1965) A new check-list of British lichens. *Lichenologist*, 3, 95–153.

James, P.W. (1978) Lichens, in Jermy, A.C. and Crabbe, J.A. (eds) *The Island of Mull, a survey of its flora and environment*. London: British Museum (Natural History), pp. 14·1–14·62.

James, P.W., Allen, A. and Hilton, B. (1996) ['1995'] The lichen flora of Lundy: I the species. *Annual Report of the Lundy Field Society*, 46, 66–86.

James, P.W., Allen, A. and Hilton, B. (1997) ['1996'] The lichen flora of Lundy: II the communities. *Annual Report of the Lundy Field Society*, 47, 93–126.

James, P.W., Hawksworth, D.L. and Rose, F. (1977) Lichen communities in the British Isles: a preliminary conspectus, in Seaward, M.R.D. (ed.) *Lichen ecology*. London: Academic Press, pp. 295–415.

Laundon, J.R. (1986) *Lichens*. [Shire Natural History No. 10] Aylesbury: Shire Publications.

Lindsay, D.C. (1980) Lichens, in Clark, M.C. (ed.) *A Fungus Flora of Warwickshire*. London, British Mycological Society, pp. 232–243.

McCarthy, P.M. and Mitchell, M.E. (1988) *Lichens of the Burren Hills and the Aran Islands*. Galway: Officina Typographica.

Mitchell, M.E. (1993) *First Records of Irish Lichens 1696–1990. Collectors and localities from published sources*. Galway: Officina Typographica.

Mitchell, M.E. (1995) 150 years of Irish lichenology: a concise survey. *Glasra* (new series), 2, 139–155.

Mitchell, M.E. (1996) *Irish Lichenology 1858–1880: Selected letters of Isaac Carroll, Theobald Jones, Charles Larbalestier and William Nylander*. [Occasional Papers No. 10] Dublin: National Botanic Gardens, Glasnevin.

Mitchell, M.E. (1998) Index of collectors in Knowles' *The Lichens of Ireland* (1929) and Porter's *Supplement* (1948), with a conspectus of lichen recording in the Irish vice-counties to 1950. [Occasional Papers No. 11] Dublin: National Botanic Gardens, Glasnevin, 11, 1–53.

Nash, T. H. III (1996) Nutrients, elemental accumulation and mineral cycling, in Nash, T. H. III (ed.) *Lichen Biology*. Cambridge: Cambridge University Press, pp. 136–153.

Obermayer, W. (1994) Die Flechtengattung *Arthrorhaphis* (*Arthrorhaphidaceae, Ascomycotina*) in Europa und Grönland. *Nova Hedwigia*, 58, 275–333.

Orange, A. (1994a) *Lichens on Trees. A guide to some of the commonest species*. Cardiff: National Museum of Wales.

Orange, A. (1994b) Lichens of Glamorgan, in Wade, A.E., Kay, Q.O.N and Ellis, R.G. (eds) *Flora of Glamorgan*. London: HMSO, pp. 249–271.

Pentecost, A.R. (1987) The lichen flora of Gwynedd. *Lichenologist*, 19, 97–166.

Phillips, R. (1980) *Grasses, Ferns, Mosses and Lichens of Great Britain and Ireland*. London: Pan Books.

Pope, C.R. (1985) A lichen flora of the Isle of Wight, *Proceedings of the Isle of Wight Natural History and Archaeological Society*. 7, 577–599.

Printzen, C. (1995) Die Flechtengattung *Biatora* in Europa. *Bibliotheca Lichenologica*, 60, 1–275.

Purvis, O.W., Coppins, B.J., Hawksworth, D.L., James, P.W. and Moore, D.M. (1992) *The lichen flora of Great Britain and Ireland*. London: Natural History Museum Publications.

Purvis, O.W., Coppins, B.J. and James, P.W. (1993) Checklist of lichens of Great Britain and Ireland. *British Lichen Society Bulletin*, 72 (Supplement), 1–75.

Purvis, O.W., Coppins, B.J. and James, P.W. (1994a) *Checklist of lichens of Great Britain and Ireland*. London: British Lichen Society.

Purvis, O.W., Gilbert, O.L. and Coppins, B.J. (1994b) Lichens of the Blair Atholl limestone. *Lichenologist*, 26, 367–382.

Purvis, O.W., Jørgensen, P.M. and James, P.W. (1995) The lichen genus *Thelotrema* in Europe. *Bibliotheca Lichenologica*, 58, 335–360.

Richardson, D.H.S. (1975) *The Vanishing Lichens*. Newton Abbott: David and Charles.

Rose, F. (1976) Lichenological indicators of age and environmental continuity in woodlands, in Brown, D.H., Hawksworth, D.L. and Bailey, R.H. (eds) *Lichenology: progress and problems*. London, Academic Press, pp. 279–307.

Rose, F. (1992) Temperate forest management: its effect on bryophyte and lichen floras and habitats, in Bates, J.W. and Farmer, A.M. (eds) *Bryophytes and lichens in a changing environment*. Oxford: Clarendon Press, pp. 211–233.

Rose, F. and Coppins, B.J. (1983) Lichens of Colonsay. *Proceedings of the Royal Society of Edinburgh, B*, 83, 403–413.

Rose, F. and James, P.W. (1974) Regional studies on the British lichen flora I. The corticolous and lignicolous species of the New Forest, Hampshire. *Lichenologist*, 6, 1–72.

Rose, F., Stern, R.C., Matcham, H.W. and Coppins, B.J. (1991) *Atlas of Sussex Mosses, Liverworts and Lichens*. Brighton: Booth Museum of Natural History.

Sandell, K.A. and Rose, F. (1996) The lichen flora, in Brewis, A., Bowman, P. and Rose, F. (eds) *The Flora of Hampshire*. Colchester: Harley Books, pp. 306–324.

Sanderson, N. (1994) An ecological survey of the lichens *Catillaria laureri* and *Parmelia minarum* in the New Forest. *BLS Bulletin*, 74, 33–34.

Scheidegger, C. (1993) A revision of European saxicolous species of the genus *Buellia* de Not. and formerly included genera. *Lichenologist*, 25, 315–364.

Seaward, M.R.D. (1975) Lichen flora of the West Yorkshire conurbation. *Proceedings of the Leeds Philosophical and Literary Society, Science Section*, 10, 141–208.

Seaward, M.R.D. (1979) Lichens, in Booth, E.M. *The Flora of County Carlow*. Dublin, pp. 149–151.

Seaward, M.R.D. (1983) The lichens of County Laois, in Feehan, J. (ed.) *Laois: an environmental study*. Ballykilcavan, pp. 130–131, 491–493.

Seaward, M.R.D. (1984) Census catalogue of Irish lichens. *Glasra*, 8, 1–32.

Seaward, M.R.D. (1988a) Progress in the study of the lichen flora in the British Isles. *Botanical Journal of the Linnean Society*, 96, 81–95.

Seaward, M.R.D. (1988b) Contribution of lichens to ecosystems, in Galun, M. (ed.) *CRC Handbook of Lichenology, Vol. II*. Boca Raton: CRC Press, pp. 107–129.

Seaward, M.R.D. (1993) Lichens and sulphur dioxide air pollution: field studies. *Environmental Reviews*, 1, 73–91.

Seaward, M.R.D. (1994a) Vice county distribution of Irish lichens. *Biology and Environment, Proceedings of the Royal Irish Academy*, 94B, 177–194.

Seaward, M.R.D. (1994b) Checklist of Yorkshire lichens. *Proceedings of the Leeds Philosophical and Literary Society, Science Section*, 11(5), 85–120.

Seaward, M.R.D. (ed.) (1995) *Lichen Atlas of the British Isles, Fascicle 1*. London: British Lichen Society.

Seaward, M.R.D. (1998) Time-space analyses of the British lichen flora, with particular reference to air quality surveys. *Folia Cryptogamica Estonica*, 32, 85–96.

Seaward, M.R.D. and Hitch, C.J.B. (1982) *Atlas of the lichens of the British Isles. Vol. 1*. Cambridge: Institute of Terrestrial Ecology.

Sérusiaux, E. (1989) *Liste rouge des macrolichens dans la communauté Européene*. Liège: Centre de Recherches sur les Lichens.

Siddiqi, M.R. and Hawksworth, D.L. (1982) Nematodes associated with galls on *Cladonia glauca*, including two new species. *Lichenologist*, 14, 175–184.

Smith, A.L. (1918) *A Monograph of the British Lichens*. Vol. 1, 2 edn, London: British Museum (Natural History).

Smith, A.L. (1926) *A Monograph of the British Lichens*. Vol. 2, 2 edn, London: British Museum (Natural History).

UK Biodiversity Group (1999) *Tranche 2 Action Plans. Volume III – plants and fungi*. Peterborough: English Nature.

van Herk, C.M. (1999) Mapping of ammonia pollution with epiphytic lichens in the Netherlands. *Lichenologist*, 31, 9–20.

Watson, M.F., Hawksworth, D.L. and Rose, F. (1988) Lichens on elms in the British Isles and the effect of Dutch elm disease on their status. *Lichenologist*, 20, 327–352.

Wolseley, P.A. and Pryor, K.V. (1999) The potential of epiphytic twig communities on *Quercus petraea* in a Welsh woodland site (Tycanol) for evaluating environmental changes. *Lichenologist*, 31, 41–61.

Woods, R.G. (1993) *Flora of Radnorshire*. Cardiff: National Museum of Wales.

Woods, R.G. and Orange, A. (1999) *A Census Catalogue of Welsh Lichens*. Cardiff: National Museums and Galleries of Wales.

World Conservation Union (1994) *IUCN Red List Categories*. Gland: IUCN.

Chapter 8

Terrestrial and freshwater eukaryotic algae

David M. John, Allan Pentecost and Brian A. Whitton

Although there are many accounts of the distribution and abundance of freshwater algae in the British Isles, and some for terrestrial species, only a few sites have been sampled over long periods. Among them are lakes in the English Lake District, Loch Leven in Scotland, Lough Neagh in Northern Ireland and a few reservoirs. Changes observed in these large water bodies are the result of changes in nutrient levels and to other factors. Eutrophication can lead to a shift from a phytobenthos to a plankton-dominated lake (for example Cheshire Meeres, Irish Midlands); a key factor is overgrowth of benthic plants by filamentous algae followed by algal blooms. Short-term studies taken together with anecdotal reports also indicate that changes have taken place at many other sites. Inevitably such studies tend to focus on nuisance or conspicuous algae. *Cladophora glomerata* has long been regarded as the principal nuisance alga, but there are now reports of *Enteromorpha*, *Hydrodictyon* and *Vaucheria* becoming more widespread and abundant. These and other filamentous forms (such as *Spirogyra* or *Oedogonium*) not only restrict water movement in rivers, ditches and canals, but also affect their aesthetic and recreational value, as well as pose problems to managers of fish farms, anglers and water-treatment works. The spread of *Hydrodictyon* is probably more a response to elevated and extended summer temperatures and low flows rather than to eutrophication. During the 1990s it has invaded many rivers (such as the Wear and Wye) and in some years forms extensive growths.

Clare Island in County Mayo and Malham Tarn in North Yorkshire are sites where long-term comparisons have been made. At these two sites there has been little change in the eukaryotic algae over several decades, although at Malham Tarn considerable changes have taken place in the prokaryotic cyanobacteria. Losses from sites are well documented for many charophytes, but little information is available for other groups of eukaryotic algae. Regrettably, algae have tended to be given low priority by UK conservation bodies, and Red Lists of endangered species published for some European countries do not exist for Great Britain and Ireland with the exception of the charophytes.

In the terrestrial environment, nuisance or unsightly growths are most often reported for golf courses, where several common genera of green algae can reach nuisance proportions as a result of poor management regimes. Similarly some terrestrial algae often form conspicuous growths on buildings and the edges of fertilised fields. There is considerable evidence for genetic adaptation to particular environmental factors such as zinc, copper and herbicides. However, it remains unclear to what extent major taxa or species differ in their ability to respond to such factors, the time period over which adaptation takes place, or the extent to which the distribution of adapted strains between sites is important. One reason for the absence of reliable baseline information against which to measure change is the lack of suitable identification guides.

The availability of a coded checklist (Whitton *et al.* 1998) and the *Freshwater Algal Flora of the British Isles* (John *et al.* 2001) should make it much easier for the reviewer of this subject 25 years from now to provide a more comprehensive overview of change.

REFERENCES

John, D.M., Whitton, B.A. and Brook, A.J. (2001) *The British Freshwater Algal Flora.* Cambridge: Cambridge University Press.

Whitton, B.A., John, D.M., Johnson, L.R., Boulton, P.N.G., Kelly, M.G. and Haworth, E.Y. (1998) *A Coded List of Freshwater Algae of the British Isles.* Wallingford: Institute of Hydrology.

Chapter 9

Cyanobacteria (blue-green algae)

Brian A. Whitton and S. J. Brierley

One obvious change since the previous review in 1973 is in the name. Research scientists have mostly followed the lead of the late R.Y. Stanier in calling these organisms cyanobacteria to emphasize their prokaryotic structure, whereas those involved in practical management continue to use 'blue-green algae'. As the binomials based on the International Code of Botanical Nomenclature are the only ones suitable for routine field survey, the latter term is used here.

Since the overview in 1973 for the freshwater algae as a whole, blue-green algal populations have continued to increase in many eutrophic lakes and have sometimes reached nuisance densities for the first time. Environmental management has in some cases reversed the trend, however, especially in the London reservoirs, as a consequence of pumping water to prevent the formation of a thermocline. An important feature of many large oligotrophic lakes, such as Wastwater, which has been realized only within the past 15 years, is that blue-green algae also form a significant component of the algal biomass here in summer, though the overall biomass is relatively low (Hawley and Whitton 1991). This is due to the presence of picoplankton, organisms < 2 µm in their largest dimension.

There have also been an increasing number of reports of mats of blue-green algae floating down rivers in early summer. However, it is much more difficult to say whether or not these reflect nutrient eutrophication or whether it is merely that more people are on the look-out for such phenomena.

Many planktonic blue-green algae appear to be cosmopolitan, at least within broad climatic zones, so spread or invasion would seem unlikely unless there has been a shift in the environment. Nevertheless, there has been an increasing number of records of single filament forms (rather than flakes) of the nitrogen-fixer *Aphanizomenon* (Whitton and Brierley, unpubl. data), so the question arises whether there have been shifts in nutrient regime or other environmental factors that favour this particular form. There are some data showing that strains of particular species of cyanobacteria may become genetically adapted to elevated zinc from mine drainage (Shehata and Whitton 1981), but no studies have been reported for the UK showing whether or not strains adapted to other potentially toxic factors such as herbicides are widespread. It seems likely that losses of calcite-depositing blue-green algae have often occurred in calcareous streams as a result of eutrophication; one example of this is the loss of *Rivularia* from a calcareous flush at Cwm Nofydd Site of Special Scientific Interest (near Cardiff) (Whitton, unpublished data). In the case of loss of calcifying aquatic organisms, phosphate enrichment seems the most likely cause (Pentecost and Whitton 2000). However, the continuing enrichment of upland catchments with combined nitrogen due to atmospheric deposition must be considered a threat to streams dominated by the nitrogen-fixing cyanobacterium *Rivularia*, such as the streams draining peat and metamorphosed limestone in Upper Teesdale (Livingstone and Whitton 1984). Such upland stream communities of cyanobacteria and eukaryotic algae are particularly well suited for monitoring long-term and seasonal changes in the nitrogen to phosphorus ratio in upland catchments (Whitton 1999).

REFERENCES

Hawley, G.R.W. and Whitton, B.A. (1991) Seasonal changes in chlorophyll-containing picoplankton populations of ten lakes in northern England. *Internationale Revue des gesamten Hydrobiologie*, **76**, 545–554.

Livingstone, D. and Whitton, B.A (1984) Water chemistry and phosphatase activity of the blue-green alga *Rivularia* in Upper Teesdale streams. *Journal of Ecology*, **73**, 405–421.

Pentecost, A. and Whitton, B.A. (2000) Limestones, in Whitton, B.A. and Potts, M. (eds) *The Ecology of the Cyanobacteria*. Dordrecht: Kluwer, 257–280.

Shehata, F.H.A. and Whitton, B.A. (1981) Field and laboratory studies on blue-green algae from aquatic sites with high levels of zinc. *Verhandlungen der Internationalen Vereinigung für Theoretische und Angewandte Limnologie*, **21**, 1466–1471.

Whitton, B.A. (1999) Perspective on the use of phototrophs to monitor nutrients in running waters. *Aquatic Conservation*, **9**, 545–549.

Chapter 10

Diatoms

Elizabeth Y. Haworth

ABSTRACT

Changes in diatom composition are good indicators of environmental change and this has been put to good use in many recent studies. Long-term datasets, both as algal records and as the study of sediment profiles, have recently proved invaluable, for without them it would be impossible to identify the changes in diatom species over recent time. A study of changing diversity requires that sites are revisited or reworked time and time again but this practice is proving less popular as science becomes more economically constrained. Windermere provides a good example of the value of long-term observation and routine monitoring. The palaeolimnological studies within the research on acid rain or lake enrichment have, however, also provided the evidence about the waxing and waning of certain species due to changes in our environment, and illustrate the pattern of successful immigration or disappearance.

There is a problem in identifying recent changes where there may not be older records for comparison or where a species may have previously been overlooked. This is further complicated by the great advances in identification, the use of electron microscopy, and the revision of species and genera in the last 25 years. Diatoms will surely continue to play a part in our environmental predictions.

1 Introduction

Whitton (1974) reviewed the then recent changes in British freshwater algae and remarked that 'it is becoming clear that some changes in diatom species composition are particularly good indicators of eutrophication'. This useful characteristic has led to a positive bloom of studies of diatom assemblages in recent sediments in Britain, and the onset of the 'acidification debate' in the 1980s also directed attention to those taxa characteristic of oligotrophic, acidic waters (Battarbee *et al.* 1990).

2 The value of long-term algal observations

How do we know that there have been changes, since not many water-bodies are studied at all, much less observed routinely? In the spring of 1973, Dr J.W.G. Lund noticed the sudden appearance of a large population of a small-sized *Stephanodiscus* in Blelham Tarn. As he had been monitoring the phytoplankton populations in this small Cumbrian lake since 1945, he could be confident that this was a new phenomenon

(Lund 1979). This provided an opportunity for monitoring the way in which the biological material of a spring diatom bloom became incorporated into the sediment surface (Haworth 1976) and was a lake-wide improvement on an experiment that added a marine diatomite (*Coscinodiscus* sp.) to one of the prototype 'Lund tubes' to provide a marker horizon (Pennington *et al.* 1977). After three seasons, numerous valves of *S. minutulus* could be found, to varying depths within the upper 10 cm of sediment depending on location within the lake (Haworth 1979: Fig. 2). This identified the material as relating to those years and showed remarkably good stratigraphic distribution, with a lack of bioturbation, as might be expected from the known paucity of benthic animals (Jones 1980). A stratigraphic profile was also analysed to see how closely the sedimentary record actually reflected the diatom plankton populations monitored by Lund (1979). The result was a diatom-based timescale where successive sedimentary peaks of different species could be assigned dates according to the long-term data (Haworth 1980). This showed how stepwise changes in the diatoms, *Cyclotella praetermissa*, *Asterionella formosa*, *Aulacoseira subarctica*, *Fragilaria crotonensis*, *F. crotonensis* var. *prolongata* and *Stephanodiscus minutulus* had followed nutrient enrichment of the lake since the 1950s (Lund 1979). Earlier there had been no form of absolute dating for core material less than 1000 years old but a new method of dating sediments was just being assessed (Pennington *et al.* 1976), using lead isotopes – a comparative analysis of ^{210}lead and ^{226}radon – and this now provides an essential palaeolimnological timescale for the last 150 years.

Over the last 20 years cores from Blelham Tarn have been studied on several occasions and the collection of a core, frozen *in situ*, in 1990 for the study of Chernobyl caesium (van der Post *et al.* 1997) showed that the most recent sediments were now visibly laminated as light and dark coloured bands. Diatom and radio-isotope analysis both indicated that there has been a considerable increase in the annual accumulation (both as cm yr^{-1} and gm cm^{-3}) since *c.* 1980 that could only partly be due to eutrophication as calculated from algal biomass. Although there has been no further change in the species composition of the diatom assemblage, *Stephanodiscus* populations have clearly increased and the examination of material from cores collected in 1997 suggests that there are changes in valve size and morphology; it is possible that this may correlate with the silica:phosphorus ratios.

It is only researchers who keep looking at samples from the same site, be it as live samples or as the sedimentary accumulation, who are able to observe changes as they happen. H. Canter first spotted the appearance of *Aulacoseira islandica* subsp. *helvetica* in the phytoplankton populations of Windermere in 1987 (Canter and Haworth 1991). Such populations must have to reach a certain minimum size before they are likely to appear in samples under the microscope. Since then this has become nearly as abundant as *A. subarctica*, which has been one of the spring diatom bloom dominants since the 1940s (Table 10.1; Sabater and Haworth 1995). This was soon followed by the observation of filaments of *A. granulata* var. *angustissima* (Canter and Haworth 1992) but this species was not seen again, suggesting that the population was not viable. The failure of *A. granulata* to establish itself in Windermere at that time may be ascribed to the timely removal of phosphorus from the treated effluent from the local sewage treatment works and declining concentrations in the water (Reynolds 1998). However, it made another and more significant appearance in the late autumn of 1999 (Parker and Dent, pers. comm.), so further monitoring is required.

Table 10.1 Comparison of the annual population maximum of *Aulacoseira islandica* (as cells l⁻¹) and the representation in the diatom assemblages in the recent sediments of the north and south basins of Windermere. The former relates only to plankton algae and the latter relates only to diatoms of wider habitats. Note that the core top predates the maximum input for that year.

S. Basin core 90-1 Pb210 date	S. Basin core 90-1 % of diatoms	W'mere SB max. cells ml⁻¹ yr⁻¹	Year	W'mere NB max. cells ml⁻¹ yr⁻¹	N. Basin core 92-2 % of diatoms	N. Basin core 92-2 Pb210 date
		379	1994	79		
		349	1993	148		
		877	1992	286	5.2	1992
					5.0	
		1348	1991	984	8.8	1991
					8.4	
1990	7.6	1273	1990	363	4.6	1990
					6.6	
1989	1.8	770	1989	4	10.6	1989
					8.4	
1988	0.0	28	1988	3	7.0	1988
					4.2	1987
1987	1.9				0.4	
					6.0	1986
1985	0.0				0.0	
					1.0	1984
					2.0	
					0.6	1982
					0.0	1981

This pattern is somewhat similar to the reported appearance of *A. ambigua* in Bassenthwaite Lake. This diatom was first noticed in samples preserved during a fisheries study of this lake in 1987 (Mubamba 1989; Canter and Haworth 1992) and now dominates the spring algal bloom there (Hall *et al.* 1996). Both of these lakes had been receiving increasing concentrations of phosphorus and analyses of the recent sediments of Windermere (north and south basins) include *A. islandica* ssp. *helvetica* in the uppermost sediments, following an increase in the proportion of *Stephanodiscus parvus* (Sabater and Haworth 1995; Haworth, unpubl. data). A slot-sequencing correlation showing the goodness of fit of the sedimentary profile to the annual population maxima reveals an extremely good match, especially for *Aulacoseira islandica* ssp. *helvetica* (Battarbee *et al.* 1996). During a survey of all the major Cumbrian lakes (Table 10.2; Hall *et al.* 1996), it became clear that this taxon had also migrated into Ullswater and from there it has moved into Haweswater in the water transferred by pipeline. As Lake District lakes are visited not only by large numbers of small pleasure craft but also by increasing numbers of Environment Agency monitors, not to say students of freshwater biology, the surprise is more that we haven't had many more biological intrusions into the ecosystem, rather than the other way about.

Another area which has undergone a period of diatomaceous scrutiny is the Severn Estuary (Underwood and Paterson 1993). Here a new *Navicula* species has been

Table 10.2 A comparison of *Aulacoseira* species abundance in four Cumbrian lakes in 1995.

	Lakes				
				Windermere	Windermere
	Bassenthwaite	Haweswater	Ullswater	(N)	(S)
Phytoplankton (as cell/ml)					
A. ambigua	816				
A. islandica ssp. helvetica		127	426	5	19
A. subarctica	145		255	10	124
Surface sediment (as diatom %)					
A. ambigua	20.6				
A. islandica ssp. helvetica		4.4	8.4	2.6	2.4
A. subarctica	16.0	16.4	20.2	10.0	16.0

described – *N. pargemina* (Underwood and Yallop 1994) – and found in considerable abundance. The authors consider this another recent and successful recruit, but it is just possible that it reflects the lack of investigation within the wider locality, or has previously been included elsewhere, in another species. For a discussion on the finding and description of several other new species see Gardner *et al.* (1995) for a list of studies on coastal material. They state that 'Many diatoms may have been overlooked in earlier studies or lost during conventional cleaning procedures owing to the fact that many of the taxa found in this [episammic] community have thinly silicified frustules.' From the listed distribution it would appear that *Stoschiella hebetata* itself may have been here a while.

The expansion of the list of centric diatoms to be found in the Norfolk Broads (Clarke 1992) may also owe as much to the advances in careful investigation, our taxonomic definition and use of the electron microscope as to new recruitment or a real expansion in the changing water quality of that region.

3 Palaeolimnological profiles

The diatom profile from Bassenthwaite Lake sediments told a rather different story. There, the sediments indicate that *A ulacoseira ambigua* first appeared in the early 1900s and again *c.* 1960 (Fig. 10.1) but didn't really gain a foothold until *c.* 1980, only becoming prominent in the late 1980s when the species was first noticed in plankton samples. By the early 1990s it formed about 10% of the diatom assemblage (Cranwell *et al.* 1995).

Many studies of lowland lakes could tell a story of pollution and (or) over-enrichment in the succession of their diatoms (Bennion *et al.* 1997). A recent example is that of the over-enrichment of three small, shallow coastal lakes in Anglesey (Haworth *et al.* 1996). Llyn Coron, near Aberffraw, close to the south-west coast, is a lake with an undulating

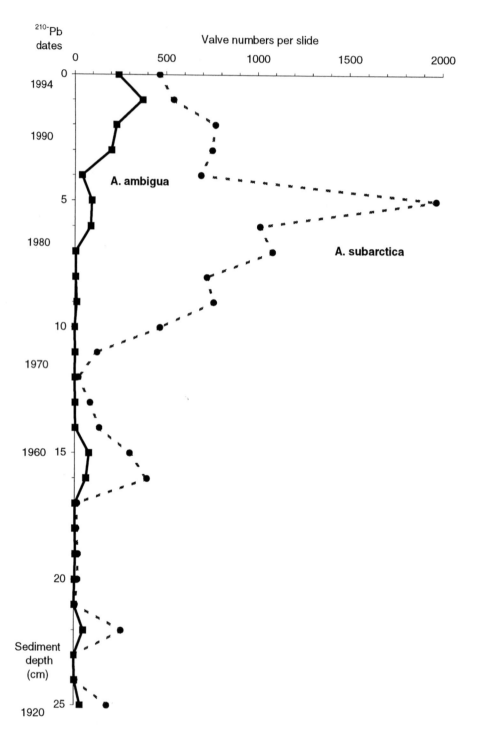

Figure 10.1 The changing concentrations of two *Aulacoseira* species in the recent sediments of Bassenthwaite Lake, Cumbria as found in a given volume of wet sediment.

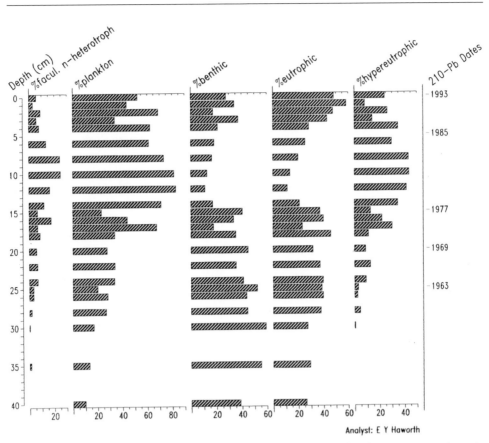

Figure 10.2 Recent ecological changes in Llyn Coron, Anglesey, reflected in the percentages of certain habitat, nutrient- and pollution-level indicators as found in the diatom assemblage profile.

agricultural catchment which, prior to the 1940s, was occupied by small mixed farms. Agriculture became more intensive in the 1960s and intensive pig- and poultry-rearing units were also introduced. The greatest application of phosphorus fertilizers to the land occurred in the mid-1970s and the pig unit closed in 1983. The sedimentary diatom assemblages indicate that the lake was mesotrophic over a long period. Several small planktonic *Stephanodiscus* and *Cyclostephanos* species, which are considered to indicate hypertrophy (Van Dam *et al.* 1994), appeared in the assemblages (Fig. 10.2) in the 1960s and dominated them from the late 1970s to the extent that there was a switch in dominance from benthic taxa to planktonic ones and many taxa were eliminated or very much reduced (Haworth *et al.* 1996). It is no surprise that *Pinnularia, Eunotia* and some of the *Navicula* species all disappear at a time when increased plankton populations were shading out even this shallow benthos. It is also quite plain from these profiles that, while there are very few incoming species additions to the floras of these lakes since changes prior to the 1970s, the increasing populations of some taxa have severely diminished those of others, by competition, over the last few decades. In extreme conditions this leads to species exclusion and, for central Europe, Lange-Bertalot (1997) has suggested that most of these are species of oligotrophic or dystrophic water-bodies.

Not all incoming taxa retain their initial foothold, as the Windermere records indicate. Whitton (1974) commented on the appearance and subsequent establishment in the upper reaches of the tidal River Thames of *Hydrosera triquetra*, a tropical and sub-tropical marine form, as noted by the amateur diatomist Gleave (1972). Although I cannot find any further published reference to this species, E. Cox (pers. comm.) assures me that it is still frequently found in samples of these reaches of the Thames. Obviously it is able to survive the British climate and water quality despite its usual preference for warmer, more saline waters. Presumably it came with a sea-going vessel, perhaps one of the many that discharge some 10 billion tonnes of water ballast annually throughout the world, before taking on cargo (Forbes and Hallegraeff 2001). This activity is now regarded as potentially risky, as it moves problematical organisms around the world. Although one would expect that London is more likely to be a potential exporter than importer, this is clearly not always the case. One wonders what other microscopic life-forms have come to our waters by the same route?

During the 1980s, collaborative Surface Waters Acidification Project studies were spearheaded by R.W. Battarbee of London and I. Renberg of Umeå in Sweden (Battarbee *et al.* 1990). The resulting suite of diatom profiles from recently accumulated sediments in lakes, especially those in our northwestern uplands, showed how diatom assemblages in differing sites have been modified by an environmental change that was finally, after much testing, shown to be correlated with acidity from the atmosphere (ibid.). During this work, attention focused on the diatoms more typical of acid, oligotrophic waters resulting in the recognition of several new taxa, for example *Aulacoseira tethera* (Haworth 1988), *Achnanthes scotica* and *A. marginulata* f. *major* (Flower and Jones 1989),now transferred to the new genus *Psammothidium*, and *Tabellaria binalis* var. *elliptica* (Flower 1989), now transferred to the genus *Oxyneis*. *Tabellaria binalis* is a good indicator of acid waters that appeared in some small, montane Lake District lakes around 1950 (Knudson 1954) and expanded to become prominent in their diatom assemblages during the early 1980s (Haworth *et al.*, 1988; Haworth 1993). Although *A. tethera* may have entered Scoat Tarn somewhat earlier, it has increased significantly in abundance since *c.*1970 (Haworth 1993; Fig. 10.3). A problem with recognizing the arrival of new species in these particular lakes is that not only has no one been monitoring these sites over the last 50 years but the sedimentary accumulation is very slow and so there are few samples within the relevant timescale to examine. A concern that *A. tethera* would disappear from Seathwaite during an experiment to increase the alkalinity of that acid water lake, by adding phosphorus to encourage plant growth to boost the carbonate levels (Davison *et al.* 1995), has proved that predictions do not always materialize. Instead, there was some indication of population enhancement, reinforcing a more rational view that the first beneficiaries of increased nutrients are the current incumbents but that taxa more capable of utilizing the new environmental conditions soon arrive or attain significant populations. Studies by Cameron (1995) of the Loch Fleet experiment, where lime was applied to the catchment to raise alkalinity, failed to show the expected response of certain *Achnanthes* species.

More recently, Gibson *et al.* (1993) found that *Skeletonema subsalsum* had invaded Lough Erne in Northern Ireland. Palaeolimnological study has now confirmed the phytoplankton monitoring and showed that *Skeletonema* arrived around 1980 (Anderson *et al.* 1996). Another species of the same genus, *S. potamos*, was found in a Shropshire Reservoir early in 1998 and considerable populations of this species were recorded from the River Thames in 1995 (Hutchings, pers. comm.).

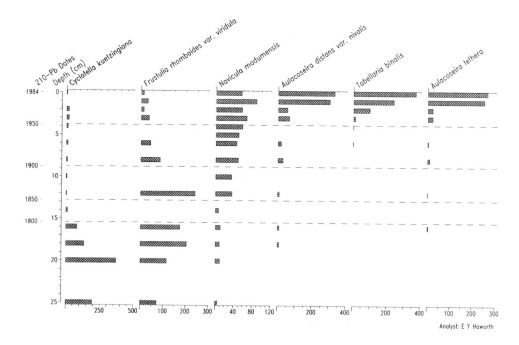

Figure 10.3 Recent changes in diatom assemblages in Scoat Tarn, Cumbria illustrating the trend from the species of a circumneutral pH towards those of increased acidity.

4 Taxonomic and other changes

Owing to the use of the electron microscope, more especially the scanning electron microscope, our understanding of the morphology of the diatom frustule has changed radically during the last three decades and there have been numerous name changes in an attempt to group taxa more naturally. Three works of consequence for British diatomists have appeared within the last decade: *The Diatoms* (Round *et al.* 1990), *An Atlas of British Diatoms* (Hartley *et al.* 1996) and the *Bacillariophyceae* volumes of the *Süsswasserflora von Mitteleuropa* (Krammer and Lange-Bertalot 1986–91). There have been many other, smaller, helpful works, including a checklist of the algae of south Wales (Benson-Evans and Antoine 1993). One of the ways in which ongoing comparisons may now be confounded is in the numerous and usually necessary name changes: Whitton (1974) was right to say that past collections may be more useful than old lists of names, for many of the latter have been found to be incorrect when compared with the type material, and species have been divided, amalgamated and moved into new genera such that one species had changed names four times in 30 years of diatom study. A new British checklist of freshwater algae has recently been prepared (Whitton *et al.* 1998) but such lists are outdated almost before they are printed and represent only the best efforts of the compilers; if scientific study is to progress, they can never be set in stone, or whatever is currently the most durable material.

The methods of studying diatom assemblages have also altered, together with our much improved understanding of the ecological significance of these assemblages (Van Dam *et al.* 1994). With specific questions to answer, such as the changing pH or nutrient levels, some studies are now focused on producing transfer functions so that diatoms can provide data on water chemistry that will help forecast or hindcast trends for the use of water-resource managers (Bennion *et al.* 1996). We have unfortunately lost several experts who not only collected and studied diatoms from a wide range of sites over many years, but also freely passed on their expertise to professionals and students alike. Science has moved on and no longer encourages the publication of 'A new species of ... from ...', for example Carter's (1987) *Navicula claytoni* from Gala Water, or the long species lists that would have made this review easier. A recent morphological study of British *Licmophora* species (Honeywill 1998) includes four species not listed in Hartley's 1986 checklist and omits two found therein, the only comment being that *L. profundeseptata* may previously have been included in *L. communis*. Many of the basic data are now stored in unpublished databases; statistical and other treatments have also taken the place of qualitative assessment, the list and the line drawing; it has become an applied science, with all the commercialism that implies. Unfortunately such treatment may only answer the specific question in hand and not supply answers for the future researcher in related fields. At least for diatoms, samples of the almost indestructible frustule can readily be retained and provide us with excellent type or reference material. Recently there has been much study, by Cox (1998) and other researchers, to validate our taxomony against the older material. There is also the sedimentary store as a measure for hindcasting long-term changes, in the hope of persuading humanity to mend its ways.

5 Global warming

Many researchers are seeking to correlate changing biotic assemblages and weather patterns over various timescales but, although we are now able to look at short-term changes in the sedimentary record, and some long-term phytoplankton data, there is as yet no clear evidence on the most recent period. Clearly algal populations are affected by both wind and temperature, and these and other seasonal effects have been shown to occur in the long-term records of one of the common planktonic diatoms, *Asterionella formosa* (Maberly *et al.* 1994), while the competitive ability of *Aulacoseira subarctica* may be considerably improved in warmer winters (Haworth 1993). The immediate reaction to the blossoming of *A. granulata* var. *angustissima* in Windermere is to speculate that the longer, warmer autumn and delayed stratification have provided the opportunity.

Zong and Horton (1999) presented a first transfer function to aid high-resolution investigations of past changes in sea level. This is based on the altitudinal range of the diatom indicators along transects of some coastal saltmarshes. This change in diatom distribution, indicative of the past patterns of sea-level change and coastal inundation, may also have implications for the future.

One possible effect of climate change could be the massive fish-kill in the Kennet and Avon Canal in March 1998 (Johnson *et al.* 1998) when, for a time, it was thought that a very small diatom species common to highly enriched waters had become a fish killer and 4 km of canal were closed while chemists and biologists hastily sought the culprit. Exceedingly high chlorophyll levels were found to relate to an almost monoculture of the planktonic diatom *Stephanodiscus hantzschii*, and the sudden death and deposition of these dense populations on the canal bed were shortly followed by further fish-kills.

Although it was demonstrated that the tiny spines and long chitinous setae of these *c*.10 μm diameter 'barrels' did not cause death by clogging the fish gills, it may well be that the unusually warm weather, linked to canal dredging, high nutrient levels and low flows all combined to encourage the build-up of these high populations and a subsequent toxin or maybe drastic, and unnoticed, oxygen depletion. Unless this was strictly a one-off occurrence, we may see similar circumstances again, if global warming is to bring higher spring temperatures.

ACKNOWLEDGEMENTS

I should like to thank David Windal for counting *Aulacoseira* in Bassenthwaite Lake (Fig. 10.1) for me, and Drs M. Robinson and J.W.G. Lund for their help and support in critically reading the manuscript. My thanks also to all those who provided me with information, especially Julie Parker, and to Ian McCulloch who searched the various library systems so diligently.

REFERENCES

Anderson, N.J., Patrick, S.T. and Appleby P.G. (1966) Palaeolimnological surveys of Lough Erne and Lough Melvin (Northern Ireland). *Environmental Change Research Centre, University of London Research Report*, **28**, 1–29.

Battarbee, R.W., Mason, Sir John, Renberg, I. and Talling, J.F. (1990) *Palaeolimnology and Lake Acidification*. London: The Royal Society.

Battarbee, R.W., Barber, K.E., Oldfield, F., Thompson, R., Stevenson, A.C., Eglinton, G., Haworth, E.Y., Brooks, S.J., Holmes, J., Cameron, N.G., Rose, N.L. and Maddy, D. (1996) *Proxy records of climatic change in the UK over the last two millennia.* Final Report NERC Special Topic Grant GST/02/70.

Bennion, H., Juggins, S. and Anderson, N.J. (1996) Predicting epilimnetic phosphorus concentrations using an improved diatom-based transfer function and its application to lake eutrophication management. *Environmental Science and Technology.* **30**, 2004–2007.

Bennion, H., Harriman, R. and Battarbee, R. (1997) A chemical survey of standing waters in south-east England, with reference to acidification and eutrophication. *Freshwater Forum*, **8**, 28–44.

Benson-Evans, K. and Antoine, R. (1993) *A Check-list of the Freshwater, Brackish and Marine Algae of South Wales, UK*. Cowbridge: D. Brown and Sons.

Cameron, N.G. (1995) The representation of diatom communities by fossil assemblages in a small acid lake. *Journal of Paleolimnology*, **14**, 185–223.

Canter, H.M. and Haworth, E.Y. (1991) The occurrence of two new planktonic diatom populations in the English Lake District; *A. islandica* ssp. *helvetica* and *A. ambigua*. *Freshwater Forum*, **1**, 39–48.

Canter, H.M. and Haworth, E.Y. (1992) Another species of *Aulacoseira* found in Windermere (English Lake District). *Freshwater Forum*, **2**, 21.

Carter, J.R. (1987) *Navicula claytoni*, a new diatom. *Microscopy*, **35**, 633–635.

Clarke, K.B. (1992) Centric diatoms in the Norfolk Broads. *Microscopy*, **36**, 692–702.

Cox, E.J. (1998) The identity and typification of some naviculoid diatoms (Bacillariophyta) from freshwater or brackish habitats. *Phycologia*, **37**, 162–175.

Cranwell, P.A., Haworth, E.Y., Lawlor, A. and Lishman, J.P. (1995) *Assessment of the Historical Changes in Bassenthwaite Lake from a deep sediment profile*. Report to National Rivers Authority, NW Region.

Davison, W., George, D.G. and Edwards, N.J.A. (1995) Controlled reversal of lake acidification by treatment with phosphate fertilizer. *Nature, 377,* 504–507.

Flower, R.J. (1989) A new variety of *Tabellaria binalis* (Ehrenb.) Grun. from several acid lakes in the UK. *Diatom Research,* **4,** 21–23.

Flower, R.J. and Jones, V. (1989) Taxonomic descriptions and occurrences of new *Achnanthes* taxa in acid lakes in the UK. *Diatom Research,* **4,** 227–239.

Forbes, E. and Hallegraeff, G. (2001) Transport of potentially toxic *Pseudonitzscha* diatom species via ballast water, in *Proceedings of the 15th International Diatom Symposium* (ed. J. John). Koenigstein: Koeltz Scientific Books.

Gardner, C., Schulz, D., Crawford, R.M. and Wenderoth, K. (1995) *Stoschiella hebetata* gen.et sp.nov. a diatom from intertidal sand. *Diatom Research,* **10,** 241–250.

Gibson, C.E., McCall, R.D. and Dymond, A. (1993) *Skeletonema subsalsum* in a freshwater Irish lake. *Diatom Research,* **8,** 65–71.

Gleave, H.H. (1972) *Hydrosera triquetra,* a diatom new to European waters. *Microscopy,* **32,** 208.

Hall, G.H., Cubby, P.R., Deville, M., Fletcher, J., Haworth, E.Y., Hewitt, D.P., Hodgson, P., James, B., Lawlor, A.J., Parker, J., Rigg, E. and Woof, C. (1996) *A survey of the limnological characteristics of the lakes of the English Lake District: Lakes Tour 1995.* Report to the Environment Agency, North West Region.

Hartley, B. (1986) A check-list of the freshwater, brackish and marine diatoms of the British Isles and adjoining coastal waters. *Journal of the Marine Biological Association of the United Kingdom,* **66,** 531–610.

Hartley, B., Barber, H.G., Carter, J.R. and Sims, P.A. (1996) *An Atlas of British Diatoms.* Bristol: Biopress.

Haworth, E.Y. (1976) The changes in the composition of the diatom assemblage profiles found in the surface sediments of Blelham Tarn in the English Lake District during 1973. *Annals of Botany,* **40,** 1195–1205.

Haworth, E.Y. (1979) The distribution of a species of *Stephanodiscus* in the recent sediments of Blelham Tarn, English Lake District. *Nova Hedwigia Beiheft,* **64,** 395–410.

Haworth, E.Y. (1980) Comparison of continuous phytoplankton records with the diatom stratigraphy in the recent sediments of Blelham Tarn. *Limnology and Oceanography,* **25,** 1093–1103.

Haworth, E.Y. (1988) Distribution of diatom taxa of the old genus *Melosira* (now mainly *Aulacoseira*) in Cumbrian waters, in *Algae and the Aquatic Environment* Round, F.E. (ed.). Bristol: Biopress, pp. 135–164.

Haworth, E.Y. (1993) The responses of aquatic biota and their use as indicator organisms, in *Biological Indicators of Global Change,* Symoens, J.J., Devos, P., Rammeloo, J. and Verstraeten, Ch. (eds). Brussels: Royal Academy of Overseas Sciences, pp. 115–127.

Haworth, E.Y., Lishman, J. P. and Tallantire, P. (1988) *A Further Report on the Acidification of Three Tarns in Wasdale, Cumbria, North-west England.* Report to the Department of the Environment.

Haworth, E.Y., Pinder, L.C.V., Lishman, J.P. and Duigan, C.A. (1996) The Anglesey Lakes, Wales, UK – A study of recent environmental change in three standing waters of nature conservation importance. *Aquatic Conservation: Marine and Freshwater Ecosystems,* **6,** 61–80.

Honeywill, C. (1998) A study of British *Licmophora* species and a discussion of its morphological features. *Diatom Research,* **13,** 221–271.

Johnson, I., Barnard, S., Sims, I., Conrad, A., James, H., Parr, W., Hedgecott, S. and Cartwright, N. (1998) *Technical Investigation of the Kennet and Avon Canal Pollution Incident (March 1998).* Water Research Centre Report to Environment Agency.

Jones, J.G. (1980) Some differences in the microbiology of profundal and littoral lake sediments. *Journal of General Microbiology* ,**117**, 285–292.

Knudson, B. (1954) The ecology of the diatom genus *Tabellaria* in the English Lake District. *Journal of Ecology*, **42**, 345–358.

Krammer, K and Lange-Bertalot, H. (1986–91). Bacillariophyceae 1–4, in *Süßwasserflora von Mitteleuropa* (eds H. Ettl, J. Gerloff, H. Heynig and D. Mollenhauer), 4 vols. Stuttgart and New York: Gustav Fischer Verlag.

Lange-Bertalot, H. (1997) A first ecological evaluation of the diatom flora in Central Europe. Species diversity, selective human interactions and the need of habitat protection. *Lauterbornia*, **31**, 117–123.

Lund, J.W.G. (1979) Changes in the phytoplankton of an English Lake, 1945–1977. *Hydrobiological Journal*, **14**, 6–21.

Maberly, S.C., Hurley, M.A., Butterwick, C., Corry, J.E., Heaney, S.I., Irish, A.E., Jaworski, G.H.M., Lund, J.W.G., Reynolds, C.S., and Roscoe, J.V. (1994) The rise and fall of *Asterionella formosa* in the south basin of Windermere: analysis of a 45-year series of data. *Freshwater Biology*, **31**, 19–34.

Mubamba, R. (1989) *Ecology of English Coregonid fishes*. University of Wales PhD thesis.

Pennington, W. (Mrs Tutin), Cambray, R.S., Eakins, J.D. and Harkness, D.D. (1976) Radionuclide dating of the recent sediments of Blelham Tarn. *Freshwater Biology*, **6**, 317–331.

Pennington, W. (Mrs Tutin), Cranwell, P.A., Haworth, E.Y., Bonny, A.P. and Lishman, J.P. (1977) Interpreting the environmental record in the sediments of Blelham Tarn. *Freshwater Biological Association Annual Report*, **45**, 37–47.

Reynolds C.S.(1998) Back from the brink: reversing the deterioration of Windermere. *Cumbrian Wildlife*, **50**, 8–13.

Round, F.E., Crawford, R.M. and Mann, D.G. (1990) *The Diatoms; biology and morphology of the genera*. Cambridge: Cambridge University Press.

Sabater, S. and Haworth, E.Y. (1995) An assessment of recent trophic changes in Windermere (England) based on diatom remains and fossil pigments. *Journal of Palaeolimnology*, **14**, 151–163.

Underwood, G.J.C. and Paterson, D.M. (1993) Seasonal changes in diatom biomass, sediment stability and biogenic stabilisation in the Severn Estuary, UK. *Journal of the Marine Biological Association of the United Kingdom*, **73**, 871–887.

Underwood, G.J.C. and Yallop, M.L. (1994) *Navicula pargemina* sp. nov. – a small epipelic species from the Severn Estuary, UK. *Diatom Research*, **9**, 473–478.

Van Dam, H., Mertens, A. and Sinkeldam, J. (1994) A coded check-list and ecological indicator values of freshwater diatoms from the Netherlands. *Netherlands Journal of Aquatic Ecology*, **28**, 117–127.

Van der Post, K.D., Oldfield, F., Haworth, E.Y., Crooks, P.J. and Appleby, P.G. (1997) A record of accelerated erosion in the recent sediments of Blelham Tarn in the English Lake District. *Journal of Palaeolimnology*, **18**, 103–120.

Whitton, B.A. (1974) Changes in the British Freshwater Algae, in Hawksworth, D.L. (ed.) *The Changing Flora and Fauna of Britain*. London: Academic Press, pp.115–141.

Whitton, B.A., John, D.M., Johnson, L.R., Boulton, P.N.G., Kelly, M.G., Haworth, E.Y., Tindall, C.I. and Moore, R.V. (1988) *A Coded List of Freshwater Algae of the British Isles*. Wallingford: Institute of Hydrology.

Zong, Y. and Horton, B.P. (1999) Diatom-based tidal-level transfer functions as an aid to reconstructing Quaternary history of sea-level movement in the UK. *Journal of Quaternary Science*, **14**, 153–167.

Chapter 11

Viruses

Roger T. Plumb and J. Ian Cooper

ABSTRACT

The study of viruses in the wildlife of Great Britain and Ireland has been much neglected. What work has been done has demonstrated not only that viruses are widespread but that the discovery of new viruses is a fertile area of study. In consequence, either comparing virus incidence now with any earlier time, or giving a remotely definitive description of viruses in even a small part of Britain's wildlife is impossible. Viruses and similarly intangible organisms, such as prions, are however the subject of everyday concern, and on occasions alarm. Changes in wildlife and the policies that affect it will inevitably have consequences for the presence of viruses and their ease of transfer, often via invertebrate vectors, between hosts. On balance, viruses seem likely to become a more obvious component of and influence on British wildlife.

1 Introduction

Viruses in Britain's wildlife have never been systematically surveyed, even in a small component of our native biota. The only plants and animals for which any detailed knowledge is available are crop plants and domesticated animals. Epidemiological studies of viruses in these hosts often lead to investigations of possible 'wild' or 'alternative' hosts in native populations so that the published host ranges of many viruses include native species. However, this does no more than identify them as hosts and gives little idea of their prevalence, distribution and importance of infection to the wild host.

Not only is there thus little knowledge of virus presence in British wildlife, but there is no baseline against which to measure change, except in a few cases usually linked to the deliberate or accidental introduction of a virus or virus strain to Britain. Part of the difficulty in working with and identifying viruses is the need for specialist equipment and skills, and studies of viruses have largely been the province of the well-equipped professional. Recent developments in simple diagnostic methods open up the possibility of more widespread surveys, however funding and human resources are a continuing constraint.

Viruses also suffer from a bad press. They are almost universally considered 'a bad thing' and, for those species that matter to the population at large – in other words ourselves and our domesticated animals and plants – much time, effort and money is devoted to the eradication of viruses that infect them. Such efforts sometimes bring success, as in the case of smallpox (variola virus); for others such as influenza and rhinoviruses (the common cold) there has, so far, been conspicuous failure.

Viruses are also often synonymous, in the public perception, with diseases for which the cause is often unknown, uncertain or for which the putative agent is even more difficult to comprehend than viruses. Thus, the recognition of a new variant of Creutzfeldt–Jakob disease, the causal agent of which appears to be a prion, that affects mainly young people, was linked in many peoples' minds with viruses and the spectre of the spread of such diseases from animals to humans became a very great cause for concern.

Thus both the nature of viruses and the concentration of specialist effort needed to recognise and identify them means that providing any comprehensive overview of viruses in wildlife, let alone their current status, is not possible. Nevertheless, the clear, but undeniable, statement can be made that viruses are ubiquitous. What we hope we have provided below are a few snapshots of particular host–virus combinations, in both plants and animals, that illustrate some of the problems in understanding viruses in the context of wildlife in Britain.

2 What are viruses?

As the properties of viruses dictate the methods that need to be used to work with them it may be appropriate to give an outline of their characteristics.

Viruses are small; indeed only the poxviruses have individual particles large enough to come within the limit of resolution of light microscopy. For the remainder electron microscopy is required to visualise particles. In size they range from the brick-shaped orthopoxviruses which are about 200 nm × 200 nm × 250 nm to the satellite viruses of <17 nm in diameter (Fig. 11.1), although the latter usually depend upon and co-exist with other larger virus. Viruses consist essentially of a nucleic acid core, either DNA or RNA but not both, usually although not invariably enveloped within a proteinaceous coat. They depend entirely upon their host for their replication and occasionally on each other. They have no motility of their own and the great majority depend on an intermediary for their spread. Dissemination may be through seeds and via pollen or semen, but is more often associated with invertebrate vectors such as aphids, beetles, ticks, mosquitoes, leaf and plant hoppers, whiteflies and thrips, and in the soil by fungi or nematodes. Some viruses multiply in their vector as well as in their principal hosts. Taxonomy and species definition, as applied to sexually reproducing eukaryotes, are inappropriate for viruses, although much effort has been devoted to generating a hierarchical taxonomy (van Regenmortel *et al.* 2000). So defining what a virus is is remains problematical.

3 Infection and disease

Visible manifestation of the presence of a virus in a host is usually a range of symptoms; many viruses, however, exist within their host in a latent or inactive form, and these are known as orphan viruses in animals and cryptic viruses in plants. Several viruses can be present in one host, thus identifying a single cause of an overt condition can be difficult, and requires lengthy isolation, identification, reinfection and reisolation. Viruses can also recombine to produce viruses with mixed properties; in this way host range and

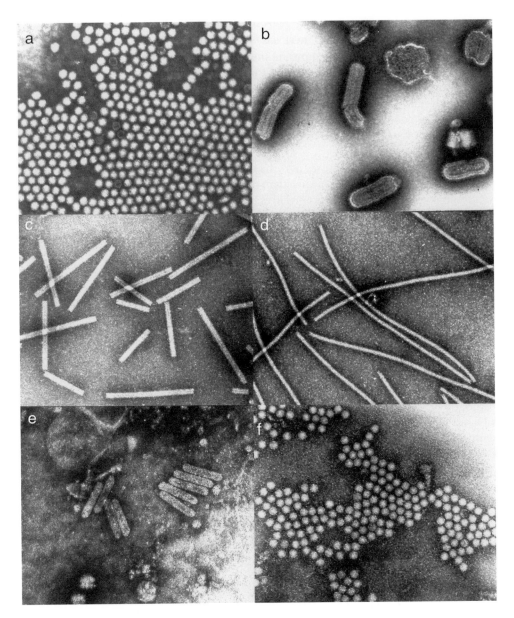

Figure 11.1 (a) Bee Paralysis Virus – 30 nm diameter; (b) Maize Mosaic Virus – 180–380 × 60–100 nm;
(c) Oat Golden Stripe Virus – 90–160 and 290 × 18 nm; (d) Bean Yellow Mosaic Virus –
680–900 × 12 nm; (e) Cocoa Swollen Shoot Virus – 120–130 × 28 nm; (f) Beet Western
Yellow Virus – 25 nm diameter.

vectors can change. Viruses can also interact with different viruses to prevent or limit infection; this process is known as immunisation in animals and cross-protection in plants.

4 Viruses in wildlife

4.1 Plants

Surveys of viruses in wild *Brassica* species growing on cliffs in Dorset (Raybould *et al.* 1999 and unpublished data) showed that wild black mustard (*Brassica nigra*) is naturally infected by *Turnip crinkle* and *Turnip rosette viruses*, but that these viruses were uncommon in wild cabbage (*B. oleracea*). Four other viruses, *Beet western yellows virus*, *Cauliflower mosaic virus*, *Turnip mosaic virus* and *Turnip yellow mosaic virus* were also present, but more frequently in *B. oleracea* than *B. nigra* (Fig. 11.2). Seventy-eight per cent of *B. oleracea* and 35% of *B. nigra* plants sampled were infected by at least one virus and multiple infections were frequent. There were also differences among sites in the frequency of infection. With these differences between species and sites it is impossible to draw any general conclusions about effects, even in a relatively localised area such as the area surveyed. In another survey of wild grasses (Edwards *et al.* 1985, 1989), 50 different viruses were isolated of which 14 were not readily identifiable, and thus likely to be 'new'. While there is limited evidence from wild species, geographical differences in virus frequency might be expected. *Cherry leaf roll virus* (CLRV) which seems to be mainly transmitted through seed via pollen and ovules was more common in birch trees in England (4%), than in Scotland (0.6%) and Wales (1.2%). By contrast, infection by CLRV in birch trees grown for amenity, and probably imported, was 17.5% (Cooper *et al.* 1984). Hammond (1981) surveyed viruses of *Plantago* species, which are common weeds in agriculture, from widely separated locations in England (Table 11.1). Eight viruses were found, only two of which were already known, and brought the total number of viruses known to infect *Plantago* species to 39.

While knowledge of viruses in wild plants is very sketchy, relative virus prevalence in cultivated crops is exploited. The production of the highest grades of seed potato is restricted to parts of Scotland where aphid-transmitted viruses such as *Potato virus Y* and *Potato leaf roll virus*, which are so damaging to potatoes, are uncommon early in the growing season. In central England, *Cucumber mosaic virus* and *Lettuce mosaic virus* were common in a wide range of wild plants (Tomlinson *et al.* 1970), but cucumber mosaic was not reported in Scotland, although *Tobacco rattle virus* was common, especially in *Stellaria media* and *Viola arvensis* (Cooper and Harrison 1973). However, the presence of viruses in cultivated plants can also differ greatly between years. The incidence of viruses in sugar beet is very closely monitored each year and wide differences are seen between years (Fig. 11.3). The fluctuations are largely the result of a combination of environmental influences on the aphids that spread the viruses and, possibly to a lesser extent, on the sources of virus available in wild species on remnants of the previous crop.

The results from these surveys demonstrate how numerous viruses are likely to be in wild plants, how little we know about the occurrence and identity of many of these viruses and how significant their presence is in influencing growth, survival and

Table 11.1 Viruses in *Plantago* spp (from Hammond 1981).

Locality	Number of viruses found
Northumberland 1	4
Northumberland 2	1
Cumbria	2
Warwickshire	1
Norwich	4
Cambridge	7
Essex	4
Reading	3
Cornwall	3

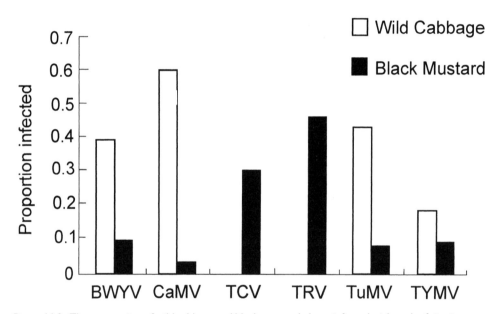

Figure 11.2 The proportion of wild cabbage and black mustard plants infected with each of six viruses.

BWYV Beet western yellows virus
CaMV Cauliflower mosaic virus
TCV   Turnip crinkle virus
TRV   Tobacco rattle virus
TuMV Turnip mosaic virus
TYMV Turnip yellow mosaic virus

competition between species. It has been suggested (Gibbs 1980) that viruses may protect their hosts from herbivores and thus could be at a competitive advantage. Other viruses may make hosts more attractive to herbivores, including those that commonly transmit viruses. The *Luteoviridae*, as the name suggests, often cause yellowing

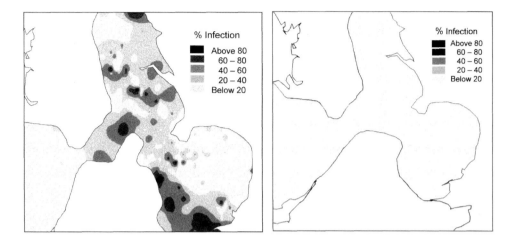

Figure 11.3 Incidence and distribution of viruses in sugar beet in 1989 (left) and 1991 (right). Virus Yellows disease is caused by a combination of *Beet yellows virus* and *Beet mild yellowing virus*.

symptoms in infected hosts and yellow is attractive to several important virus vectors (Klingauf 1987). By contrast red forms of lettuce are less attractive to aphids (Muller 1964) and thus less likely to be infected by aphid-borne viruses than conventionally coloured plants. Thus it is clear that the distribution and dissemination of viruses in wild plants is likely to be a complex of host plant, vector, virus and environment interactions.

4.2 Animals

While concern for viruses in animals is more immediately apparent to the 'person-in-the-street', outbreaks of foot and mouth disease are front-page news and nearly everyone knows of, and has a view on, rabies and the desirability or otherwise of quarantine restrictions, there have been few systematic efforts to determine the presence of viruses in wild animals. Only when large-scale deaths occur, such as those of seals in the North Sea, are any detailed studies made. The problem then is not whether or not viruses are present, but which ones are the cause of disease and death, with the possibility that none of them are. Other opportunistic testing for viruses is done for some animals killed on roads or captured for other purposes. There seems no reason why a systematic survey should not show a wide range of viruses, especially in species that have little or no direct contact with humans. For those that come into human contact, willingly or unwillingly, the concern about cross-infection has resulted in a closer examination of the viruses present. However, in addition to the often unknown movement of viruses between wild species, there is the possibility of reverse zoonotic events by which human viruses enter the wild animal population, for example through animal rescue services which depend for their income on donations and for which public access is essential. The animals, often stressed and immunologically compromised, can acquire viruses from other animals similarly afflicted as well as from their human carers and visitors. The most likely viruses to spread in this way are those which are air-borne and cause coughs, colds and influenza, but those associated with enteric diseases are also potential candidates for crossing the species barrier. Rescued animals are rarely, and sometimes for good reasons, returned to the sites from which they were collected. There is thus potential for the spread of human viruses to other wildlife and to other regions. General debilitation associated with chronic infection may be more important to wildlife and the balance between species than acute effects, but these impacts have not been addressed; neither have the consequences of virus in prey on predators and scavengers.

4.3 Other host groups

If viruses in the more obvious plants and animals are little known, knowledge of viruses in other hosts is almost non-existent. Viruses have however been recorded in all the other groups discussed in this volume. Interestingly, but undoubtedly because of their medical significance, viruses of bacteria – the bacteriophages – are probably some of the best known and most widely studied.

Little is known of viruses in freshwater organisms or in waterside animals. This is surprising considering the importance of fresh, clean water to human life, and the increased public concern about all matters related to food hygiene. Preliminary studies of Priest Pot, a small lake in Cumbria (Edwards *et al.* 1999), revealed viruses pathogenic for an indigenous green microalga and a diverse array of virus-like agents capable of lysing bacteria were cultured from the lakewater. There were also large numbers of 'orphan' virus-like particles in the water and more than half the bacterial species present contained virus-like particles. As with other examples this is only a tiny, fragmentary sample but illustrates not only the widespread distribution of viruses, but also their great diversity and our lack of knowledge of them.

5 Changes 1973–98

Given the paucity of knowledge about virus occurrence it is impossible to draw any realistic comparisons between 1973 and 1998. The most obvious and best known differences are in the presence of human viruses. Smallpox has been eradicated and measles is going the same way, but human immunodeficiency viruses (HIVs), unknown in 1973, are now widespread. As far as wildlife in the usually accepted sense is concerned there is little information, although the recent epidemic of an orthomyxo-like virus associated with infectious salmon anaemia in Scotland's farmed salmon has received much publicity. Less well reported was the mutation in the parvovirus causing distemper in dogs, which spread throughout the world in two years. This virus is now certainly present in wildlife, but has not been monitored, nor are its effects on wildlife populations known. Over the last 25 years there have been measurable successes in keeping certain viruses (for instance rabies) out of Britain, or at least preventing the import of further viruses and potentially decreasing the incidence of a virus already established (*Poplar mosaic virus* in *Populus nigra*). Viruses in domesticated vertebrates are contained by immunisation policies and, in extreme circumstances, by widespread slaughter. In 'domesticated' plants the incidence of viruses has fluctuated widely from year to year (see Fig. 11.3).

Some viruses of plants new to Britain have been reported, such as *Beet necrotic yellow vein virus* (BNYVV) first reported in 1983, and the soil-borne mosaic viruses in barley, first reported in 1980. How long these viruses had been present before they became obvious is a matter of conjecture but it seems likely that BNYVV was introduced to Britain more recently than the barley mosaic viruses. Both viruses have spread since first reported and the barley mosaic disease is now known to be caused by two distinct viruses, *Barley mild mosaic virus* and *Barley yellow mosaic virus*; the latter is now also known to occur in two forms.

Any comparison of a list of viruses in Britain in 1973, should one have existed, with one compiled in 1998, would certainly show a very large increase in their number and of distinct variants of viruses. It would be a mistake to assume that this reflected a real increase. It is much more likely to be a demonstration of the advances in techniques that have allowed the viruses to be isolated and identified.

6 Future changes

It is certain that, as a result of the development of techniques and methods of working with viruses, new ones will be identified; new viruses will also be found just by looking at previously unexplored hosts. It is more appropriate, however, to try and anticipate the effects of some outside influences on virus incidence.

6.1 Trade and policy issues

Plant seed or animal semen are known sources of some viruses and trade in them has long been the subject of scrutiny or specific exclusion policies, although these are not easy to apply against an overarching policy which aims to liberalise world trade, especially within the European Union. Most people returning to the UK from overseas know little of the restrictions on animal and plant imports, except for domestic pets and the requirement for six months' quarantine to ensure freedom from rabies and other infectious agents. Current government policy is moving to control rabies by vaccination and the veterinary profession suspects that this has the capacity to increase virus prevalence, with the possibility of rabies infecting foxes or other susceptible animals. Undoubtedly international policies and domestic pressures are likely to decrease the checks and restrictions on the inadvertent importation of viruses. It is possible that, as our knowledge of viruses in wildlife throughout the world increases, further restrictions may be appropriate but will require data on the risks and consequences of such importation to support any legislation.

6.2 National policies – land management

There is much current public concern about pesticides, many of which are used directly to restrict the spread of viruses through control of their vectors. The dipping of sheep is one use of pesticides that has received much publicity. One purpose of dipping is to control the ticks that spread louping-ill flavivirus(es) in sheep and in Britain's wildlife. Any restriction on dipping is likely to favour tick populations and this could increase louping-ill, manifest by encephalitis and temporary lameness, in sheep. Louping-ill has been known from Scotland since 1932, and there has been recent concern about tick infestations in young game birds. Many more native and introduced species have been implicated as hosts in the epidemiology of the disease (Reid 1988).

Restrictions on chemical use through a pesticide 'tax' may increase virus infection and will encourage the development of alternative methods of control. One much promoted, although relatively little exploited in the field, is biological control by viruses. The most publicly recognised of such controls is probably myxomatosis and this has led to changes in the habits of rabbits and in the virus. Recent suggestions for the use of a calicivirus to provide more effective control – up to 95% mortality can occur – raised concerns that such a release would affect other species; being stable and enteric, caliciviruses could thus be spread by predators or scavengers. Some circumstantial evidence suggests that marine mammals may be sources and that these viruses can infect reptiles, amphibians, molluscs and birds as well as mammals. However, there are already known to be one or more caliciviruses present in Britain's rabbits and hares (Chasey *et al.* 1997). Some viruses, notably the baculoviruses, are also used to control a range of lepidopteran pests and are currently the subject of efforts to enhance their effectiveness. There is concern that such modified viruses may have effects on non-target species and thus down the food chain of which they are part.

Another consequence of the pressure to decrease pesticide use and thus 'sustainability' of agriculture has been the increase in the development of transgenic crops especially, in the context of this chapter, those constructed to resist or tolerate plant viruses. While there are a range of possibilities for the production of such plants, most involve the introduction to the host plant genome of a part of the virus genome. There are a number of concerns about the method of producing such plants but the main one, for virus-tolerant plants, is the possibility that the presence of virus gene sequences in plants could lead to recombination, perhaps with unknown viruses in wild plants, to produce enhanced virus variability or prevalence, with unpredictable consequences. In animals, protective inoculation programmes may also aid the evolution of viruses of significance to intensive farm animal management.

The largely unknown status of water as a source of viruses has already been mentioned. Both population and climate pressures on water supply are likely to increase in Britain and the need for more efficient use, and hence recycling, of water will increase. Viruses are part of water quality assessments and if water is recycled even more frequently than at present it could acquire a range of picorna, hepatitis and probably rotaviruses.

6.3 Global warming

There is a generally accepted consensus that the Earth is warming, largely as a result of the emission of greenhouse gases. The extent and consequences of this warming for Britain are the subject of much debate. It seems certain that there will be additions to our wildlife as the range of various animals, especially very mobile groups such as birds and insects, extends northwards. These may well introduce new viruses, either those that infect them or, in the case of insects, are transmitted by them. It is also possible that we may lose some species who find the warmer conditions at altitude or in the north of Britain unsuitable. Certainly new arable crops, such as grain maize and sunflowers, may become viable, and existing crops will be grown further north and at greater altitude than at present, although the ability to grow many new crops in southern Britain will depend on adequate water management, with the attendant problems for viruses already mentioned.

7 Conclusions

Even from the limited knowledge of viruses affecting wildlife in Britain it seems reasonable to conclude that they are both numerous and ubiquitous. The overt presence of viruses is critical to the efficient management of crops and farm animals, and to humans. However, it seems clear that in many hosts the presence of viruses is not obvious but is often widespread. The consequences of such chronic infections are unknown but they seem likely to have influenced the ecological balance in the past and virus populations seem equally likely to influence host populations in the future. The possibility, especially in animals, of the transfer of viruses between species, including humans, seems likely to increase. Pressure on water supply as a result of industrial and domestic needs, and global warming, seem likely to both increase the number of viruses present and change their effects. What the consequences of this will be are unknown.

Novel methods of crop and animal protection may decrease the effects of viruses but may result in increased opportunities for virus 'evolution'. The future for viruses in Britain looks rosy.

REFERENCES

Chasey, D., Trout, R.C., Sharp, G. and Edwards, S. (1997) Seroepidemiology of rabbit haemorrhagic disease in wild rabbits in the UK and susceptibility to infection, in Chasey, D., Gaskell, R.M. and Clarke, I.N. (eds) *Proceedings of the 1st International Symposium on Caliciviruses.* Weybridge: European Society for Veterinary Virology, pp. 156–162.

Cooper, J.I. and Harrison, B.D. (1973) The role of weed hosts and the distribution and activity of vector nematodes in the ecology of tobacco rattle virus. *Annals of Applied Biology*, 73, 53–66.

Cooper, J.I., Massalski, P.R. and Edwards, M.-L. (1984) Cherry leaf roll virus in the female gametophyte and seed of birch and its relevance to vertical virus transmission. *Annals of Applied Biology*, 105, 55–64.

Edwards, M.-L., Cooper, J.I., Massalski, P.R. and Green, B. (1985) Some properties of a virus-like agent in *Brachypodium sylvaticum* in the United Kingdom. *Plant Pathology*, 34, 95–104.

Edwards, M.-L., Kelley, S.E., Arnold, M.K. and Cooper, J.I. (1989) Properties of a hordevirus from *Anthoxanthum odoratum*. *Plant Pathology*, 38, 209–218.

Edwards, M.-L., Timms-Wilson, T., Lilley, A. and Cooper, J.I. (1999) Bugs and their viruses in a wetland community. *NERC News*, (spring), 20–21.

Gibbs, A.J. (1980) A plant virus that partially protects its wild legume host against herbivores. *Intervirology*, 13, 42–47.

Hammond, J. (1981) Viruses occurring in *Plantago* species in England. *Plant Pathology*, 30, 237–243.

Klingauf, F.A. (1987) Biology: Host plant finding and acceptance, in Minks, A.K. and Harrewijn, P. (eds) *Aphids, Their Biology, Natural Enemies and Control*, Vol. A. Amsterdam: Elsevier, pp. 209–223.

Muller, H.J. (1964) Über die Auflingdichte von Aphiden auf farbige Salatpflanzen. *Entomologia Experimentalis Applicata*, 7, 85–102.

Raybould, A.F., Maskell, L.C., Edwards, M-L., Cooper, J.I. and Gray, A.J. (1999) The prevalence and spatial distribution of viruses in natural populations of *Brassica oleracea*. *New Phytologist*, 141, 265–275.

van Regenmortel, M.H.V., Fauquet, C.M., Bishop, D.H.L., Carstens, E.B., Estes, M.K., Lemon, S.M., Maniloff, J., Mayo, M.A., McGeoch, D.J., Pringle, C.R. and Wickner, R.B. (eds) (2000) *Virus Taxonomy, Classification and Nomenclature of Viruses*, 7th Report of the International Committee on Taxonomy of Viruses. Wien: Springer Verlag.

Reid, H.W. (1988) Louping ill, in Monath, T.P. (ed.) *The Arboviruses: epidemiology and ecology*. vol 3, Boca Raton: CRC Press, pp. 17–135.

Tomlinson, J.A., Carter, Anne L., Date, W.T. and Simpson, Carol J. (1970) Weed plants as sources of cucumber mosaic virus. *Annals of Applied Biology*, **66**, 11–16.

Chapter 12

Protozoa

Bland J. Finlay

ABSTRACT

A brief account is given of the key characteristics of free-living *Protozoa*. The point is stressed that most species are probably ubiquitous, that their global number is modest, and that a distinctive British protozoan fauna probably does not exist. The *Protozoa* found in a particular habitat type in Britain are also found in the same habitat type wherever that exists worldwide – thus it is the habitat which selects the protozoan species that will thrive.

A short historical account of the development of protozoology concludes by establishing the status of the study of *Protozoa* in Britain. British research output is relatively strong, but there are significant gaps in identification skills – especially with respect to the sarcodines, and some flagellate groups. This situation has arisen largely as a consequence of the retirement, death and emigration of specialists in recent years, and the lack of recruitment of younger staff. There has also been a significant decline in the number and breadth of taught courses in protozoology in Britain over the past 25 years. The prognosis however, is that research on the 'biodiversity' of *Protozoa* will develop, partly because of interest in *Protozoa* as experimental tools for the investigation of the meaning and importance of biodiversity, and because of the economic importance of *Protozoa* as the principal grazers of other microbes in the biosphere. The common ground between taxonomy and ecology of *Protozoa* is considered fertile territory for fundamental research to tackle key questions such as 'what is a species?', 'are molecular techniques useful in refining species concepts?', 'to what extent are genotypic and phenotypic characters correlated in *Protozoa*?' and 'what is the ecological significance of sibling species?'

1 Introduction

The *Protozoa* failed to secure an entry in *The Changing Flora and Fauna of Britain* (Hawksworth 1974). At that time there was no substantial body of relevant information, there were no mapping schemes and checklists of species existed for only a few groups. The current picture is not significantly different, although the taxonomic skills that are available – at least with respect to the ability to confidently identify the *Protozoa* living in Britain – are probably in worse shape today. On the other hand, our understanding of the key characteristics and likely dimensions of protozoan 'biodiversity' in Britain (and globally) is probably better now than it has ever been.

Protozoa are microscopic (generally 5–500 μm) organisms with animal-like features, and all free-living *Protozoa* are phagotrophic organisms that specialise in feeding on other micro-organisms. Each *protozoon* typically exists as an independent cell, but in some species these join to form colonies, as in peritrich ciliates and choanoflagellates. In other species, such as slime moulds, cells coalesce to form plasmodia. Some *Protozoa* can also photosynthesise because they have chloroplasts, or because they support photosynthetic endosymbionts, but it is the key attribute of phagotrophy that defines them as *Protozoa* and dictates their fundamental ecological role.

All *Protozoa* are aquatic, including those that live in soil. They are typically very abundant. One gram of soil will contain 10^3–10^7 amoebae; 10^5 planktonic foraminiferans will normally be found beneath 1 m^2 of oceanic water; every millilitre of fresh and sea water on the planet supports at least 100 heterotrophic flagellates, and these numbers are at least one order of magnitude greater in marine and lake sediments.

Because of their great abundance, *Protozoa* are quantitatively the most important grazers of microbes, and the heterotrophic flagellates alone can probably consume all bacterial production in the aquatic environment. protozoan grazing of bacteria and other microbes also stimulates the rate of organic matter decomposition. This may operate by increasing the rate of turnover of essential nutrients that would otherwise remain 'locked up' in microbial biomass.

Many *Protozoa* are microaerobic – that is to say they seek out habitats with a low level of dissolved oxygen. As many bacteria are microaerobic also, this behaviour tends to bring many types of *Protozoa* into contact with elevated abundances of microbial food. It also facilitates the maintenance of nutritional endosymbionts such as green algae, which benefit from being located in opposing gradients, for example oxygen and light on the one hand, and carbon dioxide, hydrogen sulphide and other reductants on the other (e.g. Finlay *et al.* 1996b). Many *Protozoa*, perhaps the majority, have symbiotic micro-organisms living inside them and on their external surfaces. These include sulphide-oxidising bacteria, photosynthetic non-sulphur purple bacteria, methanogens and sulphate reducers (see Fenchel and Finlay 1995) and complex consortia consisting of multiple bacterial types (Esteban *et al.* 1993).

Many micro-aerobic *Protozoa* are facultative anaerobes (Bernard and Fenchel 1996), but unlike the 'true' anaerobes (see below) their metabolism is fundamentally aerobic. However, a great variety of free-living *Protozoa* are true anaerobes, and for these, oxygen is toxic. These species live principally in freshwater and marine sediments. Most use hydrogen-evolving fermentations for energy generation, and in most cases the hydrogen is used as a substrate by endosymbiotic methanogenic bacteria, so that methane is released from these protozoan 'consortia'. *Protozoa* are probably the only phagotrophic organisms capable of living permanently in the absence of oxygen, so anaerobic food chains exist only because *Protozoa* exist.

One likely consequence of the extraordinary abundance of *Protozoa*, and the absence of effective barriers to their continuous geographical dispersion, is that migration rates will be relatively high. Thus rates of speciation and extinction will be low, the global number of species will be modest, and local species richness will be a relatively large proportion of global species richness (Fenchel 1993). This is now supported by much evidence (Finlay 1998). Notably, the total number of extant species of free-living *Protozoa* in the biosphere is believed to lie in the range 10 000–15 000 species (Finlay 2000) and local species richness is indeed a large proportion of global richness. The number of species of ciliated *Protozoa* recorded so far from a 1 ha pond in the English Lake District represents roughly 25% of all ciliate species ever recorded from fresh waters worldwide (Finlay 1998), and more than 75% of the global number of species in the flagellate genus *Paraphysomonas* have been found in < 0.1 cm^2 of sediment from the same pond (Finlay and Clarke 1999a, b).

Although most protozoan species are probably ubiquitous, in the natural world only a limited number of protozoan niches are available at any moment in time, so the very rare or dormant species (often accounting for approximately 90% of those present) are difficult to find until suitable conditions are created for them. These conditions are mostly created through reciprocal interactions involving a variety of biological and non-biological factors. This process can be observed in the natural environment (Finlay *et al.* 1997; Finlay and Esteban 1998) and it can be demonstrated experimentally (Finlay *et al.* 1996a; Fenchel *et al.* 1997).

We can be reasonably sure that a distinctive 'British' protozoan fauna does not exist. It is also unlikely that there are any British endemic species, for a protozoan species will thrive wherever in the world it finds a suitable habitat. The *Protozoa* living in similar habitats in England and Australia can be predicted to support the same species of *Protozoa* – indeed this has recently been shown to be the case (Lee and Patterson 1998; Finlay *et al.* 1999; Esteban *et al.* 2000).

2 Historical outline

If we seek some historical perspective on the development of knowledge of the *Protozoa* living in Britain, it is necessary in the first place to establish exactly what is meant by the term. It is actually rather recent, having crept slowly into common usage only in the latter half of the nineteenth century. There has never been unanimous agreement as to where the boundaries of the *Protozoa* might lie, and there are those who would even question the propriety of the term. It is also necessary to disentangle the '*Protozoa*' from the '*Infusoria*' – 'this intangible and omnipresent group of organisms' (Kent 1880–82). The term '*Infusoria*' was introduced by Ledermüller in Nuremberg in 1763, to include all those microscopic animals that appeared in water in which hay had been steeped (i.e. the 'infusion') for several days. Many of these organisms would originally have been present as spores and cysts, so the '*Infusoria*' that appeared typically included rotifers and other metazoans, as well as algae. At various times, and in the hands of different workers, the '*Infusoria*' have included ciliates, flagellates, amoeboid *Protozoa*, *Hydra*, rotifers, desmids, diatoms and bacteria. The two early pre-eminent works on this diverse group are the *Animalcula Infusoria* by the Danish naturalist O.F. Müller (published posthumously in 1786), and C.G. Ehrenberg's *Die Infusionsthierchen* (1838). Other large works followed. Dujardin (1841) questioned the accuracy of some of Ehrenberg's interpretations (such as that foraminifera are unicellular), and uncertainty continued to surround the unicellular character of the *Infusoria* in the works of Claparède and Lachmann (1858–61) and others. The clearest assertion that the infusoria should embrace only those organisms that can be reduced to the state of a single cell, is attributable to Carl Theodor von Siebold who, in 1845, gathered together all the unicellular animal-like organisms within a subkingdom of the *Invertebrata*. He named the subkingdom '*Protozoa*' (a term previously introduced by Goldfuß in 1818 for a group that also included some sponges and bryozoans), and divided it into the *Rhizopoda* (the amoebae), and the *Infusoria* (the ciliates and flagellates).

The second half of the nineteenth century was a period of sustained exploration of protozoan diversity in the natural environment, but particularly in continental Europe. Notable examples from this period would include Perty's (1852) account of the algae, *Protozoa* and microfauna collected in the Bernese Alps, Stein's (1867) monograph on the heterotrich ciliates, and his 'Infusionsthiere' of 1878. Ernst Haeckel, in addition to assuming the role of vociferous champion in Europe for 'Darwinism', revealed for the first time (1887) the extraordinary diversity of the *Radiolaria* from the open ocean (Haeckel 1877). He also created a new third kingdom – the *Protista* – a systematic concept based on his appreciation of evolutionary phylogeny, and designed to hold the *Protozoa*, algae and 'lower' fungi. The term '*Protista*' was originally proposed in 1866; it enjoyed less than wholehearted acceptance during the following century and is now experiencing something of a renaissance (as one of many terms and concepts currently jockeying for general acceptance). It is now considered unlikely that a single-kingdom *Protista* can contain the diversity of taxa that comprise the 'lower' eukaryotes (see Corliss 1998a).

By the beginning of the twentieth century the study of free-living *Protozoa* had become a truly international effort, although the impressive European tradition was maintained with a steady flow of both seminal and comprehensive works – notably those of Fauré-Fremiet (see Corliss 1998b) in France and, in the 1920s, by Kahl in Germany (1930–35). But outside of Europe too there was growing interest in the free-living *Protozoa*, especially in North America (Stokes 1888) and in Russia (Schewiakoff 1893). Corliss (1979a) documents much of this work, and provides biographical sketches (Corliss 1978, 1979b).

In Britain in the nineteenth century, the first notable publication was Pritchard's *A History of the Infusoria* (1861; the first edition having appeared in 1834). This, however, is largely a compilation of the views of Ehrenberg, Dujardin and others. Kent's substantial *A Manual of the Infusoria* was published between 1880 and 1882. His 'infusoria' are roughly equivalent to the modern *Protozoa*, although he gives special emphasis to the flagellates, including many photosynthetic chrysomonads, euglenids and dinoflagellates. So far as the investigation of the natural history of free-living *Protozoa* is concerned, Kent was probably the first British protozoologist of any standing. He was based at what is now the Natural History Museum in London and his explorations appear to have been fuelled by 'an abundant supply of living material for investigation', from 'Mr Thomas Bolton of Birmingham, and Mr John Hood of Dundee'. His monumental work may safely be regarded as the first substantial contribution to the study of *Protozoa* in Britain. Many of the organisms that he and other naturalists described were being found by other workers in other parts of the world, and the apparent widespread acceptance at that time of the ubiquity of protozoan species – an idea that has re-emerged only in recent years – is a point that is also stressed in the present article.

Kent was far from impressed by the contribution of 'English investigators' to 'our knowledge of infusorial life', and all of the learned works that he relied upon were written by authors in central Europe – specifically in France, Germany and Switzerland. Why should this have been so? One possibility is that advances in the study of *Protozoa* were closely related to the history of developments in microscopy, and specifically to the production of objective lenses offering high contrast and resolution. By 1830, some microscopes could resolve particles as small as one micrometre (the size of many bacteria), but not until the 1880s did lens designs correct for chromatic aberration. The first phase-contrast objectives appeared in 1932, and interference contrast microscopy dates from the early 1960s. To a large extent the quality of microscopes dictated what could be discovered about the diversity of form and function in *Protozoa*, so it is perhaps not surprising that protozoan taxonomy has been in a state of continuous revolution since Leeuwenhoek manufactured the first crude microscopes in the late seventeenth century. It is widely acknowledged that the best-quality microscopes manufactured in the nineteenth century came from continental Europe; we might imagine therefore that British microscopists experienced difficulty in getting hold of adequate equipment, and that this impeded progress, although there is little if any evidence to support this idea. On the contrary, there is abundant evidence that microscopy in Britain was alive and well in the period before the 1880s. Enthusiasts such as Henry Baker, for example, in works published in 1742 (*The Microscope Made Easy*) and in 1753 (*Employment for the Microscope*), provided the first descriptions of many of the organisms that developed in 'infusions', including the unmistakable ciliate *Lacrymaria olor* (named *Proteus* by Baker). And towards the end of the nineteenth century it is clear that many enthusiastic amateurs had access to the best equipment available at the time. Nelson (1889a) writes of 'Professor Abbé and Dr Zeiss … [that] … their new half-inch object glass may be justly termed the highest achievement of optical science'.

We may perhaps gain a more relevant insight into the status of the scientific investigation of *Protozoa* in the nineteenth century from some of Nelson's other observations. Referring to a condenser made by Andrew Ross in 1831, he says:

> We must remember by whom the microscope was used at that time. As far as this country was concerned, it was merely looked upon as a philosophical toy. It was principally to be found in the hands of a few dilettanti; science of every kind was tabooed; the microscope being placed at the lowest end of the scale.
>
> (Nelson 1889b)

And there, perhaps, we have it: enthusiasts succeeded in raising amateur interest in microscopy to a very high level, whereas professional scientists expressed little interest in what such 'toys' could reveal about free-living *Protozoa* in the natural environment. While professional protozoologists in Germany were writing the first textbooks on protozoology (Bütschli 1880–89), establishing zoological institutes (such as those founded by Bütschli in Heidelberg in 1878 and by Hertwig in Munich) and launching the first scientific journal in the area (*Archiv für Protistenkunde*, by Schaudinn in 1902), amateurs in Britain seem for the most part to have been occupied in describing in exquisite detail virtually anything they could recognise using a microscope – although their most passionate interest seems to have been reserved for the new technical developments in optics.

Whatever may have caused the slow development of scientific interest in the free-living *Protozoa* in Britain in the nineteenth century, the same cannot be said of the twentieth – a period of significant advances in Britain and elsewhere. For example, the origins of soil protozoology can be traced to the work of Russell and Hutchinson at Rothamstead around 1910 (Sandon 1927). At that time, it was suspected that *Protozoa* might influence the fertility of soil, but this was difficult to pursue because very little was known about the identities and activities of soil *Protozoa*.

Other major advances in Britain were fuelled by economic factors, such as unravelling the role of *Protozoa* as agents responsible for the clarification of liquor in sewage treatment processes (Curds 1975), and the description and quantification of methane-generating protozoan consortia in the anaerobic layers of municipal landfill sites (Finlay and Fenchel 1991). Among many other 'British' historical examples is Picken's (1937) attempt to paint the first holistic view of protozoan community structure, while Mare (1942) carried out the first study in Britain of the role of *Protozoa* in the marine benthic food chain.

The most prominent recent example of a truly international effort directed at studying *Protozoa* in the natural environment, and in which there has also been a large British contribution (e.g. Berninger *et al.* 1991; Sleigh and Zubkov 1998) is the work carried out over the last 15–20 years which identified the importance of planktonic *Protozoa* in controlling bacterial abundance – giving these microscopic grazers a pivotal role in the 'microbial loop' (see Fenchel 1987).

There have, however, been no extensive taxonomic studies of the occurrence and distribution of *Protozoa* in specific habitats in Britain, apart from the almost unbroken study of freshwater *Protozoa* in the Lake District since the 1950s, which is the nearest thing we have to a 'long-term study' (Webb 1961 ; Goulder 1974; Finlay and Esteban 1998). Apart from some isolated examples which vary greatly in their coverage of specific taxonomic groups (e.g. benthic foraminifera, Murray 1971, 1979; testate amoebae, Ogden and Hedley 1980; Heal 1964; naked amoebae, Cash *et al.* 1905–21; Page 1976, 1988; and marine dinoflagellates, Dodge 1982; and ciliates, Curds 1982; Curds *et al.* 1983) no attempts have been made to document a 'British' protozoan fauna.

3 The status of research on *Protozoa*

We have attempted to determine the current status of scientific research on free-living *Protozoa* in Britain. We have also quantified the relative importance of such research, and we have collated the opinions of 20 British protozoologists.

An analysis of all published work indicates that the quantity of recent research published in Britain is at least as great as that from any other European country (Table 12.1).

The first search was on 'Protozoa', plus each of the countries listed. Note that searching by countries is done using the address of the corresponding author of each reference; and as this information is not always included in the initial abstracting procedures, the absolute numbers shown in the Table are almost certainly lower than the real numbers. However, for the purpose of making a comparison between countries, the numbers are useful. Separate searches were performed for parasitic and for fossil *Protozoa*. The combined totals for these were then subtracted from 'All Protozoa' to yield derived estimates for 'Free-living *Protozoa*'.

In order to obtain a simple measure of the relative importance of research on free-living *Protozoa*, we searched for published work featuring 'fish' emanating from countries in the same list. In comparing the research output on *Protozoa* with that for fish, it is clear that the protozoan output is far from insignificant. In the case of the USA and UK, it accounts for about one-third of the output of fish. As expected, research on fish

Table 12.1 The number of published research articles, arranged by subject, abstracted by the Zoological Record in the period 1981–98.

Country	All Protozoa	Free-living Protozoa (a)	Fish (b)	(a/b) %
USA	1491	1049	3431	31
UK	485	323	818	39
Germany	398	300	425	71
France	334	249	377	66
Russia	264	185	475	39
Canada	252	178	962	19
Spain	223	152	303	50
Italy	201	125	209	60
Denmark	93	71	68	104
Poland	96	69	160	43

The first search was on 'Protozoa', plus each of the countries listed. Note that searching by countries is done using the address of the corresponding author of each reference; and as this information is not always included in the initial abstracting procedures, the absolute numbers shown in the table are almost certainly lower than the real numbers. However, for the purpose of making a comparison between countries, the numbers are useful. Separate searches were performed for parasitic, and for fossil Protozoa. The combined totals for these were then subtracted from 'All Protozoa' to yield derived estimates for 'Free-living Protozoa'.

Table 12.2 Analysis of published references dealing specifically with the ecology of free-living Protozoa (excluding truly marine forms), dated within the period 1910–81 (see Finlay and Ochsenbein-Gattlen 1982), arranged by geographical location of the lead author).

	Britain[1]	America[2]	Other[3]
Fresh waters	92 (74,18)[4]	153	529
Terrestrial	90 (74,16)	100	237
Inland saline waters	10 (8,2)	48	118
Waste treatment	60 (55,5)	45	127
General & reviews	78	165	128
Totals	330	511	1139

[1] Including Ireland.

[2] Includes North and South America, but USA and Canada account for the great majority of references.

[3] Including the rest of Europe (most references originate from Germany, France and central Europe) with the remainder largely from Russia, Japan and Australasia.

[4] the first number in parenthesis is the number of references to work in Britain by British workers; the second, to work in other places in the world (e.g. the tropics, or Antarctica) by British workers.

is relatively more important in Canada (consider the strength and reputation of the Fisheries Research Board of Canada). In Germany, the output from research on *Protozoa* is apparently greater than that from research on fish.

The relative strength of British research output on free-living *Protozoa* is also supported by a bibliographic survey which focused specifically on published research in the area of free-living *Protozoa*, during the period 1910–81 (Finlay and Ochsenbein-Gattlen 1982) (Table 12. 2).

A questionnaire was sent to 20 protozoologists who are, or were recently, in established positions in the UK, and working in the general area of free-living *Protozoa*. Four of them have a bias towards marine *Protozoa* and one is a micropaleontologist. They were asked to restrict their considerations to the free-living *Protozoa* living in freshwater and terrestrial environments in Britain. There were 17 returns.

Question 1   Considering only the ability to identify *Protozoa* (phagotrophic microbial eukaryotes) with the aid of a light microscope, do you think there are in the UK at the present time, individuals who can confidently identify ciliates, sarcodines and flagellates?
Answers: Numbers of respondents replying 'yes' or 'no' with respect to each of the three main taxonomic groups.

	Yes	No
Ciliates	17[a]	0
Sarcodines	8	9
Flagellates[b]	14	3

[a] All three taxonomic groups are very broad. The point was made repeatedly that many individual protozoologists are able to confidently identify some species within parts of each large taxon, while lacking confidence with the taxon as a whole.

[b] Heterotrophic only.

Question 2   Do you think there are significant gaps or strengths within any of these broad taxonomic groups?
Answers: numbers of respondents specifying particular 'gaps' and 'strengths'.

	Gaps	Strengths
Ciliates	2 (marine species)	3 (freshwater species)
Sarcodines	14 (small naked amoebae) 2 (sarcodines in general) 3 (testate amoebae)	2 (opportunistic pathogens, especially *Acanthamoeba* and *Naegleria*)
Flagellates	7 (flagellates in general – but this is largely because the taxonomic diversity is still being explored)	2 (choanoflagellates, kinetoplastids and some soil flagellates)
Others	3 (soil *Protozoa* generally)	

Question 3 Do you believe there have been significant changes in the human resources available to identify *Protozoa* using traditional (i.e. non-molecular) methods over the past 25 years?

Answers: Numbers of respondents (and their principal comments).

Decline	Increase
16 (due to retirement and death, emigration, and lack of recruitment of younger staff)	1 (in the number of micropaleontologists)

Question 4 Do you know of any taught courses in British universities and colleges, that are devoted specifically to the biology/ecology/taxonomy of *Protozoa* or protists? Do you have an opinion on how the quality and quantity of teaching in this area has changed in the past 25 years?

Answers: Eight respondents did not know of any taught courses. The other respondents provided the following information. University College London offers a course on *Protozoa* for first-year students (biologists, zoologists and microbiologists). It was felt that this course had improved in quality and scope, and had remained popular, over the period of the past 25 years. An MSc in micropalaeontology is also offered at UCL. At Surrey University, a 10-hour course on *Protozoa* is given by staff from the Natural History Museum (London). Twenty years ago, the duration of this course was 16–18 hours. At Nottingham University there is no separate course on Protozoology, but *Protozoa* feature in microbiology courses and in a microbial ecology module. At Glasgow University, protozoology teaching has been reduced from eight lectures to one lecture in the second year. Twenty-five years ago, in the Zoology department at Glasgow, there was a third-year course lasting 5 weeks, and a two-term fourth-year option on protozoology. Neither of these is currently provided.

Ten respondents offered opinions on the quality and availability of teaching in protozoology in British universities at the present time. Eight of them indicated that there had been a significant decline. The other two expressed the opinion that standards had continued to improve at UCL. The limited body of evidence available does indicate that there has been a significant decline in the teaching of protozoology in Britain over the past 25 years. What are the reasons for this?

One contributory reason may be the overwhelming chaos and continuous reorganisation of the higher level taxonomy of *Protozoa*. There is no widespread agreement on the taxonomic boundaries of the *Protozoa*, and the status of the taxon may have been further eroded by being dissolved within the 'Protista' (or even 'Protoctista'). If students believe Protozoology is ill-defined, with a declining status in the teaching curricula of universities, they cannot be blamed for doubting that it could one day provide them with a career.

Second, in the current economic climate, the continued survival of a university course is largely dependent on sufficient numbers of students taking that course. It is not obviously the case that 'protozoology' has broad appeal to students, and that new courses should be created to cater for that demand.

Third, laboratory practical classes on *Protozoa* (especially those that are parasites or opportunistic pathogens) have become difficult to conduct due to increased costs, new Health and Safety Regulations, and possibly, some ethical considerations.

Finally, one possible reason for the low demand for protozoology courses is that the study of natural history in universities now appears to be less popular than that of molecular biology, and in particular, the study of the applications of molecular technologies. Additionally, undergraduate practical classes are now much larger than they used to be, and so it may be more difficult for students to gain access to a microscope of reasonable quality, and to become acquainted with *Protozoa* (although excellent introductory illustrated texts are now available; see for example Patterson and Hedley 1992). It is quite likely that most of those protozoologists who are now close to retirement were first fascinated with *Protozoa* when a sympathetic teacher took them on one side, and showed them something remarkable (and living) on a microscope slide.

4 Prognosis for the next 25 years

1   There is likely to be continued exploration of the apparent global ubiquity of protozoan species. Much of the motivation for this research will continue to be generated in Britain, but it will be fuelled by increased international scientific collaboration. Confirmation of the absence of protozoan biogeographies will mean that discoveries about the activities of specific protozoan communities in Britain (e.g. stimulation of the carbon flux in soil, activated sludge, or in aquatic microbial food webs) may also apply directly to communities with the same species composition, living in the same habitat type, in many other parts of the world.

2   Greater use is likely to be made of protozoan communities in the experimental investigation of the importance of 'biodiversity' – in particular, the relevance of the latter to the stability and productivity of ecosystems. This research has already begun (e.g. McGrady-Steed *et al.* 1997). But protozoan (and microbial) 'biodiversity' is quite different in character to that of macroscopic animals and plants (e.g. with respect to the ubiquity of individual species) and the most fruitful explorations in this area will be those that seek better understanding of protozoan (and microbial) diversity as integral components of ecosystem function.

3   As soil *Protozoa* are now recognised as being economically important, it is likely that more attention will be devoted to obtaining solid information about their identities, population dynamics and general ecology.

4   It is unlikely that 'mapping schemes' will be initiated for free-living *Protozoa* in Britain.

5   The common ground between taxonomy and ecology of *Protozoa* is fertile territory for scientific research, and many fundamental questions need to be answered. What is a species? Can molecular techniques be useful in refining species concepts? To what extent are genotypic and phenotypic characters correlated in *Protozoa*? Can we define a protozoan 'niche', and what is the ecological significance of sibling species?

ACKNOWLEDGEMENTS

I should like to thank the following for assistance in providing information and helpful criticism: Peter Burkill, Colin Curds, John Corliss, John Darbyshire, Jackie Parry, Andrew Gooday, Fionna Hannah, Harriet Jones, Johanna Laybourn-Parry, Barry Leadbeater, Ray Leakey, John Murray, Franco Novarino, David Patterson, Terry Preston, Andrew Rogerson, Michael Sleigh, Humphrey Smith, Sue Tong, Keith Vickerman, Alan Warren; and thanks to Ian McCulloch and Carolyn Williams, the librarians at Ferry House.

REFERENCES

Bernard, C. and Fenchel, T. (1996) Some microaerobic ciliates are facultative anaerobes. *European Journal of Protistology*, **32**, 293–297.

Berninger, U.-G., Finlay, B.J. and Kuuppo-Leinikki, P. (1991) Protozoan control of bacterial abundances in fresh water. *Limnology and Oceanography*, **36**, 139–147.

Bütschli, O. (1880–89) *Protozoa*, in Bronn, H.G. (ed.) *Klassen und Ordnungen des Tierreichs*. Heidelberg: Winter.

Cash, J., Wailes, G.H. and Hopkinson, J. (1905–21) *The British Freshwater Rhizopoda and Heliozoa*, vols I–V. London: Ray Society.

Claparède, E. and Lachmann, J. (1858–61) Etudes sur les infusoires et les rhizopodes. *Mémoires de l'Institut National Genévois*, **5**, 1–260, **6**, 261–482, **7**, 1–291.

Corliss, J.O. (1978) A salute to fifty-four great microscopists of the past: a pictorial footnote to the history of protozoology. Part I. *Transactions of the American Microscopic Society*, **97**, 419–458.

Corliss, J.O. (1979a) *The Ciliated Protozoa*, 2nd edn. Oxford: Pergamon.

Corliss, J.O. (1979b) A salute to fifty-four great microscopists of the past: a pictorial footnote to the history of protozoology. Part II. *Transactions of the American Microscopic Society*, **98**, 26–58.

Corliss, J.O. (1998) Biodiversity of the cilioprotists and the contributions of Fauré-Fremiet to the field. *Protist*, **149**, 277–290.

Corliss, J.O. Haeckel's Kingdom Protista and current concepts in systematic protistology. *Stapfia*, **56**, 85–104.

Curds, C.R. (1975) *Protozoa*, in Curds, C.R. and Hawkes, H.A. (eds) *Ecological Aspects of Used-Water Treatment*. London: Academic Press, pp. 203–268.

Curds, C.R. (1982) *British and Other Freshwater Ciliated Protozoa, Part I*. Synopses of the British Fauna No. 22. Cambridge: Cambridge University Press.

Curds, C.R., Gates, M.A. and Roberts, D.McL. (1983) *British and Other Freshwater Ciliated Protozoa, Part II*. Synopses of the British Fauna No. 23. Cambridge: Cambridge University Press.

Dodge, J.D. (1982) *Marine Dinoflagellates of the British Isles*. London: The Stationery Office.

Dujardin, F. (1841) *Histoire Naturelle des Zoophytes Infusoires*. Paris: Librairie Encyclopédique de Roret.

Ehrenberg, C.G. (1838) *Die Infusionsthierchen als vollkommene Organismen*. Leipzig: Leopold Voss.

Esteban, G., Guhl, B.E., Clarke, K.J., Embley, T.M. and Finlay, B.J. (1993) *Cyclidium porcatum* n. sp.: a free-living anaerobic scuticociliate containing a stable complex of hydrogenosomes, eubacteria and archaeobacteria. *European Journal of Protistology*, **29**, 262–270.

Esteban, G., Finlay, B.J., Olmo, J.L. and Tyler, P.A. (2000) Ciliated *Protozoa* from a volcanic crater-lake in Victoria, Australia. *Journal of Natural History*, **34**, 159–189.

Fenchel, T. (1987) *The Ecology of Protozoa*. Madison: Science Tech. Publishers/Springer-Verlag.

Fenchel, T. (1993) There are more small than large species? *Oikos*, **68**, 375–378.

Fenchel T., Esteban G.F., and Finlay B.J. (1997) Local versus global diversity of microorganisms: cryptic diversity of ciliated *Protozoa*. *Oikos*, **80**, 220–225.

Fenchel, T. and Finlay, B.J. (1995) *Ecology and evolution in anoxic worlds* [Oxford Series in Ecology and Evolution]. Oxford: Oxford University Press.

Finlay, B.J. and Ochsenbein-Gattlen, C. (1982) *Ecology of Free-Living Protozoa – a Bibliography*, occasional publication no.17. Ambleside: Freshwater Biological Association.

Finlay, B.J. and Fenchel, B.J. (1991) An anaerobic protozoon with symbiotic methanogens, living in municipal landfill. *FEMS Microbial Ecology*, **85**, 169–180.

Finlay, B.J., Esteban, G.F. and Fenchel, T. (1996a) Global diversity and body size. *Nature*, **383**, 132–133.

Finlay, B.J., Maberly, S.C. and Esteban, G. (1996b) Spectacular abundance of ciliates in anoxic pond water: contribution of symbiont photosynthesis to host respiratory oxygen requirements. *FEMS Microbial Ecology*, **20**, 229–235.

Finlay, B.J., Maberly, S.C. and Cooper, J.I. (1997) Microbial diversity and ecosystem function. *Oikos*, **80**, 209–213.

Finlay, B.J. (1998) The global diversity of *Protozoa* and other small species. *International Journal of Parasitology*, **28**, 29–48.

Finlay, B.J. (2000). *Protozoa*, in Levin, S.A. (ed.) *Encyclopedia of Biodiversity*. San Diego: Academic Press.

Finlay, B.J. and Esteban, G.F. (1998) Planktonic ciliate species diversity as an integral component of ecosystem function in a freshwater pond. *Protist*, **149**, 155–165.

Finlay, B.J. and Clarke, K.J. (1999a) Ubiquitous dispersal of microbial species. *Nature*, **400**, 828.

Finlay, B.J. and Clarke, K.J. (1999b) Apparent global ubiquity of species in the protist genus *Paraphysomonas*. *Protist*, **150**, 419–430.

Finlay, B.J. , Esteban, G.F., Olmo, J.L. and Tyler, P.A. (1999) Global distribution of free-living microbial species. *Ecography*, **22**, 138–144.

Goulder, R. (1974) The seasonal and spatial distribution of some benthic ciliated *Protozoa* in Esthwaite Water. *Freshwater Biology*, **4**, 127–147.

Haeckel (1887) Report on the Radiolaria collected by HMS *Challenger* during the years 1873–1876. *Challenger Scientific Reports Zoology*, **18** (Part 1): 1–888; **18** (Part 2): 889–1893.

Hawksworth, D.L. (ed.) (1974) *The Changing Flora and Fauna of Britain*. London: Academic Press.

Heal, O.W. (1964) Observations on the seasonal and spatial distribution of *Testacea* (*Protozoa: Rhizopoda*) in *Sphagnum*. *Journal of Animal Ecology*, **33**, 395–412.

Kahl, A. (1930–1935) Urtiere, oder *Protozoa*. I. Wimpertiere oder Ciliata (Infusoria), in Dahl, F. (ed.) *Die Tierwelt Deutschlands*. Jena: Gustav Fischer.

Kent, W.S. (1880–82) *A Manual of the Infusoria*, 3 vols. London: David Bogue.

Lee, W.J. and Patterson D.J. (1998) Diversity and geographic distribution of free-living heterotrophic flagellates – analysis by PRIMER. *Protist*, **149**, 229–244.

Mare, M.F. (1942) A study of a marine benthic community with special reference to the microorganisms. *Journal of the Marine Biological Association UK*, **25**, 517–554.

McGrady-Steed, J., Harris, P.M. and Morin, P. (1997) Biodiversity regulates ecosystem predictability. *Nature*, **390**, 162–165.

Müller, O.F. (1786) *Animalcula Infusoria Fluviatilia et Marina*. Stockholm and Leipzig.

Murray, J.W. (1971) *An Atlas of British Recent Foraminiferids*. London: Heinemann.

Murray, J.W. (1979) British Nearshore Foraminiferids, in Kermack, D.M. and Barnes, R.S.K. (eds) *Synopses of the British Fauna* (new series), no.16. London: Academic Press.

Nelson, E.M. (1889a) On a method of detecting certain spurious diffraction images, and on the new half-inch apochromatic object-glass of Zeiss. *Journal of the Quekett Microscopical Club*, 2nd series, **4**, 55–58.

Nelson, E.M. (1889b) The substage condenser: its history, construction, and management; and its effect theoretically considered. *Journal of the Quekett Microscopical Club*, 2nd series **4**, 116–136.

Ogden, C.G. and Hedley, R.H. (1980) *An Atlas of Freshwater Testate Amoebae*. London: British Museum (Natural History) and Oxford University Press.

Page, F.C. (1976) *An Illustrated Key to Freshwater and Soil Amoebae*. Scientific Publications of the Freshwater Biological Association, no. 34. Ambleside: Freshwater Biological Association.

Page, F.C. (1988) *A New Key to Freshwater and Soil Gymnamoebae*. Ambleside: Freshwater Biological Association.

Patterson, D.J. and Hedley, S. (1992) *Free-Living Freshwater Protozoa – a Colour Guide*. London: Wolfe.

Perty, M. (1852) *Zur Kenntniss kleinster Lebensformen*. Berne: Jent & Reinert.

Picken, L.E.R. (1937) The structure of some protozoan communities. *Journal of Ecology*, **25**, 368–384.

Pritchard, A. (1834, 1845, 1852, 1861) *A History of the Infusoria*, 4 editions. London: Whittaker & Co.

Sandon, H. (1927) *The Composition and Distribution of the Protozoan Fauna of the Soil*. Edinburgh: Oliver & Boyd.

Schewiakoff, W. (1893) Über die geographische Verbreitung der Süsswasser-Protozoen, *Mémoires Academy Imperial Sciences, St. Pétersburg* (sér. 7), vol. 41.

Sleigh, M.A. and Zubkov, M.V. (1998) Methods of estimating bacterivory by Protozoa. *European Journal of Protistolology*, **34**, 273–280.

Stein, F.R. (1867) *Der Organismus der Infusionsthiere nach eigenen Forschungen in Systematischer Reihenfolge Bearbeitet*, II. Leipzig: Engelmann.

Stokes, A.C. (1888) A preliminary contribution towards a history of the fresh-water infusoria of the United States. *Journal of the Trenton Natural History Society*, **1**, 71–344.

Webb, M.G. (1961) The effects of thermal stratification on the distribution of benthic *Protozoa* in Esthwaite Water. *Journal of Animal Ecology*, **30**, 137–151.

Chapter 13

Freshwater invertebrates

John F. Wright and Patrick D. Armitage

ABSTRACT

Since 1974, significant advances have been made in our knowledge of the distribution of freshwater invertebrates and many species previously unknown in Great Britain or Ireland have been recorded as a result of comprehensive surveys, new keys and taxonomic expertise. We now have more comprehensive information from which to observe future change.

Some introduced species have continued their spread within Britain, for example the American flatworm *Dugesia tigrina*, while others have appeared for the first time. In Ireland, the recent appearance of the zebra mussel is being closely monitored. Unfortunately, the introduction of the signal crayfish into England in 1976 brought with it a fungal infection, the crayfish plague, which poses a major threat to native white-clawed crayfish in Britain and Ireland.

A number of other species are also under threat for a variety of reasons, despite recent legislation. There is growing concern over the status of the pearl mussel as a result of over-fishing and poor recruitment.

Our increasing knowledge of the native fauna and specific actions being taken for species protection are positive signs for the future, but must be seen against continued pressure on water resources and the potential for further problems due to climate change.

1 State of knowledge

Our present knowledge of the freshwater invertebrate fauna of Great Britain and Ireland is probably as detailed as any country in the world. This is, in part, a consequence of the limited species richness of the British fauna compared with continental Europe. More important, however, is the long-term dedication and enthusiasm of both amateur naturalists and professional biologists in collecting, identifying and studying the freshwater fauna. Interest in some taxonomic groups can be traced back to the last century, but without doubt the scientific publications of the Freshwater Biological Association (FBA) over the last 60 years have set the standard for the production of reliable keys to many groups of freshwater organisms.

T.T. Macan of the FBA, who contributed the chapter on freshwater invertebrates in *The Changing Flora and Fauna of Britain* (Macan 1974), produced keys to a wide range of taxa including freshwater and brackish gastropoda (1977), water bugs (1965), the nymphs of *Ephemeroptera* (1970) and adult *Trichoptera* (1973). This tradition within the FBA, the Institute of Freshwater Ecology (IFE) and now the Centre for Ecology and Hydrology (CEH) continues to this day, supplemented with further substantial contributions from specialists in taxonomy drawn from universities, colleges and museums.

Inevitably, some taxonomic groups attract more interest than others. Nevertheless, in the last 25 years, significant advances have been made to our knowledge of the occurrence and detailed distribution of most freshwater invertebrates in Great Britain and Ireland. Sometimes this is the result of research on a single species, taxonomic group or a particular waterbody. In other cases extensive survey work on ponds (Pond Action 1994; Williams *et al.* 1998) or river systems in Great Britain (Wright *et al.* 1996) has yielded reliable baseline data. In Scotland, Fozzard *et al.* (1994) have reviewed the invertebrate fauna of both standing and running waters, while in Ireland a number of recent publications provide a wealth of relevant information on the current status of the freshwater fauna (Ashe *et al.* 1998; Giller 1998; Reynolds 1998).

One consequence of this increase in our knowledge is that many species unknown from Great Britain and Ireland in 1973 have been added to the species list and a number of others are known to be more widespread than previously realised. In contrast, it is sometimes difficult to obtain clear evidence of the decline of individual species over the last 25 years unless they are routinely identified during extensive surveys and their decline has been abrupt. However, we now have more comprehensive information that should make the documentation of future change more reliable.

2 Data capture (including mapping schemes)

Early editions of the FBA keys to *Ephemeroptera* (Macan 1970) and *Plecoptera* (Hynes 1977) included distribution maps showing the occurrence of the individual species by vice-counties. More detailed maps prepared by the Biological Records Centre based on species occurrence in 10-km squares of the national grid were featured in one or two of the early keys to aquatic *Diptera* including *Simuliidae* (Davies 1968) and *Dixidae* (Disney 1975). These distribution maps were very incomplete and potentially misleading, and were offered as an incentive to those with an interest in species distribution. A somewhat more comprehensive series of records was provided in a provisional atlas of 16 species of freshwater leeches (Elliott and Tullett 1982) which drew on data for 1048 10-km squares in the British Isles.

At this time, however, few freshwater groups had been the subject of intensive study in order to provide the data required for reasonably comprehensive maps. A notable exception was the freshwater *Mollusca*, which have a long history of scientific study. In 1961 the Conchological Society of Great Britain and Ireland started to use the 10-km square approach for species mapping and 15 years later an *Atlas of the Non-Marine Mollusca of the British Isles* was published (Kerney 1976) with help from the Biological Records Centre for data processing and map production. Recently a new book, representing nearly 40 years of recording, has been published as an *Atlas of Land and Freshwater Molluscs of Britain and Ireland* (Kerney 1999).

There are now many more National Biological Recording Schemes for freshwater invertebrates supported by the Biological Records Centre (Table 13.1). Some of the most successful schemes are for freshwater insects, and in the case of the *Odonata*

Table 13.1 National biological recording schemes for freshwater invertebrates.

Major taxonomic group			Common name
Platyhelminthes		Tricladida	Freshwater flatworms
Mollusca		Gastropoda and Bivalvia	Freshwater snails and mussels
Arthropoda	Crustacea	Cladocera	Water fleas
		Isopoda	Water hog-lice
Arthropoda	Insects	Ephemeroptera	Mayflies
		Odonata	Dragonflies and damselflies
		Hemiptera (Heteroptera)	Water bugs
		Coleoptera	Water beetles
		Megaloptera and Neuroptera	Alderflies, lacewing and sponge flies
		Trichoptera	Caddisflies
		Diptera (several schemes)	True flies

mapping scheme the emphasis is on the adult phase of the lifecycle. The availability of informative publications on this attractive group of insects sparked widespread interest in the mapping scheme during the 1980s and culminated in the publication of a very comprehensive *Atlas of the Dragonflies of Britain and Ireland* (Merritt *et al.* 1996).

Two other schemes deserve brief mention, each one of them co-ordinated by experts who promote interest by producing newletters and organising meetings and field excursions. For over 20 years the Balfour–Browne Club has acted as a focal point for the study of water beetles and during the 1980s, preliminary editions of maps giving the recent distribution of British water beetles were published in six of their newsletters. Similarly, the Dipterists Forum now co-ordinates a substantial number of recording schemes for different families of true flies. Individual schemes relevant to freshwater include those for craneflies, mosquitoes, meniscus midges, horseflies and hoverflies.

Frequently, biological surveys are undertaken with sites (such as ponds, lakes, streams or rivers) as the primary area of interest rather than the distribution patterns of particular species. Because many individual studies are published in the literature, this section simply provides brief reference to some of the larger data-sets collated in recent years.

The latest version of the River InVertebrate Prediction And Classification System (RIVPACS) developed by the IFE is based on information from 614 running-water sites of high biological quality in Great Britain, each one sampled in three seasons, and a further 70 sites in Northern Ireland (Wright *et al.* 1997). The 614 sites in Great Britain include 637 macroinvertebrate taxa after standardisation of all samples to the same taxonomic level (Wright *et al.* 1996) while the 70 sites in Northern Ireland include 313 taxa (Wright *et al.* 2000). These datasets, plus further data collected using similar methods, have been assembled into a national database for the macroinvertebrate fauna of British rivers which currently holds information on 1630 sites.

Pond Action, based at Oxford Brookes University, has been responsible for the

National Pond Survey (1989–94) and the more recent Lowland Pond Survey 1996 (Williams *et al.* 1998). The National Pond Survey includes a reference dataset of around 200 minimally impaired ponds, also sampled in three seasons and for which mainly species-level invertebrate data are available. This has now been supplemented with a further 200 variably degraded sites, sampled in one season only (J. Biggs pers. comm.).

The most extensive macroinvertebrate surveys are those for the appraisal of the biological quality of running waters. Within the UK, identification of the fauna at each site is to BMWP family-level only (Armitage *et al.* 1983) and the RIVPACS software is used to appraise site quality. In the 1995 national survey, the Environment Agency sampled 6700 sites in England and Wales (Environment Agency 1997), the Scottish Environment Protection Agency sampled 1256 sites (Scottish Office 1995) and the Department of the Environment (Northern Ireland) sampled 283 sites (DOE(NI) 1997). These datasets, collected using a standardised procedure, may be compared with similar surveys at many of the same sites in 1990. They also form a valuable asset for documenting future change.

In the Republic of Ireland, the national biological river quality survey is operated by the Environmental Protection Agency where a separate appraisal system has been in place for many years based mainly on family-level identifications. (McGarrigle *et al.* 1992). The current network is some 3000 sites and for logistical reasons one third of the sites are sampled each year, the most recent three-year cycle of investigation being 1995–97. The use of a consistent approach over time means that, at some sites, changes in quality can be traced back through the 1980s and sometimes back to the 1970s.

Finally, the Invertebrate Site Register, originally developed by the Joint Nature Conservation Committee (JNCC), now holds biological information on over 13 000 sites for use by the statutory nature conservation bodies of England, Wales and Scotland. A majority of these sites are terrestrial, but some include freshwater fauna, where the site has a pond or running-water system.

3 Workforce

The number of specialists in freshwater invertebrates, including both professionals and amateurs, is difficult to gauge accurately, because the fauna encompasses such a wide range of taxonomic groups. It is also important to distinguish taxonomists from those skilled in the identification of the freshwater fauna using existing keys. Within the Natural History Museum in London there are six or seven scientists with taxonomic expertise in freshwater invertebrates, and of the dozen invertebrate zoologists within the Institute of Freshwater Ecology, around half have specialist taxonomic knowledge. Many other scientists, widely dispersed throughout the British Isles in universities, colleges and museums, have also made an important contribution to the taxonomy of freshwater invertebrates.

The statutory agencies with responsibilities for monitoring river quality in the British Isles employ the largest numbers of professional biologists with skills in the identification of the freshwater fauna. In England and Wales, the Environment Agency has around 100 biologists who undertake macroinvertebrate survey work. They range from scientists with many years of experience and skills in species-level identification to new recruits receiving in-house training at identification to BMWP-family level. In

Scotland, the Scottish Environment Protection Agency (SEPA) employs over 30 biologists, many of whom undertake macroinvertebrate surveys, while in Northern Ireland a team of five biologists in the Industrial Research and Technology Unit (IRTU) conduct similar surveys on behalf of the Environment and Heritage Service of the Department of the Environment (Northern Ireland). In the Republic of Ireland the Environmental Protection Agency also maintains a small specialised team of three biologists who undertake their three-year monitoring programme.

There are also many other professionals with detailed knowledge of the freshwater fauna in educational establishments (such as universities or the Field Studies Council), the statutory conservation organisations (Joint Nature Conservation Committee, English Nature, the Countryside Council for Wales and Scottish Natural Heritage) and in environmental consultancies.

Inevitably, some groups of freshwater invertebrates have proved more attractive for study by amateurs than others. In the case of molluscs, dragonflies, water beetles and some true flies, specialist societies and clubs have helped to bring professionals and amateurs together. In this way, the efforts of hundreds of amateurs throughout the British Isles have made a substantial contribution to knowledge, and in particular to species distribution. Examples may be found in other chapters of this book.

Reliable information on the freshwater fauna is crucial, whether it is being collected by professionals or by amateurs. The Natural History Museum runs an Identification Qualification (IdQ) scheme which seeks to improve standards in environmental monitoring, impact assessment and conservation work by awarding certificates of competence in animal and plant identification. Examinations to test competence in the identification of freshwater macroinvertebrates are held at species and at BMWP family level.

4 Synopsis of changes to 1973

In *The Changing Flora and Fauna of Britain*, Macan (1974) pointed out that human impact on standing and running waters over many hundreds of years must have had consequences for the distribution and abundance of the freshwater invertebrate fauna. However, apart from the molluscs, water beetles and to a lesser extent dragonflies, which enjoyed popularity in the nineteenth century, the fauna was not well known, even at the beginning of the twentieth century. It was, therefore, difficult to know whether species had changed their status, been lost or gained a foothold as a consequence of human activity. Macan did list two water beetles (*Ilybius subaeneus* and *Acilius canaliculatus*) whose current distribution is very different from that noted in the nineteenth century. He also mentioned the loss of the damselfly *Coenagrion scitulum* which was known from a restricted area in Essex between 1946 and 1953 but which then became extinct in the British Isles, following flooding by exceptionally high tides in 1953. With these exceptions, Macan focused on freshwater invertebrates that have, to varying degrees, been successful at invading British freshwaters from abroad.

In all, Macan (1974) discusses 16 species (Table 13.2), of which 13 were regarded as outstandingly or moderately successful colonisers: one was restricted to a single location and two others failed to establish. He describes the spectacular spread of *Potamopyrgus jenkinsi* (now *P. antipodarum*) from the first record in the Thames estuary in 1883 to its present distribution throughout Great Britain and Ireland. Macan

Table 13.2 Sixteen species of freshwater invertebrates listed in Macan (1974) which had colonised Great Britain (and Ireland*) prior to 1974. Species defined as outstandingly successful colonists by Macan are indicated by ++.

Taxonomic group	Species	Successful colonists	One location	Failures
Hydrozoa	Cordylophora lacustris	+		
Tricladida	Planaria torva*	+		
	Dugesia tigrina	+		
Oligochaeta	Branchiura sowerbyi*	+		
Gastropoda	Potamopyrgus jenkinsi*	++		
	Lymnaea catascopium			+
	Physa acuta	+		
	Planorbis dilatatus	+		
Bivalvia	Dreissena polymorpha	+		
Crustacea	Asellus communis		+	
	Corophium curvispinum	+		
	Crangonyx pseudogracilis[1]	++		
	Gammarus tigrinus[1]	+		
	Orchestia cavimana	+		
	Eriocheir sinensis			+
Oomycota	Aphanomyces astaci[2]	++		

[1] These species now known to have been present in Ireland in 1974.
[2] This record is now in doubt.

also records the progressive colonisation from the 1930s of the American amphipod *Crangonyx pseudogracilis* across England, Wales, and finally into Scotland, although in 1974 he had no records for Ireland. The third species regarded as a successful colonist was *Aphanomyces astaci*, which causes crayfish plague disease. This was known from many European countries in the late nineteenth century, but was less well documented in England. Macan simply drew attention to records from the 1930s in which native crayfish underwent alternate abundance and scarcity, as potential evidence of an epidemic disease. The modern view (Holdich *et al.* 1995) is that there is no firm evidence of crayfish plague in the British Isles prior to the 1980s and that earlier mortalities among the white-clawed (native) crayfish (*Austropotamobius pallipes*) were probably due to 'porcelain' disease caused by a protozoan.

5 The freshwater invertebrate fauna of the British Isles

Before considering the species currently under threat and changes in status since 1973, it is important to recognise the extent to which our knowledge of species occurrence has increased over the past 25 years. Maitland (1977) published a coded checklist of animals occurring in freshwater in the British Isles and this provides an authoritative statement for that time, based on taxonomic keys and checklists. Inevitably, judgements were required on the validity of some records and on whether to include species that could be regarded as semi-aquatic or brackish-water species. In general, species were included even when their true status was in doubt in order to minimise the need to add codes in the future. Therefore, the presence of a given species on the list should not be regarded as a definitive statement that it occurs in the British Isles. Since then, a large number of additional species have been recorded in the British Isles, and in some cases new information has clarified the status of particular species. Many taxonomists and ecologists have contributed to current knowledge and much of this information has been collated into an updated version of the coded checklist (Furse *et al.* 1998).

A comparison between the number of non-insect taxa listed in Maitland (1977) and in Furse *et al.* (1998) is given in Table 13.3. In a small minority of cases (*Porifera, Rotifera, Nematoda* and *Gastrotricha*), the number of freshwater species in the latest listing has decreased as a result of more accurate information. However, in 50% of major groups the number of species has increased, sometimes as a result of genuine additions to the fauna and at other times due to a broader viewpoint on those species

Table 13.3 Number of non-insect freshwater invertebrates in the British Isles listed in Maitland (1977) and in the updated version of the checklist compiled by Furse *et al.* (1998).

Major taxonomic group		Number of species (Maitland 1977)	Number of species (Furse et al. 1998)
Porifera		8	5
Coelenterata		8	8
Platyhelminthes	Microturbellaria	46	47
	Tricladida	11	12
Nemertini		2	2
Nematomorpha		4	4
Rotifera		511	475
Nematoda		97	69
Gastrotricha		22	12
Polyzoa		9	11
Mollusca	Gastropoda	52	64
	Bivalvia	27	28
Annelida	Polychaeta	1	4
	Oligochaeta	118	126
Hirudinea		14	16

continued on next page

Table 13.3 Number of non-insect freshwater invertebrates (cont.).

Arthropoda	Tardigrada	35	42
(others)	Hydracarina	322	324
	Oribatei	4	4
	Araneae	1	1
Arthropoda	Anostraca	2	2
(crustacea)	Notostraca	1	1
	Cladocera	88	90
	Ostracoda	88	89
	Copepoda	111	113
	Branchiura	2	2
	Malacostraca	33	43

appropriate for inclusion in the list. Overall, the total of 1594 species in the latest listing is slightly lower than the 1617 listed by Maitland (1977).

The recent checklist of Irish aquatic insects by Ashe *et al.* (1998) also includes a table listing the richness of each major taxonomic group in the British Isles and this has been used for comparisons with Maitland (1977) in Table 13.4. With the exception of the *Plecoptera* and *Neuroptera* where there have been no additions, and the *Lepidoptera*

Table 13.4 Number of aquatic insects in the British Isles as listed in Maitland (1977) and in a recent listing from Ashe *et al.* (1998).

Major taxonomic group	Number of species (Maitland 1977)	Number of species (Ashe et al. 1998)
Collembola	17	21
Ephemeroptera	47	48 (plus 3*)
Plecoptera	34	34
Odonata	45	46
Hemiptera	62	63
Hymenoptera	27	40
Coleoptera	300	364
Megaloptera	2	3
Neuroptera	4	4
Trichoptera	193	198
Lepidoptera	11	5
Diptera	1138	1542

* See text section 7.2 for details.

where the list of genuine freshwater species is lower than before, all other groups have increased in species richness. In the case of the *Coleoptera* and the *Diptera* the new listings are substantially longer than in 1977 and overall the total list of insects has increased from 1880 to 2371 species. This last figure includes three species of *Ephemeroptera* recently recorded in Great Britain which were not included in Ashe *et al.* (1998).

6 Species under threat and recent progress in protection

Boon *et al.* (1992) provide a concise review of legislation for protecting the freshwater biota, and a more recent publication by English Nature (1997) offers an account of the many programmes currently underway for protecting the biota of a wide range of freshwater ecosystems in England.

Legislation concerned with species protection is largely directed towards threatened species and the *Red Data* books, including one for non-insects (Bratton 1991) and a second for insects (Shirt 1987), are the primary sources of information. The threat categories given in these publications are RDB1 (endangered), RDB2 (vulnerable) and RDB3 (rare). Bratton (1991) and Shirt (1987) include both terrestrial and freshwater species, but not all major freshwater groups are covered and subsequent publications have filled some but not all the gaps. The number of *Red Data Book* species by major taxonomic group and threat category is given in Table 13.5, together with the relevant source reference.

Table 13.5 Number of Red Data Book species per major group of freshwater invertebrates.

Major taxonomic group	RDB 1	RDB2	RDB3	RDB5/K	Reference
Nemertea				1RDBK	Bratton (1991)
Polyzoa			1	1RDBK	Bratton (1991)
Gastropoda	6	5	3		Bratton (1991)
Bivalvia	1		2		Bratton (1991)
Hirudinea			1	1RDBK	Bratton (1991)
Crustacea	2	1	1	1RDB5	Bratton (1991)
Ephemeroptera	1	2	1		Bratton (1990)
Plecoptera		1		3RDB5	Bratton (1990)
Odonata	1	2	3		Merritt *et al.* (1996)
Hemiptera			3		Kirby (1992)
Coleoptera*	3	18	28		Foster (in press)
Trichoptera	7	7	7	7RDBK	Wallace (1991)
Diptera (incomplete)	14	34	29		Falk (1991)

RDB1 endangered; RDB2 vulnerable; RDB3 rare; RDBK insufficiently known; RDB5 endemic.
* Note that for the *Coleoptera*, the new IUCN categories of Critically Endangered, Endangered and Vulnerable apply.

Note that, as yet, no RDB species have been designated in several major freshwater groups including *Tricladida, Oligochaeta, Megaloptera, Neuroptera* or in several families of freshwater *Diptera* including *Psychodidae, Chaoboridae, Chironomidae* and *Simuliidae.*

In the last few years, the World Conservation Union (IUCN) has formally adopted revised Red List categories and criteria (IUCN 1994) which are more rigorous and place more emphasis on quantitative criteria. As yet, the new system has not been widely applied to invertebrates, but it has been adopted for aquatic *Coleoptera* (see Table 13.5).

The main instrument for species protection in Great Britain is the Wildlife and Countryside Act 1981 together with subsequent quinquennial reviews. Schedule 5 includes a select list of freshwater invertebrates, drawn mainly from the *Red Data Book* lists, where legal protection is considered likely to bring tangible conservation benefits. It also covers a few non-listed species in need of protection as a result of recent population declines including the white-clawed crayfish (*Austropotamobius pallipes*) and the freshwater pearl mussel *Margaritifera margaritifera.* As a result of the third quinquennial review, the pearl mussel now has full protection and there is a moratorium on the collection of pearls until populations have recovered.

Another important development in the protection of species and habitats under threat resulted from the Convention on Biological Diversity drawn up in Rio de Janeiro in June 1992. This led to the publication of the UK Biodiversity Action Plan (Department of the Environment 1994a) followed by the UK Steering Group Report

Table 13.6 Freshwater invertebrates on the 'priority' listing for Species Action Plans.

Major taxonomic group	Species	Major taxonomic group	Species
Nemertea	Prostoma jenningsi	Plecoptera	Brachyptera putata
Hirudinea	Hirudo medicinalis*	Odonata	Coenagrion mercuriale*
Gastropoda	Anisus vorticulus*	Hemiptera	Hydrometra gracilenta
	Catinella arenaria*	Coleoptera	Agabus brunneus
	Myxas glutinosa*		Bidessus minutissimus
	Segmentina nitida*		B. unistriatus
	Vertigo moulinsiana*		Helophorus laticollis
Bivalvia	Margaritifera margaritifera*		Hydrochara caraboides
	Pseudanodonta complanata*		Hydroporus cantabricus
	Pisidium tenuilineatum*		H. rufifrons
Crustacea	Austropotamobius pallipes*		Laccophilus poecilus
	Triops cancriformis		Graphoderus zonatus
			Paracymus aeneus

* Species on the 'short' list with Action Plans in the UK Steering Group Report (Department of the Environment 1995).

(Department of the Environment 1995) which outlined a major programme on the conservation of biodiversity within the UK. Volume 2 of the Steering Group Report (1995) gives 'short' and 'middle' lists of species which are either globally threatened or rapidly declining in the UK. It also includes detailed Action Plans for the 11 freshwater invertebrates on the 'short' list. More recently a 'priority' listing has been drawn up (Table 13.6) which includes the 'short' list species plus 14 additional freshwater invertebrates largely, but not exclusively, drawn from the original 'middle' list. Lead partners have been appointed for individual species in order to co-ordinate each Action Plan and ensure progress towards defined targets.

Finally, mention should also be made of the Species Recovery Programme initiated by English Nature in 1991, which has now become a primary means of meeting species targets within the UK Biodiversity Action Plan. Two freshwater species to receive detailed attention are the white-clawed crayfish and the freshwater pearl mussel, also protected under the Wildlife and Countryside Act 1981.

7 Changes between 1973 and 1998

7.1 Losses and declines

Some of the freshwater invertebrates under threat, and therefore in need of both legal protection and positive action, have been highlighted in the previous section. Here, the focus is on a selection of the individual species that are known to have been lost or have undergone a substantial decline in the last 25 years. As previously emphasised, it is sometimes difficult to obtain clear evidence, except in the case of large, easily identified species within routine monitoring programmes or species included in Biological Records Centre mapping schemes. Hence, some major taxonomic groups receive no mention but comments are made on a few species or groups that may warrant close attention in the future. For information on declines and losses in *Diptera* refer to Chapter 16, this volume.

Tricladida   The recent spectacular increase in the geographical distribution of the American triclad *Dugesia tigrina* in British rivers and canals (Wright 1987), and further extensions to its range in standing waters (Young and Reynoldson 1999), raises the question of whether it has the potential to affect the distribution and abundance of native flatworms. Evidence of the almost entire displacement of the native *Polycelis tenuis* and *P. nigra* from a well-studied lake on Anglesey following the arrival of *D. tigrina* (Gee and Young 1993) – and additional evidence of its detrimental effects on native triclads in Colemere, England (Young and Reynoldson 1999) – suggests that this concern may be well founded in some circumstances.

Gastropoda   A number of freshwater gastropods have continued their decline and are now regarded as endangered or vulnerable. The glutinous snail *Myxas glutinosa* was last recorded in England in 1993 and is now only known from Llyn Tegid in North Wales (Environment Agency, 2000), having disappeared from several previously documented sites. Similarly, the shining ram's-horn snail *Segmentina nitida* and *Anisus vorticulus*, which both frequent drains within grazing marshes, are believed to have declined through a combination of over-frequent ditch clearance, nutrient enrichment and conversion of grazing levels to arable farming, with associated lowering of the water table.

Bivalvia   The freshwater pearl mussel (*Margaritifera margaritifera*) is a species of global conservation concern and throughout the British Isles strenuous efforts are being made to document and understand the causes of its decline. It now appears that many populations of this long-lived species may not have produced young for over 30 years. In England and Wales just four rivers have large or medium populations while in other rivers the species is close to extinction. Poor water quality, changes to the physical habitat, a decline in host fish populations, conifer planting, sedimentation and amateur pearl fishing have all been proposed as reasons for their decline. Scotland is a major stronghold, and a recent survey has revealed recruitment to some populations, but confirmed that destructive pearl fishing is a major cause of decline in this region. In Ireland, there are similar concerns over recruitment although substantial populations do occur in some widely spaced rivers (Reynolds 1998). The Irish hardwater subspecies (*M. margaritifera durrovensis*), which is confined to a 10-km section of the River Nore in south-east Ireland, is also the subject of recent research.

Crustacea   Undoubtedly, one of the most worrying changes in the status of a native member of the freshwater fauna has been the recent decline of the white-clawed crayfish. Although still widespread on mainland Europe, it has a patchy distribution as a result of disease, over-fishing and pollution, and the British Isles now holds the greatest concentration of this endangered species in Europe (Holdich *et al.* 1995). It prefers clean alkaline waters and is therefore most common in northern, central and southern England, eastern Wales and central Ireland.

However, since the early 1980s the crayfish plague (*Aphanomyces astaci*) has been responsible for eliminating a number of populations in central and southern England. Fortunately, the species remains widespread in northern England. Crayfish plague is carried by the North American signal crayfish, *Pacifasticus leniusculus*, which was introduced into Great Britain from Sweden in the 1970s for culture in crayfish farms. Subsequently it has been released into the wild at many river locations, particularly in southern England. Whereas the signal crayfish can harbour crayfish plague and survive, the native white-clawed crayfish is highly susceptible and can be eliminated very rapidly. In addition, plague-free populations of the signal crayfish are very aggressive and where they initially co-existed with the native crayfish, by the 1990s all known mixed populations had become monospecific in favour of signal crayfish (Holdich *et al.* 1995). The species action plan for the white-clawed crayfish summarises the factors causing loss and decline, and the measures proposed to conserve it, including the designation of 'no-go' areas for the keeping of non-native crayfish.

In Ireland, despite the absence of non-native crayfish, stocks of white-clawed crayfish were brought to extinction in some Irish lakes by crayfish plague in the 1980s. Reynolds (1988) suggests that the most probable route for the infection was on the waders or fishing equipment of anglers arriving from Britain or the continent.

Ephemeroptera   The mayfly *Paraleptophlebia werneri* has RDB3 status and is known from a relatively small number of calcareous streams and winterbournes. When it occurs in winterbournes, the larvae must complete their larval stage in spring and emerge before the stream dries up in summer. Resistant eggs in the stream-bed then survive the period of drought through the autumn and early winter. A series of major droughts in the 1990s, abstraction schemes within chalk catchments, and the potential for agricultural pollution in these small streams, all suggest that the status of this rare species should be monitored.

Odonata   Macan (1974) mentioned the loss of one species, *Coenagrion scitulum*, prior to 1973, but two additional species, *C. armatum* and *Oxygastra curtisii*, were also lost to the British list prior to 1973, although at the time of writing the hope of finding relict populations was still present. For more details on these species and of current actions to conserve threatened species, see Brooks (Chapter 17, this volume).

Hemiptera   Despite the efforts of a small group of recorders in Great Britain, it is difficult to comment on changes in status in this group since 1973. In Ireland, the flightless water bug *Microvelia pygmaea* was first discovered in the 1970s, but after establishing itself in several lakes in County Cork, the populations crashed and it now appears to be extinct (Walton 1985).

Trichoptera   The relatively small number of specialists capable of identifying the larvae and adults of this important group make it difficult to assess genuine changes in status. An example of the problem is the RDB1 (endangered) species *Hydropsyche saxonica* which was only known from a single stream near Oxford in the 1950s. Subsequent housing development caused its demise at this site through pollution and siltation, and during the 1970s and 1980s the lack of an adequate description of the larvae in British publications, and limited sampling of small streams, delayed a true assessment of its status. However, more recently, it has been found at no less than 34 sites from 30 different 10-km squares in England, Wales and Scotland (Blackburn and Forrest 1995); a lower threat category may now be appropriate.

7.2 Additions and increases

As previously indicated, many additions to the freshwater fauna of the British Isles since 1973 have been a result of the discovery of previously unrecognised species which were, nevertheless, present 25 years ago. In most cases, these species are already known from mainland Europe, but in a few cases they are new to science. Aliens and introduced species form a small but sometimes important component of the total increase in the faunal listings as given in Tables 13.3 and 13.4. Only a selection of the additions and increases in the fauna can be highlighted here, with the focus on interesting and important changes in Great Britain and in Ireland. For comments on additions and increases in the *Diptera* refer to Chapter 16 in this volume.

Tricladida   Prior to 1973, the American immigrant *Dugesia tigrina* was known only from a few flowing-water sites in Great Britain, but by 1985 it had colonised a large number of lowland rivers and canals in England and Wales and was even reported in the River Tweed in Scotland (Wright 1987). Its ability to exploit a wide range of food resources, asexual mode of reproduction and toleration of enriched waters may help to explain its success. The need to monitor the impact of this species on the native triclad fauna has already been mentioned. An additional American species, *Phagocata woodworthi*, found in Loch Ness (Ball and Reynoldson 1981) was probably brought over accidentally by American scientists engaged in the search for the 'Loch Ness Monster'.

Gastropoda One American immigrant, the freshwater limpet *Ferrissia wautieri* has colonised a number of sites in England recently, while in Northern Ireland, another American immigrant *Physa gyrina*, first seen in 1995, has now spread through Lough Neagh and the Lower Bann (Giller 1998). Additionally, in 1996, *Viviparus viviparus* was recorded for the first time in an alkaline lake in the Republic of Ireland (Cotton 1996).

Bivalvia The zebra mussel, *Dreissena polymorpha*, which originates from Eastern Europe has been known in Great Britain since 1824 and Macan (1974) regarded it as a successful colonist. However, in the 1980s, when it colonised the North American Great Lakes, it caused major problems. Hence, its appearance in Lough Ree and Lough Derg in Ireland, probably resulting from an initial introduction into the lower Shannon in 1994, was of concern (McCarthy and Fitzgerald 1997). Since then, it has been spread by boat traffic from Lough Derg to the Lough Erne system in County Fermanagh, Northern Ireland via the Shannon–Erne canal (Rosell, Maguire and McCarthy 1999).

Annelida Not surprisingly, several oligochaetes have been added to the British list since 1974. In addition, the leech *Boreobdella verrucata* has been confirmed in both Britain and Ireland, while a second species, *Haementaria costata*, which is new to the British Isles, has been recorded (Elliott *et al.* 1979).

Crustacea Macan (1974) regarded the Chinese mitten crab, *Eriocheir sinensis*, as a failed immigrant to British waters (Table 13.2), for although it had colonised a number of European estuaries and rivers, only a single specimen was taken in the River Thames in 1935. In the last 25 years, this species has reappeared in the River Thames and there have been further records in the Trent, Ouse and Humber (Gledhill *et al.* 1993). A specimen of the American Blue Crab *Callinectes sapidus*, which is also widespread in European coastal waters, was taken in the River Trent in 1982 (Clark 1984).

The importation of signal crayfish into Great Britain in the 1970s and the negative consequences for the native white-clawed crayfish have already been mentioned. By 1994, there were 99 registered crayfish farms, mainly in southern and central England and wild populations were documented in at least 40 rivers, most notably in the Rivers Great Ouse, Kennet, Loddon and the Thame (Holdich *et al.* 1995). Unfortunately, wild populations of three additional non-native crayfish have also established themselves although at present they are much less widespread than the signal crayfish. The narrow-clawed or Turkish crayfish, *Astacus leptodactylus*, was imported in the 1970s and is now well established in lakes and canals in the London area, including the Serpentine in Hyde Park. Following its introduction into lakes within the Stour catchment, it then colonised the River Stour in Essex. The noble crayfish *Astacus astacus* from mainland Europe, itself a victim of crayfish plague in Scandanavia and elsewhere, is now being farmed at one location in the Mendips, but has recently escaped into the wild (Holdich *et al.* 1995). Finally, the red swamp crayfish *Procambarus clarkii*, a native of the southern USA, was originally introduced into Britain for aquaculture and for the aquarist trade. To date only one wild population is known: from a lake on Hampstead Heath (Holdich *et al.* 1995).

A number of amphipod crustacea continue to colonise new locations, particularly in Ireland, where the native freshwater fauna was, until fairly recently, limited to just two species, *Gammarus lacustris* and *G. dueboni.* Macan (1974) regarded *Crangonyx pseudogracilis* as an outstandingly successful colonist and this species is now more widely distributed within Scotland, including Loch Ness and some rivers in north-east Scotland. The first record for Ireland was in Phoenix Park, Dublin in 1969, but since then it has been found in the River Boyne, River Liffey, Lough Erne, Lough Neagh and their associated rivers (Giller 1998). Whereas the North American *Gammarus tigrinus* is thought to have been accidentally introduced into Lough Neagh via the River Bann during the First World War, *G. pulex*, was introduced into several rivers in Northern Ireland in connection with fish farming and is now also present in Lough Neagh. The impact of these three non-native amphipods in Lough Neagh on the Irish freshwater subspecies *Gammarus dueboni celticus* has been the subject of detailed investigations by Dick (1996); whereas the single native species was dominant, there is now a patch-work of mixed species assemblages.

Ephemeroptera and Plecoptera   In view of the importance of mayflies in freshwaters and the widespread interest shown in them by both fishermen and scientists, it is sur-prising that no less than four species new to Great Britain have been discovered in the last 15 years. Admittedly, three species (*Caenis pusilla, C. pseudorivulorum* and *C. beskidensis*) are small and unobtrusive (Gunn and Blackburn 1997, 1998), but the first two are quite widely distributed in flowing waters. More spectacular was the recent dis-covery of *Electrogena affinis* in the lower reaches of the River Derwent in Yorkshire (Blackburn *et al.* 1998). In contrast, there have been no additions to the list of *Plecoptera* (stoneflies) for the British Isles, although it is only in the last 25 years that *Capnia atra* and *Leuctra nigra* have been reported from Ireland.

Odonata   Another exciting find was the discovery of the north European damselfly *Coenagrion lunulatum* in Ireland (Cotton 1982). It is now known to be a widespread but uncommon member of the fauna of mesotrophic lakes and pools, and the Irish pop-ulation is one of the largest in Western Europe; this species is absent from Great Britain. Further information on changes in the status of native species and invasions from conti-nental Europe may be found in Brooks (Chapter 17, this volume).

Megaloptera   The first record of the alderfly *Sialis nigripes* in the British Isles was in Ireland in 1976 (Elliott *et al.* 1979), but since then it has been found at numerous loca-tions in England, Wales and southern Scotland.

Trichoptera   Two RDB1 (endangered) species in Great Britain, *Limnephilus pati* and *L. tauricus*, have recently been found in Ireland (O'Connor and Bond 1995).

8 Factors affecting change

The character of streams and rivers depends largely on the topography, geology and cli-mate of the region in question. This intimate association with the surroundings means that any changes or disturbances in the catchment area will affect the watercourses draining it and this may be reflected in the biota which live in them. The main impacts are attributable to agriculture, forestry and other catchment disturbance, urbanisation and associated effects on the management of water resources. Biotic interactions may also affect changes in community composition (see above).

8.1 Agricultural impacts – polluting effects

In a study (National Rivers Authority 1992) the impacts of agricultural pollution were reviewed and the results are summarised here.

- *Organic pollution (silage/slurry)*   Acute incidents following spills plus both chronic and diffuse inputs from inadequate storage facilities result in invertebrate kills, due primarily to low dissolved oxygen concentrations.
- *Nutrient input (nitrogen/phosphorus)*   Application of fertilisers leads to chronic and diffuse problems resulting in increased eutrophication of receiving waters and the development of large filamentous algal populations.
- *Herbicides*   These may impact non-target species in rivers and kill stands of instream vegetation thereby affecting the communities of invertebrates that live in them.
- *Pesticides including sheep dips*   Acute pollution can occur but most problems are of a chronic or diffuse nature resulting from surface run-off and aerial drift. Most modern pesticides are less persistent in the environment but their sub-lethal effects are unknown.
- *Oil and other contaminants*   Oil pollution is relatively rare but can cause severe effects over many kilometres of river. Acidic runoff from drained soils containing sulphides lowers pH and mobilises metals which may reach toxic levels. Pharmaceuticals including antibiotics and disinfectants are commonly used in agriculture and may also enter streams and rivers resulting in the evolution of resistant strains of bacteria and increased mortality of invertebrates.

8.2 Agricultural impacts – physical effects

- *Channel engineering*   This involves the straightening and dredging of rivers and the removal of riparian strips. These disturbances may result in lowered bed stability, the physical destruction of prime bankside and instream habitat resulting in a reduction in the diversity of habitats and major changes to the original pre-disturbance faunal assemblages. The removal of riparian buffer strips may increase the flow of sediment into the stream and accelerate bank erosion. The cutting of trees and shrubs will reduce resting places and remove landmarks used as swarm markers for adult insects and may lead to reduced shading and increased growth of macrophytes, resulting in modifications to the faunal community (RSPB, NRA and RSNC 1994; Brookes and Shields Jr 1996)
- *Abstraction of water for irrigation*   This practice reduces flows and affects habitat availability (Armitage and Petts 1992).
- *Drainage of wetlands*   Drainage schemes destroy wetland habitat and reduce

connectivity between floodplain and river. They can also result in the removal of ponds and the culverting of ditches or their conversion into underground field drains (Acreman and Adams 1998).

- *Mobilisation of silt*   Cross-contour ploughing leads to concentrated runoff and removal of topsoil into rivers and streams. This in turn reduces faunal diversity and modifies existing faunal assemblages (Wood and Armitage 1997).

8.3 Forestry and other disturbances

Afforestation may exacerbate the acidification of streams and lakes resulting in a diminished faunal community. Reduced water yield has also been associated with afforestation through transpiration and increased evaporation from the tree canopy. Soil disturbance, drainage off new plots and the use of heavy machinery may lead to sediment input and increase the risk of oil pollution.

Other large scale activities such as quarrying, strip mining and peat-cutting will mobilise sediment and in some cases lower the pH of receiving waters.

8.4 Urbanisation

The process of catchment urbanisation can lead to increased atmospheric pollution through the burning of fossil fuels and this may result in the acidification of poorly buf-fered waters. The construction and establishment of the built environment disturbs topsoil, alters drainage patterns and radically alters the flood hydrograph. Runoff from the streets may carry high sediment loads and a cocktail of toxic hydrocarbons, all of which eventually enter streams and rivers. In addition, sewage effluent may comprise a major proportion of the total discharge in large conurbations, and leachates from old industrial sites can enter the drainage system. This combination of stresses impover-ishes the fauna directly (House *et al.* 1993) and physical changes to the river channel result in a reduction in instream habitat diversity and poor bankside and riparian habi-tat (see Channel engineering above).

8.5 Water resource management

The storage and transport of water for industrial and domestic use and the generation of power can have major impacts on rivers and streams. The regulation and modifica-tion of natural flow patterns through the construction of reservoirs, river transfer schemes and abstractions may result in a variety of effects which will radically alter faunal communities in the stored water and in the receiving water course (Armitage 1979; Petts 1984). In recent years the combination of low summer flows and abstrac-tion of water has resulted in severe impacts to both macrophytes and freshwater inver-tebrates, particularly in chalk streams where extremes of discharge are not a natural event (Mainstone *et al.* 1998).

8.6 Checks and balances

The impacts listed above are known to be potentially harmful to freshwater habitats and many are still occurring, but the major change in the last 25 years is the recognition of this fact and the implementation of a series of check and balances which mitigate against disturbance. This, in association with increased public awareness of wildlife issues through the media and the higher profile of environmental matters in schools curricula, has altered the populations view of wildlife and its conservation. Wildlife Trusts, the RSPB, conservation agencies and angling associations have all contributed to the environmental debate and have been instrumental in the rehabilitation, restoration and construction of wildlife habitat (Andrews and Kinsman 1990; Biggs *et al.* 1994; Williams *et al.* 1997).

Rivers, streams and standing waters are monitored nationally and the development of catchment management plans, and local environment agency plans have identified pressures on the environment which can then be acted on with the support of environmental legislation, initiatives and codes of good practice. These include the Wildlife and Countryside Act, the SSSI system, Biodiversity Action Plans, the EU Habitats Directive and other European Union initiatives (Department of the Environment 1994b).

9 Prognosis for the next 25 years

Most of the positive changes over the past 25 years have been attributable to two main causes. First, the publication and improvement of taxonomic keys together with the increase in the number of both professional and amateur freshwater biologists in the British Isles. Secondly, the establishment of agencies with responsibilities for the assessment and restoration of aquatic habitats and their biota. Targets are being set for river quality and specific actions are being taken for the protection of individual species and communities. These factors are likely to play a major role in the future as our knowledge-base improves and the environment is protected with increasing vigour.

There is also greater public interest in and awareness of the biota of the British Isles and increasing pressure to conserve our natural heritage for the enjoyment of present and future generations. Mapping schemes continue to develop and projects such as the National Biodiversity Network (Chapter 22, this volume) should increase the general level of awareness and availability of data. National and international legislation for protecting habitats and species should help to counteract the pressures placed upon the countryside and its fauna by the human population.

Against this positive background there is continued pressure on land and associated water resources due to intensive arable and livestock production and through urbanisation. Inevitably, agricultural and industrial processes generate a wide range of anthropogenic compounds including pesticides. Some of these compounds may interfere with the endocrine system of aquatic organisms resulting in changes in faunal composition.

The effects of climate change may vary with location but would involve changes in rainfall patterns, water temperatures and water usage with potentially severe consequences for wetlands and rivers. Changing conditions may favour alien species that could impact on the distribution and abundance of native species. Any future need to move water from one area to another via river transfer also has the potential to introduce species and diseases into the recipient river catchment. The increasing mobility of the population means that fewer areas are left undisturbed and management of water resources has to consider the multiple use of the aquatic environment with some inevitable compromises.

In general, the prognosis for the next 25 years is good provided that economic stability is maintained. Increased communication and knowledge will mitigate against increased stress on the environment. However, it is essential that taxonomic work on poorly known groups of freshwater invertebrates is encouraged and funded in order to increase our knowledge and understanding of the aquatic fauna. Finally, although information on the distribution of species and faunal assemblages is growing, we need more information on the processes controlling colonisation, response to disturbance, and the amount of habitat needed to sustain a particular species. This knowledge is essential to the implementation of sustainable environmental management.

REFERENCES

Acreman, M.C. and Adams, B. (1998) *Low Flows, Groundwater and Wetland Interactions – Scoping Study*. Report to Environment Agency (W6–013), UKWIR (98/WR/09/1) and NERC (BGSWD/98/11), R&D Technical report W112.

Andrews, J. and Kinsman, D. (1990) *Gravel Pit Restoration for Wildlife – A Practical Manual*. Sandy: Royal Society for the Protection of Birds.

Armitage, P.D. (1979) Stream regulation in Great Britain, in Ward, J.V. and Stanford, J. (eds) *The ecology of regulated streams*. New York: Plenum, pp. 165–181.

Armitage, P.D., Moss, D., Wright, J.F. and Furse, M.T. (1983) The performance of a new biological water quality score system based on macroinvertebrates over a wide range of unpolluted running-water sites. *Water Research*, **17**, 333–347.

Armitage, P.D. and Petts, G.E. (1992) Biotic score and prediction to assess the effects of water abstractions on river macroinvertebrates for conservation purposes. *Aquatic Conservation: Marine and Freshwater Ecosystems*, **2**, 1–17.

Ashe, P., O'Connor, J.P. and Murray, D.A. (1998) *A Checklist of Irish Aquatic Insects*. Occasional publication, no. 3. Dublin: The Irish Biogeographical Society.

Ball, I.R. and Reynoldson, T.B. (1981) *British Planarians*, [Synopses of the British fauna, no. 19], Linnean Society of London. Cambridge: Cambridge University Press.

Biggs, J., Corfield, A., Walker, D., Whitfield, M. and Williams P. (1994) New approaches to the management of ponds. *British Wildlife*, **5**, 272–287.

Blackburn, J.H. and Forrest, M.B. (1995) New records of *Hydropsyche saxonica* McLauchlan (*Trichoptera, Hydropsychidae*) from small streams in Great Britain. *Entomologist's Monthly Magazine*, **131**, 71–76.

Blackburn, J.H., Gunn, R.J.M. and Hammett, M.J. (1998) *Electrogena affinis* (Eaton, 1885) (*Ephemeroptera, Heptageniidae*), a mayfly new to Britain. *Entomologist's Monthly Magazine*, **134**, 257–263.

Boon, P.J., Morgan, D.H.W and Palmer, M.A. (1992) Statutory protection of freshwater flora and fauna in Britain. *Freshwater Forum*, **2**, 91–101.

Bratton, J.H., (1990) *A Review of the Scarcer Ephemeroptera and Plecoptera of Great Britain*, Research and Survey in Nature Conservation, no. 29. Peterborough: Nature Conservancy Council.

Bratton, J.H. (ed.) (1991) *British Red Data Books: 3. Invertebrates other than Insects*. Peterborough: Joint Nature Conservation Committee.

Brookes, A. and Shields Jr, F.D. (eds) (1996) *River Channel Restoration: Guiding Principles for Sustainable Projects*. Chichester: John Wiley.

Clark, P. (1984) Recent records of alien crabs in Britain. *The Naturalist*, **109**, 111–112.

Cotton, D.F.C. (1982) *Coenagrion lunulatum* (Charpentier) (Odonata: Coenagriidae) new to the British Isles. *Entomologist's Gazette*, **33**, 213–214.

Cotton, D. (1996) First Irish living record of *Viviparus viviparus* (L.). *Irish Naturalists' Journal*, **25**, 278.

Davies, L. (1968) *A Key to the British Species of Simuliidae (Diptera) in the Larval, Pupal and Adult Stages*, Scientific Publications no. 24. Ambleside: The Freshwater Biological Association.

Department of the Environment (1994a) *Biodiversity. The UK Action Plan*. London: HMSO.

Department of the Environment (1994b) *Planning Policy Guidance: Nature Conservation*. London: Department of the Environment PPG9.

Department of the Environment (1995) *Biodiversity: The UK Steering Group Report*. London, HMSO.

Department of the Environment (Northern Ireland) (1997) *River Quality in Northern Ireland 1995*. London: Environment and Heritage Service.

Dick, J.T.A. (1996) Animal introductions and their consequences for freshwater communities, in Giller, P.S. and Myers, A.A. *Disturbance and Recovery in Ecological Systems*. Dublin: Royal Irish Academy, pp. 47–58.

Disney, R.H.L. (1975) *A key to the Larvae, Pupae and Adults of the British Dixidae (Diptera), the Meniscus Midges*, Scientific Publications no. 31. Ambleside: The Freshwater Biological Association.

Elliott J.M. and Tullet, P.A. (1982) *Provisional Atlas of the Freshwater Leeches of the British Isles*, Occasional Publication no. 14. Ambleside: The Freshwater Biological Association.

Elliott, J.M., O'Connor, J.P. and O'Connor, M.A. (1979) A key to the larvae of Sialidae (Insecta: Megaloptera) occurring in the British Isles. *Freshwater Biology*, **9**, 511–514.

English Nature (1997) *Wildlife and Freshwater. An agenda for sustainable management*. Peterborough: English Nature.

Environment Agency (1997) *The Quality of Rivers and Canals in England and Wales 1995*. Bristol: Environment Agency.

Environment Agency (2000) *Focus on Biodiversity*. Bristol: Environment Agency.

Falk, S.J. (1991) *A Review of the Scarce and Threatened Flies of Great Britain. Part 1*, Research and Survey in Nature Conservation, no. 39. Peterborough: Joint Nature Conservation Committee.

Foster, G.N. (in press) *A Review of the Scarce and Threatened Coleoptera of Great Britain. Part 3. Aquatic Coleoptera*. Peterborough: Joint Nature Conservation Committee.

Fozzard, I.R., Doughty, C.R. and Clelland, B.E. (1994) Invertebrates, in Maitland, P.S., Boon, P.J. and McLusky D.S. (eds) *The Fresh Waters of Scotland: A national resource of international significance*. Chichester: John Wiley, pp. 171–190.

Furse, M.T., McDonald, I. and Abel, R. (1998) *A Revised Coded Checklist of Freshwater Animals Occurring in the British Isles*. Biological Dictionary Determinand Working Group. Unpublished software, Version 1.1. London: Department of the Environment.

Gee, H. and Young, J.O. (1993) The food niches of the invasive *Dugesia tigrina* (Girard) and indigenous *Polycelis tenuis* Ijima and *P. nigra* (Muller) (*Turbellaria; Tricladida*) in a Welsh lake. *Hydrobiologia*, **254**, 99–106.

Giller, P.S. (1998) (ed.) *Studies in Irish Limnology*. Dublin: The Marine Institute.

Gledhill, T., Sutcliffe, D.W. and Williams, W.D. (1993) *British Freshwater Crustacea Malacostraca: A Key with Ecological Notes*. Scientific Publications no. 52. Ambleside: The Freshwater Biological Association.

Gunn, R.J.M. and Blackburn, J.H. (1997) *Caenis pseudorivulorum* Kieffermuller (Ephem., Caenidae), a mayfly new to Britain. *Entomologist's Monthly Magazine*, **133**, 97–100.

Gunn, R.J.M. and Blackburn, J.H. (1998) *Caenis beskidensis* Sowa (*Ephemeroptera, Caenidae*), a mayfly new to Britain. *Entomologist's Monthly Magazine*, **134**, 94.

Holdich, D.M., Rogers, W.D. and Reader, J.P. (1995) *Crayfish Conservation*, R & D Note 466. Bristol: National Rivers Authority.

House, M.A., Ellis, J.B., Herricks, E.E., Hvitved-Jacobsen, T., Seager, J., Lijklema, L., Aalderink, H. and Clifforde, I.T. (1993) Urban drainage – impacts on receiving water quality. *Water Science and Technology*, 27, 117–158

Hynes, H.B.N. (1977) *A Key to the Adults and Nymphs of the British Stoneflies (Plecoptera) with Notes on their Ecology and Distribution*, Scientific Publications no.17. Ambleside: The Freshwater Biological Association.

IUCN (1994) *IUCN Red List Categories. Prepared by the IUCN Species Survival Commission. As approved by the 40th Meeting of the IUCN Council, Gland, Switzerland*. Gland: The World Conservation Union.

Kerney, M.P. (1976) *Atlas of the Non-Marine Mollusca of the British Isles*. Cambridge: Institute of Terrestrial Ecology.

Kerney, M.P. (1999) *Atlas of Land and Freshwater Molluscs of Britain and Ireland*. Colchester: Harley Books.

Kirby, P. (1992) *A Review of the Scarce and Threatened Hemiptera of Great Britain*, UK Nature Conservation, no. 2. Peterborough: Joint Nature Conservation Committee.

Macan, T.T. (1965) *A Revised Key to the British Water Bugs (Hemiptera-heteroptera) with Notes on their Ecology*, Scientific Publications no. 16. Ambleside: The Freshwater Biological Association.

Macan, T.T. (1970) *A Key to the Nymphs of the British Species of Ephemeroptera with Notes on their Ecology*. Scientific Publications no. 20, Ambleside: The Freshwater Biological Association.

Macan, T.T. (1973) *A Key to the Adults of the British Trichoptera*, Scientific Publications no. 28. Ambleside: The Freshwater Biological Association.

Macan, T.T. (1974) Freshwater Invertebrates, in Hawksworth D.L. (ed.) *The Changing Flora and Fauna of Britain*, London: Academic Press, pp. 143–155.

Macan, T.T. (1977) *A Key to the British Fresh- and Brackish-Water Gastropods with notes on their Ecology*, Scientific Publications no. 13. Ambleside: The Freshwater Biological Association.

Mainstone, C.P., Holmes, N.T., Armitage, P.D., Wilson, A.M., Marchant, J.H., Evans, K., Solomon, D. and Westlake, D. (1999) *Chalk Rivers-Nature Conservation and Management*. Report to English Nature and the Environment Agency.

Maitland, P.S. (1977) *A Coded Checklist of Animals occurring in Fresh Water in the British Isles*. Edinburgh: Institute of Terrestrial Ecology.

McCarthy, T.K. and Fitzgerald, J. (1997) The occurrence of the zebra mussel *Dreissena polymorpha* (Pallas, 1771), an introduced biofouling freshwater bivalve in Ireland. *Irish Naturalists' Journal*, 25 (11/12), 413–416.

McGarrigle, M.L., Lucey, J. and Clabby, K.C. (1992) Biological assessment of river quality in Ireland in Newman, P.I., Piavaux, M.A. and Sweeting, R.A. (eds) *River Water Quality: Ecological Assessment and Control*. Brussels: Commission of the European Communities, pp. 371–393.

Merritt, R., Moore, N.M. and Eversham, B.C. (1996) *Atlas of the Dragonflies of Britain and Ireland*, Natural Environment Research Council. London: HMSO.

National Rivers Authority (1992) *The Influence of Agriculture on the Quality of Natural Waters in England and Wales*, Water Quality Series no. 6. Bristol: National Rivers Authority.

O'Connor J.P. and Bond, K.G.M. (1995) *Limnephilus pati* O'Connor rediscovered and *L. tauricus* Schmid (Trichoptera: Limnephilidae) new to Ireland. *Entomologists's Gazette*, 46, 207–208.

Petts, G.E. (1984) *Impounded Rivers – Perspectives for ecological management*. Chichester: Wiley-Interscience.

Pond Action (1994) *National Pond Survey: 1989–1993. Interim Report to the World Wide Fund for Nature*. Oxford: Pond Action.

Reynolds, J.D. (1988) Crayfish extinctions and crayfish plague in central Ireland. *Biological Conservation*, **45**, 279–285.

Reynolds, J.D. (1998) *Ireland's Freshwaters*. Dublin: The Marine Institute.

Rosell, R.S., Maguire, C.M. and McCarthy, T.K. (1999) First reported settlement of zebra mussels *Dreissena polymorpha* in the Erne system, Co. Fermanagh, Northern Ireland. *Proceedings of the Royal Irish Academy*, **98** B, no. 3, 191–193.

RSPB, NRA and RSNC 1994 *The New Rivers and Wildlife Handbook*. Sandy: RSPB, NRA and Wildlife Trusts.

Scottish Office (1995) *Water Quality Survey of Scotland 1995*, Scottish Office, Agricultural, Environmental and Fisheries Department. Edinburgh: The Stationery Office.

Shirt, D.B. (1987) *British Red Data Books. 2. Insects*. Peterborough: Nature Conservancy Council.

Wallace, I.D. (1991) *A Review of the Trichoptera of Great Britain*, Research and Survey in Nature Conservation, no. 32. Peterborough: Nature Conservancy Council.

Walton, G.A. (1985) *Microvelia pygmaea* (Dufour) (Hemiptera: Veliidae) in Ireland. *Irish Naturalists Journal*, **21**, 493–495.

Williams, P., Biggs, J., Corfield, A., Fox, G., Walker, D. and Whitfield M. (1997) Designing new ponds for wildlife. *British Wildlife*, **8**, 137–150.

Williams, P.J., Biggs, J., Barr, C.J., Cummins, C.P., Gillespie, M.K., Rich, T.C.G. *et al.* (1998) *Lowland Pond Survey 1996*. London: Department of the Environment, Transport and the Regions.

Wood, P.J. and Armitage, P.D. (1997) Silt and siltation in the lotic environment. *Environmental Management*, **21**, 203–217.

Wright, J.F. (1987) Colonization of rivers and canals in Great Britain by *Dugesia tigrina* (Girard) (Platyhelminthes: Tricladida). *Freshwater Biology*, **17**, 69–78.

Wright, J.F., Blackburn, J.H., Gunn, R.J.M., Furse, M.T., Armitage, P.D., Winder, J.M. and Symes, K.L. (1996) Macroinvertebrate frequency data for the RIVPACS III sites in Great Britain and their use in conservation evaluation. *Aquatic Conservation: Marine and freshwater ecosystems*, **6**, 141–167.

Wright, J.F., Moss, D., Clarke, R.T. and Furse, M.T. (1997) Biological assessment of river quality using the new version of RIVPACS (RIVPACS III), in Boon, P.J. and Howell, D.L. (eds) *Freshwater Quality: Defining the Indefinable?*, Edinburgh: The Stationery Office.

Wright, J.F., Gunn, R.J.M., Blackburn, J.H., Grieve, N.J., Winder, J.M. and Davy-Bowker, J. (2000) Macroinvertebrate frequency data for the RIVPACS III sites in Northern Ireland and some comparisons with equivalent data for Great Britain. *Aquatic Conservation: Marine and Freshwater Ecosystems*, **10**, 371–389.

Young, J.O. and Reynoldson, T.B. (1999) Continuing dispersal of freshwater triclads (Platyhelminthes; Turbelleria) in Britain with particular reference to lakes. *Freshwater Biology*, **42**, 247–262.

Chapter 14

Nematodes

Brian Boag and David J. Hunt

ABSTRACT

Nematodes are very numerous, worm-like animals, found in soil, plants, rivers, lakes, vertebrates and invertebrates. They are usually microscopic and require considerable expertise to identify. Little is known about their biology, distribution or even how many species there are in Britain except for a few which are economically important parasites of plants or animals. Nematodes can also be an important constituent of the soil nutrient web and have considerable potential as bio-indicators of soil processes and soil 'health'. Changes in agricultural practice and land use are not likely to cause extinction of any of the British soil nematode fauna but some animal parasitic species could be lost if the host itself becomes extinct. A variety of new species, mostly plant or insect parasitic, has been found and described from Britain in the last 25 years and this trend is likely to continue. Although new molecular techniques are being introduced into nematology and parasitology they cannot yet replace the expertise of traditional taxonomists, many of whom have retired in the last 25 years.

1 Introduction

Nematodes are small, multicellular worm-like animals and are found in virtually all environments – marine, freshwater and terrestrial – some species even being adapted to survive the climatic extremes found in the dry valleys of Antarctica. So numerous are they that N.A. Cobb (1914) remarked 'if all the matter in the universe except the nematodes were swept away, our world would still be dimly recognisable … We would find its mountains, hills, valleys, rivers, lakes and oceans represented by a film of nematodes'. Their trophic specialisms are diverse – many species are microbivorous, yet others are predatory (either on nematodes or other invertebrates), mycetophagous, or parasitic on higher plants, invertebrates or vertebrates. The diversity of nematodes is undoubtedly much greater than our current knowledge indicates, yet estimates of the total number of species remain nebulous, varying from hundreds of thousands to many millions (see for example Nielsen 1998).

In Great Britain the predominant nematologist of the nineteenth century was Henry Bastian who, in his *Monograph of the Anguillulidae*, described 100 new species belonging to 30 genera of which 23 were new (Bastian 1865). Bastian was also acutely aware that the contemporary generic scheme was far too conservative for the degree of morphological variation recorded within the *nematoda* and he should be regarded as one of the founding fathers of nematode systematics. Other seminal publications in British nematology include *Plant Parasitic Nematodes and the Diseases they Cause* by T. Goodey (1933) and his later *Soil and Freshwater Nematodes* (1951) a text subsequently revised by J.B. Goodey (1963). As an example of how even large soil nematodes could be readily overlooked, Hooper (1961) described five new species of large phytoparasitic nematodes from the UK, the existence of which in British soils had not been recognised until the appropriate extraction methodology was utilised.

Nematology within Britain and throughout the world has been mainly led by researchers investigating control measure for the most economically important species such as the potato cyst nematodes *Globodera rostochiensis* and *G. pallida* and parasitic species associated with domestic farm animals. This has resulted in some groups being better known than others. More recently, however, interest has widened to include other nematodes since their abundance and diversity provide them with the potential for exploitation as bio-indicators of environmental quality in both soil and aquatic ecotypes. Animal parasitic species may also play a part, as an imperfectly density-dependent factor controlling the populations of wild animals found in the countryside.

To understand why so little is known about nematodes – a mostly cryptic, microscopic and taxonomically difficult group – it is worth briefly considering some of the problems associated with their study.

2 Soil and plant nematodes

2.1 Sampling

A major problem when sampling qualitatively for soil nematodes relates to the fact that a negative result, rather than indicating absence, may only mean that a species was not detected. A properly constituted sampling strategy allied to appropriate extraction procedures will contribute to minimising such errors, but the sheer volume of the soil phase relative to that of the sample will usually preclude certainty as to the absence of a particular species. Soil-inhabiting nematodes have an aggregated distribution (Boag and Topham 1984; Webster and Boag 1992) and distinct assemblages can be detected (Boag and Topham 1985; Topham, Boag and McNicol 1991; Hominick *et al.* 1996). Sampling to detect the size of plant-parasitic nematode populations has received considerable attention (Schomaker and Been 1990) but the sampling procedures required to obtain a representative assemblage of species in an area have received less attention. Ettema (1998) reviewed the literature and reported that the number of species per single core sample could be as little as 3% of the total nematodes present (Orr and Dickerson 1966) while Johnson *et al.* (1972) found a single core sample could have half of the total species occurring in 32 core samples. Although differences in soil nematode species diversity have been recorded from as few as six sub-samples (Freckman and Ettema 1983) the comprehensive list by Hodda and Wanless (1994) of 154 species from English chalk grassland was obtained from 96 bulked soil samples.

Because of the aggregated distribution pattern shown by nematodes in soil, a stratified random sampling procedure has been recommended as the optimum method for quantifying their population size and a similar procedure should probably be used for investigating nematode diversity (Boag *et al.* 1992). Tools used to collect samples from the soil can also affect the efficiency with which nematodes are recovered, narrow sample cores having an adverse effect on numbers of the larger nematodes recovered from soil samples (Boag and Brown 1985).

2.2 Extraction

Another variable significantly influencing both the size and species composition of a sample is the extraction procedure. Many of these procedures are dependent on the mobility of the nematode to separate it from inert soil debris (Whitehead and Hemming 1965; Flegg 1967; Brown and Boag 1988). Others, dependent upon a rising column of water (Trudgill *et al.* 1973), specific gravity gradient (Griffiths *et al.* 1991) or centrifugation (Harrison and Green 1976), are generally considered to give a more representative sample of the community structure of nematode populations but are more labour intensive and time consuming. These extraction techniques, or modifications of them, are probably those most commonly used by nematologists in the British Isles, but more comprehensive coverage of extraction techniques, including those utilised by researchers in Europe and America, is given by Hooper (1985a) and Hunt and De Ley (1996). Nematodes in plant material, usually located within the roots but sometimes in the aerial parts, are typically extracted using the same principles as those for soil nematodes, albeit with certain modifications. Hooper (1985b) gives details of extraction techniques.

2.3 Identification

Nematology is still a relatively young science with new species and genera, particularly from tropical soils and the marine environment, being described in large numbers. In this situation both the taxonomy and systematics of the group as a whole are in such a state of continuous change that any identification keys rapidly become out of date. New species of terrestrial nematodes described from the British Isles since 1973 are listed in Table 14.2.

One of the first attempts to produce a comprehensive textbook on soil nematodes was that of Goodey (1933), and later Goodey (1951) and J.B. Goodey (1963) as noted above in section 1. This standard work stood the test of time until the mid-1980s when the impracticality of having a single comprehensive book covering all soil and plant nematodes became apparent. Siddiqi (1986) produced a monograph *Tylenchida, Parasites of Plants and Insects* while Jairajpuri and Ahmad (1992) produced the complementary *Dorylaimida*. Andrássy (1984) also prepared a treatise covering the major groups of free-living soil nematodes. Since then more specialist books have become available, for example *Aphelenchida, Longidoridae and Trichodoridae: Their Systematics and Bionomics* (Hunt 1993). There have also been identification guides confined to a certain country, for example The Netherlands (Bongers 1987) and South Africa (Heyns 1971), but these are of limited use for identifying the British nematode species. The lack of any book dedicated to the British soil nematode fauna has meant that progress in studying nematode ecology and the use of nematodes as bio-indicators of environmental quality has been adversely affected. Recently, F. Wanless (pers. comm.) has produced a provisional checklist of the nematode species from soils and freshwater in the British Isles.

Knowledge of the British nematode fauna is based on the relatively few scientific papers that attempt to give species/taxa listings for the habitat surveyed. The first

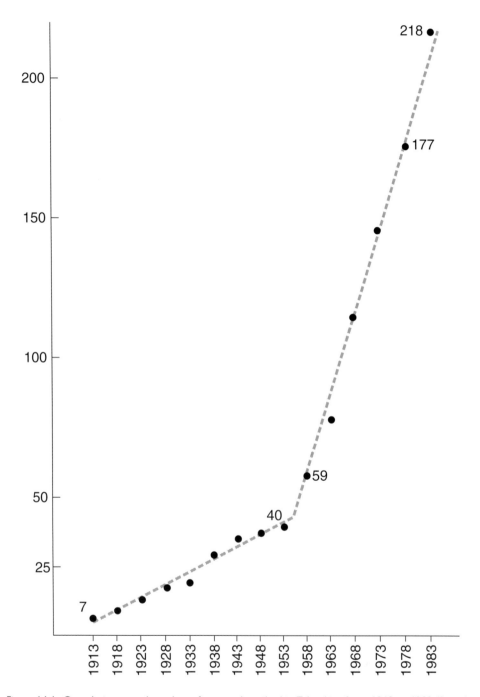

Figure 14.1 Cumulative annual number of genera described in *Tylenchina* from 1913 to 1983 (from Luc *et al.* 1987).

attempt was probably that of Robertson (1929) who reported the species present in arable soil in the north of Scotland. Banage (1962) studied nematodes from Moor House National Nature Reserve while Winslow (1964) and Yuen (1966) investigated the nematode fauna at Rothamsted Experimental Station, Hertfordshire. Yeates (1970, 1971) described the nematode assemblages from some of the Orkney Islands and Wicken Fen while Boag (1974) reported on the genera of nematodes associated with forest and woodland trees in Scotland.

More recently, intensive sampling by Hodda and Wanless (1994) at the Site of Special Scientific Interest at Porton Down listed 154 species from 96 samples while Ruess et al. (1966) found 79 nematode species from an experimental Sitka spruce site near Edinburgh. The composite list of all soil and fresh water nematodes compiled by Wanless (see above), although far from being complete, is the first attempt at a comprehensive assessment of the soil and freshwater nematode species in the British Isles and will be of value in a number of fields of British nematology. The production of such a British nematode species list points out the extensive deficiencies in our knowledge and expertise of many of the major soil-inhabiting groups including rhabditids, cephalobids, enoplids, plectids, dorylaimids and so on, particularly when compared with that in other countries such as The Netherlands, where the nematode fauna is far better documented (see Bongers 1987).

2.4 Nematode abundance

One of the contributing factors making nematodes potentially useful bio-indicators of soil and water quality is that they are very numerous organisms and therefore amenable to statistical analysis. Actual numbers vary considerably between sites, soil type and over time (Boag and Yeates 1998). Counts of between 500 to 10 000 per 200 g soil of migratory nematodes are quite common in British soils and even higher counts of sedentary semi-endoparasites are possible, for example *Globodera* sp. egg counts can be as high as 20 000 per 200 g soil (Jones and Kempton 1982). Another useful characteristic for using nematodes as environmental bio-indicators is that they are relatively immobile, migration of even large nematodes e.g. *Longidorus elongatus* in the soil phase being a maximum of 10 cm per month under laboratory conditions (Thomas 1981).

3 Nematodes in freshwater estuaries

Intertidal areas around the British coast have been studied in some detail with complete species lists produced for some of them, the numbers varying between 37 to over 100 species per site (Capstick 1959; McIntyre and Murison 1973; Platt 1977; Warwick and Price 1979). Analysis of the community structure of these nematodes has shown them to be bio-indicators of sewage pollution in the Tay estuary (Neilson 1993).

4 Nematodes of invertebrates

Nematodes are known to be associated with or infect a broad spectrum of insects and other invertebrate animals (see Poinar 1975), but very little is known about either the diversity and distribution of such species in Britain or their ecology. The knowledge we do have is largely confined to nematodes being exploited as biological control agents of slugs and insects. Several surveys in the British Isles (Blackshaw 1988; Hominick and Briscoe 1990; Griffin *et al.* 1991; Boag *et al.* 1992; Hominick *et al.* 1995; Gwynn and Richardson 1996) have yielded a relatively restricted list of entomopathogenic species. Gwynn and Richardson found *Steinernema affine, S. feltiae, S. kraussi* and three undescribed *Steinernema* species. However, the morphology and morphometrics of these species are so close in this highly conserved group as to necessitate the use of restriction fragment length polymorphisms (RFLPs) to reliably differentiate between them (Reid and Hominick 1992). *Heterorhabditis* sp. was also recorded from the British Isles (Hominick and Briscoe 1990). The nematode *Phasmarhabditis hermaphrodita*, a naturally occurring nematode in Britain, has been cultured and used as a biological control agent against the slug *Deroceras reticulatum* (Wilson *et al.* 1993) and is now commercially available.

5 Nematodes of mammals, birds and fish

Although mammals and birds can harbour large numbers of nematodes, very little is known about either the species or size of populations in wild animals in Britain when compared with those in domestic animals. It is not intended to deal with these parasites in any detail in this chapter other than that necessary to bring them to the attention of the reader.

The techniques used to collect nematodes of vertebrates and the identification of the more common species in domestic animals are given in the Ministry of Agriculture, Fisheries and Food (MAFF) Technical Bulletin no.18 (1987). The nematode populations found in domestic animals were dramatically reduced when new drugs were introduced in the 1960s (Michel 1976) especially when integrated with other management practices (Paton and Boag 1987). However, since then the development of resistance to anthelmintics has meant that new control strategies have had to be developed, including the use of nematophagous fungi (Walker and Larsen 1993), breeding resistant/tolerant animals (Prichard 1990) and altering animal nutrition (Van Houtert and Sykes 1996).

The nematode fauna of most of the 55 wild mammal listed in *The Handbook of British Mammals* (Corbet and Harris 1991) is still very poorly known. There are a number of reasons for this, not least the fact that in order to collect and identify most parasitic nematodes the animals have to be dead. Many mammal species are rare and protected by law so that a knowledge of their nematode fauna can often only be gleaned from fresh road kills, for example. What little knowledge we do have about the nematode fauna of most of the British mammals can be obtained from reading the paragraph on parasites in *The Handbook of British Mammals*. Some wild animals, however, are classed as pests and are therefore not protected to the same extent by law (e.g. deer, rabbit, fox and mink). Consequently, more knowledge exists about the nematode fauna of these species. A list of the number of nematodes associated with a range of British mammals is given in Table 14.1. Some of the mammals can harbour the same nematode species; for example, rabbits, brown hares and mountain hares all have the same native fauna of *Graphidium strigosum, Trichostrongylus retortaeformis* and *Passaburus ambiguus* (Boag 1985, 1987; Boag and Jason 1986). Lagomorphs and artiodactyls can also harbour many of the nematode parasites found in domestic cattle

Table 14.1 Number of nematode parasites recorded from selected British mammals.

Mammal	Number of parasites	Reference
Hedgehog (*Erinaceus europeus*)	5	Cameron and Parnell (1933), Boag and Fowler (1988)
Common shrew (*Sorex araneus*)	3	Lewis (1987)
Pygmy shrew (*S. minutus*)	2	Lewis (1987)
Rabbit (*Oryctolagus cuniculus*)	3	Boag (1972), (1985)
Brown hare (*Lepus europaeus*)	3	Irvin (1970), Boag (1987)
Mountain hare (*L. timidus*)	3	Irvin (1970), Boag and Jason (1986)
Bank vole (*Clethrionomys glareolus*)	7	Alibhai and Gipps (1991)
Field vole (*Microtus agrestis*)	6	Gipps and Alibhai (1991)
Wood mouse (*Apodemus sylvaticus*)	6	Flowerdew (1991)
House mouse (*Muscus domesticus*)	4	Berry (1991)
Fox (*Vulpes vulpes*)	9	Beresford-Jones (1961), Lloyd (1980)
Badger (*Meles meles*)	6	Jones *et al.* (1980), Hancox (1980)
Otter (*Lutra lutra*)	10	Stephens (1957) Weber (1991)
Wildcat (*Felis sylvestris*)	1	Burt Pike *et al.* (1980)
Red deer (*Cervus elaphus*)	10[1]	Dunn (1964), Sleeman (1982), Hollands (1985)
Fallow deer (*Dama dama*)	18[1]	Batty and Chapman (1970)

[1] Numbers of nematodes can include those from domestic animals and could be even greater (Hawkins 1988).

and sheep and may help to spread parasites between farm animals (Hawkins 1988). No nematode parasites have been reported from British bats (Corbet and Harris 1991).

The nematode fauna of birds in Britain has received little attention except for *Trichostrongylus tenuis* which infects red grouse (*Lagopus lagopus scoticus*) where it can cause considerable loss of condition and affect bird numbers (Wilson 1983; Hudson 1986). Similarly, little is known about nematodes found in freshwater fish although a high proportion may be infected (Brattey 1979). Kennedy (1974) gives a checklist of the British and Irish freshwater fish parasites and their distribution.

6 Nematological expertise

The Natural History Museum is the only museum in the UK that currently has expertise specifically in nematode taxonomy. It is also an internationally recognised depository for type material from around the world. However, much of their work involves research on deep-sea nematodes with relatively little effort going into identifying and studying terrestrial nematodes. The only other organisation that now has any in-depth expertise in this field is CABI Bioscience's UK Centre (Egham). This centre also acts as an international depository for type material, has an extensive slide collection of plant and soil nematodes and also runs identification courses. The Institute of Arable Crop Research, Rothamsted Experimental Station (RES), although currently lacking a nematode taxonomist, also has an extensive collection of nematode type material and much other slide-mounted material amassed over a long period of time.

To aid identification computer programs have been written for some of the more common nematode groups such as the genera *Helicotylenchus* (Boag *et al.* 1988) and *Longidorus* (Rey *et al.* 1988) but these are not widely used. An attempt to develop an expert system, NEMISYS (Diederich and Milton 1988), failed. New computer programs, aided by spectacular advances in processor power, may be more user friendly and be taken up more readily in the future (Ryss 1997; Boddy *et al.* 1998; Dallivity 1998).

Nematology and parasitology are taught at a number of universities throughout Great Britain although always as a minor part of a degree course. These courses are generally, although not always, part of an applied degree in agricultural, environmental or veterinary science and reflect the economic importance nematodes have in the agricultural, horticultural and veterinary fields. Apart from the universities, nematological expertise associated with economically important species may be found at government-funded agencies including research institutes. Even at these establishments (such as the Central Science Laboratory, Scottish Agricultural Science Agency, Scottish Agricultural College, ADAS, Institute of Grassland and Environmental Research Aberystwyth Research Centre, Silsoe Research Institute, Institute for Animal Health Compton, Moredun Research Institute and the Scottish Crop Research Institute) nematological/parasitological expertise is usually focused on only a very limited number of phytoparasitic nematodes or animal parasites. The thrust at these sites is nearly always aimed at developing better, more economic and environmentally friendly ways of controlling a relatively narrow spectrum of nematodes.

There have recently been several attempts to use nematodes as bio-indicators of environmental change, a role for which they have considerable potential (e.g. Bongers 1990; de Goede 1993; Yeates *et al.* 1993). Nematodes were monitored at the Scottish Crop Research Institute during a MAFF commissioned research and development experiment 'Towards a lower input system minimising agrochemicals and nitrogen' (TALISMAN) and a comparable long-term experiment on the effects of setting land aside out of agricultural production as required by European Union regulations. Nematodes have also been monitored by commercial companies at pollution sites as bio-indicators of soil and sediment quality.

7 Data capture

Because nematodes are mostly small, cryptic and difficult to study they have mostly been avoided by enthusiastic amateurs and thus the majority of the research is dependent on professional nematologists employed at museums or other research establishments, particularly those involved in agriculture. Data capture has been largely haphazard and often related to certain crops or projects rather than a planned accumulation for a data bank.

There is no national scheme co-ordinating the nematological data produced from the different centres in Britain. The Natural History Museum, CABI Bioscience's UK

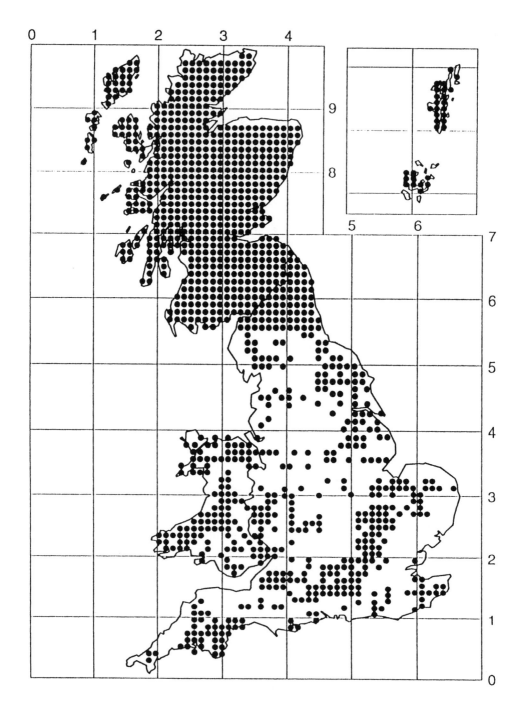

Figure 14.2 Distribution of 10-km squares sampled.

Centre and Rothamsted Experimental Station are all recognised depositories for nematological type material but do not actively seek data on UK nematodes and parasites. The most comprehensive set of distribution data was that produced from a combination of a North Atlantic Treaty Organisation (NATO) financed survey of virus-vector nematodes (Taylor and Brown 1976) and a Natural Environment Research Council (NERC) financed survey of nematodes associated with trees in Scotland (Boag 1974). A number of maps were produced for the *Provisional Atlas of the Nematodes of the British Isles* (Heath *et al.*1977). Over 5400 samples were collected from throughout the British Isles and stored at the Scottish Crop Research Institute (Fig. 14. 2). Initially they were examined for economically important virus-vector nematodes belonging to

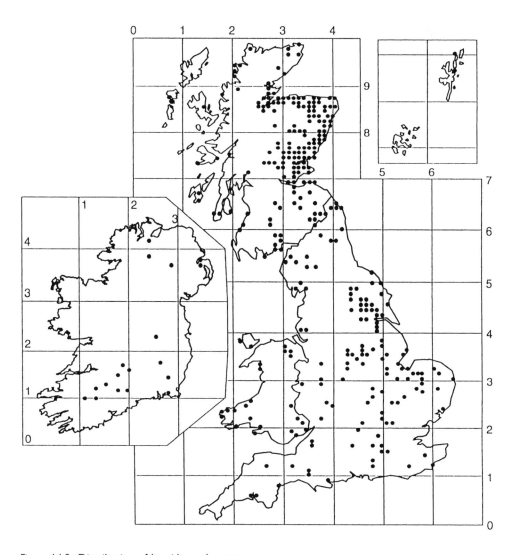

Figure 14.3 Distribution of *Longidorus elongatus*.

the genera *Longidorus, Xiphinema, Trichodorus, Paratrichodorus* and *Paralongidorus* but subsequently slides were made of other nematode groups (i.e. criconematid, hoplolaimid and mononchid nematodes) before the samples had to be disposed of, the nematodes having disintegrated as a result of being stored in TAF (Hooper *et al.* 1983).

The differences in geographical distribution of the nematodes mapped during the surveys were not just of academic interest. Whereas *Longidorus elongatus*, the vector of raspberry ringspot virus (RRV), had a cosmopolitan distribution throughout the British Isles (Fig. 14.3) that of *Xiphinema diversicaudatum*, the vector of strawberry latent ringspot virus (SLRV), was restricted to the south of Scotland and England (Fig. 14.4) (Taylor and Brown 1976). This meant that land planted to strawberries in the

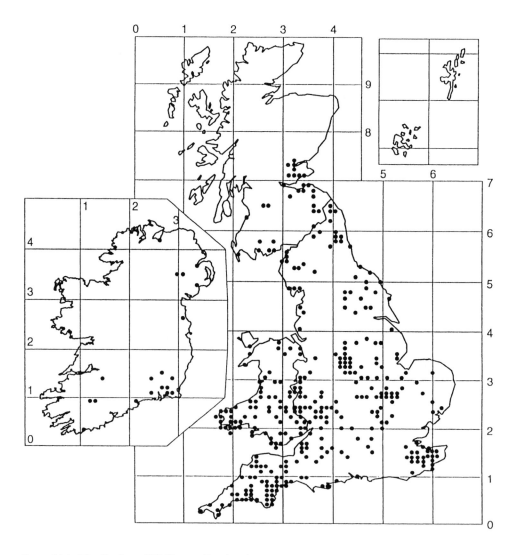

Figure 14.4 Distribution of *Xiphinema diversicaudatum*.

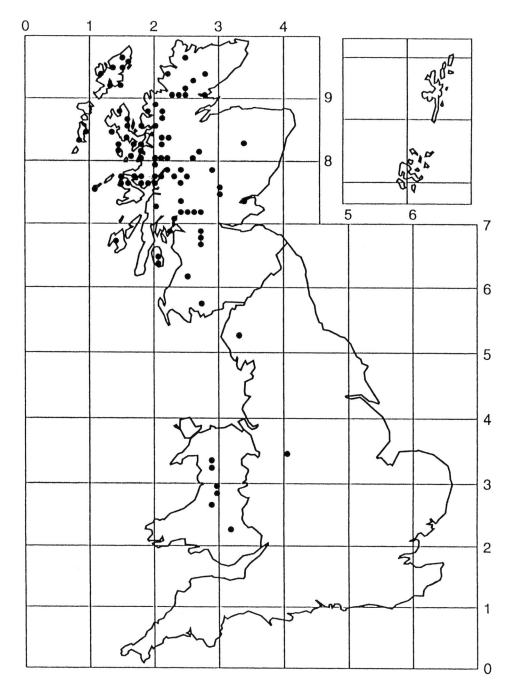

Figure 14.5 Distribution of *Truxonchus dolichurus*.

northern part of Scotland did not need to be tested for virus-vector nematodes. These data also revealed other distinct distribution patterns, thus providing an insight into the ecology of certain nematode species. For instance, *Truxonchus dolichurus*, a predatory nematode, was mainly found in the north-west of Scotland and was associated with wet acidic soils (Boag *et al.* 1992c; Fig. 14.5).

8 Changes since 1973

It is difficult to objectively estimate how many new nematode species have become established in Britain in the last 25 years since there is no definitive baseline set of data for comparison. This state of affairs is due to the size, cryptic nature and lack of biosystematic coverage of nematodes. Most phytoparasitic species are polyphagous and are therefore not affected by the increased rarity or loss of a particular higher plant from the flora. Many species (including both new records and new species) have been added to the British list of plant and soil nematodes in the last 25 years (Table 14.2) and a number of new entomopathogenic nematodes (*Steinernematidae*: *Steinernema*) have been recorded from British soils using molecular techniques (RFLP banding profiles), although these species have yet to be formally named and described by traditional morphological methods.

Potential new introductions entering the UK via official channels are often intercepted by quarantine authorities, although some alien species have become established in recent years e.g. *Trichodorus sparsus*, with some becoming relatively widespread e.g. *Pratylenchus bolivianus*.

It is difficult to state with any degree of certainty if or when a nematode species becomes 'extinct'. This difficulty also relates to the cryptic nature and habitat of nematodes plus the absence of any nationwide monitoring scheme. *Anguina tritici*, a seed-borne nematode – and the first phytoparasitic species to be discovered (Needham 1743) – used to be common in wheat fields but is now apparently extinct in Britain,

---

Table 14.2 New species of soil-dwelling, plant-parasitic, lichenicolous and insect-parasitic nematodes described from Britain since 1973 (the species epithets are in combination with their original designated genus).

Amplimerlinius globigerus	*Nothotylenchus adasi*
Aphelenchoides lichenicola	*Ottolenchus cabi*
A. richardsoni	*Pratylenchus wescolagricus*
A. varicaudatus	*Pseudhalenchus leevalleyensis*
Ektaphelenchoides winteri	*P. siddiqii*
Filenchus recisus	*Rotylenchus ouensensis*
Gastromermis metae	*Safianema lutonense*
Helicotylenchus scoticus	*Seinura paynei*
Heterodera pallida	*Trichodorus hooperi*
Howardula husseyi	*Tylenchus rex*
Meloidogyne maritima	*Zanenchus salmae*
Merlinius processus	

---

although still common in Mediterranean countries, eastern Europe and the Indian subcontinent. Extinction was due to a combination of efficient seed-cleaning techniques and a restricted host range and probably took place sometime after the mid-1950s. Nematodes that are obligate parasites of endangered vertebrates or invertebrate hosts may also be in danger of co-extinction, but once again few data are available as so few potential hosts, particularly invertebrates, have been examined for nematodes.

Other changes since 1973 concern an alarming decline in the number of nematologists with the relevant expertise to identify and classify the ever-increasing number of new species discovered. In the last 25 years nematode taxonomists have retired from establishments such as the International Institute of Parasitology (now merged into CABI Bioscience) and Rothamsted Experimental Station and not been replaced. This phenomenon is not just restricted to Britain but affects many other countries such as The Netherlands, France, Germany and the USA. Some of the posts previously held by traditional taxonomists have gone to molecular biologists who, using new techniques, can reliably identify individual nematode juveniles, something traditional taxonomists often have difficulty doing. However, molecular techniques are expensive and the results obtained are not necessarily clearcut; they may also suffer from interpretation difficulties. Traditional taxonomists are still required to characterise the 'new species' indicated by molecular methodologies. In this respect, most of the new entomopathogenic nematodes isolated using RFLP have still not been formally named and described.

9 Factors affecting change in the nematode fauna since 1973

The major potential impact on many terrestrial nematodes in Britain is change in agricultural practice and land use. Since 1973 the hectarage of oilseed rape has increased from virtually nothing to an estimated 467 000 ha in 1997 (Anon. 1997) and 'set-aside' land now covers up to 15% of the formerly arable land in the United Kingdom. However, evidence from experiments financed by both the MAFF and the Pesticide Safety Directorate (PSD) into the effects of set-aside and different levels of pesticide respectively have found that neither have had any significant effect on soil-inhabiting nematodes (Boag *et al.* 1998a, b). Afforestation of land previously used for agriculture could also affect soil nematode communities on a national scale but it is too soon to be able to predict to what extent this might occur. Research on the total nematode taxa inventory fauna of the relict Caledonian Scots pine forest near Aviemore in Scotland has shown that a large number of the total nematode community was not recorded in the soil at all, but occurred in the lichen, mosses and other types of secondary vegetation (V. Peneva, pers. comm.). Pollution, including raised levels of heavy metals, is another factor likely to adversely affect numbers and diversity and while effects on soil nematodes have been observed abroad (Kosthals *et al.* 1996) there is no evidence, as yet, of this happening to any extent in Britain.

10 Prognosis for the next 25 years

To the best of our rather scant knowledge no plant, soil, insect or animal nematode is directly at risk in Britain, but secondary factors, such as the loss of an associated animal host, may have an impact: for example, the endangered capercaille in which the senior author has recorded *Capillaria caudinflata* and *Heterakis gallinae* (L.F. Khalil, pers. comm.). Modern agricultural practices and crops are unlikely to have any adverse affect on nematodes and the introduction of more diverse crops and reduction in the use of pesticides may eventually increase nematode biodiversity in agricultural land. The government policy towards helping farmers to manage environmentally sensitive areas in a more extensive way and the protection of sites of special scientific interest (SSSIs) to conserve the habitats of both plants and animals may also incidentally protect certain niche-specific nematode species.

It is envisaged that research will continue on management strategies for economically important species e.g. the development of resistant potato varieties against *Globodera pallida*. Other aspects currently under consideration involve the status of nematodes in the soil community, their role in the soil nutrient cycle and their use as bio-indicators of pollution and/or soil health.

Future advances in molecular techniques may complement those of traditional taxonomists, for example in identifying individual juvenile stages of nematodes and molecular barcoding. Molecular techniques are increasingly used to investigate nematode biodiversity and evolutionary relationships occurring within the phylum nematoda (Blaxter *et al.* 1998). There will still be a need in Britain for traditional nematode taxonomists to underpin all the other activities occurring in nematology, although at present these experts are, to the detriment of both nematology and parasitology, neither being trained nor recruited.

ACKNOWLEDGEMENTS

The work of the senior author was financed by the Scottish Office Agriculture Environment and Fisheries Department.

REFERENCES

Alibhai, S.K. and Gipps, J.H.W. (1991) Bank vole, in Corbet, G.B. and Harris, S. (eds) *The Handbook of British Mammals*. Oxford: Blackwell Scientific Publications, pp. 192–203.

Andrássy, I. (1984) *Klasse Nematoda: Ardnungen Monhysterida, Desmoscolecida, Araeolaimida, Chromadorida, Rhabditida*. Stuttgart: Gustav Fischer Verlag.

Anon. (1997) *Agriculture in the United Kingdom*. London: The Stationery Office.

Banage, W.B. (1962) Some nematodes from the Moor House National Nature Reserve, Westmoorland. *Nematologica* 7: 32–36.

Bastian, H.C. (1865) Monograph on the *Anguillulidae*, or free nematoids, marine, land, and freshwater; with descriptions of 100 new species. *Transactions of the Linnean Society of London* 25: 73–184.

Batty, A.F. and Chapman, N. (1970) Gastro intestinal parasites of wild fallow deer (*Dama dama* L.). *Journal of Helminthology* 44: 57–61.

Beresford-Jones, W.P. (1961) Observations on the helminths of British wild red foxes. *Veterinary Record* 73: 882–883.

Berry, R.J. (1991) House mouse, in Corbet, G.B. and Harris, S. (eds) *The Handbook of British Mammals*. Oxford: Blackwell Scientific Publications, pp. 239–247.

Blackshaw, R.P. (1998) A survey of insect parasitic nematodes in Northern Ireland. *Annals of Applied Biology* 113: 561–562.

Blaxter, M.L., De Ley, P., Garey, J.R., Liu, L.X., Schelderman, P., Vierstraete, A., Vanfleteren, J.R., Mackey, L.Y., Dorris, M., Frisse, L.M., Vida, J.T. and Thomas, W.K. (1998) A molecular evolutionary framework for the phylum Nematoda. *Nature* 392: 71–75.

Boag, B. (1972) Helminth parasites of the wild rabbit *Osyctolagus cuniculus* (L.) in north east England. *Journal of Helminthology* 46: 73–79.

Boag, B. (1974) Nematodes associated with forest and woodland trees in Scotland. *Annals of Applied Biology* 77: 41–50.

Boag, B. (1985) The incidence of helminth parasites from the wild rabbit *Oryctolagus cuniculus* (L.) in eastern Scotland. *Journal of Helminthology* 59: 61–69.

Boag, B. (1987) The helminth parasites of the wild rabbit *Oryctolagus cuniculus* and the brown hare *Lepus capensis* from the Isle of Coll, Scotland. *Journal of Zoology* (London) 212: 352–355.

Boag, B. and Brown, D.J.F. (1985) Soil sampling for virus-vector nematodes. *Aspects of Applied Biology* 10: 183–189.

Boag, B. and Fowler, P.A. (1988) The prevalence of helminth parasites from the hedgehog *Erinaceus europaeus* in Great Britain. *Journal of Zoology* (London) 215: 379–382.

Boag, B. and Iason, G. (1986) The occurrence and abundance of helminth parasites of the mountain hare *Lepus timidus* (L.) and the wild rabbit *Oryctolagus cuniculus* (L.) in Aberdeenshire, Scotland. *Journal of Helminthology* 60: 92–98.

Boag, B. and Topham, P.B. (1984) Aggregation of plant-parasitic nematodes and Taylor's Power Law. *Nematologica* 30: 348–357.

Boag, B. and Topham, P.B. (1985) The use of associations of nematode species to aid the detection of small numbers of virus-vector nematodes. *Plant Pathology* 34: 20–24.

Boag, B. and Yeates, G.W. (1998) Soil nematode biodiversity in terrestrial ecosystems. *Biodiversity and Conservation* 7: 617–630.

Boag, B., Helden, P.M., Neilson, R. and Rodger, S.J. (1998b) Observations on the effect of different management regimes of set-aside land on nematode community structure. *Applied Soil Ecology* 9: 339–343.

Boag, B., Neilson, R. and Brown, J.D.F. (1992) Seminar: Nematode sampling and prediction. *Nematologica* 38: 459–465.

Boag, B., Neilson, R. and Gordon, S.C. (1992) Distribution and prevalence of the entomopathogenic nematode *Steinernema feltiae* in Scotland. *Annals of Applied Biology* 121: 355–360.

Boag, B., Rodgers, S.J., Wright, G.McN., Neilson, R., Helden, P., Squire, G.R. and Lawson. (1998a) Influence of reduced agrochemical inputs on plant-parasitic nematodes. *Annals of Applied Biology* 133: 81–89.

Boag, B., Small, R.W., Neilson, R., Gould, J.H. and Robertson, L. (1992) The *Mononchida* of Great Britain: Observations on the distribution and ecology of *Anatonchus tridentatus*, *Truxonchus dolichurus* and *Miconchoides studeri* (Nematoda). *Nematologica* 38: 502–513.

Boag, B., Topham, B.B., Smith, P. and Fong San Pin, G. (1988) Advances in computer identification of nematodes. *Nematologica* 34: 238–245.

Boddy, L., Morris, C.W. and Morgan, A. (1998) Development of artificial neural networks for identification, in Bridge, P., Jefferies, P., Morse, D.R. and Scott, P.R. (eds) *Information Technology, Plant Pathology and Biodiversity*. Wallingford: CAB International.

Bongers, T. (1987) *De Nematoden van Nederland*. Utrecht: Koninklijke Nederlandse Natuurhistorische Verendging.

Bongers, T. (1990) The Maturity Index: an ecological measure of environmental disturbance based on nematode species composition. *Oecologia* 83: 14–19.

Bratty, J. (1979) Intestinal helminths from the fish of the Forth and Clyde canal at Temple, Glasgow. *Glasgow Naturalist* 19: 475–479.

Brown, D.J.F. and Boag, B. (1988) An examination of methods used to extract virus-vector nematodes (*Nematoda: Longidoridae* and *Trichodoridae*) from soil samples. *Nematologia Mediterranea* 16: 93–99.

Burt, M.D.B., Pike, A.W. and Corbett, L.K. (1980) Helminth parasites of wild cats in north-east Scotland. *Journal of Helminthology* 54: 303–308.

Cameron, T.W.M. and Parnell, I.W. (1933) The internal parasites of land mammals in Scotland. *Proceedings of the Royal Physiological Society of Edinburgh* 22: 133–154.

Capstick, C.K. (1959) The distribution of free-living nematodes in relation to salinity in the middle and uppper reaches of the Blyth estuary. *Journal of Animal Ecology* 28: 189–210.

Cobb, N.A. (1914) Nematodes and their relationships, in *United States of America Department of Agriculture Yearbook*, pp. 457–490.

Corbet, G.B. and Harris, S. (1991) *The Handbook of British Mammals*. London: Blackwell Scientific.

Courtney, W.D., Polley, D. and Miller, V.L. (1955) TAF, an improved fixative in nematode technique. *Plant Disease Reporter* 39: 570–571.

Dallivity, M.J., Paine, T.A. and Zurcher, E.J. (1998) Interactive keys, in Bridge, P., Jefferies, P., Morse, D.R. and Scott, P.R. (eds) *Information Technology, Plant Pathology and Biodiversity*, Wallingford: CABI.

Diederich, J. and Milton, J. (1998) Nemisys: an expert-system for nematode identification, in Fortuner, R. (ed.) *Nematode Identification Systems Technology*. New York: Plenum Press, pp. 45–63.

Dunn, A.M. (1964) *Trichostrongylus askivali* n. sp. (*Nematoda: Trichostrongylidae*) from the Red Deer. *Nature* 201: 841.

Ettema, C.H. (1998) Soil nematode diversity: species co-existence and ecosystem function. *Journal of Nematology* 30: 159–169.

Flegg, J.J.M. (1967) Extraction of *Xiphinema* and *Longidorus* species from soil by a modification of Cobb's decanting and sieving technique. *Annals of Applied Biology* 60: 429–437.

Flowerdew, J.R. (1991) Wood mouse, in Corbet, G.B. and Harris, S. (eds) *The Handbook of British Mammals*. Oxford: Blackwell Scientific, pp. 220–229.

Freckman, D.W. and Ettema, C.H. (1993) Assessing nematode communities in agroecosystems of varying human intervention. *Agriculture, Environment and Ecosystems* 45: 239–261.

Gipps, J.H.W. and Alibhai, S.K. (1991) Field vole, in Corbet, G.B. and Harris, S. (eds) *The Handbook of British Mammals*. Oxford: Blackwell Scientific, pp. 203–208.

Goede, de, R.G.M. (1993) *Terrestrial nematodes in a changing environment*. PhD thesis, Agricultural University, Wageningen.

Goodey, J.B. (1963) *Soil and Freshwater Nematodes*. London: Methuen.

Goodey, T. (1933) *Plant Parasitic Nematodes and the Diseases they Cause*. London: Methuen.

Goodey, T. (1951) *Soil and Freshwater Nematodes*. London: Methuen.

Griffin, C.T., Moore, J.F. and Downes, M.J. (1991) Occurrence of insect-parasitic nematodes (*Steinernematidae, Heterorhabditidae*) in the Republic of Ireland. *Nematologica* 37: 92–100.

Griffiths, B.S., Boag, B., Neilson, R. and Palmer, L. (1990) The use of colloidal silica to extract nematodes from small samples of soil or sediment. *Nematologica* 36: 465–473.

Gwynn, R.L. and Richardson, P.N. (1996) Incidence of entomopathogenic nematodes in soil samples collected from Scotland, England and Wales. *Fundamental and Applied Nematology* 19: 427–431.

Hancox, M. (1980) Parasites and infectious diseases of the Eurasian badger (*Meles meles* L.): a review. *Mammal Review* 10: 151–162.

Harrison, J.M. and Green, C.D. (1976) Comparison of centrifugal and other methods for standardization of extraction of nematodes from soil. *Annals of Applied Biology* 82: 299–308.

Hawkins, D. (1998) A parasitic interrelationship of deer and sheep on the Knelworth Park Estate, near Stevenage. *Deer* 7: 296–300.

Heath, J., Brown, D.J.F. and Boag, B. (1977) *Provisional Atlas of the Nematodes of the British Isles*. Monks Wood: Institute of Terrestrial Ecology.

Heynes, J. (1971) *A Guide to the Plant and Soil Nematodes of South Africa*. Cape Town: AA Balkema.

Hodda, M. and Wanless, F.R. (1994) Nematodes from an English chalk grassland: species distribution. *Nematologica* 40: 116–132.

Hollands, R.D. (1985) *Elaphostrongylus cervi cervi* in the central nervous system of red deer (*Cervus elaphus*) in Scotland. *Veterinary Record* 116: 584–585.

Hominick, W.M. and Briscoe, B.R. (1990) Occurrence of entomopathogenic nematodes (*Rhabditida*: *Steinernematidae* and *Heterorhabditidae*) in British soils. *Parasitology* 100: 295–302.

Hominick, W.M., Reid, A.P. and Briscoe, B.R. (1995) Prevalence and habitat specificity of steinernematid and heterorhabditid nematodes isolated during surveys of the UK and the Netherlands. *Journal of Helminthology* 69: 27–32.

Hominick, W.M., Reid, A.P., Bohan, D. and Briscoe, B. (1996) Entomopathogenic nematodes: biodiversity, geographical distribution and the convention on biological biodiversity. *Biocontrol Science and Technology* 6: 317–331.

Hooper, D.J. (1961) A redescription of *Longidorus elongatus* (de Man, 1876) Thorne and Swanger, 1936, (*Nematoda, Dorylaimidae*) and descriptions of five new species of *Longidorus* from Great Britain. *Nematologica* 6: 237–257.

Hooper, D.J. (1985a) Extraction of free living stages from soil, in Southey, J.F. (ed.) *Laboratory Methods for Work with Plant and Soil Nematodes*. London: HMSO, pp. 5–30.

Hooper, D.J. (1985b) Extraction of nematodes from plant material, in Southey, J.F. (ed.) *Laboratory Methods for Work with Plant and Soil Nematodes*. London: HMSO, pp. 51–58.

Hooper, D.J. (1985c) Handling, fixing, staining and mounting nematodes, in Southey, J.F (ed.) *Laboratory Methods for Work with Plant and Soil Nematodes*. London: HMSO, pp. 59–80.

Hooper, D.J., Cowland, J.A. and Spratt J. (1983) Nematode specimen collection. *Rothamsted Experimental Station Report for 1982*: p.167.

Hudson, P.J. (1968) The effect of a parasitic nematode on the breeding production of red grouse. *Journal of Animal Ecology* 55: 85–92.

Hunt, D.J. (1993) *Aphelenchida, Longidoridae and Trichodoridae: Their Systematics and Bionomics*. Wallingford: CAB International.

Hunt, D.J. and De Ley, P. (1996) Nematodes in soils, in Hall, G.S. (ed.) *Methods for the Examination of Organismal Diversity in Soils and Sediments*. Wallingford: CAB International, pp. 227–240.

Irvine, A.D. (1970) A note on the gastro-intestinal parasites of British hares (*Lepus europaeus* and *L. timidus*). *Journal of Zoology* (London) 162: 544–546.

Jairajpuri, M.S. and Ahmad, W. (1992) *Dorylaimida: Free-living, Predaceous and Plant-parasitic Nematodes*. London: E.J. Brill.

Johnson, S.R., Ferris, V.R. and Ferris, J.M. (1972) Nematode community structure of forest woodlots 1 Relationships based on similarity coefficients of nematode species. *Journal of Nematology* 4: 175–183.

Jones, G.W., Neal, C. and Harris, E.A. (1980) The helminth parasites of the badger (*Meles meles*) in Cornwall. *Mammal Review* 10: 163–164.

Jones, F.G.W. and Kempton, R.A. (1982) Population dynamics, population models and integrated control in Southey, J.F. (ed.) *Plant Nematology*. London: HMSO.

Kennedy, C.R. (1974) A checklist of British and Irish freshwater fish parasites with notes on their distribution. *Journal of Fish Biology* 6: 613–644.

Korthals, G.W., van de Ende, A., van Megen, H., Lexmond Theor M., Kammenga J.E. and Bongers T. (1996) Short-term effects of cadmium, copper, nickel and zinc on soil nematodes from different feeding and life-history strategy groups. *Applied Soil Ecology* 4: 107–117.

Lewis, J.W. (1987) Helminth parasites of British rodents and insectivores. *Mammal Review* 17: 81–93.

Lloyd, H.G. (1980) *The Red Fox*. London: Batsford.

Luc, M., Maggenti, A.R., Fortuner, R., Raski, D.J. and Geraert, E. (1987) A reappraisal of *Tylenchina* (Nemata) 1. For a new approach to the taxonomy of *Tylenchina*. *Revue de Nématologie* 10: 127–134.

McIntyre, A.D. and Murison, D.J. (1973) The meiofauna of a flatfish nursery grounds. *Journal of the Marine Biology Association of the UK* 53: 93–118.

Michel, J.F. (1976) The epidemiology and control of some nematode infections in grazing animals. *Advances in Parasitology* 14: 355–397.

Ministry of Agriculture, Fisheries and Food (1987) *Manual of Veterinary Parasitology Laboratory Techniques*. Technical Bulletin 18. London: Ministry of Agriculture, Fisheries and Food.

Needham, J.T. (1743) Concerning certain chalky tubulous concretions, called malm: with some microscopical observations on the farina of the red lilly, and of worms discovered in smutty corn. *Philosophical Transactions* 42: 634.

Neilson, R. (1993) *An investigation into the horizontal spatial distribution and ecology of marine nematodes in an intertidal estuarine biotope*. MSc thesis, Dundee University.

Nielsen, C. (1998) Sequences lead to tree of worms. *Nature* 392: 25–26.

Orr, C.C. and Dickerson, O.J. (1966) Nematodes in true prairie soils of Kansas. *Transactions of the Kansas Academy of Science* 69: 317–334.

Paton, G. and Boag, B. (1987) A model for predicting parasitic gastroenteritis in lambs subject to mixed nematode infections. *Research in Veterinary Science* 43: 67–71.

Platt, H.M. (1977) Ecology of free-living marine nematodes from an intertidal sandflat in Strangford Lough Northern Ireland. *Estuarine and Coastal Marine Science* 5: 685–693.

Poinar, G.O. (1975) *Entomogenous Nematodes. A Manual and Host List of Insect-nematode Associations*. Leiden: E.J. Brill.

Pritchard, R.K. (1990) Anthelmintic resistance in nematodes: extent, recent understanding and future directions for control and research. *International Journal for Parasitology* 20: 515–523.

Reid, A.P. and Hominick, W.M. (1992) Restriction fragment length polymorphisms within the ribosomal DNA repeat unit of British entomopathogenic nematodes (Rhabditida: Steinernematidae). *Parasitology* 105: 317–323.

Rey, M.J., Andres, M. Fe. and Arias, M. (1988) A computer method for identifying nematode species. 1. Genus *Longidorus* (Nematoda: Longidoridae). *Revue de Nématologie* 11: 129–135.

Robertson, D. (1929) Free-living nematodes occurring in arable soil in the north of Scotland. *Proceedings of the Royal Physiological Society* 21: 253–263.

Ruess, L., Sandbach, P., Cudlin P., Dighton J. and Crossley A. (1996) Acid deposition in a spruce forest soil: effects on nematodes, mycorrhizas and fungal biomass. *Pedobiologia* 40: 51–66.

Ryss, A. 1997. Computerized identification of species of the genus *Radopholus* (Tylenchida: Pratylenchidae). *Russian Journal of Nematology* 5: 137–142.

Schomaker, C.H. and Been, T.H. (1990) Sampling strategies for potato cyst nematodes; developing and evaluating a model, in Gommers, F.J. and Maas, P.W.T. (eds) *Nematology from Molecule to Ecosystem*. The Netherlands: Dekker and Huisman Wildervank, pp. 182–200.

Siddiqi, M.R. (1986) *Tylenchida Parasites of Plants and Insects*. Wallingford: CAB International.

Sleeman, D.P. (1983) Parasites of deer in Ireland. *Journal of Life Sciences Royal Dublin Society* 4: 203–209.

Stephens, M.N. (1957) *The Natural History of the Otter*. Potters Bar: UFAW.

Taylor, C.E. and Brown, D.J.F. (1976) The geographical distribution of *Xiphinema* and *Longidorus* nematodes in the British Isles and Ireland. *Annals of Applied Biology* 84: 383–402.

Thomas, P.R. (1981) Migration of *Longidorus elongatus*, *Xiphinema diversicaudatum* and *Ditylenchus dispsaci* in soil. *Nematologia Mediterranea* 9:75–81.

Topham, P.B., Boag, B. and McNicol, J.W. (1991) An assessment of some measures of association between species based on presence/absence and applied to plant-parasitic nematode data. *Nematologica* **37**: 470–480.

Trudgill, D.L., Evans, K. and Faulkner, G. (1973) A fluidising column for extracting nematodes from soil. *Nematologica* **18**: 469–475.

Van Houtert, M.F.J. and Sykes, A.R. (1996) Implications of nutrition for the ability of ruminants to withstand gastrointestinal nematode infections. *International Journal for Parasitology* **26**: 1151–1168.

Waller, P.J. and Larsen, M. (1993) The role of nematophagous fungi in the biological control of nematode parasites of livestock. *International Journal of Parasitology* **23**: 539–546.

Warwick, R.M. and Price, R. (1979) Ecological and metabolic studies on free-living nematodes from an estuarine mud-flat. *Estuarine and Coastal Marine Science* **9**: 257–271.

Weber, J.M. (1991) Gastrointestial helminths of the otter, *Lutra lutra*. Shetland. *Journal of Zoology* **224**: 341–346.

Webster, R. and Boag, B. (1992) The spatial distribution of cyst nematodes in soil. *Soil Science* **153**: 583–595.

Whitehead, A.G. and Hemming, J.R. (1965) A comparison of some quantitative methods of extracting small vermiform nematodes from soil. *Annals of Applied Biology* **55**: 25–38.

Wilson, G.R. (1983) The prevalence of caecal threadworms (*Trichostrongylus tenuis*) in red Grouse (*Lagopus legopus scoticus*). *Oecologia* **58**: 265–268.

Wilson, M.J., Glen, D.M. and George, S.K. (1993) The rhabditid nematode *Phasmarhabditis hermaphrodita* as a potential biological control agent for slugs. *Biocontrol Science and Technology* **3**: 503–511.

Winslow, R.D. (1964) Soil nematode population studies. I. The migratory root Tylenchida and other nematodes of the Rothamsted and Woburn six-course rotations. *Pedobiologia* **4**: 65–76.

Yeates, G.W. (1970) The diversity of soil nematode fauna. *Pedobiologia* **10**: 104–107.

Yeates, G.W. (1971) Plant and soil nematodes of Wicken Fen. *Nature in Cambridgeshire* **14**: 23–25.

Yeates, G.W., Bongers, T., de Goede, R.G.M., Freckman, D.W. and Georgieva, S.S. (1993) Feeding habits in nematode families and genera – an outline for soil ecologists. *Journal of Nematology* **25**: 315–331.

Yuen, P.H. (1966) The nematode fauna of the regenerated woodland and grassland of Broadbalk Wilderness. *Nematologica* **12**: 195–214.

Chapter 15

Mites and ticks

Anne S. Baker

ABSTRACT

The number of mite species recorded in Great Britain and Ireland to date is estimated at 2100, with 65 species new to science and 64 faunal additions reported since 1973. Because few taxa have been studied in detail and a wide range of habitats have yet to be closely examined for mites, there are undoubtedly many more species and much distributional information still to discover. Changes in the fauna are extremely difficult to detect when our knowledge is so incomplete; the few that have been documented concern species that affect humans through their medical or veterinary impact. Progress in data recording is hindered by the poor taxonomy of many groups, the lack of up-to-date checklists and user-friendly identification keys, and the small number of specialists able to identify material, carry out revisionary work and train professional or amateur workers. In contrast to mites, ticks have been extensively sampled and the 22 species known to occur are considered to be a complete inventory of the established fauna. Potential changes to populations as a result of threats to mammal hosts or off-host habitats have not, however, been investigated.

1 Introduction

Mites and ticks make up the subclass *Acari*, the most morphologically and ecologically diverse division of the class *Arachnida*. Representatives of four of the six mite orders occur in Great Britain and Ireland, and the present author estimates that 2100 species have been recorded here to date (order *Astigmata* 278, *Mesostigmata* 501, *Oribatida* 311 and *Prostigmata* 910). Mites have colonized just about every category of terrestrial and aquatic habitat; soil-dwelling forms play an important role as secondary decomposers of decaying plant matter, while many of those living in association with animals and plants are of great medical, veterinary or horticultural importance. The order *Ixodida*, which comprises the ticks, also has a wide geographical distribution, but is represented by just 22 species and all its members are blood-feeding ectoparasites of vertebrates. Ticks readily attack humans and domestic animals and, in addition to debilitating their host through exsanguination, they can transmit a wide range of pathogens and cause paralysis through the introduction of neurotoxins during feeding.

Current faunistic and distributional knowledge of the *Acari* is far from complete and, in most instances, is insufficient to allow changes which may be occurring to be identified. The following account attempts to put this state of knowledge into context by reviewing the main sources of faunistic data and outlining the problems in detecting change. The concluding section considers the requirements necessary for obtaining a more complete picture of the fauna, which will enable effective monitoring to be carried out in future.

2 Sources of faunistic data

2.1 Mites

The identification and recording of the British fauna began in earnest in the late 1800s when Michael began to study the soil-dwelling *Oribatida*. His classic works on the order contain descriptions and intricate illustrations of 104 species (Michael 1884, 1888), and he later produced a similarly detailed treatment of the free-living *Astigmata* (Michael 1901, 1903). Discoveries of species from all four orders that were new to science or to the fauna increased greatly from Michael's time. Many were due to the work of Hirst and Hull, who specialized in parasitic taxa and in the soil mites of northern England respectively (see for example Hirst 1916, 1920, 1922; Hull 1918, 1931). The prostigmatid freshwater mites were also the subject of early attention, with descriptions of the 216 species known in Britain and Ireland at the time being published in three monographic volumes (Soar and Williamson 1925, 1927, 1929). Initial recording in Ireland was carried out virtually single-handedly by Halbert, who covered a wide range of both aquatic and terrestrial taxa, and whose research remains the most significant single body of work on the Irish acarine fauna (e.g. 1915, 1920, 1944).

The publications of the above-mentioned and other pioneering acarologists contain much information that is still of value, but many of their records need to be reassessed in the light of subsequent revisionary work and because the morphological descriptions given are now insufficiently detailed for species recognition. Identities can be clarified when original specimens have survived; for example, the collections of Michael and Halbert (housed in the Natural History Museums of London and Dublin, respectively) have proved invaluable for such work (e.g. Luxton 1987c, 1989, 1998). The proliferation in taxonomic and distributional data resulting from this early work motivated Turk to compile the first and, to date, only catalogue of British and Irish *Acari* (Turk 1953a,b). He included some 1600 species, but subsequent studies have shown a significant number to be synonyms or of doubtful identity; for example, only 238 of the 308 listed for the freshwater mites are now considered to be valid. In spite of this, Turk's checklist remains a useful basis for investigating taxa that have not been the subject of subsequent study.

The documentation of the *Oribatida* begun by Michael continued throughout the twentieth century, with particularly important contributions being made by Seyd, who specialized in the fauna of upland areas (Seyd 1988, 1992; Seyd *et al.* 1996), and by Luxton. Luxton's studies of material from a wide range of localities yielded descriptions of species new to science, identifications of new faunal records, and much new information concerning geographical distribution and habitat relationships (e.g. Luxton 1987a,b, 1990, 1998). He also provided a great service by reinterpreting many of the taxa described or identified by early workers (Luxton 1987c,d, 1989, 1998), and by compiling a checklist of the species recorded from Great Britain and Ireland (Luxton 1996).

A significant increase in faunistic knowledge of the *Mesostigmata* occurred in the 1950s when Evans and colleagues began taxonomic revisions of the major free-living and ectoparasitic families (e.g. Evans 1958; Evans and Till 1966; Hyatt 1980; Hyatt and Emberson 1988; Skorupski and Luxton 1996). As well as clarifying nomenclature, this work resulted in the addition of many new species and locality records. Evans and Till (1979) prepared a general account of the British and Irish Mesostigmata which included identification keys to suborders, families and genera, as well as a comprehensive bibliography to direct users to papers containing descriptions and keys to species. The only modern checklist for the order covers the 235 species recorded from Ireland (Luxton 1998).

The *Astigmata* and *Prostigmata* are the most neglected of the orders and few families or genera have been the subject of taxonomic review or detailed collecting. The most studied astigmatids are the free-living species which infest stored foods and the dust of homes and workplaces (e.g. Griffiths 1970; Cusack *et al.* 1975; Hughes 1976). In the *Prostigmata*, aquatic groups are the best documented, and nomenclatorially up-to-date checklists are available for both freshwater and marine taxa (Gledhill and Viets 1976; Green and Macquitty 1987). Records of animal parasitic *Astigmata* and *Prostigmata* mostly occur in taxonomic revisions by overseas workers (e.g. Fain and Elsen 1967; Southcott 1992).

After taxonomic studies, the richess source of mite records are faunistic and ecological investigations of particular localities or biotopes (e.g. Block 1965; Curry 1976; Curry and Momen 1988; Hyatt 1990, 1993; Skorupski and Luxton 1998). Such studies began to increase in the 1950s when it was realized that plant-dwelling mites could cause serious damage to crops and that soil-dwelling species were potentially important contributors to the decomposition process. This upsurge of interest in mites was also the stimulus for a plan to publish a series of five volumes describing the biology, morphology and systematics of the terrestrial *Acari* of the British Isles. Unfortunately, only the introductory volume was ever published (Evans *et al.* 1961), but it is still a useful information source because of the extensive bibliographies and many examples of native species it contains.

2.2 Ticks

The importance of accurately identifying ticks when studying their role as vectors of disease has long been recognized, with the result that they are well worked both taxonomically and faunistically. Detailed descriptions of species found in Great Britain and Ireland were included in an early monographic series on the world fauna (Nuttall *et al.* 1908; Nuttall and Warburton 1911, 1915). Over the following decades, research into the medical and veterinary impact of native ticks increased, and with it the need for an updated account of the fauna. In response, Arthur (1963) published an identification aid and information source which comprised descriptions of the life stages of each species, together with details of geographical distribution, host ranges and diseases transmitted. The taxonomic and biological data contained in Arthur's work were updated some 30 years later in a synopsis of the ticks of north-west Europe (Hillyard 1996). Many samples of British and Irish ticks have been collected, and it is not expected that there are any undiscovered established species to add to the 22 already identified. Distribution maps, largely based on material deposited in the

Natural History Museum, London, have been compiled for each species (Martyn 1988) but, because of the limited sample range and lack of updating, they are regarded as provisional.

3 Problems in detecting change

3.1 Mites

The lack of periodic collation of new records and nomenclatorial developments means that, apart from that for the *Oribatida*, there is no current catalogue of valid species on which to base investigations of change. Furthermore, while knowledge of the mite fauna is still at the stage of identifying the component species and determining their distributions, it is extremely difficult to detect changes which may be occurring. Even the more well-worked taxa are incompletely known and each new taxonomic study yields its quota of undescribed or newly recorded species (Table 15.1). Similarly, surveys usually result in huge increases in the number of local records; for example, when collections from the Isle of Man and the Isles of Scilly were identified, their species lists for the *Oribatida* rose from 2 to 59 and from 11 to 44, respectively (Luxton 1987a, 1990). Distributional ignorance is further demonstrated by Luxton's observations that 70 of the 303 oribatid species in his checklist are represented by a single record, while 35 have not been collected within the last 40–50 years (Luxton 1996).

Because there are so many localities and habitats still to be explored, the status of apparently rare species cannot be gauged definitively. Hyatt (1980), for example, stated that *Parasitus consanguineus* 'is never found abundantly', but this view had later to be revised when the mite was found in huge numbers in cattle bedding (Fox *et al.* 1989). At present, no species is classed as endangered, but it is possible that

Table 15.1   Numbers of native mite species recorded in selected taxa before and after the most recent revision or taxonomic survey.

Taxon	Study	Before	After
Prostigmata			
Freshwater mites (*Hydrachnellae*)	Gledhill and Viets (1976)	226	273
Halacaridae	Green and Macquitty (1987)	27	65
Eupodidae	Baker (1987)	5	24
Rhagidiidae	Baker (1987)	4	11
Mesostigmata			
Parasitinae	Hyatt (1980)	21	36
Pergamasus	Bhattacharyya (1963)	13	32
Macrochelidae	Hyatt and Emberson (1988)	23	32
Phytoseiidae	Baker (in prep.)	25	39
Oribatida	Luxton (1996)	296	303

unrecorded taxa confined to threatened habitats or hosts may themselves be facing extinction.

A fundamental hindrance to documenting more of the fauna is the difficulty in identifying specimens in the first instance. Many of the genera and families which occur in Great Britain and Ireland are in need of worldwide revision before species can be reliably identified. Also, there is a general lack of comprehensive identification keys (user-friendly or otherwise) for both the native and worldwide faunas, and attempts to determine material may necessitate time-consuming trawls through the literature.

3.2 Ticks

Interest in changes to the tick population traditionally centres on the effect they may have on the health of host animals, and not on the conservation of the parasites themselves. At least three species are potentially vulnerable to documented reductions in host numbers or off-host habitats, but, because their populations are not being monitored, predictions of change are speculative. The water vole (*Arvicola terrestris*), the principal host of *Ixodes apronophorus*, is under threat from human manipulation of its habitats and predation by mink (Jefferies *et al.* 1989). If host numbers continue to fall, *I. apronophorus* may itself become a rarity or disappear from the fauna. Similarly, *Haemaphysalis punctata* and *Dermacentor reticulatus* both need marshy habitats in which to moult and oviposit, but the continued draining of such land may result in their distribution becoming even more restricted than it is at present (Martyn 1988).

4 Synopsis of changes

4.1 Additions

Since 1973, 65 species of mites new to science and 64 new to the fauna have been recorded, the majority originating from investigations of poorly studied taxa, localities or hosts. A number were found on captive exotic birds or imported items (see for example Fain and Laurence 1979; Haines 1988), but no follow-up surveys have been carried out to determine whether populations have established here.

Varroa jacobsoni (the honey bee brood mite) was recorded for the first time in 1992. It was initially collected in a hive in Devon, but has since migrated northwards into Scotland and is now considered to be established (Paxton 1992). *V. jacobsoni* has devastated hives of the European honey bee (*Apis mellifera*) in most countries of the world, damaging honey production and reducing the prospect of crop pollination. Its spread has been attributed to the transport of infested hives, but its precise route into the British Isles is uncertain.

Psorobia bos (the cattle-itch mite) was reported for the first time in 1996 (Baker *et al.* 1996). This species is a skin parasite which is transmitted during physical contact between hosts; it has only previously been recorded from southern Africa and the USA. The single record of a native infestation was found in England on a UK-bred cow which apparently had not been in contact with any potentially infested imported animal. It is possible, therefore, that *P. bos* has existed in the British Isles for some time, but clinical symptoms were either overlooked or attributed to another cause.

Rhipicephalus sanguineus (the kennel- or brown-dog tick) originates from Africa, but has spread to many parts of the world as a result of the transportation of its primary host, the domestic dog. Specimens have periodically been removed from dogs imported into the British Isles, but, in 1983, a breeding population was found in a London house (Fox and Sykes 1985). *R. sanguineus* is now included in accounts of the native fauna (Martyn 1988; Hillyard 1996), but its occurrence is expected to be restricted to kennels or houses where dogs are kept, because the climate is too cold for it to survive out of doors.

4.2 Losses

Psoroptes equi, a mange-causing skin parasite of horses, was a common find during the period when horses and ponies were routinely used in industry and agriculture. Compulsory chemical treatment was effective in controlling infestations and, in 1948, the mite was considered to have been eradicated (P. G. J. Bates, pers. comm.). To date, there is no evidence that *P. equi* has been reintroduced (Meleney 1985).

Hyalomma aegyptium (the tortoise tick) parasitizes tortoises (*Testudo* spp.) living in the Mediterranean region and North Africa. It used to be found frequently on animals imported for the pet trade, but there have been no records since this practice was banned in the 1970s.

5 Prognosis for the future

In order to generate the faunistic data that will enable changes in the mite population to be recognized, it will be necessary to instigate a programme of rigorous collecting and taxonomic revisions. The most urgent action is needed in the latter category, because, until the confusion in many of the groups which occur here is resolved, reliable species identifications cannot be made. Such work is also a prerequisite for devising user-friendly keys and compiling up-to-date and nomenclatorially sound checklists. Considering that there are currently only six employed and four retired professional mite taxonomists in the British Isles (compared with a total of about 20 in 1973), and that only a small amount of total research time is spent documenting the native fauna, progress in the immediate future will be slow. Furthermore, overseas workers cannot compensate for the decrease in the workforce because a similar loss through redundancy, retirement or transfer to other areas of research is occurring throughout the world.

The dwindling number of expert taxonomists is also preventing the potential involvement of amateur workers. Three of the important early acarologists (Michael, Hull and Turk) carried out research in their spare time, but the increasing complexity of mite taxonomy has earned it the reputation of being too difficult for non-professionals. It is evident, however, that amateurs with an eye for detail and access to sufficiently high-quality microscopes and acarological advice can make a useful contribution; for example, the careful work of Monson is yielding new records and distributional data for the *Oribatida* (Monson 1997, 1998). Periodic interest in identifying mites is also expressed by volunteers who find specimens during the course of studies on more high-profile animals, such as bats, but it is difficult to exploit this interest because of the lack of identification aids and experts with the time to act as mentors.

Ticks have the advantage over mites of being relatively large (up to 20 mm long) and of being the subject of identification keys which enable inexperienced workers to accurately determine species (Hillyard 1996). However, there will have to be an enormous change in public perception of these parasites before they are valued as an important part of the native fauna, and are the subject of surveys and conservation programmes in the way that, for example, mammals and butterflies are.

ACKNOWLEDGEMENTS

Particular thanks are due to the following UK colleagues for valuable acarological discussions: Dr Malcolm Luxton (Department of Zoology, National Museum of Wales, Cardiff); Dr Don Griffiths (Acarology Consultants, Middlesex); Mr Joe Ostoja-Starsweski (Central Science Laboratory, MAFF, York) and Mr Peter Bates (Central Veterinary Laboratory, Weybridge, Surrey). I am also grateful to Drs Malcolm Luxton and Rory Post (Department of Entomology, The Natural History Museum) for commenting on the manuscript.

REFERENCES

Arthur, D.R. (1963) *British Ticks*. London: Butterworth.
Baker, A.S. (1987) *Systematic studies on mites of the superfamily Eupodoidea (Acari: Acariformes) based on the fauna of the British Isles*. PhD Thesis, University of London.
Baker, A.S., Andrews, A.H. and Fox, M.T. (1996) First record of the cattle itch mite, *Psorobia bos (Prostigmata: Psorergatidae)*, in the Palaearctic region, *Systematic and Applied Acarology*, 1, 213–216.
Bhattacharyya, S.K. (1963) A revision of the British mites of the genus Pergamasus Berlese s. lat. (*Acari: Mesostigmata*). *Bulletin of the British Museum (Natural History) (Zoology)*, 11, 133–242.
Block, W.C. (1965) Distribution of soil mites (*Acarina*) on the Moor House National Nature Reserve, Westmorland, with notes on their numerical abundance. *Pedobiologia*, 5, 244–251.
Curry, J.P. (1976) The arthropod fauna of some common grass and weed species of pasture. *Proceedings of the Royal Irish Academy*, 76B, 1–35.
Curry, J.P. and Momen, F.M. (1988) The arthropod fauna of grassland on reclaimed cutaway peat in Central Ireland. *Pedobiologia*, 32, 99–109.
Cusack, P.D., Evans, G.O. and Brennan, P.A. (1975) A survey of the mites of stored grain and grain products in the Republic of Ireland. *Scientific Proceedings of the Royal Dublin Society (Series B)*, 3, 273–329.
Evans, G.O. (1958) A revision of the British *Aceosejinae* (*Acarina*: Mesostigmata). *Proceedings of the Zoological Society of London*, 131, 177–229.
Evans, G.O. and Till, W.M. (1966) Studies on the British *Dermanyssidae* (*Acari*: Mesostigmata). Part II. Classification. *Bulletin of the British Museum (Natural History) (Zoology)*, 14, 107–370.
Evans, G.O. and Till, W.M. (1979) Mesostigmatic mites of Britain and Ireland (*Chelicerata: Acari: Parasitiformes*). An introduction to their external morphology and classification. *Transactions of the Zoological Society of London*, 35, 139–270.
Evans, G.O., Sheals, J.G. and Macfarlane, D. (1961) *Terrestrial Acari of the British Isles*. London: British Museum (Natural History).
Fain, A. and Elsen, P. (1967) Les *Acariens* de la famille Knemidokoptidae producteurs de gale chez les oiseaux. *Acta Zoologica et Pathologica Antverpiensia*, 45, 3–145.

Fain, A. and Laurence, B.R. (1979) Two new species of *Neottialges* Fain (*Acari*, *Hypoderatidae*) under the skin of birds, with a key to the hypopi of this genus. *Journal of Natural History*, **20**, 849–856.

Fox, M.T. and Sykes, T.J. (1985) Establishment of the tropical dog tick, *Rhipicephalus sanguineus*, in a house in London. *Veterinary Record*, **116**, 661–662.

Fox, M.T., Baker, A.S. and Fisher, M.A. (1989) Bovine and human infestation with *Parasitus consanguineus* Oudemans and Voigts (*Mesostigmata: Parasitinae*). *Veterinary Record*, **124**, 64.

Gledhill, T. and Viets, K.O. (1976) A synonymic and bibliographic check-list of the freshwater mites (*Hydrachnellae* and *Limnohalacaridae*, *Acari*) recorded from Great Britain and Ireland. Occasional publication no. 1. Freshwater Biological Association 1–59.

Green, J. and Macquitty, M. (1987) *Halacarid Mites*. Shrewsbury: Field Studies Council.

Griffiths, D.A. (1970) A further systematic study of the genus *Acarus* L., 1758 (*Acaridae*, *Acarina*), with a key to species. *Bulletin of the British Museum (Natural History) (Zoology)*, **19**, 83–118.

Haines, C.P. (1988) A new species of predatory mite (*Acarina*: *Cheyletidae*) associated with bostrichid beetles on dried cassava. *Acarologia*, **29**, 361–375.

Halbert, J.N. (1915) Clare Island Survey. 39. *Acarinida*. ii – Terrestrial and marine *Acarina*. *Proceedings of the Royal Irish Academy*, **31**, 45–136.

Halbert, J.N. (1920) The *Acarina* of the seashore. *Proceedings of the Royal Irish Academy*, **35**, 106–152.

Halbert, J.N. (1944) List of Irish freshwater mites (*Hydracarina*). *Proceedings of the Royal Irish Academy*, **50**, 39–104.

Hillyard, P.D. (1996) *Ticks of North-West Europe*. Shrewsbury: Field Studies Council.

Hirst, S. (1916) Notes on parasitic *Acari*. A. On some species of *Acari* parasitic on mammals and birds in Great Britain. *Journal of Zoological Research*, **1**, 59–76.

Hirst, S. (1920) Revision of the English species of red spider (genera *Tetranychus* and *Oligonychus*). *Proceedings of the Zoological Society of London*, 1920, 49–60.

Hirst, S. (1922) *Mites Injurious to Domestic Animals (with an Appendix on the Acarine Disease of Hive Bees)*. London: British Museum (Natural History).

Hughes, A.M. (1976) *The Mites of Stored Food and Houses*. London: HMSO.

Hull, J.E. (1918) Terrestrial *Acari* of the Tyne province. *Transactions of the Natural History Society of Northumberland*, **5**, 13–88.

Hull, J.E. (1931) Terrestrial *Acari* of the Tyne province. 4. *Tyroglyphidae*. *Transactions of the Northern Naturalists Union*, **1**, 37–44.

Hyatt, K.H. (1980). Mites of the subfamily *Parasitinae* (*Mesostigmata: Parasitidae*) in the British Isles. *Bulletin of the British Museum (Natural History) (Zoology)*, **38**, 237–378.

Hyatt, K.H. (1990) Mites associated with terrestrial beetles in the British Isles. *Entomologist's Monthly Magazine*, **126**, 133–147.

Hyatt, K.H. (1993) The acarine fauna of the Isles of Scilly. *Cornish Studies*, **1**, 120–161.

Hyatt, K.H. and Emberson, R.M. (1988) A review of the *Macrochelidae* (*Acari*: *Mesostigmata*) of the British Isles. *Bulletin of the British Museum (Natural History) (Zoology)*, **54**, 63–125.

Jefferies, D.J., Morris, P.A. and Mulleneux, J.E. (1989) An inquiry into the changing status of the water vole, *Arvicola terrestris*, in Britain. *Mammal Review*, **19**, 111–131.

Luxton, M. (1987a) Oribatid mites from the Isle of Man. *Naturalist*, **112**, 67–77.

Luxton, M. (1987b) Mites of the genus *Malaconothrus* (*Acari*: *Cryptostigmata*) from the British Isles. *Journal of Natural History*, **21**, 199–206.

Luxton, M. (1987c) The oribatid mites (*Acari*: *Cryptostigmata*) of J.E. Hull. *Journal of Natural History*, **21**, 1273–1291.

Luxton, M. (1987d) The British oribatid mites (*Acari*: *Cryptostigmata*) of Warburton and Pearce. *Journal of Natural History*, **21**, 1359–1365.

Luxton, M. (1989) Michael's British damaeids (*Acari*: *Cryptostigmata*). *Journal of Natural History*, **23**, 1367–1372.

Luxton, M. (1990) Oribatid mites (*Acari: Cryptostigmata*) from the Isles of Scilly. *Naturalist*, 115, 7–12.

Luxton, M. (1996) Oribatid mites of the British Isles. A checklist and notes on biogeography (*Acari, Oribatida*). *Journal of Natural History*, 30, 803–822.

Luxton, M. (1998) The oribatid and parasitiform mites of Ireland, with particular reference to the work of J.N. Halbert (1872–1948). *Bulletin of the Irish Biogeographical Society*, 22, 2–72.

Martyn, K.P. (1988) *Provisional Atlas of the Ticks (Ixodoidea) of the British Isles*. Grange-over-Sands: Institute of Terrestrial Ecology.

Meleney, W.P. (1985) Mange mites and other parasitic mites, in Gaafar, S.M., Howard, W.E. and Marsh, R.E. (eds) *Parasites, Pests and Predators*. Amsterdam: Elsevier, pp. 317–346.

Michael, A.D. (1884) *British Oribatidae*, Vol. 1. London: Ray Society.

Michael, A.D. (1888) *British Oribatidae*, Vol. 2. London: Ray Society.

Michael, A.D. (1901) *British Tyroglyphidae*, Vol. 1. London: Ray Society.

Michael, A.D. (1903) *British Tyroglyphidae*, Vol. 2,.London: Ray Society.

Monson, F. (1997) Two moss mites new to the British Isles (*Acari, Oribatida*). *Entomologist's Monthly Magazine*, 133, 9–11.

Monson, F. (1998) Oribatid mites (*Acari: Cryptostigmata*) from Slapton Wood and the vicinity of Slapton Ley. *Field Studies*, 9, 325–336.

Nuttall, G.H.F. and Warburton, C. (1911) *Ticks: A Monograph of the Ixodoidea*. Part II *Ixodidae*. London: Cambridge University Press.

Nuttall, G.H.F. and Warburton, C. (1915) *Ticks: A Monograph of the Ixodoidea*. Part III *The genus Haemaphysalis*. London: Cambridge University Press.

Nuttall, G.H.F., Warburton, C., Cooper, W.F. and Robinson, L.E. (1908) *Ticks: A Monograph of the Ixodoidea*. Part I *The Argasidae*. London: Cambridge University Press.

Paxton, P. (1992) The mite marches on: *Varroa jacobsoni* found in the UK. *Bee World*, 73, 94–99.

Seyd, E.L. (1988) The moss mites of the Cheviot (*Acari: Oribatei*). *Biological Journal of the Linnean Society*, 34, 349–362.

Seyd, E.L. (1992) The moss mites of Yes Tor, Dartmoor, Devon (*Acari: Oribatida*) and their evolutionary significance. *Zoological Journal of the Linnean Society*, 106, 115–126.

Seyd, E.L., Luxton, M.S. and Colloff, M.J. (1996) Studies on the moss mites of Snowdonia. 3. Pen-y-Gadair, Cader Idris, with a comparison of the moss mite faunas of selected montane localities in the British Isles (*Acari: Oribatida*). *Pedobiologia*, 40, 449–460.

Skorupski, M. and Luxton, M. (1996) Mites of the family *Zerconidae* Canestrini, 1891 (*Acari: Parasitiformes*) from the British Isles, with descriptions of two new species. *Journal of Natural History*, 30, 1815–1832.

Skorupski, M. and Luxton, M. (1998) Mesostigmatid mites (*Acari: Parasitiformes*) associated with yew (*Taxus baccata*) in England and Wales. *Journal of Natural History*, 32, 419–439.

Soar, C.D. and Williamson, W. (1925) *The British Hydracarina*. Vol. I. London: Ray Society.

Soar, C.D. and Williamson, W. (1927) *The British Hydracarina*. Vol. II. London: Ray Society.

Soar, C.D. and Williamson, W. (1929) *The British Hydracarina*. Vol. III. London: Ray Society.

Southcott, R.V. (1992) Revision of the larvae of *Leptus* Latreille (*Acarina: Erythraeidae*) of Europe and North America, with descriptions of post-larval instars. *Zoological Journal of the Linnean Society*, 105, 1–153.

Turk, F.A. (1953a) A synonymic catalogue of British *Acari*: Part I. *Annals and Magazine of Natural History*, (series 12) 6, 1–26.

Turk, F.A. (1953b) A synonymic catalogue of British *Acari*: Part II. *Annals and Magazine of Natural History*, (series 12) 6, 81–99.

Chapter 16

Flies

Alan E. Stubbs

ABSTRACT

The British fly list has increased from approximately 5728 to about 6668 species, reflecting high levels of taxonomic revision and field work. The Cranefly Recording Scheme, begun in 1973, gave rise to the Dipterists Forum, which runs eight national recording schemes and five study groups, and promotes the study of *Diptera* overall. Conservation bodies have also generated much new data.

Elements of the fauna have spread, especially in conifer plantations and tephritid flies of 'weeds'. Various sun-loving species that suffered population crashes in the 1960s have recovered in phase with hotter summers. Land-use intensification must, however, have led to declines in most species, notably – among other reasons – as a result of the massive scale of field under-drainage during the 1970s–80s, lowland river improvement and groundwater abstraction.

Flies now have a higher profile in conservation. As the pressures and complexity of conservation increase, better training is essential for those making decisions on the ground; much depends on better co-ordination of science and action so that sound measures are taken and gaffes avoided. The single most worrying impact is likely to be water supply, especially underground aquifer exploitation.

The greatest impediment to assessing change over the last 25 years has been the lack of resources to provide and process data. The National Biodiversity Network has the potential to aid a solution, but time will tell.

1 State of knowledge

Knowledge of our fly fauna has advanced considerably over the last 25 years, reflecting high levels of activity in both taxonomic revision and field work. While Smith's (1974) figure for the total number of known species was an estimate, firmer figures are provided by a 1976 checklist which had about 6000 species (Kloet and Hincks 1976); another for late 1998 had 6668 named species, plus acknowledgment of 50 unnamed (Chandler 1998); see Fig. 16.1.

Even some supposedly well-studied groups, such as chironomid midges and fungus gnats, have been shown to have many extra species, and little-studied families such as *Sciaridae* have shown major increases. The study of early stages and ecology has been a growing trend, leading to the recognition of extra species in a number of families.

The Irish list, with 29 species not known from Britain, stands at about 2900 species

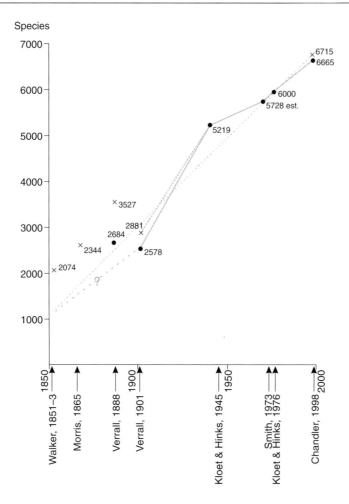

Figure 16.1 The growth in the British list of *Diptera* over the last 150 years. The higher and lower figures for **1888** and **1901** allow for species yet to be confirmed during a period of re-evaluation: earlier checklists also contained unconfirmed species.

Based on Smith (1974 Table 16.1), Kloet and Hincks (1976) and Chandler (1998).

(Chandler 1998), reflecting greater biogeographic post-glacial isolation, the immense loss of woodland and very few recorders; however, there has been steady progress over last 25 years. An initial compilation for Wales totalled 3219 species (Fowles 1985), but for Scotland lists have only been assembled for selected families (Ward 1997). About 250 species are classified as pseudoendemics (Henshaw and White 1997), some of which may yet be found in under-worked parts of Europe.

2 Data capture

The embryonic Cranefly Recording Scheme gained formal status in 1973, and encouraged the launch of schemes for hoverflies, larger *Brachycera*, and some other groups in 1976 (Stubbs 1990). In 1988 the *Dipterists Digest* was launched and the Dipterists

(a)

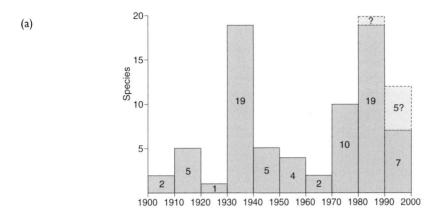

(b)

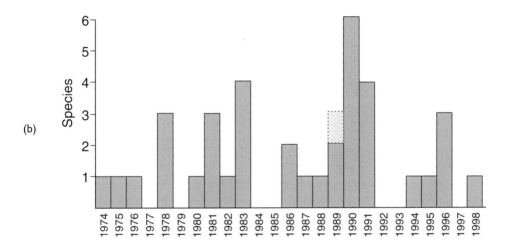

Figure 16.2 The growth of the British hoverfly fauna. (a) **1901** to present based on first publication date; despite the Syrphidae being one of the most popular and best collected fly families, the last 25 years have seen many additions resulting from more critical taxonomy, new colonisation and greater numbers of interested amateurs. Already there have been seven additions during the 1990s and at least five others are under consideration. After Stubbs (1995) with later revisions. (b) The annual additions over the last 25 years, based on first publication date; the 36 additions (plus one query) show a steady incremental rate.

Forum, a society covering all *Diptera*, was founded in 1995. The Forum currently runs eight national recording schemes (1476 species) plus five study groups (1672 species).

Knowledge of the distribution and status of species has increased enormously. Districts that were previously little studied have been subject to week-long field meetings of 20–35, often visiting over 100 sites. A recent meeting in Wales visited 132 sites in 45 10 km squares through eight vice-counties, and recorded over 1487 species of which 204 were in RDB or Notable categories (Howe 1998). Cumulatively, 25 years of such field meetings, together with smaller-scale meetings for the autumn fauna, have made a huge difference to knowledge of the distribution of many species of *Diptera* (Stubbs 1999). Local initiatives have been spawned, with various county atlases in preparation or published for hoverflies (see Stubbs 1996a; also Levy 1998; Morris 1998) and some other groups.

This increased effort has led to new insights in field-craft, which have resulted in greater success in locating otherwise little known species. This applies to both adults and larvae. Some soldierflies are now much easier to find as larvae than adults and their distribution is much better understood (Stubbs *et al.* in press). Surveys of larvae of *Xylophagus ater* (Clements and Alexander 1987), and particularly of the hoverfly *Callicera rufa* (Rotheray and MacGowan 1990), show that species rarely seen as adults are more widespread than historic recording had revealed.

Additionally, the conservation bodies have generated surveys adding significantly to the state of knowledge. The Nature Conservancy Council (NCC) Field Unit's surveys (Palmer 1991) included *Diptera* in such areas as the Lizard, New Forest bogs, West Dorset valleys, Blackdown Hills, Culm grasslands and grazing levels in Gwent, Somerset, Kent, Sussex and the Suffolk coast. The NCC issued a contract to survey invertebrates of moorland in northern England in the late 1970s. There were in-house invertebrate surveys of Welsh wetlands in the late 1980s, and East Anglian wetlands in the early 1990s, including volunteer specialists. The National Trust Biological Survey Team has included elements of *Diptera* in its surveys since the 1980s. There have been some more localised projects, including the City of London survey of Burnham Beeches, Buckinghamshire, a *Diptera* survey of Kenfig NNR organised by the National Museum of Wales, a major survey of Merseyside organised by Liverpool Museum, Warwickshire by Coventry Museum, and also work on Biodiversity Action Plans. Many individual dipterists have advanced county and site surveys, including the raising of the Wicken Fen list to approximately 1730 species, the largest list for any site in Britain so far (Perry in Friday and Harley, 2000).

Data capture has increased enormously, but data processing remains a fundamental problem. There are now home computers and record-handling packages, although few dipterists have these. The Biological Records Centre (BRC) has published two provisional atlases using processed data (*Sepsidae* and selected larger *Brachycera*) but the cranefly atlases were hand-plotted. Some other maps have been prepared for publication at BRC (such as *Dixidae*), and they were funded to process some hoverfly data. However, other atlases, such as those for mosquitoes, *Sciomyzidae*, *Tephritidae* and the bulk of the processed data for hoverflies (approximately 300 000 records), have been prepared independently of BRC (all cited in references below).

Until recently the conservation movement was only interested in very recent data (i.e. the last five or at most 10 years), especially to validate SSSIs. The new IUNC *Red Data Book* (Mace *et al.* 1992), and in turn the *Biodiversity Action Plan* (BAP) process, has put a parallel emphasis on rate of decline as a criterion for species status and thence priority for action. Suddenly historic data gains high profile, but much of the required information is locked in the literature and museum collections. No cohesive plan for the extraction of such data is in sight.

As the volume of recent recording far outweighs that of earlier periods, even for many species whose habitat has seriously declined, there are now far more recent than historic data. This distortion in data has to be interpreted to ascertain meaningful trends.

International status is a criterion for BAP action. The European Invertebrate Survey has made little headway in collating the necessary data on *Diptera*, even for a popular family such as hoverflies. Consultations have revealed that many of our declining

Figure 16.3 The distribution of *Diptera* recording scheme field meetings 1973–98. Black star summer
residential field meetings(7–8 days); open star autumn residential field meetings (4–5 days);
open star in black circle both summer and autumn meetings; small open circle day field
meetings (some masked by other symbols).

Reproduced from Stubbs (1999).

species are also declining over much of Europe for the very same reasons, in some cases
even faster. Status evaluation on a European scale is highly desirable.

Twenty-five years ago the conservation movement had no interest in data on
Diptera. The Wildlife and Countryside Act 1981 then required a more refined statement to owners of SSSIs about the exact scientific interest and the basis of listed 'damaging operations'. Schedules with comments such as 'interesting insects also occur' had
to be spelt out, with confirmation that they had been seen recently. Fortunately the
Invertebrate Site Register (ISR), which begun as a pilot in the mid-1970s, started full
operation at NCC in 1979 and was poised to provide answers.

The statutory system became locked into a process whereby species data had quasi-
legal status. This depended, and still depends, on efficient data capture followed by
putting that data into context. In 1998 the *Diptera* recording schemes were still without the level of support they need. The ISR now scarcely functions in the rapid routing
of vital data, and major help via BRC or the new National Biodiversity Network is still
over the horizon. Additionally, four national reviews on the status of groups of flies are
becoming out of date awaiting publication. The BAP process nationally, and at local
Agenda 21 level (local Biodiversity Action Plans), is deprived of essential information.

3 Workforce

Amateurs have always made a major contribution to the study of *Diptera*, indeed they do most of the fieldwork to produce data and reveal additional species: the partnership between amateur and professional has been and remains strong.

One of the biggest changes since 1973 has been severe staff reductions at what is now the Natural History Museum. The national museums in Edinburgh, Cardiff, Belfast and Dublin each have a dipterist, and about 10 provincial or county museums have strong to partial expertise. Some museums are acting as a catalyst for study and recording of flies, and Liverpool Museum runs workshops.

Following the split of the Nature Conservancy Council in 1991, the England Field Unit was axed. The superseding agencies have fewer contract entomologists in-house. The Biodiversity Action Plan has however started to generate new contract projects deploying dipterists, several of whom now work for environmental consultancies or independently. The National Trust had two entomologists with knowledge of *Diptera* for much of the 1990s. In Eire the government Wildlife Service entomologist is a leading dipterist. The Freshwater Biological Association has carried out some work on *Diptera*, especially *Chironomidae* and *Simulidae*. Sadly, few universities now have taxonomists or ecologists with a specialist understanding of *Diptera*.

Amateur dipterists have had to become more self-contained. The recording schemes, and corporately the Dipterists Forum, have been successful in encouraging more people to study *Diptera*: over 200 people currently belong to the Forum, and others occasionally submit data. The national entomological societies have also played a valuable part. The Royal Entomological Society has published further key works and has an excellent library. The British Entomological and Natural History Society has also published key works, tailored for the non-specialist (Stubbs and Falk 1983; Stubbs 1996a; Stubbs *et al.* in press), run workshops, talks and field meetings, provided a library and maintained reference collections. The Amateur Entomological Society has published an introduction to *Diptera* (Stubbs and Chandler 1978) and a handbook on insect conservation (Fry and Lonsdale 1991). The Freshwater Biological Association has published several key works on Diptera over the last 25 years, including mosquitoes for which there was a recording scheme (Cranston *et al.* 1987). The amateur work force is encouraged by these and local societies, but the recording schemes are the major focus in monitoring change.

4 Changes 1973–98

Despite severe land-use intensification, extinctions within well-known groups have been few. However, within such a large fauna, a small incremental loss within *Diptera* as a whole must still represent many species.

4.1 Species in formal status categories

The *Red Data Book* (Shirt 1987) encapsulates evaluations made in the late 1970s, with updating prior to publication. Partial revision was made and Notable species added (Falk 1991). There was major revision in the mid-1990s but four documents remain within JNCC unpublished and so hence without formalisation of categories. The

Table 16.1 Diptera species placed in formal status categories.

	Shirt (1987)	Falk (1991)
Endangered	270	203
Vulnerable	266	213
Rare	328	292
Insufficiently known (RDB)	—	102
Endemic	?	?
Extinct (not seen in the twentieth century)	3	14
Total RDB	827	810
% RDB	—	13.8
Notable (scarce)	—	644

'Insufficiently Known' category was introduced to accommodate those species whose placing within RDB categories was uncertain.

The revised IUCN status categories (Mace *et al.* 1992) employ criteria differing from those previously used: a complete overhaul is required, embracing our improved knowledge.

There are no British flies covered by international legislation, although some of our Endangered and declining species are in a perilous state over much of Europe.

The BAP Species Action Plans for *Diptera* have been very restricted, although a huge number deserve action. Priority has been to representative species that highlight particularly neglected conservation issues, such as dung, dead wood in streams, sand shoals along rivers, bare ground, and water abstraction affecting seepage habitats. The first tranche contained three *Diptera*, and the second 18 species; the published Long List will be revised before further tranches.

4.2 Losses and declines

It is now fairly certain that some species are extinct. The bee-fly *Villa venusta* (*V. circumdata*) was confined to sites on the Dorset heaths by the 1950s but after 1959 the cumulative effects of heathland loss, deterioration and fragmentation seemingly led to extinction, the balance tipped by either a decline in disturbed ground and verge heath with flower nectar sources (for example by cessation of traditional grazing) or the poor summers of the early 1960s that would have been difficult for a sunny climate species reduced to small populations (Stubbs 1996b).

Warble flies of cattle have been subject to an eradication programme. Legislation requiring compulsory treatment goes back to 1936, and infection became notifiable in 1982 in a final push to eradicate the pest (Tarry 1981). In November 1992, a submission was made to the CEC to declare the UK free of warble-fly infection. *Hypoderma bovis* and *H. lineatum* are officially extinct through deliberate action. A third species, *H. diana*, infects deer.

In Kent *Volucella zonaria* and *V. pellucens*, commensals of social wasps nests, have declined since the 1970s. *V. bombylans*, a commensal of bumble-bee nests, has shown a similar but less severe decline (Clemons 1998). Any relationship to changing status of the hosts is unclear.

Callicera spinolae, a rare European species, declined by the mid-1990s to a known breeding population in two water-filled rot holes in only two beech trees at a site in Cambridgeshire; one tree was blown down in a gale and propped up again. It has a Biodiversity Action Plan and was rediscovered at an historic site in 1998.

The switch from mainly cool and wet summers of the 1960s and early 1970s to the drought and hot summers of 1975–76 caused a dramatic reduction in the abundance of craneflies. Most species live in damp or wet places so declined as their habitat became too dry. The snail-killing *Sciomyzidae* also displayed a severe decline in the early 1990s.

The sequence of drought years severely reduced population levels of fungus gnats and other fungus-dependent species; some may have locally died out. To compound the problems for this large fauna, since about 1972 there has been a huge increase in public harvesting of mushrooms for food. A pickers' code (English Nature 1998) is likely to have only limited effect. Removal equates to destruction of essential dipteran habitat, and the eggs and larvae of the next generation.

Gales bring down trees and break off trunks and branches, regularly creating saproxylic habitat, but in some districts they have cumulatively reduced the prospect of a viable continuity of saproxylic habitats of large timber.

4.3 Additions and increases

Smith (1974) gave prominence to *Volucella zonaria* as a species that had colonised southern Britain, but despite the hot summers of the 1990s, this spectacular hoverfly has shown only slight expansion in range and only reached south Hertfordshire in 1996 (Aldridge 1998).

The *Tephritidae*, a phytophagous group of picture-winged flies, perhaps best exemplify rapidly changing distributions (Clemons 1997). *Tephritis cometa* (Fig. 16.4) breeds in the flower heads of creeping thistle and used to be only known from south-east England (White 1988). The first specimen came from south Hampshire in 1907 (introduction via Southampton cannot be ruled out), but its centre was the London area where Niblett (1956) was only able to give three records despite his diligent specialisation in this family. By the late 1970s the species had become frequent around London, and has now spread to the Mersey–Humber line, with a disjunct occurrence near Inverness and Lairg. Two species on mugwort, *Campiglossa absinthii* and *C. misella*, have become frequent over much of England to a similar extent. *Tephritis bardanae* of burdock, and *Trupaena stellata* of various 'weed' composites, are further distinctive examples showing huge expansions since 1973.

The conspicuous hoverfly *Sericomyia silentis* was not known on the Cambridgeshire Fens in the 1970s until an apparent vagrant was seen at Lode in 1977. Several were found at Chippenham Fen in 1988 and at Wicken Fen in 1991: the species is now well established on these sites but it has only been seen in late summer (I. Perry pers. comm.). Elsewhere in its predominantly northern and western range it normally appears as from May. A wetland species with aquatic larvae, it is surprising that it should only thrive now.

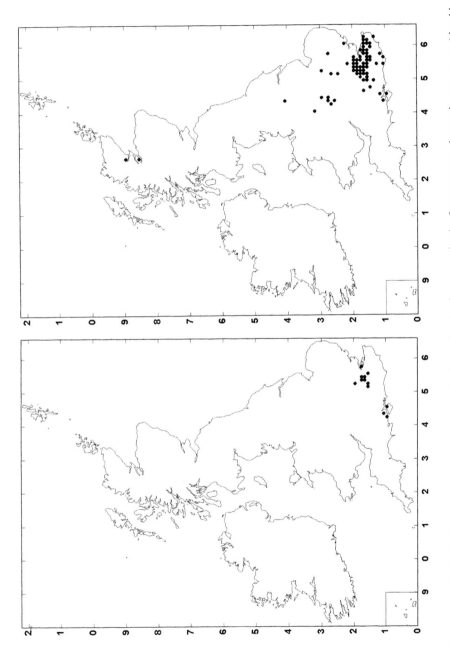

Figure 16.4 The tephritid fly *Tepthritis cometa*, whose larvae live in the flower heads of creeping thistle, *Circium arvense*, has undergone a considerable expansion in range. (a) Range up to 1973; (b) Current range.

Reproduced with permission in advance from Clemons (in prep.).

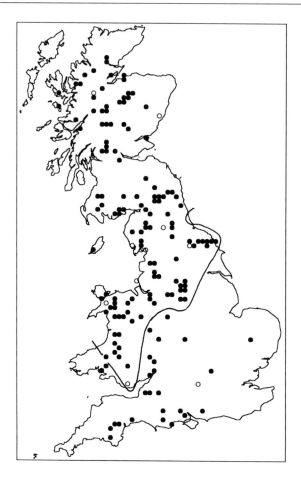

Figure 16.5 The saproxylic hoverfly *Xylota coeruleiventris* spread south in the twentieth century from its native habitat in the ancient Caledonian pine woods. The proliferation of conifer plantations reaching maturity and felling has enabled this extension of range (reproduced with permission in advance from Ball and Morris in prep.). Line indicates southern boundary of range given in Stubbs and Falk (1983). O pre-1980; ● post-1980.

Most tachinid flies are southern, requiring hot summers. Some highly distinctive ones sit on flowers as if saying 'please record me'. Numbers of *Phasia hemiptera* and *Subclytia rotundiventris* had fallen in the poor summers of the early 1960s but the series of hot summers from 1975, and especially in the 1990s, has led to a remarkable comeback, these species now being locally common within their previous range. Even great rarities, such as *Gymnosoma nitens*, are being reported again. These examples are parasites of shield bugs (*Acanthosomidae* and *Pentatomidae*), which thrive in warm summers, and climatic fluctuations may have affected the abundance of the hosts as much as their parasitic flies. The soldierfly *Stratiomys potamida* has also been more frequently seen since the late 1970s.

The dipteran fauna of conifers has been spreading quickly as post-war plantations mature and even progress to replanting. The hoverfly *Xylota coeruleiventris*, only known from the ancient pinewoods of the Scottish Highlands in the nineteenth century,

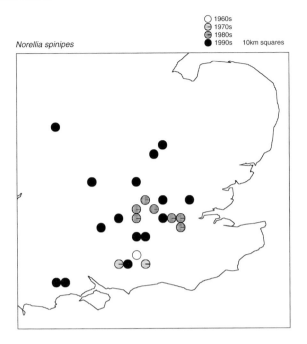

Figure 16.6 The scathophagid fly *Norellia spinipes* is phytophagous on *Narcissus*. First found in mid-Surrey in 1965, it has now substantially extended its range.

Reproduced with permission in advance from Ball (in prep.).

has spread quite widely in northern England and Wales (Stubbs and Falk 1983); it has now reached the south coast and the breckland (Fig. 16.5). The larvae occur under the bark of stumps of recently felled conifers and adults can be abundant; this is an interesting case of a native species spreading south.

The aphid-feeding hoverflies of conifers are mostly very mobile. The Highland speciality *Eupeodes neilseni* was found in Oxfordshire in the late 1970s, and other formerly rare species, such as *Parasyrphus malinellus* and *P. lineola,* have become widespread. *Epistrophe melanostoma* is associated with deciduous trees, and following the first sighting in Surrey in 1986, it became established there (Morris 1998) and in some adjacent districts, reaching Dorset.

The hoverfly *Sphegina siberica*, whose early stages are in dead wood, has been spreading rapidly across Europe; it was found in 1991 in western Scotland (Stubbs 1994), and shortly afterwards in Shropshire. In 1997 it was discovered in south-east Wales (Howe 1998) and in the Spey Valley (Perry 1998).

It is less clear whether some new migrants have established or not. *Eupeodes lundbecki* was first found in Aberdeenshire in 1976 and subsequently in Fair Isle and Dumfries (Watt and Robertson 1990) and Norfolk. Amazingly, another hoverfly new to Britain, *Helophilus affinis*, was also collected on Fair Isle in 1982 (Stuke 1996). These two species are mainly Scandinavian in distribution, so may have had a long sea crossing.

4.4 Aliens and introductions

The scathophagid *Norellia spinipes* (Fig. 16.6) is moderately large and noticeable, with larvae feeding on *Narcissus*. It was first found in 1965 at West Horsley, Surrey, and about 37 km away at Burnham Beeches in 1967, so it may have already been well established (Chandler and Stubbs 1969). By 1998 it had reached east Dorset, Oxfordshire and Huntingdonshire, and even to the south of Birmingham (Ball in prep.).

Another apparent introduction is the platypezid *Agathomyia wankowiczii*, a remarkable fly uniquely forming galls on the underside of the tough perennial bracket fungus *Ganoderma applanatum* s. lat., first found in Britain in 1990 on Ockham and Wisley Common, Surrey, followed by four sites in Kent (P. Chandler pers. comm.). The fly has been spreading from Eastern Europe, all records for Denmark, Belgium and Holland being post-1980.

5 Factors affecting change

5.1 Agriculture

Figures for the loss of semi-natural habitats can be extrapolated to declines of *Diptera* as with other wildlife. However, it is not just bald figures that count, for it is often the unquantified loss of habitat mosaic, including flowery verges, that contributes to habitat deterioration and faunal loss.

Government grant-aided field under-drainage of the 1970s–80s was probably the single most significant impact on *Diptera* in the last 25 years because of its massive scale. Since the late 1960s, extensive districts and regions have virtually lost all their wet meadows. Such drainage is not new (Philips 1989) but as farm machinery and techniques became more advanced during and after the 1940s, an era was reached where huge tracts of farmland could be efficiently drained, and field under-drainage became especially severe (Green 1976; Robinson and Armstrong 1988; Robinson 1990; Robinson *et al.* 1990). By 1973, the annual area affected in England and Wales had risen to over 100 000 ha (Robinson 1990), with peaks of 110 000 ha in 1976 and 1979, and 10 000 ha per year in lowland Scotland in the 1970s. Mole- and tile-drainage often required deepening of ditches and modification of water courses. After a peak in about 1980, government grants ceased, and accurate statistics were not maintained, but it appears to have declined to approximately 10 000 ha by 1990 and continued at that level (M. Robinson pers. comm.). The cumulative total may have been in excess of one million hectares in England and Wales since 1973, and surface drainage of moorland has also affected hill and upland wet habitats.

The decline in the quality of farmland, conservation sites and the loss of hedgerows, has resulted in a decline of nectar sources and foodplants, together with fewer shelter options for 3D plant structure, vegetation mosaics and plant community edges. Overgrazing, including disruptive management by blitz grazing from 'flying flocks' of sheep or outbreaks of rabbit abundance, and the latest BSE crisis have all caused problems for the continuity of grassland faunas; in the uplands the problems include over-grazing in gully woodland remnants.

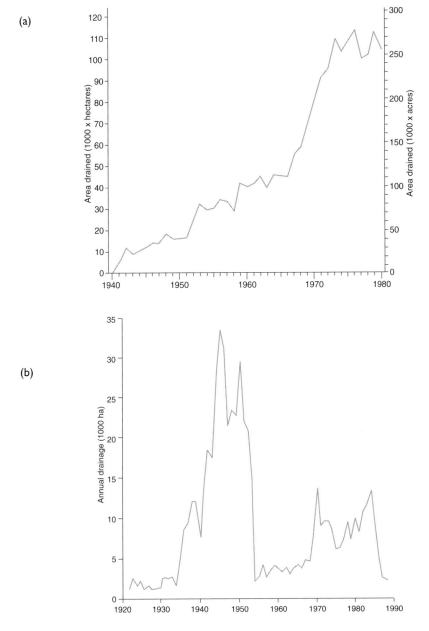

Figure 16.7 Annual area of grant-aided under-drainage. (a) In England and Wales, 1940–80; (b) In Scotland 1922–88. Fig. 16.7 (a) is reproduced from Robinson and Armstrong, 'The extent of agricultural field drainage in England and Wales 1971–80', p. 22, *Transactions of the IBG*, **18,** with permission from the Royal Geographic Society. Fig. 16.7 (b) is reproduced from Robinson *et al.* 'The extent of agricultural field drainage in Scotland 1983–86', p. 142, *Scottish Geographical Magazine*, **106**, with permission from the Royal Scottish Geographical Society.

5.2 Grazing levels

In the late 1970s the NCC started to address the knowledge gap on grazing levels faced with conversion from grazing to cereals. The main invertebrate interest was shown to be the aquatic and semi-terrestrial fauna associated with grazing-level ditches. Major concentrations of *Red Data Book* species were found at some sites, especially where cattle trampled berms occur; notably at Cliff Marshes, Kent, where the transition from freshwater to brackish conditions supported the most rarity-rich fauna. By the time the significance of the site for *Diptera* had been recognised, planning permissions had been granted for an oil refinery, a power station and the dumping of dredgings from the River Thames; fortunately, the planning permissions have not yet been taken up.

Site management has often been unsatisfactory for *Diptera* which in any case requires more than one strategy. While 3–4 year rotation ditch clearance may be appropriate traditional management in some circumstances, a modern mechanical digger making sharp-sided ditches without engineering a berm generally benefits only the open water fauna. Other elements of the fauna require shallow water or choked conditions.

5.3 The dung fauna

In the late 1980s and especially the early–mid-1990s, avermectins for veterinary treatment of gut parasites made dung lethal for elements of the dung fauna, including many *Diptera*. Timing of application can be harmful (during insect activity) or minimal (late autumn) but the new trend is towards bolas application which may lead to lethal dung production for very long periods. The hoverfly *Rhingia campestris* is widely thought to have declined in the avermectin era, though there may be an overlay of cyclic variations in annual abundance (Morris 1998). Fears for the future of the hornet robberfly, *Asilus crabroniformis*, one of the largest and most spectacular British flies, resulted in it being one of the first BAP Action Plan species, since its larvae are believed to feed on dung beetle larvae. Its ecology seems to depend on maintaining viable metapopulations since dependance on one field or even one farm could not be relied upon.

5.4 Spray drift

Spray drift must also have had consequences for *Diptera* on small conservation sites. With respect to ground sprays of herbicides, Davis (1992) noted a significant effect up to 16 m in light winds, and for orchard sprayers 30–40 m downwind. As regards pesticides, Davis (1994) recommended a statutory 250 m buffer zone around SSSIs to provide acceptable protection from downwind drift and over-spraying in the majority of cases, although the sensitivity of the majority of *Diptera* is unknown.

5.5 Woodland management

Major post-war forestry programmes have had a considerable impact on *Diptera*. Though much woodland is neglected, there has been a growing commitment from local authorities and the conservation movement to acquire and manage woodland, with varying effect.

The New Forest has substantially declined in quality for *Diptera* since the early 1950s due in part to the felling of veteran trees and over-grazing. Replacing the Forest

Nature Reserves of 1959 by a wider SSSI in 1971 did not halt the problems (Tubbs 1986). Since 1973, over-grazing has caused a serious decline in habitat quality in much of the woodland as well as open habitats, heightened by drainage, coniferisation, extensive fires and public pressure. Herbaceous nectar sources have become limited for *Diptera*, and habitat lost for phytoghagous and other herb-layer species, more so when fences for enclosures were not maintained. The massive decline of butterflies in this area (Oates 1996) is mirrored by the impoverishment of the dipteran fauna. Some older trees left as fringes around felled areas succumbed to wind throw, and the 1976 drought killed many beeches. The stock of ancient trees has been declining, and much fallen timber, even in former Forest Nature Reserves (FNRs) such as Mark Ash, was removed for firewood. Many beech trees are over-mature and it is uncertain whether there will be viable cohorts of replacement trees.

Windsor Forest lost many of its ancient trees to coniferisation before 1973, and some that were left became shaded out. Every year gales have reduced the number of over-mature ancient beeches, which are especially important to *Diptera*, and fallen trees have been removed for firewood. In the mid-1990s, a more effective conservation policy was implemented and there is now far more fallen-tree habitat for *Diptera* than any living dipterist has previously seen. However, as the number of mature live trees dwindles and replacements seem thin on the ground, future continuity problems are foreseen.

In Epping Forest, long said to be over-tidied, there is now a proper concern for conservation of veteran trees and saproxylic fauna. Pollarding was re-established in the 1990s at Burnham Beeches, which has led to the practice being tried out on other sites.

The cessation of coppice management must also have had a strong impact though this has not been fully evaluated. Laurence (1997) studied the *Diptera* of different age-class coppice in Bradfield Woods, Suffolk, but that only gives a limited perspective. The element most vulnerable is that requiring sunny rides and other canopy openings. In the East Midlands the hoverfly genus *Sphaerophoria* (apart from *S. scripta*) is absent from most woods where a period of closed canopy has occurred, even if rides have been recently opened up. These species are seemingly confined to very large woods where management has provided continuity of open-ride conditions.

Deer grazing has increased over the last 25 years in many woods. In Monks Wood NNR, muntjac, first noticed in the early 1970s, are now plentiful, and other deer species have become common. The wood has become more dominated by grasses and sedges, bramble has been reduced, and various other plants have declined; the overall aspect of the rides is a sparsity of flowers suitable for nectaring *Diptera*. Much of the blame is placed on muntjac deer (Crampton *et al.* 1998).

Dutch elm disease fungus, *Ophiostoma novo-ulmi*, was gaining a hold in limited districts in 1971 but by 1973 it was spreading fast (Rackham 1987). By the late 1970s it was well on the way to eliminating mature elms over extensive districts of southern Britain in both woodland and hedgerows. Sadly, elm was especially important for saproxylic *Diptera* and other insects (Archibald and Stubbs 1980).

While aspen is almost ubiquitous in Britain, sites with historic continuity of mature and dying trees are not: there are none in England or Wales. The Scottish Highlands has about eleven sites, mainly small and fragile, with nationally significant saproxylic faunas. The hoverfly *Hammerschmidtia ferruginea* is special, occurring with other rare saproxylic *Diptera*. As European beaver love aspen, proposals to re-establish it in the Highlands are worrying (Kitchener and Conroy 1996). The public consultation report lacks mention of the entomological and management constraints that need to be addressed (Scott Porter 1998).

The ancient Caledonian pine and birch forests have become a cause for concern where over-grazing by deer or domestic stock has been inhibiting regeneration. As mature trees die and decay, the woodland opens up and slowly disappears, while modern forestry results in dense planting and harvesting before over-maturity. Some sites are now safer and better managed (for example the RSPB areas at Abernethy).

5.6 Industry

In the last 25 years many peatland sites have been damaged or destroyed before the *Diptera* had been surveyed. At Thorne Moors, Yorkshire, a survey was instigated and revealed an important fauna (Skidmore *et al.* 1985).

The importance of post-industrial land to *Diptera* has become evident from studies in Warwickshire (S. J. Falk pers. comm.) and around Sheffield (D. Whiteley pers. comm.). Disused quarries and landfill sites had distinctive faunas from conserved sites with ancient vegetation types. In the Grays district of south Essex important *Diptera* are being found in old quarries and rough grassland now under immense development pressure (Harvey 2000).

5.7 Water pollution

Water authorities include some *Diptera* in monitoring, but the results are rarely public. While it is difficult to comment on impacts of water pollution on *Diptera*, most of the species of concern are probably not monitored, nor the semi-terrestrial fauna of river banks and standing-water margins.

The *Sea Empress* oil spill outside Milford Haven in 1996 damaged much of the Pembrokeshire coastline. There was concern about sea birds and marine life, but not important flies of the intertidal zone, the drift-line, freshwater seepages and saltmarsh. There were no follow-up studies on impacts and recovery, such as the effect of oil and chemical sprays on *Diptera*, including detergents that deprive larvae and pupae of air to breathe. It is not known if local extinction occurred nor if the species dependent on scarce isolated habitats can recolonise.

5.8 Rivers and streams

The 1960s–80s saw ruination of virtually whole catchments in parts of southern Britain through 'river improvement'. Many *Diptera* are confined to river and stream banks with natural features such as sand and shingle shoals. Canalisation of water-courses removes or grossly alters the natural river and stream profiles, resulting in a simplification of ecological features and so of the fauna. Even minor engineering can eliminate critical niches. Only by recreating meanders, creating berms at average water level and other such cosmetic measures, does the dipteran fauna stand a chance of some degree of recovery for species within recolonisation range.

Apprehension arose in 1997 over measures to remove fallen trees and branches from the River Wye and its tributaries, even along small streams. While it was said that salmon would benefit, there was no prior assessment of the implications for the saproxylic *Diptera*.

English Nature (1997) issued a list of proposed river SSSIs, and the candidate SAC (Special Areas of Conservation) list will be distilled from these. The list omits the River Monnow on the southern Welsh borders, confirmed as of national importance for *Diptera* in the early 1980s; it has since suffered from local 'improvements'.

The River Feshie has been diverted within an SSSI at its junction with the River Spey, altering the hydrology over the river confluence shingle fan, the biggest remaining intact river confluence fan in Europe. The Feshie Fan was the only known recent British site for the RDB1 cranefly *Nephrotoma aculeata* and the site contained the greatest known concentrations of rare and scarce *Diptera* on the River Spey system (Stubbs 1991). The reason for engineering the Feshie was to prevent flooding of riverside fields. Confidence over the future for the Spey has been shaken by this decision, and subsequent other channelling and building on an eroding river terrace, on Scotland's top river for *Diptera*.

5.9 Public pressure

While more areas of the countryside and town are now 'safe' from destruction, and accessible to wildlife recording, for much of the last 25 years there has been a cult of town-park tidiness. Dead wood was tidied up, trees which were not pristine were removed, damp ground ditched, grassland treated as lawns, car parks constructed and miscellaneous 'facilities' erected.

Dipteran faunas often suffer 'death from a thousand pricks'. Ockham and Wisley Common SSSI, Surrey, a top national site for craneflies up to the mid-1970s, has been reduced to a poor site by inappropriate management, including the siting of car parks, drainage, failure to control trampling and the impact of an informal lido: added to which were upsets to hydrology from road construction and droughts. At Moat Pond on Thursley Common, Surrey, the margins were wrecked by the removal of floating rafts of vegetation by people sailing model yachts and a misguided naturalist! At Denny Wood, one of the former FNRs in the New Forest, an official campsite was designated under a magnificent stand of ancient oaks. After a branch fell off in a gale and killed someone in 1980, the ancient oaks were felled.

The 1970 Save the Village Pond Campaign, and some subsequent initiatives, have probably wrecked the dipteran fauna of as many ponds as have benefited; the gentle shelving margins where so many *Diptera* are concentrated were often dug away to create sharp sides or buried under debris.

A culture that trees are good has developed, but the converse is that other tatty habitats are 'bad' so should be planted. Many corners of the countryside with residual populations of open-habitat *Diptera* have been vanishing as a result.

It is tempting to suggest that as much dipteran habitat has been destroyed or degraded by inappropriate decisions by amenity and conservation organisations as by developers, farmers and foresters. Though over the last 10 years there has been a growing professionalism and understanding of conservation and recreation management, with few specialists to advise, blunders will continue.

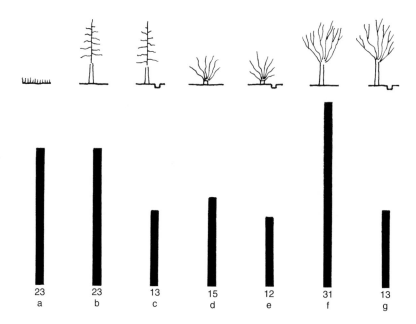

Figure 16.8   Effect of minor ditching on craneflies at Ockham and Wisley Commons, Surrey, in the mid-
1960s, showing the number of species occupying different habitats. a wet open habitat; b
alder carr without ditch; c alder carr with minor ditch; d sallow coppice without ditch; e
sallow coppice with minor ditch; f mature sallow carr without ditch; g mature sallow carr
with ditch. The ditches were only approximately 15 cm deep. Even before the droughts of
the 1990s, minor ditching reduced faunal richness during relatively wet summers.

Reproduced from Stubbs (1999).

5.10 National and international policies

In 1974 the Nature Conservancy Council started developing a national invertebrate
conservation strategy. The Invertebrate Site Register (ISR), launched in 1979, provided
a mechanism for identifying significant sites for *Diptera* and species of special note. The
Wildlife and Countryside Act (WCA) 1981 placed a premium on defining the interest
and damaging operations for SSSIs. Conservation needs of *Diptera* were incorporated
into various guideline documents, and the profile raised for invertebrate conservation
as a whole. Regrettably, the resources to progress the ISR declined steadily during the
1990s. There has been no requirement as yet to place *Diptera* on Schedule 5 of the Act
and there are no species on international legislation that concern Britain.

The WCA has measures that make it illegal to release non-native species without a
licence, partly to ensure proper deliberation before any new biocontrol agents are
released into the wild. However, this measure does not embrace bacteria and viruses,
and in the 1980s *Bacillus thuringiensis* ssp. *israelensis* (Bti) came to the fore as an insect
pest-control agent. A genotype, originally said to attack only mosquitoes, was subject
to minor trials before release and has been used with varying results in some districts
where mosquitoes have been a public nuisance (Ramsdale and Snow 1995). Since mosquito larvae are prey for other aquatic wildlife, knocking out mosquitos in grazing-
level ditches and other such environments must have food web implications, which

have yet to be properly evaluated. It is now known that Bti affects some other aquatic flies, including meniscus midges, *Dixidae*, which include species of conservation concern. In flowing water, Bti has been used against biting black fly, *Simulium* (e.g. Welton and Ladle 1993). A review is provided by Drake (1994).

The Council of Europe has made significant moves in invertebrate conservation. A Charter on Invertebrates (Pavan 1986) was adopted as an Appendix to Recommendation R(86)10 of the Committee of Ministers of the Council of Europe in 1986. More significantly, Recommendation R(88)10 of the Committee of Ministers of the Council of Europe in 1988, concerns the protection of saproxylic organisms and their biotopes. This was based on a review (Speight 1989a) and accords international status to Abernethy (pine), Epping Forest, Moccas Park, the New Forest and Windsor Forest and Park, together with some key habitat-quality indicator *Diptera*. There is a large saproxylic dipteran fauna in Britain, including many *Red Data Book* and other species of conservation concern. Regrettably, the conservation movement has largely failed to maximise the impact of the Report and Recommendation.

The Rio Earth Summit of 1992 led to a consortium of non-governmental organisations (NGOs) producing *Biodiversity Challenge* (1993, 1994) which set out issues and policies that took account of priorities for *Diptera*. The agencies produced the official Biodiversity Action Plan in 1994. There are some curious anomalies in the BAP, such as not acknowledging the priority of veteran trees or saproxylic biota in the habitat action plan for native pine woods. Much depends on the steering groups' developing habitat action plans, but the available expertise to guide implementation policy for invertebrates, let alone more specifically for *Diptera*, is a limitation. Similarly, Agenda 21 is on a scale and speed that outpaces the ability of dipterists to play a part, as is evident from reports so far produced which in some cases lack a focus on key species or habitats for *Diptera*.

6 Prognosis for the next 25 years

Much depends on government and EC policy and, crucially, public attitude; the effectiveness of BAP has yet to emerge. Further, it is still too difficult to forecast climatic change or global politics which could rapidly affect the economy and agriculture. The following factors will, however, be crucial:

1 Water supply policy: all water sources will be under further pressure. Many sites rich in *Diptera* are not properly surveyed, yet alone protected. Further tapping into underground aquifers will lead to even more widespread impoverishment and loss of dipteran faunas of springs, seepages and other wetlands.
2 The *Diptera* faunas of lowly regarded habitats (e.g. dung, post-industrial environments) will decline substantially unless the conservation movement shows greater determination.
3 Climatic change will have a significant impact on Britain's *Diptera* fauna, including resultant changes in sea level. Apart from implications for saltmarshes and coastal levels, there will be growing demands for coastal protection of soft-rock cliffs where vulnerable dipteran faunas occur.
4 The problems of data processing for *Diptera* must be resolved.

5 To stop habitat management gaffes, better training of those involved in conservation and countryside management is essential.

6 BAP presents great opportunities but also strains already limited professional and amateur expertise. In 25 years, and hopefully sooner, BAP should have become a significant influence on the changing fortunes of the *Diptera* in the UK. University departments, for instance, need to develop the will to play a larger role.

7 Will a new organisation for invertebrate conservation be necessary to co-ordinate, resource and steer?

ACKNOWLEDGEMENTS

Thanks are due to those who provided unpublished information, including maps. Laurence Clemons (organiser of the *Tephritidae* Recording Scheme) provided detail from which to select examples cited. Stuart Ball (co-organiser with Roger Morris of the Hoverfly Recording Scheme) produced the hoverfly maps and also the initial results of his analysis of data on *Norellia spinipes*.

The Institute of Hydrology, Institute of British Geographers and Royal Scottish Geographical Society have kindly given permission to reproduce graphs on field drainage. I am grateful to Dr Adrian Armstrong (IH, Wallingford) and Dr Mark Robinson (MAFF, Mansfield) for advice on recent trends in field drainage. Mrs P A Nikolaides (MAFF, Surbiton) was helpful in seeking out information on the warble-fly eradication programme.

A number of dipterists have willingly given advice, including the author of the 1973 review on *Diptera*, Ken Smith. In particular, I thank Peter Chandler, Roy Crossley and Ivan Perry.

REFERENCES

Note: Harrington and Stork (1995) have relevant papers not cited in this chapter.

Aldridge, M. C. (1998) *Volucella zonaria (Dipt., Syrphidae)* – a recent northward extension of range into Hertfordshire. *Entomologist's Monthly Magazine*, **134**, 292.

Archibald, J. F. and Stubbs, A. E. (1980) The effects of Dutch elm disease on wildlife. *Quarterly Journal of Forestry*, **74**, 30–37.

Ball, S. G. and McLean, I. F. G. (1986) Preliminary atlas. *Sciomyzidae Recording Scheme Newsletter*, no. 2, 1–36.

Biodiversity Challenge Group (1993) *Biodiversity Challenge: an agenda for conservation in the UK*. Sandy: RSPB.

Biodiversity Challenge Group (1994) *Biodiversity Challenge: an agenda for conservation in the UK*, 2nd edn, Sandy: RSPB.

Biodiversity Steering Group (1995) *Biodiversity: the UK Steering Group Report*, 2 vols, London: HMSO.

Chandler, P. J. (1998) Checklists of insects on the British Isles (new series). Part 1: *Diptera*. *Handbooks for the Identification of British Insects*, **12** (1), i–xix, 1–234.

Chandler, P. J. and Stubbs, A. E. (1969) A species of *Norellia* R.-D. (*Dipt., Scathophagidae*) new to Britain. *Proceedings of the British Entomological and Natural History Society*, **2**, 120–124.

Clements, D. K. and Alexander, K. N. A. (1987) The distribution of the fly *Xylophagus ater* Meigen (*Diptera: Xylophagidae*) in the British Isles, with some notes on its biology. *Proceedings and Transactions of the British Entomological and Natural History Society*, **20**, 141–146.

Clemons, L. (1997) A provisional atlas of the *Tephritidae* (*Diptera*) of Britain and Ireland. *Tephritidae Newsletter*, no.6, 1–42.

Clemons, L. (1998) Further notes on the genus *Volucella* (*Diptera: Syrphidae*) in Kent. *Bulletin of the Kent Field Club*, **43**, 77–84.

Crampton, A. B., Stutter, O., Kirby, K. J. and Welsh, R.C. (1998) Changes in the composition of Monks Wood National Nature Reserve (Cambridgeshire, UK) 1964–1996. *Arboricultural Journal*, **22**, 229–245.

Cranston, P. S., Ramsdale, C. D., Snow, K. R. and White, G. B. (1987) *Adults, Larvae and Pupae of British Mosquitoes* [Scientific Publication, no. 44]. Windermere: Freshwater Biological Association.

Davis, B. N. K. (ed.) (1992) *Environmental Impact of Pesticide Drift* [English Nature Research Report no. 11]. Peterborough: English Nature.

Davis, B. N. K. (ed.) (1994) *Environmental Impact of Pesticide Drift: Aerial Spraying* [English Nature Research Report, no. 112]. Peterborough: English Nature.

Disney, R. H. L. (1975) *A Key to British Dixidae* [Scientific Publication no. 31]. Windermere: Freshwater Biological Association.

Drake, C. M. (1991) *Provisional Atlas of the Larger Brachycera (Diptera) of Britain and Ireland*. Huntingdon: Biological Records Centre.

Drake, M. (1994) *Bacillus thuringiensis israelensis* (Bti) *and its Use on Conservation Sites* [Species Conservation Handbook]. Peterborough: English Nature.

English Nature (1996) *Impact of Water Abstraction on Wetland SSSIs* [English Nature Freshwater Series, no. 4]. Peterborough: English Nature.

English Nature (1997) *Wildlife and Fresh Water: an Agenda for Sustainable Management*. Peterborough, English Nature.

English Nature (1998) *The Conservation of Wild Mushrooms: The Wild Mushroom Pickers Code of Conduct*. Peterborough: English Nature.

Entwistle, P. F. and Stubbs, A. E. (1983) *Preliminary Atlas of Hoverflies (Diptera: Syrphidae) of the British Isles*. Huntingdon: Biological Records Centre.

Falk, S. (1991) *A Review of the Scarce and Threatened Flies of Great Britain (Part 1)*, [Research and survey in Nature Conservation, no. 39]. Peterborough: Nature Conservancy Council.

Fowles, A. P. (1985) *Working Checklist of the Diptera Recorded from Wales*. Bangor: Countryside Council for Wales.

Friday, L. and Harley, B. (eds) (2000) *Checklist of the Flora and Fauna of Wicken Fen*. Colchester: Harley Books.

Fry, R. and Lonsdale, D. (1991) Habitat conservation for insects – a neglected green issue. *The Amateur Entomologist*, **21**, i–xvi, 1–262, pl. 1–32.

Goldie-Smith, E. K. (1990) Distribution maps for *Dixidae* in Great Britain and Ireland. *Dipterists Digest*, **3**, 8–26.

Green, F. H. W. (1976) Recent changes in land use and treatment. *The Geographical Journal*, **124**, 12–26.

Hancock, E. G. and Horsfield, D. (1997) *Ormosia fascipennis* a short-palped cranefly (*Diptera, Limoniidae*) new to the British Isles. *Dipterists Digest*, **4**, 68–71.

Harrington, R. and Stork, N. E. (eds) (1995) *Insects in a Changing Environment*. London: Academic Press.

Harvey, P. (2000) The East Thames Corridor: a nationally important invertebrate fauna under threat. *British Wildlife*, **12**, 19–98.

Henshaw, D. J. and White, I. M. (1997) *Taxonomic review of possibly endemic British Diptera* [Contract no. V17.1E]. Peterborough: English Nature.

Howe, M. A. (1998) Field meeting of the Dipterists Forum at Abergavenny, June 1997 [*Natural Science Report no. 98/5/2*]. Bangor: Countryside Council for Wales.

Kitchener, A. C. and Conroy, J. (1996) The history of the beaver in Scotland and the case for its reintroduction. *British Wildlife*, 7, 156–161.

Kloet, G. S. and Hincks, W. D. (1945) *A Check List of British Insects*. Stockport: privately published.

Kloet, G. S. and Hincks, W. D. (1976) A check list of British Insects (2nd edn). Part 5: *Diptera and Siphonaptera. Handbooks for the Identification of British Insects*, 11 (5), 1–139.

Laurence, B. R. (1997) Flies from ancient coppiced woodland in Suffolk. *Dipterists Digest*, 4, 78–91.

Levy, E. T. and Levy, D. A. (1998) *Somerset Hoverflies*, privately published with support from Somerset Wildlfe Trust.

Mace, G. M. *et al*. (1992) The development of new criteria for listing species on the IUCN Red List. *Species*, 19, 16–22.

Morris, F. A. (1865 [–67]) *A Catalogue of British Insects, in all the Orders*. London: Longman.

Morris, R. K. A. (1998) *Hoverflies of Surrey*. Pirbright: Surrey Wildlife Trust.

Niblett, M. (1956) The flies of the London area. III. *Trypetidae. London Naturalist*, **1955**, 82–88.

Oates, M. (1996) The demise of butterflies in the New Forest. *British Wildlife*, 7, 205–216.

Palmer, M. (1991) *A Summary of the Work of the England Field Unit 1979–1991* [England Field Unit Project no. 139]. Peterborough: English Nature.

Pavan, M. (1986) *A European cultural revolution: the Council of Europe's Charter on Invertebrates*. Strasbourg: Council for Europe.

Perry, I. (1998) *Sphegina siberica* Stackelberg (*Diptera, Syrphidae*) in the Central Highlands of Scotland. *Dipterists Digest*, 5, 8–9.

Philips, A. D. M. (1989) *The Underdrainage of Farmland in England during the Nineteenth Century* [Cambridge Studies in Historical Geography, no. 15]. Cambridge: Cambridge University Press.

Pont, A. (1987) *Provisional Atlas of the Sepsidae (Diptera) of the British Isles*. Huntingdon: Biological Records Centre.

Rackham, O. (1987) *The History of the Countryside*. London: J.M. Dent and Sons.

Ramsdale, C. D.and Snow, K. R. (1995) *Mosquito Control in Britain*. Dagenham: University of East London.

Robinson, M. (1990) *Impacts of Improved Drainage on River Flows* [Report no. 113]. Wallingford: Institute of Hydrology.

Robinson, M. and Armstrong, A. C. (1988) The extent of agricultural field drainage in England and Wales, 1971–80. *Transactions of the Institute of British Geographers*, new series 13, 19–28.

Robinson, M., Clayton, M. C. and Henderson, W. C. (1990) The extent of agricultural field drainage in Scotland, 1983–6. *Scottish Geographical Magazine*, 106, 141–147.

Rotheray, G. E. and MacGowan, I. (1990) Re-evaluation of the status of the hoverfly *Callicera rufa* (Diptera, Syrphidae). *Entomologist*, 109, 35–42.

Scott Porter Research and Marketing (1998) *Re-introduction of the Beaver to Scotland: results of a public consultation* [Research, Survey and Monitoring Report, no. 121]. Perth: Scottish Natural Heritage.

Shirt, D. B. (1987) *British Red Data Books: 2. Insects*. Peterborough: Nature Conservancy Council.

Skidmore, P., Limber, M. and Eversham, B. C. (1985) The insects of Thorne Moors. *Sorby Record*, 23 (Suppl.), 89–153.

Smith, K. G. V. (1974) Changes in the British dipterous fauna, in D. L. Hawksworth (ed.), *The Changing Flora and Fauna of Britain*. London: Academic Press, pp. 371–391.

Snow, K. R., Rees, A. T. and Bulbeck, S. J. (1998) *A Provisional Atlas to the Mosquitoes of Britain*. Dagenham: University of East London.

Speight, M. C. D. (1989a) *Saproxylic Invertebrates and their Conservation* [Nature and Environment Series, no. 42.]. Strasbourg: Council for Europe.

Speight, M. C. D. (1989b) The Council for Europe and the conservation of Diptera. *Dipterists Digest*, **2**, 3–7.

Stubbs, A. E. (1972) A review of information on the distribution of the British species of *Ptychoptera* (*Dipt: Ptychopteridae*). *Entomologist*, **105**, 23–38, 308–312.

Stubbs, A. E. (1982) Conservation and the future for the field entomologist. *Proceedings and Transactions of the British Entomological and Natural History Society*, **15**, 55–67.

Stubbs, A. E. (1990) The beginning of Diptera Recording Schemes in Britain. *Dipterists Digest*, **6**, 2–10.

Stubbs, A. E. (1991) *Flood alleviation in Upper Strathspey: Terrestial Entomology*. Contract report to Institute of Hydrology.

Stubbs, A. E. (1992) *Provisional Atlas of the Long-palped Craneflies (Diptera: Tipulinae) of Britain and Ireland*. Huntingdon: Biological Records Centre.

Stubbs, A. E. (1993) *Provisional Atlas of the Ptychopterid Craneflies (Diptera: Ptychopteridae) of Britain and Ireland*. Huntingdon: Biological Records Centre.

Stubbs, A. E. (1994) *Sphegina (Asiosphegina) siberica* Stackelberg, a new species and subgenus of hoverfly (*Diptera, Syrphidae*) in Britain. *Dipterists Digest*, **1**, 23–25

Stubbs, A. E. (1995) Advances in the British hoverfly list: 1901 to 1990. *Dipterists Digest*, **2**, 13–23.

Stubbs, A. E. (1996a) *British Hoverflies: Second Supplement*. Reading: British Entomological and Natural History Society.

Stubbs, A. E. (1996b) *The Status of the Bee Flies Bombylius minor and Villa circumdata, with particular reference to the East Dorset heaths* [Contract no. VT11(B): Bee Flies]. Peterborough: English Nature.

Stubbs, A. E. (1999) The effect of drainage on *Diptera* faunas. *Bulletin of the Dipterists Forum*, **47**, 11.

Stubbs, A. E. (1999) A review of field meetings held by the *Diptera* recording schemes, including an analysis of species added to the British list. *Bulletin of the Dipterists Forum*, **47**, 21–26.

Stubbs, A. E. and Chandler, P. J. (1978) A dipterist's handbook. *The Amateur Entomologist*, **15**, i–ix, 1–255.

Stubbs, A. E., Drake, C. M. and Wilson, D. (in press) *British Soldierflies and their Allies*. Reading: British Entomological and Natural History Society.

Stubbs, A. E. and Falk, S. J. (1983) *British Hoverflies: an illustrated identification guide*. London: British Entomological and Natural History Society.

Stuke, J. S., (1996) *Helophilus affinis* new to the British Isles (*Diptera, Syrphidae*). *Dipterists Digest*, **3**, 44–45.

Tarry, D. W. (1981) Distribution of cattle warble flies in Great Britain: a larval survey. *Veterinary Record*, **108**, 69–72.

Tubbs, C. R. (1986) *The New Forest*. London: Collins.

Verrall, G. H. (1901) *A List of British Diptera*. London: Pratt.

Verrall, G. H. (1901) *A List of British Diptera*, 2nd edn. Cambridge: Cambridge University Press.

Walker, F. (1851–56) *Insecta Britannica Diptera*, 3 vols. London: Reave and Bentham.

Ward, S. (1997) *The number of Terrestrial and Freshwater Species in Scotland* [Scottish Natural Heritage Review no. 84]. Perth: Scottish Natural Heritage.

Watt, K. R. and Robinson, D. M. (1990) *Eupeodes lundbecki* (Soot-Ryen) (*Diptera: Syrphidae*) new to Britain and its separation from related species. *Dipterists Digest*, **6**, 23–27.

Welton, J. S. and Ladle, M. (1993) The experimental treatment of the blackfly *Simulium posticatum* in Dorset using the biologically produced insecticide *Bacillus thuringiensis* var *israelensis*. *Journal of Applied Biology*, **30**, 772–782.

White, I. M. (1988) Tephritid flies: *Diptera: Tephritidae*. *Handbooks for the Identification of British Insects*, **10**(5a), 1–134.

Chapter 17

True bugs, leaf- and planthoppers, and their allies

Peter Kirby, Alan J. A. Stewart and Michael R. Wilson

ABSTRACT

The *Hemiptera* is the fifth largest insect order in Britain, with around 1600 species comprising the *Heteroptera* (true bugs) and the *Homoptera*, divided into the *Auchenorrhyncha* (including leafhoppers, planthoppers and cicadas) and the *Sternorrhyncha* (aphids, whiteflies and scale insects). While less well documented than some other insect groups in the UK there is active recording of the *Heteroptera* and of the *Auchenorrhyncha* sufficient to show changes in the past 25 years. Species in these groups have been divided into several categories:

1 Species which have been known in Britain for many years, but where recent field-work has shown them to be more widely distributed than previously thought;
2 Species which have been recently described as new to Britain, but are likely to have been overlooked in the past;
3 Species that appear to have arrived in Britain recently, have become established, and have perhaps expanded their range;
4 Species known for many years that appear to have recently expanded their range in Britain;
5 Species that appear to have become rarer or gone extinct in recent years.

Examples of *Heteroptera* and *Auchenorrhyncha* corresponding to each category are discussed.

1 Introduction

The fifth largest insect order in Britain, with around 1600 species, the *Hemiptera* are a very diverse group, unified by the possession of sucking mouth parts. A distinction is generally made between the *Heteroptera* (often termed 'true bugs') and the *Homoptera* (the remainder) on the basis of the structure of the forewings; in the *Heteroptera*, the forewings comprise a thickened basal two-thirds and a membranous distal third, whereas there is no such division in the *Homoptera*. The latter group in turn is generally divided into two series, the *Auchenorrhyncha* (leafhoppers, planthoppers, treehoppers, cicadas and froghoppers) and the *Sternorrhyncha*: this includes *Aphidoidea* (aphids), *Coccoidea* (scale insects), *Psylloidea* (jumping plant lice) and *Aleyrodidae* (whiteflies). All *Homoptera* feed on plant sap, the majority from phloem sap or the contents of mesophyll cells and a few from xylem sap. Most *Heteroptera* are also herbivorous, but some species feed additionally or exclusively on animal tissue. *Hemiptera* may be found in all major habitats in Britain; leafhoppers are often among the most abundant insects in grassland.

The *Hemiptera* were not covered in the 1973 symposium, so some historical background and landmarks are needed to place current work in context. The *Heteroptera* are the best-known group (both taxonomically and biologically) of *Hemiptera* in Britain, followed by the *Auchenorrhyncha*, with other groups less well known. *Heteroptera* and *Auchenorrhyncha* are usually dry-mounted while the *Sternorrhyncha* groups are usually examined on microscope slides. The latter groups therefore tend to be more the realm of the professional entomologist or dedicated amateur. Nevertheless all groups were 'monographed' by the early part of the twentieth century and around 70% of the species on the current British list were found by these early workers – quite an achievement. These works summarised all that was known about the biology and distribution of the species but subsequent updates have been sporadic. Modern guides to the identification of these groups in the UK are given in Barnard (1999).

The conservation status, habitat requirements and general ecology of the rarer British *Hemiptera* were reviewed by Kirby (1992a). Attention was confined to the *Heteroptera* and *Auchenorryncha*, among which 242 species were designated with either *Red Data Book* or Nationally Notable status (Table 17.1). No *Sternorrhyncha* were assigned rarity status since distributional information on these groups was perceived to be inadequate, although the distribution of certain better-known species would suggest that there may be suitable candidates. Most *Hemiptera* in the UK are not pests although the aphids contain some important cereal pests and since the 1960s the incidence of a range of pest species has been carefully monitored over much of Britain using a series of suction traps. The changes in a range of species over this time period are discussed by Woiwod (1991) and by Woiwod and Harrington (1994).

Table 17.1 Hemiptera (Heteroptera and Auchenorrhyncha) of conservation interest in Britain: Numbers of species in each category of rarity (from Kirby 1992a).

Category	Definition	Heteroptera	Auchenorrhyncha
RDB1	Endangered: in danger of extinction if causal factors continue to operate	8	1
RDB2	Vulnerable, likely to move into RDB1 category in near future	3	0
RDB3	Rare, but not presently vulnerable or engangered	38	0
RDBK	Insufficiently known but suspected to fall within one of the RDB categories	4	22
Notable A	Thought to occur in less than fifteen 10-km squares	15	20
Notable B	Thought to occur in less than 100 10-km squares	66	55
Extinct	Former populations now thought to have died out	7	0
Total		143	98
Total species in Great Britain		533	376

Since the *Sternorrhyncha* are not covered in detail in this chapter two recent introductions that have rapidly become widespread should be mentioned here. The North American aphid, *Macrosiphon albifrons*, appeared in Britain around 1981 and within a few years colonised much of England and Wales, causing serious die-back of lupin plants (Carter *et al.* 1984). The horse chestnut scale, *Pulvanaria regalis*, has spread from initial records in eastern Britain to towns all over southern England, the Midlands and South Wales (Speight and Nicol 1985) where it may be very common on amenity trees such as horse chestnut, limes and maples.

2 Categories of change

This chapter deals separately with the *Heteroptera* and the *Auchenorrhyncha*. The size of each section broadly reflects the amount of information available for each group. For the purposes of describing changes in status, species will be discussed in the following five categories:

1 Species which have been known in Britain for many years, but where recent field work has shown them to be more widely distributed than previously thought.
2 Species which have been recently described as new to Britain, but are likely to have been overlooked in the past.
3 Species that appear to have arrived in Britain recently, have become established, and have perhaps expanded their range.
4 Species known for many years that appear to have recently expanded their range in Britain.
5 Species that appear to have become rarer or extinct in recent years.

3 Changes in the *Heteroptera* 1973–98

3.1 Historical background

The history of *Heteroptera* recording in Britain is fairly long, but confident and systematic recording really began with the publication of the comprehensive work by Douglas and Scott (1865). Subsequent landmark publications are by Saunders (1892), Butler (1923) and Southwood and Leston (1959). Aquatic and semi-aquatic groups are additionally served by a separate key (Macan 1965, revised by Savage 1989). Butler (1923) began a systematic record of the distribution of species by county, a process continued by Bedwell and Massee (1945) and Massee (1955). The latter work provides a particularly useful basis for assessment of subsequent changes although it is somewhat marred by the treatment of Scotland as a single unit, and by probable errors resulting from mistakes in transcription and from an uncritical acceptance of records.

Despite being well served by identification guides and a considerable number of additional papers the number of entomologists devoting substantial amounts of time to the Heteroptera at any one time has generally been rather small. This not only limits the amount of information which could be assembled, but also means that the information gathered has been heavily dependent of the areas of activity of the key workers. This has meant that much more information has been gathered on south-east England than elsewhere. This is in some ways fortunate as the *Heteroptera* are an overwhelmingly southern group. Also, and of particular relevance in the present context, the south-east contains many scarce species at the edge of their European range, and is the first stop-off point for most new colonists: it is therefore likely to be the area showing most change.

Table 17.2 lists all species for which there appears to have been a significant change in distribution and/or abundance during the whole of the period of recording, as well as species which have been recognised as British, or for which knowledge of the distribution of the species in Britain has significantly increased in the period since 1973. Most of the known changes have been within the twentieth century, not necessarily because changes have been especially great, but because only then was sufficient recording done and enough known of the group to say with confidence that the changes are likely to be real. Most of the declines in the nineteenth century are identified on the basis of absence of subsequent records.

Several species without recent records are omitted from Table 17.2 either because there is no strong evidence that they were ever established British species (for example, *Jalla dumosa*, *Hadrodemus m-flavum*, *Halticus saltator*), or because they are so difficult to find and records are so few that it is impossible to determine their status (*Pygolampis bidentata*, *Aradus corticalis*, *Acompus pallipes*, *Tingis angustata* and *Micronecta minutissima*). Also excluded are a few species of such scattered distribution that new records, even at great distances from those previously known, do not greatly affect the knowledge of their range (*Adelphocoris seticornis*, *Holcostethus vernalis* and *Eysarcoris aeneus*). There are specific reasons for the exclusion of apparent change in several other species. *Nabis brevis* has almost certainly been erroneously recorded in the past and its status at any time is undeterminable without a critical re-examination of specimens; *Cimex columbarius*, not known with certainty to have bred outside artificial dovecotes and with most old records rendered uncertain by taxonomic confusion, is omitted; and the varied fortunes of the entirely synanthropic bedbug *Cimex lectularius* are also ignored. Any possible changes in the status and distribution of *Lygus* species are masked by taxonomic changes and misidentifications. *Monosynamma bohemani*, *M. maritima* and *M. sabulicola* are considered, at least so far as British specimens are concerned, to be dubiously distinct. For current purposes, only *M. sabulicola* (the name applied to most British specimens) is listed. *Pilophorus confusus* is recorded from only two British sites at widely separated times; one of these is now unsuitable for the species, implying a decline, but the status of the species in Britain at any time is essentially unknown.

There has inevitably been a degree of subjectivity in the selection of species for Table 17.2. The decision as to what constitutes 'significant' is not always easy, and is likely to differ between individuals. It would be possible to increase considerably the number of species in the 1973–98 period for which significant change in the known distribution could be said to be shown. The relatively high activity of heteropterists, an active recording scheme and a newsletter mean that relatively minor changes are more likely to be noticed and the information transmitted than in the past. We attempted to weed out both trivial increases in known distributions and more widespread changes in apparent frequency within the known range of a species. Nevertheless, many of these changes will be real – and reflect a broad trend which, though it might prove transitory, might equally herald the start of greater changes. However, Table 17.2 includes all major changes and at least a sufficiently representative set of the remainder to illustrate general trends.

Two species of *Globiceps* have been left with queries for the most recent period.

Table 17.2 Changes in the status of British *Heteroptera* throughout the period of recording.

Species (alphabetical by genus)	Family	Change		
		1800–99	1900–73	1973–98
Acalypta nigrina	Tingidae	.	=	+
Agnocoris reclairei	Miridae	.	+	<
Anthocoris amplicollis	Anthocoridae	.	.	+
A. minki	Anthocoridae	.	.	+
Aphanus rolandri	Lygaeidae	=?	>	+?
Arenocoris falleni	Coreidae	=	=	<
A. waltli	Coreidae	.	>	–
Atractotomus parvulus	Miridae	.	+?	+?
Bathysolen nubilus	Coreidae	=	<	<
Berytinus hirticornis	Berytidae	.	+	<
Buchananiella continua	Anthocoridae	.	.	<
Campylomma annulicornis	Miridae	.	.	<
Capsodes flavomarginatus	Miridae	.	>	=?
Capsus wagneri	Miridae	.	+	+
Chlorochroa juniperina	Pentatomidae	>	>	–
Chorosoma schillingi	Rhopalidae	=	<	<
Corixa iberica	Corixidae	.	.	+
Deraeocoris flavilinea	Miridae	.	.	<
D. olivaceus	Miridae	.	<	<
Dichrooscytus gustavi	Miridae	.	.	+?
Dicranocephalus medius	Stenocephalidae	=?	>	=?
Drymus pumilio	Lygaeidae	=?	=?	+
Elasmostethus tristriatus	Acanthosomatidae	>	<	<
Elasmucha ferrugata	Acanthosomatidae	.	>	–
Emblethis denticollis	Lygaeidae	.	.	<
Eremocoris fenestratus	Lygaeidae	?	>	–
E. podagricus	Lygaeidae	=	=	<?
Eurydema dominulus	Pentatomidae	=	>?	=?
Eurygaster austriaca	Scutelleridae	>	–	–
Globiceps flavomaculatus	Miridae	=?	>	?
G. fulvicollis	Miridae	=?	>	?
G. juniperi	Miridae	.	+	+
Gonocerus acuteangulatus	Coreidae	=	=	<
Graptopeltus lynceus	Lygaeidae	?	>?	+?
Heterogaster artemisiae	Lygaeidae	=?	=?	+
Hydrometra gracilenta	Hydrometridae	.	+	+
Ischnodemus sabuleti	Lygaeidae	=	<	<
Liorhyssus hyalinus	Rhopalidae	.	.	<?
Lygocoris populi	Miridae	.	+	+

continued on next page

Table 17.2 Changes in the status of British *Heteroptera*. (cont.)

Species (alphabetical by genus)	Family	Change 1800–1899	1900–73	1973–98
Macroplax preyssleri	Lygaeidae	.	+	+
Megalonotus emarginatus	Lygaeidae	.	.	+
M. sabulicola	Lygaeidae	?	?	<
Metopoplax ditomoides	Lygaeidae	.	<?	<
Micracanthia marginalis	Saldidae	.	.	+
Microvelia pygmaea	Veliidae	.	+	+?<?
Miridius quadrivirgatus	Miridae	.	.	<
Monosynamma sabulicola	Miridae	.	.	<
Notostira erratica	Miridae	.	.	+?
Nysius senecionis	Lygaeidae	.	.	<
N. graminicola	Lygaeidae	.	.	<?
Odontoscelis fuliginosa	Scutelleridae	.	>	=?
Orius laticollis	Anthocoridae	.	+	+
Orsillus depressus	Lygaeidae	.	.	<
Ortholomus punctipennis	Lygaeidae	.	+	+
Orthotylus virens	Miridae	.	.	+
Pachybrachius luridus	Lygaeidae	.	+	+
Pachycoleus waltli	Dipsocoridae	.	+	+
Peritrechus gracilicornis	Lygaeidae	.	.	<?
Physatocheila harwoodi	Tingidae	.	>	−
Placochilus seladonicus	Miridae	.	.	<
Polymerus unifasciatus var. lateralis	Miridae	.	.	+
P. vulneratus	Miridae	.	>?	−?
Prostemma guttula	Nabidae	>	−	−
Pyrrhocoris apterus	Pyrrhocoridae	=?	=?	<?
Saldula arenicola	Saldidae	=	=	<
Scolopostethus pictus	Lygaeidae	=?	>	+
Sehirus biguttatus	Cydnidae	.	>	=?
S. dubius	Cydnidae	>	>	=?
Spathocera dalmani	Coreidae	=?	=?	<
Stephanitis rhododendri	Tingidae	.	<>	+
Stictopleurus abutilon	Rhopalidae	?	?	<
S. punctatonervosus	Rhopalidae	>	.	<
Trapezonotus ullrichi	Lygaeidae	.	.	+
Trigonotylus caelestialium	Miridae	.	.	<?
Tupiocoris rhododendri	Miridae	.	<	<
Tuponia carayoni	Miridae	.	.	<?+
Xylocoris formicetorum	Anthocoridae	.	>	=?

notes on next page

They have certainly declined during the twentieth century, but the decline certainly began long before 1973, and there is little evidence of substantial change in their fortunes since. *G. fulvicollis* has been extensively lost from wet heathland, and *G. flavomaculatus* from scrub transitions at the margins of wetlands and damp grassland. Recent (and past) changes are confused by the addition of *G. juniperi* to the British list – a species of scrubby heathland with habitat preferences overlapping those of the other species. There are enough confirmed material and associated habitat details to be confident that the declines occurred, but the likelihood of past confusions, taxonomic confusion and name-changes, means that the scale and timing of changes in the distribution and abundance of these species are impossible to determine with certainty.

3.2 Changes 1973–78

The following data summarise the species for which there is evidence of significant change in known distribution or frequency since 1973, and provide notes of the nature and timing of the change. No claim is made that the lists of references and records relevant to these changes are exhaustive, but we trust that they are sufficient to provide a reasonably complete and accurate picture. For species recently added to the British list, a distinction has been made between those that were added before 1973, but so short a while before that their status was not then known (in square brackets) from those which have been added to the British list since 1973 (unadorned).

3.2.1 Species known for many years but now shown to be more widely distributed

Acalypta nigrina   Long regarded as an exclusively upland Scottish species in Britain, this small and easily overlooked lacebug was recorded from an upland site in Caernarvon in 1988 by the Welsh Peatland Invertebrate Survey, undertaken by the Nature Conservancy Council.

Aphanus rolandri   Always a rather scarce species of southern Britain, with records almost entirely restricted to south-coast counties, a record of a single individual from Warwickshire in 1987 (Price 1996) raises the possibility (though no more) of scattered populations elsewhere.

Notes to Table 17.2

.   not known, or not long known, as British during the period in question, or for which there is no evidence for either change or constancy of status;

=   the evidence suggests reasonable constancy of status;

>   there is good evidence of decline;

<   there is good evidence of increase;

+   an increase in the known distribution is known or believed to be the result of increased or improved recording (not used in the table for the nineteenth century, where such increases are assumed, and used rather conservatively in the 1900–73 date range, chiefly for species recognised as British relatively late in the period and for species where new discoveries forced redrawing of distributions previously thought to be well established; no species are listed in the table solely on the grounds of improvements in the knowledge of the distribution prior to 1973).

Drymus pumilio   Always regarded as scarce, and with a strongly southern distribution, and emphasised by Southwood and Leston (1959) and more recently by Judd and Hodkinson (1988) as a species whose scarcity, both in Britain and elsewhere in Europe, make it a subject for particular concern. Records have continued to accumulate, albeit rather slowly, from a quite wide area. Since 1973, it has been recorded from Surrey (Alexander 1998a), Sussex, Kent, Hampshire, Huntingdonshire, Monmouthshire (Kirby 1992a), Hertfordshire and Essex (a single site on the border: Kirby 1996, 1998b), most of these being new county records. Records are mostly of single individuals or small numbers, and the habitat range is, for such a scarce insect, interestingly large, including neutral grassland, fen, mossy calcareous grassland and a woodland ride. The species is certainly not as rare as once thought, but it is difficult, in view of the distribution and nature of the records, to guess with any certainty what the true status of the species might be.

Globiceps juniperi   Though added to the British list almost 40 years ago as *G. salicicola* (Woodroffe 1959), records have accumulated rather slowly. It is now apparent that the species is widely distributed, though local, in scrubby heathlands from the south coast north to Ross and Inverness and west to Carmarthen (Kirby 1992a), though a large proportion of the records have not been published in detail. *G. juniperi* appears significantly more frequent than its previously known congeners *G. fulvicollis* and *G. flavomaculatus*, but it is possible that this is, in part, as a result of *G. juniperi* being identified as one of the other species.

Capsus wagneri   Closely similar to the very common *Capsus ater*, this species was not recorded in Britain until 1954 (Leston 1954), and then gained a reputation as an extreme rarity with records only from Wicken Fen, Cambridgeshire, and Askham Bog, Yorkshire. This reputation, rather than actual rarity, seems to have led to it going unrecorded elsewhere for some time. Records from additional localities in Yorkshire had already been made by 1973, but the greater part of the known distribution of *C. wagneri* has been elucidated since 1980, with records coming from Huntingdonshire (A.C. Warne pers. comm.), Somerset (Hodge 1990b); Middlesex, Northumberland and Dumfriesshire (Kirby 1994), Sussex (Hodge 1991), Norfolk (personal records, and recorded by the East Anglian fens survey of the Nature Conservancy Council), Northamptonshire (Kirby 1998a; J.H. Bratton pers. comm.) and Lincolnshire (Bratton *et al.* 1989).

Hydrometra gracilenta   First recorded from the Norfolk Broads in 1938, and despite long periods without records it was recorded there in the late 1980s, when it was found by Nature Conservancy Council surveys, and in 1998 (Shardlow 1999). In 1988 it was found on the Pevensey Levels, East Sussex, in a drainage ditch of decidedly different character to its Broadlands locations (Kirby 1992b). Clearly an elusive species, it may be a widely distributed but very local species of old wetlands, in which case yet more localities may await discovery, or the Pevensey find may represent a second colonisation from mainland Europe.

Heterogaster artemisiae   With a concentration of records on and near the south coast from Sussex to Cornwall, and on the north coast of Cornwall, this species was additionally known, in 1973, from rather isolated and scattered inland locations in Surrey,

Oxfordshire, Somerset, and Gloucestershire. Since 1973 its southern coastal range has been unsurprisingly extended to Glamorgan: it has been recorded on the Cumbrian coast (Parsons 1987), refound in Oxfordshire (J. Campbell pers. comm.), and specimens from Suffolk collected by B. J. MacNulty in 1972 have been identified in the collection of the National Museum of Wales. These add considerably to the scatter of records, and leave open the possibility of additional locations on calcareous dunes and grasslands.

Lygocoris populi   Described as a new species by Leston (1957), and recorded by Southwood and Leston (1959) only from Berkshire, Buckinghamshire and Surrey, there are still rather few published records of this species. It is nonetheless quite widespread and not uncommon, at least in southern England.

Microvelia pygmaea   Listed as a *Red Data Book* species (Shirt 1986), and proposed for demotion to nationally scarce by Kirby (1992a), it seems likely that even this reduced status overstates the scarcity of this species. In 1973 it was recorded only from a few locations, mostly in south-coast counties. It has now been recorded quite widely in southern counties of England and Wales (Kirby 1992a). It is not clear how far this results from spread and how far from improved recording. Until 1939, *M. pygmaea* was not recognised as separate from the common *M. reticulata* with which it can occur in mixed populations; it is generally more retiring, preferring the cover of vegetation, and is therefore more likely to be overlooked.

Micracanthia marginalis   In 1973 known chiefly from heaths in Dorset and Surrey, with a single old record for Norfolk, this species has since been found in Yorkshire (Crossley 1980), Hampshire, Shropshire (records from the England Field Unit of the Nature Conservancy Council), Denbighshire, Cardiganshire (Welsh Peatland Invertebrate Survey, Nature Conservancy Council) and Cumbria (S. Hewitt pers. comm.). It is usually found on bare wet peat, sometimes in very small areas and especially close to water margins or in very damp hollows. Small, inconspicuous and forming small highly localised populations, this species has almost certainly been widely overlooked.

Orthotylus virens   Regarded as a considerable rarity of upland peat mosses in northern England for much of its history in Britain, this bug feeds on bay willow *Salix pentandra*. Records from bay willow away from peat mosses in Yorkshire (Crossley 1979, 1981) opened up the possibility of a rather more widespread and less specialised distribution, and a rapid survey in 1988 produced a number of additional sites (Nau 1990). The details of the distribution and status of *O. virens* are still unclear; its host plant is not uncommon in the north, and there are many locations where it might occur.

Pachycoleus waltli   In 1973 this bug was known only from a few southern English counties; Southwood and Leston (1959) list Devon, Dorset, Hampshire, Berkshire and Surrey. Its range has been significantly increased by recent records, and it is now know as far north as Norfolk (Irwin 1982), Northamptonshire (Kirby 1998b) and Wales (from the Welsh Peatland Invertebrate Survey undertaken by the Nature Conservancy Council).

Pachybrachius luridus   The British stronghold of this peatland species has long been regarded, and may still be, the New Forest. Other well-known localities were

Chobham, Surrey and Hothfield, Kent. It was discovered in Wales (Cardiganshire) in 1972, and found at a second location (in Caernarvonshire) in 1993 (Alexander and Foster 1996). It may prove more frequent in the west than in the south-east.

Saldula opacula   Identification difficulties within the genus, and especially the differentiation between this species and a form of the very common and widely distributed *S. saltatoria*, have probably led to past under-recording. The species has been recorded from old fens in east Anglia, the margins of peat pools in the north (Horsfield 1989), and from saline conditions (especially grazing marsh ditches) down the east coast of England. Recent records do not greatly alter the spread of habitats or the known distribution, but do make it clear that it is a characteristic and often common component of the marginal fauna of brackish drainage ditches in grazing levels in the south-east (Drake 1988; personal records) There is also a recent record from an inland clay pit (personal records). This is a species of diminishing reputation as a rarity; listed as a Red Data Book species in Shirt (1986), proposed for demotion to nationally scarce by Kirby (1992a), it becomes increasingly clear that even this status cannot be justified, and that the species is merely local.

Scolopostethus pictus   Long regarded as a species of southern England, most frequent on the chalk, and found in haystacks and other agricultural debris, it appears to have declined, possibly to extinction, during the twentieth century, although before 1973. Crossley (1975) recorded it from the south-west coast of Scotland among strandline litter. It has been found at least twice in the same area since, and has also been found among riverine debris in Wales (Kirby 1992a; S. Judd pers. comm.) and Worcestershire (Whitehead 1991a,b). Such situations are similar to the usual habitat of the bug elsewhere in Europe.

Stephanitis rhododendri   First recorded in 1901, this species was widespread in southern England by 1922, and achieved pest status on rhododendrons locally (Judd and Rotheram 1992). Southwood and Leston (1959) reported it as present in much of England, southern Wales and southern Scotland. Between then and the early 1980s, the species appears to have declined almost to vanishing point, though it continued to be included in guides to pests into the late 1980s. It is not clear whether the decline took place before or within the last 25 years, or partly in each period. It was thought for some years to be extinct in Britain, but it has been recently recorded from Oxfordshire (Campbell 1993) and metropolitan Surrey (Jones 1993, 1998).

Trapezonotus ullrichi   A cliff-top species, probably associated with ox-eye daisy. Although known in Britain since the late nineteenth century, it was long known only from Cornwall; it has since been recorded from Devon (Alexander and Grove 1991), Pembrokeshire and Carmarthenshire (Alexander and Foster 1996). There is no obvious reason to suspect an increase in range; almost certainly it has long been present quite widely along western coasts, but has been overlooked.

3.2.2 Species recently noticed or described but likely to have been overlooked

Anthocoris amplicollis   Recorded as new to Britain from several localities along the southern edge of the North York Moors in 1978–80 (Crossley 1982), and found in the same area since (B.S. Nau pers. comm.), but seemingly nowhere else in Britain. This appears to be a well-established but highly localised and overlooked British native.

Anthocoris minki   First correctly recorded in Britain in 1983 – earlier records refer to *A. simulans* (Jessop 1983) – and rarely recorded since. Adults and nymphs live inside galls made by *Pemphigus* aphids on poplars, and there appears to be a low infestation rate even within this specialised niche, so the species is easily overlooked. There is no way to decide whether this is a recently arrived species or a long-established alien.

[*Atractotomus parvulus*]   First found in Britain in 1970 (Woodroffe 1971b), and then known only from a single locality in Surrey, it has since proved to be widespread in southern England, though published records are few (Nau 1995). Since this species has in the past been confused with *A. nigricans* it is difficult to say whether it is increasing or really being more widely recognised.

Corixa iberica   A recently recognised species, formerly confused with the very common *Corixa punctata* (Jansson 1981), it appears to be a geographical replacement for that species in parts of the extreme west: recorded from the western coasts and off-shore islands of Scotland and Ireland (Savage 1989).

[*Orius laticollis*]   Added to the British list in 1971 (Woodroffe 1971a) on the basis of specimens taken in Kent in 1956 and 1958. The species has probably been long established in Britain, but overlooked among others of this rather uniform genus. The females, how-ever, are rather distinct in the form of the thorax, and it is somewhat surprising that the species should have been missed for so long. There are still rather few published records for the species, although it is in fact of quite wide distribution in southern England and, if somewhat local, is not uncommon, being most frequent on willows and poplars.

[*Macroplax preyssleri*]   Recognised in Britain in 1968 by Dolling (1971b), reported it from two localities in Somerset with the suggestion that it might be found quite widely in the Mendips, South Wales and South Devon. This prediction has proved largely true, and it is now known from Gloucestershire (Alexander 1989, 1998b; Askew 1985), and Glamorgan (Askew 1985), but not as yet in Devon. It remains an apparently scarce and local insect.

Megalonotus emarginatus   Previously unrecognised through confusion with *M. chiragra*, although actually added to the British list in the 1960s (Roubal 1965), this went unremarked in the British literature until very recently when Aukema and Nau recorded the species from Berkshire, Dorset, Essex and Kent (Aukema and Nau 1992). It is now known to be widely distributed in southern Britain, as far west as Cornwall (Judd 1998). It is probably roughly as common as *M. chiragra*, perhaps utilising a rather wider range of soils (Allen 1993b).

Megalonotus sabulicola   Until a critical re-examination of the status of this species by Southwood (1963), it was generally regarded as a form of *M. chiragra* and therefore rather erratically recorded. Southwood noted it from Suffolk, Kent, and Devon. Subsequent records, many of them made since 1973, have largely filled in the gaps in this distribution, showing it to be a species of sandy ground along southern and eastern coasts and inland in the breckland of East Anglia. In recent years, however, it appears to have become rather more frequent and more catholic in its habitats, having been recorded inland in non-sandy districts, for example in derelict land and gardens in Northamptonshire (Kirby 1994, 1998b).

Notostira erratica   Woodroffe (1977) reported that Irish specimens of *Notostira* belonged to *N. erratica*, the English being *N. elongata*. However, two females from Wiltshire in the Butler collection in the Natural History Museum were apparently *N. erratica*. No further possible specimens of *N. erratica* from mainland Britain have been reported since, and the identity of the Wiltshire specimens must remain in doubt.

Placochilus seladonicus   First recorded in Britain from Bedfordshire in 1977 (Nau 1978), it is not completely clear whether this bug is a recent arrival or an overlooked species. The former seems more likely, since it is a distinctive species feeding openly on the flowerheads of field scabious, *Knautia arvensis*. It has since been reported from Sussex (Hodge 1990a), Oxfordshire (Hawkins 1989b), Berkshire, Hertfordshire, and at an additional site in Bedfordshire (Nau 1994).

Polymerus unifasciatus   In 1973 no distinction was made between the forms of *P. unifasciatus* in different areas. Crossley (1981) recorded var. *lateralis* from Yorkshire. Var. *unifasciatus* and var. *lateralis* appear to be distinct in both distribution and foodplants in Britain, the first occurring in southern counties north to Cambridgeshire and Norfolk and feeding chiefly on *Galium verum*; var. *lateralis* occurring from Yorkshire northwards and feeding chiefly on *Galium saxatile*. Considering the situation in Britain in isolation, it would seem that there are good grounds for re-examining the taxonomic status of var. *lateralis*.

Trigonotylus caelestialium   Overlooked among *T. ruficornis* until recently, although specimens are known from museum collections as far back as 1900 (Aukema and Nau 1992). In some places it is the commoner of the two species (Allen 1993b), and seems to be more characteristic of disturbed ground and improved grassland. Though clearly overlooked, there may have been a recent increase and its occurrence in improved and disturbed grassland make such an increase quite likely. Probably widespread in England, at least as far north as Yorkshire (Crossley 1993).

3.2.3 Species recently arrived, established and perhaps expanded in range

Agnocoris reclairei   Feeding on mature white willow, *Salix alba*, it was long regarded as a scarce fenland species, restricted to a limited area of Suffolk, Cambridgeshire and Huntingdonshire. From the 1970s it has become clear that it is now rather frequent over a broad area of Cambridgeshire, Huntingdonshire and Bedfordshire, just reaching into adjoining counties, and isolated colonies have been found in Essex (Kirby 1997), Kent (Kirby 1992a; M. Newcombe pers. comm.) and Warwickshire (Lane 1989). How far these recent discoveries reflect past under-recording or recent spread is uncertain; probably both are involved. There has certainly been a recent increase, since the species is now characteristic of white willows around older gravel workings, largely a habitat of recent origin.

Buchananiellea continua   Added to the British list by Kirby (1999), this introduced species has been found established in Buckingham Palace gardens where it seems to have been resident over several seasons.

Campylomma annulicorne   Added to the British list by Nau (1978) on the basis of specimens found in Bedfordshire, though captured a little earlier in Kent (Allen 1984b) this distinctive species has since been recorded rather widely and quite commonly in the south-eastern counties of England on osier willows, *Salix viminalis*, though there are rather few published records. Since it is a distinctive species which has spread rapidly on a plant to which hemipterists give considerable attention, it is almost certain that this is a recent arrival.

Deraeocoris flavilinea   First discovered in Essex in 1995 (Miller in prep.) this species has spread across France and Germany in recent years from Western Europe. It may now be found commonly on sycamore, *Acer pseudoplatanus*, in parts of suburban London (see for example Jones 1999).

Dichrooscytus gustavi   This is presently a scarce species of juniper on the chalk of south-east England (Sussex, Surrey, Berkshire, Buckinghamshire, Oxfordshire and Wiltshire). A single record of this as a breeding species on a cultivated juniper, *Juniperus chinensis*, in Worcestershire (Whitehead 1989) represents a considerable extension of range of the species as well as a new host.

Emblethis denticollis   First recorded in Britain in 1991 in Hampshire, this species appears to have spread rapidly through south-eastern England, with additional records from Hampshire, and then Kent, Sussex, Bedfordshire and Cambridgeshire in the following six years (Denton 1997b; Nau 1997; Hodge 1998a,b; Judd and Straw 1998), but has remained decidedly local.

Metopoplax ditomoides   Recorded on Hounslow Heath in 1952 (Woodroffe 1953), it then went unrecorded for several decades, before appearing in Oxfordshire in 1992 (Campbell 1993), and then rather widely in southern England in the following years, with records of breeding colonies from, at least, Bedfordshire, Berkshire, Hampshire, Kent, Surrey and Sussex (Denton 1997a; Nau 1997; Clemons 1998; Hodge 1998a,b; Kirby 1998; Porter 1998). It seems likely that the species was genuinely absent for most of the period between Woodroffe's capture and the recent records; that the Hounslow Heath record was of a transitory colony; and that the recent records represent a new wave of immigration. In the late 1990s the bug was seen in prodigious numbers in some places, especially field margins, abandoned arable fields and along recently created road verges.

Nysius graminicola   Reported by Allen (1984b) from Studland, Dorset, this species appears not to have since become widely established, in contrast to the following species. The Studland record may have represented only a transitory colony.

Nysius senecionis First recorded in mainland Britain in 1992 (Hodge and Porter 1997), this species spread rapidly in south-eastern England and by 1997 was often common under mayweeds on derelict land and field margins in many south-eastern counties (Denton 1997; Nau 1997; Hodge 1998a) extending north to Yorkshire (Dolling 1997).

Orsillus depressus A distinctive groundbug, first recorded in Britain in Surrey in 1987 (Hawkins 1989c), which feeds on members of the Cupressaceae, especially Lawson cypress, *Chamaecyparis lawsoniana*. It has since become quite common in parts of the south-east, spreading as far north as Northamptonshire (Kirby 1993) and Leicestershire (personal records).

Ortholomus punctipennis In 1973 known from only three localities in Norfolk, Suffolk and Hampshire, this bug has since been refound in Suffolk (in additional locations in the brecks), and also recorded from Kent (Judd 1998), Lincolnshire (Kirby 1998b) and Sussex (Porter 1998). Judd and Hodkinson (1998) suggest that such additional, and rather widely scattered, records are the result of immigration, but this is not unambiguously certain. It is perhaps likelier for the Kent and Sussex locations, which are in relatively well-worked areas, than for the Lincolnshire site, which in character is very like the brecklands, the headquarters of the species in Britain.

[*Tupiocoris rhododendri*] Described as a new species by Dolling (1971 as *Dicyphus rhododendri*) who recorded it from Surrey, Berkshire, Bedfordshire, Hertfordshire and Buckinghamshire. Clearly well established on rhododendrons in central southern England by 1973, it has spread since to Kent (Allen 1977) and Derbyshire (Kirby 1983). It was well established in southern and midland counties by the late 1980s, although its northern and western limits are not entirely clear. It is believed to have originated in North America.

Tuponia carayoni First reported in Britain by Nau (1980), from Hampshire and the Isle of Wight, this very distinctive bug feeds on tamarisk, *Tamarix gallica*. It remains very local on the south coast, and is absent even from many quite extensive south coast stands of tamarisk, suggesting that it is at the edge of its climatic tolerance.

Liorhyssus hyalinus This insect has strong powers of dispersal, and British records have generally been presumed to be vagrants or temporarily established colonies. Records tend to be clustered in particular years (Southwood and Leston 1959). The mid-1990s appear to constitute a favourable period for the species, with records from Bedfordshire (Nau 1997), Hampshire (Denton 1997a) and Pembrokeshire (Alexander and Foster 1996). It certainly bred in Britain during this period: from personal observation the Bedfordshire colony was large and included many nymphs. There is, as yet, no evidence that *L. hyalinus* has established anything more than transitory populations, but in view of the spread and establishment of other xerophilous insects in recent years, the possibility must be borne in mind.

Stictopleurus abutilon Though there are old records from Britain and it has certainly bred here in the past (Dolling 1978), it was almost certainly long extinct (perhaps for more than a century) before further records were made in the 1990s which indicated that it was once more a breeding species. It has been reported in recent years from Bedfordshire (Denton 1997b; Nau 1997), Essex (Kirby 1997), Hampshire (Denton 1997a; Hodge 1998a; Porter 1998) and Kent (personal records). Breeding has been confirmed in Hampshire (Denton 1997).

Stictopleurus punctatonervosus   There are old records from southern England, and it may have bred here, but there is little reason to believe that it has, in the past, had more than transitory colonies at best (Dolling 1978). There was only one previous twentieth-century record prior to its recent discovery in the south of England; there are records from Essex (Bowdrey 1999) and Kent (personal records). Repeated records from the Essex site suggest an established population.

Saldula arenicola   In 1973, *S. arenicola* was known with certainty as a breeding species only from seepages on soft-rock cliffs in the south-west (Hampshire, Dorset and Devon). It was recorded from Dungeness in 1972 (Kirby 1992a; E.G. Philp pers. comm.), but it was not clear whether this represented an established population. Further records from Dungeness in the 1980s and 1990s (Morris and Parsons 1992; personal records) demonstrated large breeding populations beside gravel pits. It has since been recorded beside a small pool in an active sand quarry in Surrey (personal records). It seems almost certain that these south-eastern records result from recent immigration from mainland Europe, particularly since the ecology of the populations in the east and west of England seem different: those in the east are more similar to those of mainland Europe, while the specialised ecology of the south-west populations may imply a long isolation and perhaps significant genetic divergence from the mainland populations.

3.2.4 Species that appear to have recently expanded their range

Arenocoris falleni   A ground-dwelling species feeding on storksbill *Erodium cicutarium*, and tending to be locally rather frequent in parts of southern Britain, especially in the brecklands of East Anglia and in areas with extensive coastal dunes. A number of new county records in recent years suggest that this species has participated in the spread of xerophile *Heteroptera*. In 1996 it was recorded for the first time in Bedfordshire (Nau 1997), Essex (Kirby 1997) and Northamptonshire (personal records); it was found at several additional Bedfordshire sites in 1997 (Nau 1998).

Bathysolen nubilus   This species showed a genuine increase in the twentieth century, but exactly how much and when is more difficult to determine. It has long been known from Kent, but for a long time as a rarity recorded only from Deal. Since 1945 it has become a quite frequent and locally common species in that county, aided by a liking for derelict land and quarries where its foodplants, medicks (*Medicago* sp.) are common. A substantial part of the increase occurred prior to 1973, but it seems likely that the expansion has continued and it is now quite frequent in parts of metropolitan Essex; these populations may be an expansion from its Kentish base. There are old records for Norfolk, suggesting that it is a long-established resident on the East Anglian coast, but there are no records suggesting an increase in this area. A single rather old record for Buckinghamshire has not been repeated, and could have been a transitory colony, but a first record from Bedfordshire in 1997 (Nau 1998) may indicate a continued expansion.

Berytinus hirticornis   First recorded in 1943, and for some time known only from the south Devon coast, then recorded in Cornwall in 1960, Dorset in 1968 and the Isle of Wight in 1971 (Dolling 1983): a set of records suggesting either a gradual spread along the south coast or the discovery of previously overlooked colonies. It was recorded in Kent in 1982 (Dolling 1983) and has since been found quite frequently in south-east England, particularly in Sussex, Kent and Essex, and has spread quite widely over this fairly limited area of the south-east. It was recorded in Bedfordshire in 1996 (Nau 1997) and may still be spreading. The pattern of records would be consistent with a recent arrival establishing in the south-west and spreading along the coast until it reached climatic conditions more favourable to it in the south-east; a second immigration into the south-east is, however, entirely possible.

Chorosoma schillingi   Historically, and probably still, a predominantly coastal species, with an inland stronghold in the East Anglian breckland, there have been occasional reports of other inland colonies since 1967 (Russell 1967). Inland colonies were reported in the 1980s from Kent, Lincolnshire, Huntingdonshire, Cambridgeshire and Warwickshire (Kirby 1984a, 1991a; Lane 1988). The number of such colonies appears to have gradually increased and, though still local, it is no longer an unexpected species in south-eastern and midland counties in fairly tall grass on dry soils, especially on sand, chalk and in ruderal habitats. Nau (1997) reports the spread of the species in Bedfordshire.

Eremocoris podagricus   There is some suggestion of a northward spread; it has only in recent years been recorded from Leicestershire (D. Lott pers. comm.) and Northamptonshire (Kirby 1998); in the latter county it is now present in at least two locations which were well-worked by W. E. Russell in the 1960s (Russell 1969), suggesting that the spread, or at least increase, may be genuine. On the other hand, Judd (1998) reports an old record from Staffordshire, the most north-westerly record by far, so it may be that the bug has gone unnoticed towards the edges of its range.

Elasmostethus tristriatus   At one time confined in Britain to juniper *Juniperus communis* on the southern chalk, this species spread to introduced Cupressaceae, especially Lawson cypress *Chamaecyparis lawsoniana*, in the middle years of the twentieth century and has since spread widely (Carter and Young 1974; Allen 1984a). This spread was already well in progress by 1973, but it has continued since, and the species has increased within at least some of its known areas of past expansion. It occurs north to Yorkshire (e.g. Hawkins 1995) and west to Wales (Askew 1983). It was recorded in Ireland on Lawson cypress in 1995, having been previously known in the nineteenth century (O'Connor and Ashe 1995).

Graptopeltus lynceus   Recorded from north-west England by Haughton and Bell (1996): this represents a very dramatic increase in range from its previously known distribution in south-east England, but it is not clear whether this represents the discovery of a long-established but overlooked colony, or a part of the general spread and increase of lygaeids in the late 1990s.

Deraeocoris olivaceus   First recorded in 1951 in Berkshire (Sands 1954), it was reported within the next few years from a number of locations in Buckinghamshire, Middlesex and Surrey, as well as more sites in Berkshire (Woodroffe 1954, 1956; Groves 1968), Essex (Tomlinson 1971) and from Sussex in the 1980s (Kirby 1992a; P.J. Hodge pers. comm.). It has recently been recorded in Kent (Allen 1993a), Hampshire (Denton 1997b) and Hertfordshire (J. Widgery pers. comm.). This is a large and conspicuous insect which feeds on hawthorn *Crataegus monogyna* and is presumably a fairly recent colonist, probably still spreading in south-eastern counties.

Gonocerus acuteangulatus   Known for many years from Box Hill, Surrey, this species apparently fed exclusively on box, *Buxus sempervirens*. In recent years it has spread, though not as yet very far, from this classic locality, and apparently now feeds on hawthorn, *Crataegus monogyna* (Menzies 1994a,b; Hawkins and Menzies 1998). It is a polyphagous species in other parts of its range, and its restriction to box in England always seemed a little odd; it is difficult to know whether its sudden spread to hawthorn makes the British population less or more odd.

Ischnodemus sabuleti   Spreading dramatically through the early decades of the century, and already occupying the greater part of its current range by 1973, for example reaching Yorkshire in 1971 (Crossley 1976), it has continued to spread and increase in frequency in the outlying portions of its range.

Miridius quadrivirgatus   The finding of this previously exclusively coastal insect in two inland localities in Surrey in 1986 and 1987 (Hawkins 1989a) suggests the possibility of a more general spread inland, which might mirror that of *Chorosoma schillingi* (see above). However, there appear to be no additional records to date that indicate the inland spread has continued.

Monosynamma sabulicola   A characteristic species of creeping willow, *Salix repens*, in coastal dune slacks, *M. sabulicola* appeared on other willow species in gravel pits as far inland as Derbyshire (Kirby 1984b) and Bedfordshire (Nau 1981, 1982) in the 1970s and 1980s, occurring chiefly on young scrubby growth in largely bare sand, a habitat closely mimicking its usual dune habitats.

Peritrechus gracilicornis   Old records are of rather isolated and scattered individuals, suggesting an occasional vagrant. Allen (1980) reported finding a number of individuals over a long stretch of coast at Studland Bay, Dorset, in 1977, suggesting an established breeding colony. There have been additional records since from southern England, but some are rendered uncertain by identification difficulties and it remains unclear whether there is now a permanent British population or whether the reliable records are of further vagrants and temporary colonies. In the light of other additions to the British fauna in recent years, establishment seems likely.

Pyrrhocoris apterus   Although long regarded as restricted – as a breeding species in

Britain – to a single small island off Devon, a well-established colony was recently reported in Surrey (Ashwell and Denton 1998). There is a possibility, however, that *P. apterus* has had colonies elsewhere in the past. There are records of adults on several fairly recent occasions, and older reports at various other locations in circumstances which suggest established colonies (Kirby 1992a).

Spathocera dalmani   Evidence for an increase is more ambiguous for this than for most species in this section, chiefly because its occurrence has always been rather erratic; it is rare to find a long history of the species at a single site. Nau (1997) recorded it in Bedfordshire in 1996: not only is this the first record of the species in this well-recorded county, but it was recorded in two locations, and in considerable numbers. There are also recent records from Hampshire and Surrey in 1994–96 (Denton 1997a,b). Together, these constitute the largest cluster in the recording of the species in Britain and, since they coincide with a considerable increase and spread of other species of dry and sparsely vegetated ground, it seems likely that they reflect a genuine increase.

3.3 Discussion

Interestingly, almost all the changes certainly known to have occured in the period 1973–98 are increases. Heteroptera have suffered significant losses historically, but most occurred prior to 1973. Two species are listed in the UK Biodiversity Action Plan of 1995 as having declined by at least 50% in the last 25 years, but although this is possibly the case there is no apparent evidence. *Hydrometra gracilenta* was found at two new localities in the 1980s, and there is no evidence of loss from any locality; *Orthotylus rubidus* has, quite contrary to the report of decline, been recorded since the mid-1980s in several localities on the east and south coasts, after an absence of records for over a decade. Although it seems unreasonable to suppose that some species have not declined in the face of habitat loss and change since 1973 – and many specific examples of loss of particular sites and populations could be assembled – the increase in recording effort over the same period means that the number of known locations for the potentially affected species, and usually also the total known area of distribution, has nonetheless often increased. There are species which have not been seen in part or all of their previously known range for some years, but they appear not to have been systematically searched for during this period.

A number of species have proved to be significantly more widespread or more frequent than was previously thought, apparently as a result of improved recording rather than genuine spread and increase. Some of these increases in known distribution result from the filling in and expansion of known locations and areas of distribution (e.g. *Trapezonotus ullrichi* on south-western coasts); others result from the discovery of populations (or possible populations) rather far-flung from those previously known (e.g. *Aphanus rolandri*, *Heterogaster artemisiae* and *Graptopeltus lynceus*). It is probably a general rule that it is easy to over-estimate the rarity of invertebrates.

Throughout the twentieth century and probably for longer, there was an erratic but continuous influx of new breeding species, usually establishing themselves somewhere in southern England and often spreading from their initial base. The frequency of such arrivals appears to have been especially great in the early 1990s. Only time will tell whether, and how many of, these species will remain part of the British fauna and how many will prove to be merely transitory residents. The history of the rhododendron lacebug *Stephanitis rhododendri* shows very clearly how apparently well-established species can decline as well as increase. Indeed, the considerable number of new arrivals in recent years, the apparent ease and speed with which some, at least, are able to infiltrate the British fauna, and the likelihood that they might vanish as rapidly as they appeared if the climate became once more adverse to them, gives cause to wonder at the significance of some historically recorded species. For example, the speed with which *Stictopleurus abutilon* and *S. punctatonervosus* have appeared in the 1990s, the former especially establishing itself over a wide front in a few years, gives good reason to suspect that past records of these species as breeding species in Britain may have been of transitory populations.

New arrivals form rather a broad camp, ecologically. Not surprisingly, they include a number of species associated with introduced plants, although in fact the number of such species is quite small, as is the range of plant species they occupy. Only three such bugs have been added to the British list in the last 25 years, each with a single and different plant host: rhododendron (*Tupiocoris rhododendri*), Lawson cypress (*Orsillus depressus*) and tamarisk (*Tuponia carayoni*). Considering the immense range of exotics grown in Britain, this is a small total. A further group are xerophiles, occupying a rather broad range of open, well-drained and incompletely vegetated land. These include the two *Stictopleurus* species, *Emblethis denticollis*, *Nysius senecionis* and *Metopoplax ditomoides*. The arrival of these species has coincided with a widespread increase in long-established native xerophiles, so presumably results from a change in the nature of Britain rather than (or as well as) a high rate of immigration. Climate change, or at least the particular weather conditions experienced over recent years, is the obvious and most fashionable candidate for the cause of change. Two species, apparently recently arrived (although the possibility that they spread from a previously very restricted and un-surveyed area cannot be ruled out) are associated with established British plants and are not xerophiles: *Campylomma annulicornis* on osier willow and *Placochilus seladonicus* on field scabious.

The speed with which new arrivals can spread once established is impressive. *Nysius senecionis*, for example, appears to have spread throughout south-east England, at least from Hampshire to the Humber, within five years of its first reported finding. Other species have not spread so far, but rates of spread of at least a county a year are fairly commonplace in the early years following establishment.

There has been a significant and widespread increase in xerophilic species among long-established British residents in the mid-1990s. Some non-xerophiles appear to have shared in the increase. As noted above, the most obvious explanation is that they have benefited from recent climatic trends; for many species, this increase may yet prove transitory. In part, it takes the form of northward (and for some species perhaps westward) spread, but also of increase within the previously known area of distribution. Nau (1997, 1998) summarises recent changes in Bedfordshire, a county with good and consistent coverage spanning almost exactly the 25 years covered by this chapter (1973–1997). Kirby (1997) reports the increases of some species in Essex; Denton (1997b) reports increases in usually scarce species in Surrey and Hampshire in 1996.

Some long-established species have been spreading for rather longer, several throughout the 25-year period now under review. Some are responding to increased habitat availability. In the case of the juniper shieldbug *Elasmostethus tristriatus* this is the consequence of spreading to a new host plant (Lawson cypress) and thereby transforming from a scarce species of juniper on the southern chalk to a familiar suburban insect. The increase in the willow-feeding *Agnocoris reclairei* may stem, at least in part, from the increase in gravel workings along the river valleys in and near its formerly restricted fenland area – the white willows growing around such pits are particularly favoured by the species. An interesting feature of the spread of this species is that a spread to occupy a broad and continuous area of distribution around its former range appears to have been accompanied by the formation of additional colonies at considerable distances, isolated populations being discovered in Essex, Kent and Warwickshire. Although the possibility of linking populations existing between the main area of distribution and these satellites cannot be ruled out, it does not seem likely that all links should have been overlooked for two decades.

Other species which have increased in recent years may be showing a longer-term response to climatic change. The rhopalid bug *Chorosoma schillingi* was formerly strongly maritime, its only inland stronghold being in the brecklands of East Anglia. Its coastal and breckland distribution appear to be unchanged, but it additionally occurs at many inland locations. The similarly maritime *Miridius quadrivirgatus* has shown a similar, if less widespread and convincing, tendency. Neither are exclusively, or even strongly, maritime elsewhere in Europe, and their past restriction to the coastal zone in Britain suggests a response to being at the limit of climatic tolerance.

4 Changes in *Auchenorrhyncha* 1973–98

4.1 Historical background

Serious study of the British and Irish *Auchenorrhyncha* dates from James Edwards' monograph (Edwards 1896). This work included 257 species, nearly 70% of the 376 now recognised in Britain. Given the logistic difficulties faced by Victorian entomologists when working with relatively small insects, and the restricted geographical coverage of collectors at that time, this represents a considerable achievement. Many of the species known then are now considered to be extremely rare. This prompts the interesting question as to whether Edwards and his contemporaries were simply very good collectors, or whether the species concerned have since declined in status.

The next landmark in the development of knowledge of the fauna was the publication over a 20-year period of four Royal Entomological Society (RES) handbooks (Le Quesne 1960, 1965, 1969; Le Quesne and Payne 1980). These keys made the group more accessible to amateur entomologists and attracted professional biologists interested in the ecology of the group. While some revision of these handbooks is now needed in the light of advances in our knowledge of distribution and ecology, changes in nomenclature and the addition of new species, they remain the standard identification works for the British fauna.

The influence of continental European workers in updating the taxonomy of the British fauna has been considerable. Pre-eminent among these has been Henri Ribaut, whose monographs of the French *Typhlocybinae* (1936) and the remainder of the *Cicadellidae* (1952) set a standard of illustration that has rarely been surpassed. More recently, handbooks to the Scandinavian fauna by Ossiannilsson (1978, 1981, 1983) have helped to cover the north European species not included in Ribaut's coverage of France.

There is no doubt that the sound taxonomic background provided by these various works supplied the necessary impetus for very considerable advances during the last 40 years in our knowledge of the ecology and general biology of this group in Britain. Detailed ecological studies have covered a number of community types, including species associated with trees (Claridge and Wilson 1978), herbaceous plants (Stewart 1988) and various types of semi-natural grassland (Whittaker 1969; Morris 1971; Waloff 1980). A considerable amount is also now known about many other aspects of their ecology, including population dynamics (Waloff and Thompson 1980), dispersal (Waloff 1973), parasitoids (Waloff and Jervis 1987) and acoustic signals (Claridge 1985).

Even though the RES handbooks provide virtually complete coverage, identification of species in this group is still perceived to be difficult by many entomologists. The necessity for dissection and examination of the male genitalia of species in a few groups has undoubtedly contributed to this impression. Consequently, the *Auchenorrhyncha* received far less attention by early recorders and collectors than the *Heteroptera*. There has however been a significant increase in the popularity of the group in recent years.

4.2 Data capture

Surveying and monitoring of the distribution and status of British species is co-ordinated by the *Auchenorrhyncha* Recording Scheme, one of several similar recording schemes under the aegis of the Institute of Terrestrial Ecology's Biological Records Centre. The scheme was initiated in 1979 with the primary objective of collating biogeographical data from a wide variety of sources: published literature, museum collections, site and species surveys and records from individual entomologists (Stewart 1999). These data have provided much useful information on the rarity status of individual species, as well as ecological characteristics such as seasonal phenology, voltinism, host plants and habitat associations. Such data will also provide a baseline against which to monitor future changes in national status and distribution, for example as a result of climate change.

The Recording Scheme mailing list currently includes approximately seventy amateur and professional entomologists, although not all of these are fully active in regular recording. At least six university departments have ecological research projects focused on *Auchenorrhyncha* and there are taxonomic specialists in several of the larger museums. The Scheme currently holds an estimated 40 000 records, of which approximately one quarter have been computerised. A large number of additional records have yet to be gathered from other sources, including some individual recorders and important museum collections. The existence of the Recording Scheme, together with regular field and indoor meetings, and the distribution of a newsletter, has encouraged a considerable extra recording effort. Data from individual recorders accrue at a rate of approximately 1000 records per year. Although important gaps in geographical coverage remain, sufficient data now exist to enable distinct geographical patterns to be discerned for certain species. Preliminary maps of the distribution of individual species are starting to appear (Stewart and Wilson 1999) and the production of a national distribution atlas for all species is a long-term objective.

4.3 Conservation status

Historically, *Auchenorrhyncha* have received far less attention from conservationists than many other invertebrate groups. However, the recent review of scarce and threatened species in Great Britain (Kirby 1992a) has done much to raise the profile of the group. Table 17.1 shows that 98 species (approximately 26% of the total fauna) were assigned Nationally Notable or Red Data Book (RDB) status in that review. Within the latter category, all except one are currently defined as RDBK or Insufficiently Known (that is, suspected of falling within one of the RDB categories but where there is insufficient information to be confident of this). This is a revealing indication of the state of knowledge of many rare species in this group and contrasts markedly with the situation in the *Heteroptera*; among the species treated in Kirby's (1992) review, 22% of the *Auchenorrhyncha* were assigned RDBK status, while the comparative figure for *Heteroptera* was 3%. However, greater and more focused recording effort in the eight years since Kirby's review (ibid.) now means that more informed statements can be made about many of these species.

Only one species falls unequivocally within the RDB category: the New Forest cicada *Cicadetta montana*. It is classified as endangered (RDB1) and has full protection under Schedule 5 of the 1981 Wildlife and Countryside Act. This species is at the northern edge of its natural range in Britain with only two known surviving populations, both in the New Forest. There are old records from other parts of the New Forest where the habitat remains suitable, so it is possible that the species survives in a handful of small, isolated and highly vulnerable populations. However, there are considerable problems in assessing its true status. The adults – present for less than a month in May and June – require special weather conditions for flight: high temperatures and little wind. Also, the male courtship song, which is so easily detected in other cicada species, is both faint and has a high frequency which is inaudible to the human ear in people beyond a certain age. The nymphs are subterranean, feeding on roots, their presence being most easily detected by 'emergence turrets': funnels of soil moulded by the final instar nymph immediately before the final ecdysis. Uncertainties about the exact length of the lifecycle (probably between five and eight years) create problems for assessing total population size and also make it difficult to interpret years when no adults are seen. No evidence of the species has been detected at either of the known sites since 1994. The species has been the subject of an English Nature Species Recovery Programme in an attempt to secure the survival and enhancement of any extant populations. An annually revised species action plan (UK Biodiversity Group 1999) outlines current and future action for the conservation of the species.

The other species currently receiving conservation attention is *Aphrodes duffieldi* which is a medium priority species within the UK Biodiversity Action Plan (ibid. 1999). This species is apparently confined to Dungeness and has been treated as endemic to Britain. However, recent comparisons with continental European material suggest that it may be synonymous with *A. assimilis*, also a rare species (Wilson and Stewart in prep.). This species demonstrates how the perceived status of a species may be dependent on the collection technique used. It was recorded at Dungeness several times by Duffield between 1919 and 1957 (Duffield 1960). For 30 years thereafter, a lack of records was interpreted as possible evidence of extinction, even though the shingle habitat had not perceptibly changed. However, in 1988 the species was rediscovered at a number of locations across Dungeness using pitfall traps (Morris and Parsons, 1992). This is almost certainly a reflection of the species' epigeal or possibly subterranean microhabitat and explains why standard sweep netting and hand-searching techniques did not reveal it.

4.4 Changes: 1973–98

In common with other hemipteran groups, the auchenorrhynchan fauna in Britain has experienced both gains and losses of species. The following account summarises the species for which there is evidence of significant change in known distribution or frequency since 1973 and provides notes on the nature and timing of change. Table 17.3 lists 19 species which have either been recorded as new to Britain since 1973 or were not included in the relevant RES Handbook. In the account that follows, the same categories of change in status are used as for the *Heteroptera*. However, as the *Auchenorrhyncha* are a less well-worked group, it is important to distinguish between species for which there is good evidence for a change in status, compared to those where the paucity of records mean that meaningful statements about any change in status are impossible.

4.4.1 Long-established members of the British fauna that have recently been shown to be more widely distributed than previously thought

This category refers to species whose rarity status has been revised, or requires revision, in the light of a significant increase in the number of records. In such cases, the suggestion is that the true status of the species has not changed, but the detection rate has improved. This may have resulted either from greater survey effort being directed at the species' habitat or from the application of new or improved sampling and collection techniques. The increasingly widespread use of a cheap lightweight suction sampler, as an alternative to the expensive and heavy D-Vac, has facilitated better access to the lower and more inaccessible parts of vegetation which are passed over by more general sweep netting techniques (Stewart and Wright 1995). This has generated many new records for certain grassland and bog species that live close to the soil surface.

Similarly, pitfall trapping has recently been more widely adopted for recording this group of insects, as it is known to be highly effective in trapping the species that live close to the ground (Stewart in press). Certain pitfall trap surveys, often targeted primarily at other invertebrate groups such as *Carabidae*, have greatly increased the number of *Auchenorrhyncha* records, and in some cases extended the known range, of several local and scarce species. Good examples include the East Anglian Fen Survey (1985–87) and the Welsh Peatland Invertebrate Survey (1987–89).

The latter survey, encompassing 118 sites across Wales, revealed several species to be considerably more widespread than previously thought, particularly the rarer delphacid species that inhabit the very wet parts of bogs such as *Tyrphodelphax distinctus*, *Oncodelphax pullulus* and *Delphacodes capnodes*. Additionally, recent use of suction sampling equipment on Welsh mires has revealed further sites for these species (M.R. Wilson, unpubl. data). Other rare wetland delphacids that have been recorded more widely through suction sampling include *Florodelphax paryphasma*, *Megamelodes lequesni* and *Paradelphacodes paludosus*.

Table 17.3 Species of *Auchenorrhyncha* recorded as new to Britain since 1973 or since publication of associated RES Handbook.

Species	Year discovered	Original collector	Reference where first added to British list
Trigonocranus emmeae	1959	Duffield	Le Quesne (1964)
Eupteryx origani	1974	Le Quesne	Le Quesne (1974)
*Alebra coryli**	1977	Le Quesne	Le Quesne (1977)
Edwardsiana rosaesugans	1977	Wilson	Claridge and Wilson (1978)
Kyboasca bipunctata	1978	Wilson	Wilson (1979)
Iassus scutellaris	1978	Wilson	Wilson (1981)
Issus muscaeformis	1979	Payne	Payne (1979)
Cicadula flori	1979	Straw	Le Quesne (1983a)
Psammotettix maritimus	1980	Ely	Payne (1981)
Empoasca pteridis	1982	Davis	Le Quesne (1983b)
*Muellerianella extrusa**	1982	Booij	Le Quesne (1983b)
Cicadella lasiocarpae	1985	Le Quesne	Le Quesne (1987)
Chlorita dumosa	1987	Kirby	Kirby (1991b)
Cicadula ornata	1987	Flint	Flint (1990)
Idiocerus ustulatus	1991	Stewart	Stewart (1992)
Eurysa brunnea	1992	Whitehead	Stewart (1997)
Balclutha saltuella	1993	Ashmole	Ashmole and Ashmole (1995)
Elymana kozhevnikovi	1996	Woodward	Stewart (1996)
Fieberiella florii	1998	Hodge	Hodge (1999)

Species listed in order of discovery as new to Britain. * denotes that novel status is due to clarification of taxonomic status of closely related species.

Other species for which more extensive fieldwork has modified our understanding of status and distribution include:

Struebingianella litoralis   Until the 1980s, this rare planthopper had been recorded only from the edge of lochs near Aviemore and Perth in Scotland. Outside Britain, the species is known only from Finland, where it has not been recorded since 1940. It was recorded from a range of sites in the Welsh Peatland Invertebrate Survey and has since also been recorded from a limited number of sites in northern England. It appears to be a species that is characteristic of the very wettest, and therefore least accessible, parts of raised and blanket mires in upland Britain. It is undoubtedly rare, but further searching may reveal it to be quite widely distributed.

Doratura impudica   This species is found on the seaward colonising edge of coastal sand dunes. Early records were from only two sites on the north Norfolk coast. More recently, the species has been confirmed from historical and additional sites in Norfolk and also recorded from various sand dune sites along the coastline of south-eastern England, including in Suffolk, Essex, Kent and East Sussex.

Calligypona reyi   A species of coastal marshes where it seems to be most closely associated with glaucous bulrush *Schoenoplectus tabernaemontani* and sea club-rush *Scirpus maritimus*. Early records from four widely spaced sites have been augmented by a further four sites in recent years following more focused searching on patches of host plant. Neither of the host species are rare, although they tend to occur in small and comparatively isolated patches, so continued targeting of known sites for the plants may well generate further records.

Reptalus panzeri   A limited number of historical records scattered across south-east England have recently been supplemented by the discovery that this species is both widespread and sometimes very abundant in the Greater London area (Jones and Hodge 1999). While its precise ecological requirements are still somewhat puzzling, it has been suggested that the known rural localities share the feature that the soils tend to crack upon summer drying. This may provide ovipositing females with access to plant roots on which the nymphs are known to develop. It may also help to explain the widespread occurrence of the species on early successional grasslands and waste ground in London, which commonly have coarse soil substrates and extensive patches of bare ground.

4.4.2 Recent additions to the British list that may have been overlooked

Certain species that have been added to the British list recently may nevertheless have been previously resident in Britain for a considerable period of time but remained undetected. The most likely reason for this would be that they occur at naturally low population densities and below the level of normal detection by field entomologists. Alternatively, species may be associated with habitats that are not routinely sampled by entomologists who are interested in this group, or they may belong to taxonomic groupings that are problematic or require determination by experts. Finally, this category is used to cover species that have arisen through taxonomic revision of critical groups.

Trigonocranus emmae   The few records for this species, usually of singletons, are widely scattered from Kent to Lancashire and Pembrokeshire. Recent captures of five individuals in underground pitfall traps placed in shingle on the north-west coast of Norfolk (A. Irwin, pers. comm.) strongly suggest that at least part of the lifecycle of this species may be subterranean. Furthermore, captures of macropterous individuals may represent a small fraction of populations that are mostly brachypterous (Remane and Frohlich 1994).

Edwardsiana rosaesugans   This leafhopper species was first found in 1977 (Claridge and Wilson 1978) having been recorded only from a single *Rosa* sp. bush in Breconshire, South Wales. Further records in the UK have not yet been forthcoming and efforts to confirm the continued presence of the species at the single known site have been unsuccessful. In continental Europe it is widely known in the central Alpine regions. The species belongs to a taxonomically challenging group in which specific identification requires dissection of the male genitalia; this may partly explain why it has not been recorded from more localities.

Eurysa brunnea   Recently added to the British list (Stewart 1997) from material collected in Worcestershire and subsequently discovered at additional sites in central England (Whitehead 1998; M.R. Wilson and A.Taylor, unpubl. data). This planthopper is normally brachypterous and probably lives at the base of grasses, assuming it utilises the same microhabitat as its congeners. It may therefore have been previously overlooked in grassland habitats.

Cicadella lasiocarpae   Confusion between this rare species and the very common *C. viridis* may account for the comparatively recent recognition of the former as British (Le Quesne 1987). Consequently, some old records of *C. viridis* may actually refer to *C. lasiocarpae* and should be checked. The distribution of the host plant, *Carex lasiocarpa*, is scattered with concentrations towards the north and west of Britain. The leafhopper has been widely reported in Ireland following recent survey work.

Muellerianella extrusa   This species was separated from *M. fairmairei* by Booij (1981) on the basis of differences in morphology of the male genitalia, external morphometric parameters and host plants. Old records attributed to *M. fairmairei* may in fact be *M. extrusa*, especially where collected off the latter species' host plant, *Molinia caerulea*.

Elymana kozhevnikovi   This species can only be distinguished from the common *E. sulphurella* by examination of the male genitalia. The recent discovery of this species in southern Scotland (Stewart 1996) raises the interesting possibility that historical records for *E. sulphurella* may, on further examination, turn out to be *E. kozhevnikovi*. Its predominantly northern and eastern distribution within Europe (Ossiannilsson 1983) and the location of the British record both suggest that it is likely to be a predominantly upland species within Britain.

4.4.3 Recent additions to the British list that are probably genuine new arrivals

As indicated, it is impossible to prove that a species is genuinely a new addition to the British fauna, rather than simply one that has been overlooked because of its cryptic habits or low population density. However, certain species are sufficiently large or distinctive in appearance to make it unlikely that they have been present in Britain for a long time without being detected. Information on the expansion or northward extension of a species' range in continental Europe (see for example Remane and Frohlich 1994) can sometimes be used to support the proposition that a species has recently colonised Britain. Some of these species appear to be spreading quite rapidly since colonisation.

Idiocerus ustulatus   First recorded in 1991 (Stewart 1992), this species has now been reported from a wide area of southern Britain (Stewart unpubl. data). The idiocerines are a popular group among collectors and the distinctive appearance of this species makes it unlikely that it was previously present but overlooked. Furthermore, a rapid northward spread in continental Europe has been reported in recent years (Remane and

Frohlich 1994). Its spread within Britain may have been greatly assisted by the widespread planting of its usual hosts, *Populus alba* and *P. canescens*, as amenity trees.

Iassus scutellaris   First discovered in 1978 (Wilson 1981) in suburban south London, this is another highly distinctive species that is unlikely to have been overlooked by earlier recorders. It is superficially similar to the common oak-feeding *I. lanio*, but lives on *Ulmus* species. It has subsequently been widely recorded in, but so far not outside, the south-eastern counties of England. Further spread may have been hindered by the scarcity of host trees following the Dutch elm disease epidemic. It will be interesting to see whether the subsequent recovery of old elms and the introduction of new varieties will encourage further expansion of this species' range in future.

4.4.4 Long-established species that have recently expanded their range

It is difficult to prove that a species has expanded its range when the geographical coverage of historical records for so many species is comparatively poor. There are, however, certain large and distinctive species which it is hard to believe could have been overlooked earlier, even given the limited number of field entomologists who would have had a knowledge of the group. In one case, the history and geographical source of expansion is known with remarkable accuracy.

Athysanus argentarius   This species was well documented in the first half of the nineteenth century as being strictly confined to south-eastern coastal sites (Salmon 1959). However, since the 1960s, it has undergone a dramatic expansion of its range, both inland and in a northerly direction (Fig. 17.1a; Salmon and Chapman 2000). It is interesting to speculate on why this species should have remained confined for so long to coastal localities before colonising sites further inland. Its history over the last 40 years invites comparison with the geographical spread over a similar time period of the long-winged conehead *Conocephalus discolor* (Haes and Harding 1997) and the rhopalid *Chorosoma schillingi* (Kirby 1991a).

Graphocephala fennahi   The so-called rhododendron leafhopper was first reported (as *G. coccinea*) from Chobham, Surrey in 1933 (China 1935). An accidental introduction from North America, this species had become widespread in southern England by the 1960s (Le Quesne 1965). The current distribution (Fig. 17.1b) shows some encroachment into central England, but it has not spread as far as perhaps might have been expected given the distribution of its host plant *Rhododendron ponticum*; this may be due to limited climatic tolerance. Alternatively, the restriction of *R. ponticum* to acidic soils may create large gaps in the geographic coverage of suitable host plants which natural dispersal by the leafhopper cannot bridge. Colonies in north-west England represent interesting outliers in this distribution, which may have arisen through eggs being transported in nursery stock rather than through the natural dispersal of the insect.

Idiocerus herrichi   Up until about 1980, there were only a handful of scattered records for this species. Since then however, it appears to have undergone considerable range expansion, having been widely reported across eastern and southern England. The Idiocerines are an attractive and popular group among collectors and *I. herrichi* is one of the more distinctive species; it seems very unlikely that it could have been over-looked previously. Its host plants, usually either *Salix alba* or *S. fragilis*, are common along waterways and are good colonists of recently created reservoirs and lakes, features which may have aided expansion of the insect's range.

4.4.5 Range contractions or extinctions in the past 50 years

While range expansions are hard to demonstrate because of the poor coverage of historical data, more confidence can perhaps be attached to data that suggest that a species' range has contracted. This is mainly because more recording effort is now being directed towards the *Auchenorrhyncha*. Also, although it is virtually impossible to prove absence, failure to re-record a species at its historical sites adds weight to the evidence for range contraction. There are a number of species that have not been recorded for considerable periods of time, in some cases despite concerted searching of historical sites and other ostensibly suitable habitats, which must therefore be considered possibly extinct.

Asiraca clavicornis   There are old records for this delphacid scattered right across southern and eastern England from Norfolk to Devon (Fig. 17.1c). However, in the last 30 years it has not been recorded outside the London and Thames basin (Jones and Hodge 1999). Reasons for this are still obscure, particularly because the species tends to be associated with early successional and often disturbed grassland habitats.

Platymetopius undatus   This is a highly distinctive arboreal leafhopper that could not easily be mistaken for any other species even by a novice to the group. There are old records from a range of widely separated sites across southern and eastern England (Fig. 17.1d), but there have been no confirmed sightings since the 1940s. Its association with mature, often oak, woodland provides no clue to the reason for its disappearance, since this habitat type is still plentiful and indeed many of the insect's historical sites have remained substantially unchanged.

Laodelphax striatellus   This is almost certainly a species for which Britain is right at the northern edge of its range. Historical records probably represent long-distance migrants rather than established populations. It is known to be a highly migratory species; it was widely recorded across France in aerial suction traps (della Giustina and Balasse 1999). It is a widespread pest of cultivated rice in south-east Asia (Wilson and Claridge 1985).

4.4.6 Species for which there are very few records

In most such cases, it is likely that the species occurs in populations with naturally low densities. Alternatively, the species may be highly specific to a habitat which is itself local or scarce. To a large extent, these tend to be species whose precise microhabitat requirements remain obscure. In many cases, there are no obvious reasons why the species occurs at one site but not at another in which the habitat appears equally suitable.

Tettigometra impressopunctata   This unusual planthopper (the only British member of the *Tettigometridae*) is found in dry calcareous grasslands (mostly on chalk substrate, but also in sand-dune slacks). Records are generally of single individuals, and known

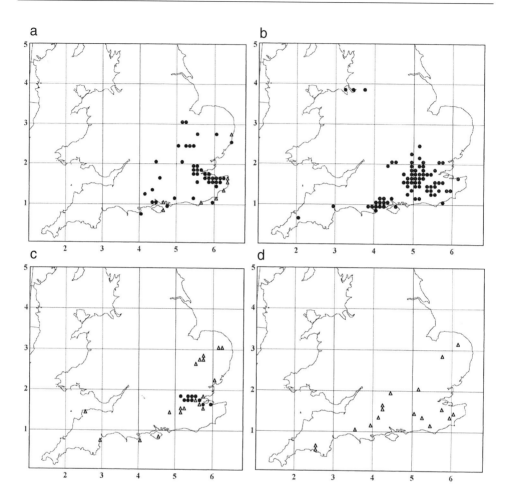

Figure 17.1 10-km square distributions based on data in the Auchenorrhyncha Recording Scheme: (a) *Athysanus argentarius*; (b) *Graphocephala fennahi*; (c) *Asiraca clavicornis*; and (d) *Platymetopius undatus*. Grey triangles refer to records before 1960; black circles refer to records from 1960 onwards. Numbers along axes refer to 100-km intervals on the national grid. The species have not been recorded in Britain outside the areas mapped.

Maps produced by DMAP (copyright Alan Morton).

sites tend to be floristically rich grasslands in warm sheltered locations. For this reason, it is regarded as a good indicator of diverse and undisturbed calcareous grasslands.

Euscelis venosus   A species recorded from only three sites, in each case from tall grassland on a chalk substrate. The only recent record (Hawes and Stewart 1997) suggests that, under favourable conditions, population densities may be quite high. The precise habitat requirements are obscure, however, since such grasslands are quite widespread, yet the insect is absent.

Issus muscaeformis   This species may be genuinely confined to a rather restricted area of north-west England, where it has been reported from only three sites. Its congener, *I. coleoptratus*, is local but widespread across England and Wales south of this area. The precise limits to the ranges of these two species, and whether they overlap, is unclear but would be well worth investigating.

A subcategory of this grouping concerns species whose status is based on very old records from a few sites. These are the species whose status is perhaps the most difficult to evaluate, since they may be surviving in small isolated populations at densities below the normal level of detection. Alternatively, historical records may simply represent transient populations that have long since disappeared. Progress in checking the true status of these species is often hindered by almost complete ignorance of the species' life history and habitat requirements. All have been assigned RDBK status.

Macrosteles cyane   A species recorded from only four sites, all in southern England. The species has not been recorded in the last thirty years and it has been impossible to trace the exact locations of historical records. It is associated with *Potamogeton natans*, but precisely which life-history stages utilise this plant, and how, are unknown.

Metalimnus formosus   Recorded from only a single locality in East Anglia in 1906. The original lowland fen site where the species may have persisted unnoticed survives and there are other sites that may be suitable. The considerable attention given to wetland sites in recent years does however suggest that this species may be extinct in Britain.

Mocuellus collinus   Represented by a single record from the Isle of Wight in the nineteenth century. The extreme southern location in Britain suggests that this species is right at the limit of its climatic tolerance and that any populations may therefore be both vulnerable and transitory.

4.5 Factors affecting change in Auchenorrhyncha

In the case of many rare *Auchenorrhyncha*, so little is known of their general ecology and basic habitat requirements that it is difficult to be certain about what factors have influenced past changes in status and what will constitute the major threats in future, although it would be safe to assume that such species have been affected by many of the same factors as have affected other invertebrate groups. The severity of human-induced factors grades from inappropriate habitat management, through modification of habitats, to wholesale habitat destruction. For example, *Auchenorrhyncha* communities on grassland of all types are threatened by ploughing or other agricultural 'improvements' (such as fertilising or reseeding) that change the species composition and structure of the vegetation. Often however an insidious but more serious threat is posed by alteration to the grazing regime or removal of grazing altogether. The successional changes that follow lead to progressive dominance by coarse grasses and eventually to invasion by scrub, with the resultant loss of faunal communities associated with open grassland.

Species associated with wet habitats, including wet grassland, fen, wet heath and bog, are particularly vulnerable to artificial drainage or any lowering of the water table. Both lowland and upland bogs support a characteristic suite of planthoppers (*Delphacidae*) many of which are rare and highly localised; their vulnerability probably stems from a combination of poor dispersal ability (most populations of these species are dominated by brachypterous individuals) and the widespread destruction and fragmentation of mire systems.

Extensions in geographical range are dependent upon an ability to disperse. While data from aerial suction traps show that many species can be highly dispersive, at least when the right meteorological conditions prevail (Waloff 1973; della Giustina and Balasse 1999), many of the rarer *Auchenorrhyncha* probably have rather limited powers of dispersal. Many planthopper species are most frequently brachypterous, only occasionally producing macropterous forms. In certain species, there is good evidence of a trade-off between dispersal ability (putting resources into wings and flight muscle) and fecundity (allocating resources to reproduction) (Denno 1994). The switch from brachyptery to macroptery is apparently most often triggered by deteriorating host-plant quality, allowing long-winged forms to abandon patches of poor food resources and disperse to more promising areas. The numerous long-winged specimens of the normally brachypterous delphacid planthopper *Conomelus anceps* that were recorded during the hot summers of 1996 and 1997 may have been an example of this. Although such mass dispersal events are probably rather infrequent, their effects may be profound. The ability to disperse will dictate whether species can take advantage of newly created habitats or respond to large-scale climatic change (Masters *et al.* 1998; Whittaker and Tribe 1998). If rare species do indeed have generally poor powers of dispersal, they may be badly equipped to cope with the natural and anthropogenic changes in the British landscape that are likely to take place in future.

REFERENCES

Alexander, K. N. A. and Foster, A. P. (1996) Heteroptera recording in Wales during 1993 and 1994. *British Journal of Entomology and Natural History*, **9**, 3–6.

Alexander, K. N. A. and Grove, S. J. (1991) Heteroptera recording in Cornwall and Devon during 1989 and 1990. *British Journal of Entomology and Natural History*, **4**, 119–121.

Alexander, K. N. A. (1989) A second Gloucestershire locality for *Macroplax preyssleri* (Fieber) (Heteroptera, Lygaeidae). *British Journal of Entomology and Natural History*, **2**, 92–93.

Alexander, K. N. A. (1998a) *Drymus pumilio* Puton (Hemiptera: Lygaeidae) in Surrey. *British Journal of Entomology and Natural History*, **10**, 202.

Alexander, K. N. A. (1998b) Exhibit at the annual exhibition of the British Entomological and Natural History Society 1997. *British Journal of Entomology and Natural History*, **11**, 110.

Allen, A. A. (1977) *Dicyphus rhododendri* Dolling (Hem., Miridae) in Kent. *Entomologist's Monthly Magazine*, **112**, 176.

Allen, A. A. (1980) *Peritrechus gracilicornis* Puton (Hem., Lygaeidae) well established in the Studland area, Dorset. *Entomologist's Monthly Magazine*, **116**, 65.

Allen, A. A. (1984a) First occurrences in Britain of two plant bugs (Hem.-Het.). *Entomologist's Record and Journal of Variation*, **96**, 66–67.

Allen, A. A. (1984b) *Cyphostethus tristriatus* F. (Hem., Acanthosomatidae) on cypress in E. Surrey. *Entomologist's Record and Journal of Variation*, **96**, 187.

Allen, A. A. (1993a) *Deraeocoris olivaceus* (F.) (Hem.: Miridae) in Kent. *Entomologist's Record and Journal of Variation*, **105**, 30.

Allen, A. A. (1993b) Records of two species of Heteroptera (Plant bugs) recently recognised as British. *Entomologists Record and Journal of Variation*, **105**, 37–38.

Ashmole, P and Ashmole, M. (1995) Arthopod fauna of a cave on Tresco, Isles of Scilly, English Channel. *The Entomologist*, **114**, 79–82

Ashwell, D. and Denton, J. (1998) Firebugs *Pyrrhocoris apterus* (L.) (Hemiptera, Pyrrhocoridae) breeding in Surrey. *British Journal of Entomology and Natural History*, **10**, 219.

Askew, R. R. (1983) *Cyphostethus tristriatus* (F.) (Hem., Acanthosomatidae) on *Chamaecyparis* in South Wales,. Entomologist's Monthly Magazine, **119**, 220.

Askew, R. R. (1985) *Macroplax preyssleri* (Hem., Lygaeidae) in Gloucestershire and South Wales. *Entomologist's Monthly Magazine*, **121**, 8.

Aukema, B. and Nau, B. S. (1992) *Megalonotus emarginatus* (Rey) (Lygaeidae) and *Trigonotylus caelestialium* (Kirkaldy) (Miridae) (Hem.-Het.) new to Britain. *Entomologist's Monthly Magazine*, **128**, 11–14.

Barnard, P. C. (1999) *Identifying British Insects and Arachnids. An annotated bibliography of key works.* Cambridge: Cambridge University Press.

Bedwell, E. C. and Massee, A. M. (1945) The county distribution of the British Hemiptera–Heteroptera. *Entomologist's Monthly Magazine*, **81**, 253–273.

Booij, C. J. H. (1981) Biosystematics of the *Muellerianella* complex (Homoptera, Delphacidae), Taxonomy. Morphology and Distribution. *Netherlands Journal of Zoology*, **31**, 572–595.

Bowdrey, J. P. (1999) *Stictopleurus punctatonervosus* (Goeze 1778) (Hem.: Rhopalidae) rediscovered in Britain and new to Essex. *Entomologists Record and Journal of Variation*, **111**, 135–136.

Bratton, J. H., Brierley, S. J. and Kirby, P. (1989) Additional records and amendments to the Heteroptera of Lincolnshire and South Humberside. *Transactions of the Lincolnshire Naturalists' Union*, **22**, 115–116.

Butler, E. A. (1923) *A biology of the British Hemiptera–Heteroptera*. London: Witherby.

Campbell, J. M. (1993) True bugs. *Oxfordshire Biological Records Centre Newsletter*, **17** (1992), 6.

Campbell, J. M. (1994) Hemiptera, bugs. *Oxfordshire Biological Records Centre Newsletter*, **18** (1993), 7.

Carter, C. I. and Young, C. W. T. (1974) *Chamaecyparis lawsoniae* (Murray) Parlatore another host plant for *Cyphostethus tristriatus* (F.) (Hem., Acanthosomatidae). *Entomologist's Monthly Magazine*, **109**, 180.

Carter, C. I. Fourt, D. F. and Bartlett, P. W. (1984) The Lupin Aphid's arrival and consequences. *Antenna*, **8**, 129–132.

China, W. E. (1930) The origin of the British Heteropterous fauna, in Peuplement des Iles Britanniques. *Société de Biogeographie*, **3**, pp. 79–90.

China, W. E. (1935) A North American Jassid (Homoptera) in Surrey. *Entomologist's Monthly Magazine*, **71**, 277–279.

Claridge, M. F. and Wilson, M. R. (1976) Diversity and distribution patterns of some mesophyll-feeding leafhoppers of temperate woodland canopy. *Ecological Entomology*, **1**, 231–250.

Claridge, M. F. and Wilson, M. R. (1978) Observations on new and little known species of Typhlocybine leafhoppers (Hemiptera: Cicadellidae) in Britain. *Entomologist's Gazette*, **29**, 247–251.

Claridge, M. F. (1985) Acoustic signals in the Homoptera: behaviour taxonomy and evolution. *Annual Review of Entomology*, **30**, 297–317.

Clemons, L. (1998) *Metopoplax ditomoides* (Costa) (Hem., Lygaeidae) in north-west Kent. *Entomologist's Record and Journal of Variation*, **110**, 255.

Crossley, R. (1974) *Scolopostethus pictus* (Schilling) (Hem., Lygaeidae) in Scotland. *Entomologist's Monthly Magazine*, **110**, 226.

Crossley, R. (1976) Notes on some Yorkshire Hemiptera. *Entomologist's Monthly Magazine*, **112**, 238.

Crossley, R. (1979) Entomological reports for 1977–78. *Naturalist*, **104**, 126

Crossley, R. (1980) Notes on Yorkshire shorebugs. *Entomologist's Monthly Magazine*, **116**, 40.

Crossley, R. (1981) *Polymerus unifasciatus* (F.) var. *lateralis* (Hahn) (Hem., Miridae) in Britain. *Entomologist's Monthly Magazine*, **116**, 155–156.

Crossley, R. (1982) *Anthocoris amplicollis* Horváth (Hem., Anthocoridae), new to Britain/ *Entomologist's Monthly Magazine*, 118, 111.

Crossley, R. (1993) A further county record for *Trigonotylus caelestialium* (Kirk.) (Hem., Miridae). *Entomologist's Record and Journal of Variation*, 105, 176.

della Giustina, W. and Balasse, H. (1999) Gone with the wind: Homoptera Auchenorrhyncha collected by the French network of suction traps in 1994. *Marburger Entomologische Publicationen*, 3(1), 7–42.

Denno, R. F. (1994) Life history variation in planthoppers, in *Planthoppers: Their Ecology and Management* R. F. Denno and T. J. Perfect (eds). London: Chapman and Hall. pp. 163–215.

Denton, J. S. (1997a) Recent records of rare and notable Hemiptera in Surrey and North Hampshire. *Entomologist's Monthly Magazine*, 133, 79–80.

Denton, J. S. (1997b) Rare and notable Hemiptera in Surrey, Middlesex and North Hampshire 1995–6. *Entomologist's Monthly Magazine*, 133, 175–176.

Dolling, W. R. (1971) *Macroplax preyssleri* (Fieber) (Hem., Lygaeidae) new to Britain. *Entomologist's Monthly Magazine*, 106, 155–156.

Dolling, W. R. (1972) A new species of *Dicyphus* Fieber (Hem., Miridae) from southern England. *Entomologist's Monthly Magazine*, 107, 244–245.

Dolling, W. R. (1978) The British species of *Stictopleurus* Stål (Hemiptera: Rhopalidae). *Entomologist's Gazette*, 29, 261–264.

Dolling, W. R. (1983) Two new county records for *Berytinus hirticornis* (Brullé) (Hem., Berytidae). *Entomologist's Monthly Magazine*, 119, 70.

Dolling, W. R. (1997) Two recent immigrants found north of the Humber (Hym., Vespidae and Hem., Lygaeidae). *Entomologist's Monthly Magazine*, 133, 130.

Douglas, J. W. and Scott, J. (1865) *The British Hemiptera. Vol. 1*. Hemiptera–Heteroptera. London: Ray Society.

Drake, C. M. (1988) *A Survey of the Aquatic Invertebrates of the Essex Grazing Marshes*. Peterborough: Nature Conservancy Council.

Duffield, C. A. W. (1960) The Auchenorrhyncha (Homoptera) of Kent. *Transactions of the Kent Field Club*, 1, 161–170.

Edwards, J. (1896) *The Hemiptera–Homoptera (Cicadina and Psyllina) of the British Islands*. London: L. Reeve.

Flint, J. R. (1990) *Cicadula ornata* (Melichar) (*Hemiptera: Cicadellidae*) new to Britain. *Entomologist's Gazette*, 40, 345–246.

Groves, E. W. (1968) Hemiptera–Heteroptera of the London area. Part V. *London Naturalist*, 47, 50–80.

Haes, E. C. M. and Harding, P.T. (1997) *Atlas of grasshoppers, crickets and allied insects in Britain and Ireland*. London: HMSO.

Haughton, A. and Bell, J. (1996) *Graptopeltus lynceus* (F.) (Hem., Lygaeidae) in north-west England. *Entomologist's Monthly Magazine*, 132, 16.

Hawes, C. and Stewart, A. J. A. (1997) Notes on the habitat preference of *Euscelis venosus* (Kbm) (Hem., Auchenorrhyncha). *Entomologist's Monthly Magazine*, 133, 63–64.

Hawkins, R. D. (1989a) *Miridius quadrivirgatus* (Costa) (Hem., Miridae) found inland in Surrey. *Entomologist's Monthly Magazine*, 125, 174.

Hawkins, R. D. (1989b) A further record of *Placochilus seladonicus* (Fallén) (Hem., Miridae)'. *Entomologist's Monthly Magazine*, 125, 205.

Hawkins, R. D. (1989c) *Orsillus depressus* Dallas (Hem., Lygaeidae) an arboreal groundbug new to Britain. *Entomologist's Monthly Magazine*, 125, 241–242.

Hawkins, R. D. (1995) *Elasmostethus tristriatus* and related shieldbugs in the Sheffield area. *Sorby Record*, 31, 66–67.

Hawkins, R. D. and Menzies, I. S. (1998) Exhibit at the annual exhibition of the British Entomological and Natural History Society 1997. *British Journal of Entomology and Natural History*, 11, 110.

Hodge, P. J. (1990a) *Placochilus seladonicus* (Fln) (Hem., Miridae) in East Sussex. *Entomologist's Monthly Magazine*, **126**, 211.

Hodge, P. J. (1990b) *Capsus wagneri* Remane (Hem., Miridae) in North Somerset. *Entomologist's Monthly Magazine*, **126**, 260.

Hodge, P. J. (1991) Exhibit at the annual exhibition of the British Entomological and Natural History Society. *British Journal of Entomology and Natural History*, **4**, 43.

Hodge, P. J. (1998a) Exhibit at the annual exhibition of the British Entomological and Natural History Society 1996. *British Journal of Entomology and Natural History*, **10**, 178.

Hodge, P. J. (1998b) Four species of Hemiptera from south-east England. Exhibit at the annual exhibition of the British Entomological and Natural History Society 1998. *British Journal of Natural History*, **11**, 110.

Hodge, P. J. (1999) Exhibit at the annual exhibition of the British Entomological and Natural History Society 1998. *British Journal of Entomology and Natural History*, **12**, 180.

Hodge, P. J. and Porter, D. A. (1997) *Nysius senecionis* (Schilling) (Hemiptera: Lygaeidae) new to the British Isles. *British Journal of Entomology and Natural History*, **10**, 1–2.

Horsfield, D. (1989) Two records of *Saldula opacula* (Zett.) (Hem., Saldidae) from the Scottish Highlands. *Entomologist's Monthly Magazine*, **125**, 249.

Irwin, A. G. (1982) *Pachycoleus waltli* Fieber (Hemiptera, Dipsocoridae) new to East Anglia. *Entomologist's Monthly Magazine*, **118**, 194.

Jansson, A. (1981) A new European species and notes on synonymy in the genus *Corixa* Geoffroy (Heteroptera, Corixidae). *Annales Entomologici Fennici*, **47**, 65–68.

Jessop, L. (1983) The British species of *Anthocoris* (Hem., Anthocoridae). *Entomologist's Monthly Magazine*, **119**, 221–223.

Jones, R. A. (1993) The rhododendron lacebug, *Stephanitis rhododendri* Horváth, rediscovered in south-east London. *British Journal of Entomology and Natural History*, **6**, 139–140.

Jones, R. A. (1998) Exhibit at the annual exhibition of the British Entomological and Natural History Society 1996. *British Journal of Entomology and Natural History*, **10**, 177.

Jones, R.A. (1999) Entomological surveys of vertical river flood defence walls in urban London-brownfield corridors; problems, practicalities and some promising results. *British Journal of Entomology and Natural History*, **12**, 193–213.

Jones, R.A. and Hodge, P.J. (1999) Notes on the occurrence of the planthoppers *Reptalus panzeri* (Low) (Hemiptera: Cixiidae) and *Asiraca clavicornis* (Fab.) (Hemiptera: Delphacidae). *British Journal of Entomology and Natural History*, **12**, 239–240.

Judd, S. and Rotheram, I. D. (1992) The phytophagous insect fauna of *Rhododendron ponticum* L. in Britain. *Entomologist*, **111**, 134–150.

Judd, S. (1998) New national, regional and county records of British seed bugs (Hem., Lygaeidae). *Entomologist's Monthly Magazine*, **134**, 311–314.

Judd, S. and Hodkinson, I. (1998) The biogeography and regional diversity of the British seed bugs (Hemiptera: Lygaeidae). *Journal of Biogeography*, **25**, 227–249.

Judd, S. and Straw, N. A. (1998) A new seed bug, *Emblethis denticollis* Horváth (Heteroptera, Lygaeidae) for Britain, with a key to nymphs of *Emblethis*. *British Journal of Entomology and Natural History*, **10**, 220–225.

Kirby, P. (1983) *Neodicyphus rhododendri* (Dolling) (Hem., Miridae) in Derbyshire. *Entomologist's Monthly Magazine*, **119**, 116.

Kirby, P. (1984a) Inland records of *Chorosoma schillingi* (Schummel) (Hem., Rhopalidae). *Entomologist's Monthly Magazine*, **120**, 177.

Kirby, P. (1984b) *Monosynamma sabulicola* Wagner (Hem., Miridae) in Derbyshire. *Entomologist's Monthly Magazine*, **120**, 222.

Kirby, P. (1991a) Further inland records of *Chorosoma schillingi* (Schummel) (Hem., Rhopalidae). *Entomologist's Monthly Magazine*, **127**, 250.

Kirby, P. (1991b) *Chlorita dumosa* (Ribaut) (Hemiptera: Cicadellidae) in Britain, *Entomologist's Gazette*, **42**, 41–43.

Kirby, P. (1992a) A review of the scarce and threatened Hemiptera of Great Britain. Peterborough: Joint Nature Conservation Committee (UK Nature Conservation, no. 2).

Kirby, P. (1992b) *Hydrometra gracilenta* Horváth (Hemiptera, Hydrometridae) in East Sussex. *British Journal of Entomology and Natural History*, 5, 128.

Kirby, P. (1993) News digest. *Heteroptera Study Group Newsletter*, 12, 2.

Kirby, P. (1994) A selection of interesting Heteroptera taken in 1991 and 1992, Exhibit at the annual exhibition of the British Entomological and Natural History Society 1993. *British Journal of Entomology and Natural History*, 7, 174–5.

Kirby, P. (1996) New and interesting records of Essex Hemiptera. *Essex Naturalist (new series)*, 13, 13–24.

Kirby, P. (1997) Essex Heteroptera: report for 1996. *Essex Naturalist (new series)*, 14, 18–20.

Kirby, P. (1998a) Exhibit at the annual exhibition of the British Entomological and Natural History Society 1996. *British Journal of Entomology and Natural History*, 10, 177–178.

Kirby, P. (1998b) An annotated list of the Heteroptera of Lincolnshire and South Humberside. *Transactions of the Lincolnshire Naturalists Union*, 22, 41–70.

Kirby, P. (1999) *Buchananiella continua* (B.White) (Hemiptera: Anthocoridae) established in Britain. *British Journal of Entomology and Natural History*, 12, 221–223

Lane, S. A. (1988) *Chorosoma schillingi* (Schummel) (Hem., Rhopalidae) in Warwickshire. *Entomologist's Monthly Magazine*, 124, 80.

Lane, S. A. (1989) *Agnocoris reclairei* (Wagner) (Hem., Miridae) in Warwickshire. *Entomologist's Monthly Magazine*, 125, 36.

Le Quesne, W. J. (1960) Hemiptera: Fulgoromorpha. *Handbooks for the Identification of the British Insects*, 2(3), London: Royal Entomological Society.

Le Quesne, W. J. (1964) *Trigonocranus emmeae* Fieber (Hem., Cixiidae) new to Britain. *Entomologist's Monthly Magazine*, 100, 117.

Le Quesne, W. J. (1965) Hemiptera Cicadomorpha (excl. Deltocephalinae and Typhlocybinae), *Handbooks for the Identification of the British Insects*, 2(a). London: Royal Entomological Society.

Le Quesne, W. J. (1969) Hemiptera Cicadomorpha Deltocephalinae, *Handbooks for the Identification of the British Insects*, 2(b). London: Royal Entomological Society.

Le Quesne, W. J. (1974) *Eupteryx origani* Zakhvatkin (Hem., Cicadellidae) new to Britain. *Entomologist's Monthly Magazine*, 109, 203–206.

Le Quesne, W. J. (1977) A new species of *Alebra* Fieber (Hem., Cicadellidae). *Entomologist's Monthly Magazine*, 112, 49–52.

Le Quesne, W. J. (1983a) *Cicadula flori* (Sahlberg), new to Britain (Hem., Cicadellidae). *Entomologist's Monthly Magazine*, 119, 177.

Le Quesne, W. J. (1983b) Leafhopper news. *Auchenorhyncha Recording Scheme Newsletter*, 3, 2–3.

Le Quesne, W. J. (1987) *Cicadella lasiocarpae* Ossiannilsson (Hemiptera: Cicadellidae). *Entomologist's Gazette*, 38, 87–89.

Le Quesne, W. J. and Payne, K. R. (1980) Cicadellidae (Typhlocybinae) with a check list of the British Auchenorhyncha (Hemiptera, Homoptera), *Handbooks for the Identification of the British Insects*, 2(c). London: Royal Entomological Society.

Leston, D. (1954) *Capsus wagneri* Remane (Hem., Miridae), a plant-bug new to Britain. *Entomologist's Monthly Magazine*, 90, 1.

Leston, D. (1957) The British *Lygocoris* Reuter (Hem., Miridae), including a new species. *Entomologist*, 90, 128–135.

Macan, T. T. (1965) *A Key to British Water Bugs*, 2nd edn. Scientific Publications of the Freshwater Biological Association, 16. Ambleside: Freshwater Biological Association.

Massee, A. M. (1955) The county distribution of the Hemiptera-Heteroptera, 2nd edn. *Entomologist's Monthly Magazine*, 91, 7–27.

Masters, G. J., Brown, V. K., Clarke, I. P., Whittaker, J. B., and Hollier, J. A. (1998) Direct and indirect effects of climate change on insect herbivores: Auchenorrhyncha (Homoptera). *Ecological Entomology*, 23, 45–52.

Menzies, I. (1994a) Hemiptera-Heteroptera. *London Naturalist*, 72, 171.

Menzies, I. (1994b) Heteroptera from Surrey. Exhibit at the annual exhibition of the British Entomological and Natural History Society 1993. *British Journal of Entomology and Natural History*, 7, 175.

Morris, M. G. (1971) Differences between the invertebrate faunas of grazed and ungrazed chalk grassland. IV. Abundance and diversity of Homoptera: Auchenorrhnycha. *Journal of Applied Ecology*, 8, 37–52.

Morris, R. K. and Parsons, M. S. (1992) *A survey of invertebrate communities on the shingle of Dungeness, Rye Harbour and Orford Ness*, JNCC Report 77. Peterborough: Joint Nature Conservation Committee.

Nau, B. S. (1978) Two plant bugs new to Britain, *Placochilus seladonicus* (Fall.) and *Campylomma annulicornis* (Sig.) (Heteroptera, Miridae). *Entomologist's Monthly Magazine*, 114, 157–159.

Nau, B. S. (1980) *Tuponia carayoni* Wagner (Hem., Miridae) new to Britain. *Entomologist's Monthly Magazine*, 116, 83–84.

Nau, B. S. (1990) The status of *Orthotylus virens*. *Heteroptera Study Group Newsletter*, 9, 4–5.

Nau, B. S. (1994) Notes on *Placochilus seladonicus* (Fall.) (Hem., Miridae) in Britain. *Entomologist's Monthly Magazine*, 130, 209–210.

Nau, B. S. (1995) Status of *Atractotomus parvulus* Reuter (Hem., Miridae) in Britain. *Entomologist's Monthly Magazine*, 131, 64.

Nau, B. S. (1997) Range-changes in some Hemiptera-Heteroptera in Bedfordshire. *Entomologist's Monthly Magazine*, 133, 261–262.

Nau, B. S. (1998) Further notes on range-changes of Hemiptera-Heteroptera in Bedfordshire. *Entomologist's Monthly Magazine*, 134, 52.

O'Connor, J. P. and Ashe, P. (1996) *Elasmostethus tristriatus* (F.) (Hem., Acanthosomatidae) confirmed as an Irish species. *Entomologist's Monthly Magazine*, 133, 314.

Ossiannilsson, F. (1978) The Auchenorrhyncha (Homoptera) of Fennoscandia and Denmark. Part 1. Introduction, infraorder Fulgoromorpha. *Fauna Entomologica Scandinavica*, 7(1), Klampenborg: Scandinavian Science Press.

Ossiannilsson, F. (1981) The Auchenorrhyncha (Homoptera) of Fennoscandia and Denmark. Part 2. The Families Cicadidae, Cercopidae, Membracidae and Cicadellidae (excl. Deltocephalinae). *Fauna Entomologica Scandinavica*, 7(2), Klampenborg: Scandinavian Science Press.

Ossiannilsson, F. (1983) The Auchenorrhyncha (Homoptera) of Fennoscandia and Denmark. Part 3. The Family Cicadellidae: Deltocephalinae, Catalogue, Literature and Index. *Fauna Entomologica Scandinavica*, 7(3), Klampenborg: Scandinavian Science Press.

Parsons, M. S. (1987) *Review of invertebrate sites in England: Cumbria – interim report (A and B sites only)*. Peterborough: Nature Conservancy Council (Invertebrate Site Register report no. 102, CSD report no. 788).

Payne, K. R. (1979) Auchenorrhyncha (Homoptera) of Gait Barrows National Nature Reserve near Arnside, Lancs. (SD/47). *Entomologist's Monthly Magazine*, 114, 210.

Payne, K. R. (1981) Leafhopper news. *Auchenorhyncha Recording Scheme Newsletter*, 2, 2.

Porter, D. (1998) Exhibit at the annual exhibition of the British Entomological and Natural History Society 1996. *British Journal of Entomology and Natural History*, 10, 178–179.

Price, J. M. (1996) *A provisional atlas of the true bugs of Warwickshire (Insecta: Hemiptera: Heteroptera: 6433)*. Warwick: Warwickshire Museum Service (Warwickshire County Council).

Remane, R. and Frohlich, W. (1994) Beiträge zur chorologie einiger Zikaden-arten (Homoptera Auchenorrhyncha) in der Westpaläarktis. *Marburger Entomologische Publikationen*, 2(8), 131–188.

Ribaut, H. (1936) Homopteres Auchenorhynches. I. Typhlocybidae. *Faune de France*, **31**, Paris: Paul Lechevalier et Fils.

Ribaut, H. (1952) Homopteres Auchenorhynches. II. Jassidae. *Faune de France*, **57**, Paris: Paul Lechevalier et Fils.

Roubal, J. (1965) *Chiragra*-Komplex unter der Lygaeiden-Gattung *Megalonotus* Fieber 1860 aus dem Europaischen Festland. – Ein Versuch um die taxonomische Losung. *Acta entomologica Musei nationalis Pragae*, **36**, 555–588.

Russell, W. E. (1967) A record of *Chorosoma schillingi* Sch. (Heteroptera, Rhopalidae) from Peterborough. *Entomologist's Gazette*, **19**, 8.

Russell, W. E. (1969) A preliminary list of plant bugs (Hemiptera, Heteroptera) recorded for the Peterborough district. *Entomologist's Gazette*, **20**, 125–135.

Salmon, M. A. (1959) On the rediscovery of *Athysanus argentarius* (Fab.) (Hom. Cicadellidae) in Britain. *Entomologist's Gazette*, **10**, 51–53.

Salmon, M. A. and Chapman, H. (2000) On the history and distribution of *Athysanus argentarius* Metcalf (hem.: Cicadellidae) in Britain. *British Journal of Entomology and Natural History*, **13**, 91–93.

Sands, W. A. (1954) *Deraeocoris olivaceus* (F.) (Hem., Miridae) new to Britain. *Entomologist's Monthly Magazine*, **90**, 301.

Saunders, E. (1892) *The Hemiptera-Heteroptera of the British islands*. London: L. Reeve.

Savage, A. A. (1989) *Adults of the British aquatic Hemiptera Heteroptera: a key with ecological notes*. Scientific publication no. 50, Ambleside: Freshwater Biological Association.

Shardlow, M. (1999) Exhibit at the annual exhibition of the British Entomological and Natural History Society. *British Journal of Entomology and Natural History*, **12**, 180.

Shirt, D. B. (ed.) (1986) *British Red Data Books: 2. Insects*. Peterborough: Nature Conservancy Council.

Southwood, T. R. E. and Leston, D. (1959) *Land and water bugs of the British Isles*. London: Warne.

Southwood, T. R. E. (1963) *Megalonotus sabulicola* (Thomson, 1870) in Britain. *Entomologist*, **96**, 124–126.

Speight, M.R. and Nicol, M. (1985) Horse chestnut scale – another problem for urban trees? *Antenna*, **9**, 176–178.

Stewart, A. J. A. and Wright, A. F. (1992) A new inexpensive suction apparatus for sampling arthropods in grassland. *Ecological Entomology*, **20**, 98–102.

Stewart, A. J. A. (1988) Patterns of host plant utilisation by leafhoppers in the genus *Eupteryx* (Hemiptera: Cicadellidae) in Britain. *Journal of Natural History*, **22**, 357–379.

Stewart, A. J. A. (1992) A new species to Britain – *Idiocerus ustulatus*. *Auchenorrhyncha recording Scheme Newsletter*, **15**, 5.

Stewart, A. J. A. (1996) Three new species to Britain. *Auchenorrhyncha Recording Scheme Newsletter*, **15**, 1–2.

Stewart, A. J. A. (1997) *Eurysa brunnea* Melichar, 1896 (Hemiptera: Delphacidae) new to Britain. *British Journal of Entomology and Natural History*, **10**, 3–5.

Stewart, A. J. A. (1999) Twenty years of recording the distribution of Auchenorrhyncha (Hemiptera: Homoptera) in Britain and Ireland: progress, achievements and prospects. *Reichenbachia*, **33**, 207–214.

Stewart, A. J. A. and Wilson, M. R. (1999) Notes on the distribution of *Ledra aurita* (L.) (Homoptera: Cicadellidae) in Britain. *British Journal of Entomology and Natural History*, **11**, 167–168.

Stewart, A. J. A. (in press) Field sampling techniques, in *Leafhopper and Planthopper Technology: Techniques for Research on the Homoptera*, Auchenorrhyncha. M.A. Jervis and M.R. Wilson (eds). Andover: Intercept.

Tomlinson, R. (1971) Field meetings. Stanford-le-Hope, Essex, 14 June 1970. *Proceedings and Transactions of the British Entomological and Natural History Society*, **4**, 25.

UK Steering Group. (1995) *Biodiversity: the UK Steering Group report*. Vol. 2: *Action Plans*. London: HMSO.

UK Biodiversity Group (1999) *Tranche 2 Action Plans. Volume IV. Inverterbrates*. Peterborough: English Nature.

Waloff, N. (1973) Dispersal by flight of leafhoppers (Auchenorrhyncha: Homoptera). *Journal of Applied Ecology*, **10**, 705–730.

Waloff, N. (1980) Studies on grassland leafhoppers (Auchenorrhyncha, Homoptera) and their natural enemies. *Advances in Ecological Research*, **11**, 81–215.

Waloff, N. and Thompson, P. (1980) Census data and analyses of populations of some leafhoppers (Auchenorrhyncha, Homoptera) of acidic grassland. *Journal of Animal Ecology*, **49**, 395–416.

Waloff, N. and Jervis, M. A. (1987) Communities of parasitoids associated with leafhoppers and planthoppers in Europe. *Advances in Ecological Research*, **17**, 281–402.

Whitehead, P. F. (1998) British records of *Eurysa brunnea* Melichar, 1896 (Hemiptera: Delphacidae). *Entomologist's Monthly Magazine*, **134**, 307–310.

Whitehead, P. F. (1989) The most northerly breeding colony of *Dichrooscytus valesianus* (Hemiptera: Miridae) Meyer-Dür in Britain. *Entomologist's Monthly Magazine*, **125**, 49–51.

Whitehead, P. F. (1991a) *Scolopostethus pictus* (Schilling) (Hem., Lygaeidae) new to Worcestershire. *Entomologist's Record and Journal of Variation*, **103**, 82.

Whitehead, P. F. (1991b) A further note on *Scolopostethus pictus* (Schilling) (Hem., Lygaeidae). *Entomologist's Record and Journal of Variation*, **103**, 262.

Whittaker, J. B. (1969) Quantitative and habitat studies on the frog-hoppers and leaf-hoppers (Homoptera: Auchenorrhyncha) of Wytham Woods, Berkshire. *Entomologist's Monthly Magazine*, **105**, 27–37.

Whittaker, J. B. and Tribe, N. P. (1998) Predicting numbers of an insect (*Neophilaenus lineatus*: Homoptera) in a changing climate. *Journal of Animal Ecology*, **67**, 987–991.

Wilson, M. R. (1979) *Kyboasca bipuncta* (Oshanin) (*Homoptera: Auchenorhyncha: Typhlocybinae*); a new species to Britain. *Entomologist's Record and Journal of Variation*, **91**, 194.

Wilson, M. R. (1981) Identification of European *Iassus* species (Homoptera: Cicadellidae) with one new species to Britain. *Systematic Entomology*, **6**, 115–118.

Wilson, M. R. and Claridge, M. F. (1985) The leafhopper and planthopper fauna of rice fields, in *The Leafhoppers and Planthoppers*, L. R. Nault and J.G. Rodriguez (eds). Chichester: John Wiley.

Woiwod, I. P. (1991) The ecological importance of long-term synoptic monitoring, in Firbank, L.G. *et al.* (eds) *The Ecology of Temperate Cereal Fields*. 32nd Symposium of the British Ecological Society. Oxford: Blackwell Scientific.

Woiwod, I. P. and Harrington, R. (1994) Flying in the face of change: the Rothamsted Insect Survey, in Leigh, R.A. and Johnstone, A.E. (eds) *Long-term Experiments in Agricultural and Ecological Sciences*. Wallingford: CAB International.

Woodroffe, G. E. (1953) On the occurrence at Hounslow, Middlesex, of *Metopoplax ditomoides* Costa (Hem., Lygaeidae) new to Britain. *Entomologist*, **86**, 224–225.

Woodroffe, G. E. (1954) Further notes on the Hemiptera-Heteroptera of Hounslow Heath, Middlesex. *Entomologist*, **87**, 15–18.

Woodroffe, G. E. (1956) Miscellaneous records of Hemiptera-Heteroptera captured during 1955. *Entomologist's Monthly Magazine*, **92**, 47–48.

Woodroffe, G. E. (1971a). The identity of the British *Orius* (Wolff) (Hem., Anthocoridae). *Entomologist*, **104**, 258–259.

Woodroffe, G. E. (1971b) An undescribed British *Atractotomus* Fieber (Hem., Miridae). *Entomologist*, **104**, 265–267.

Woodroffe, G. E. (1977) *Notostira erratica* (L.) and *N. elongata* (Geoffroy) in the British Isles. *Entomologist's Gazette*, **28**, 123–126.

Chapter 18

Butterflies and moths

Richard Fox

ABSTRACT

The past 25 years have witnessed a transformation of the state of knowledge of the distribution, abundance and ecology of butterflies and moths. Many thousands of volunteers have participated in two major national surveys of butterflies, monitoring schemes have been set up to measure relative abundance and many local atlases have been published.

Habitat destruction and deterioration have continued apace and many species have declined dramatically, some to extinction, despite the introduction of protective legislation, species recovery programmes and biodiversity action planning. The charity Butterfly Conservation considers about 40% of British butterflies to be threatened and about a third of all resident macro-moth species are classified as nationally scarce. Most of these species have suffered major distribution declines in recent decades. By contrast, a number of relatively widespread species continue to expand their ranges and new species are colonising from continental Europe, in parallel with the climatic amelioration attributed to global warming.

The crisis facing our native butterflies and moths was recognised 25 years ago and remains today. However, we are now in a much better position to tackle the underlying problems. Enormous improvements have been made in our knowledge of species requiring conservation action and the location of core populations. We also now have a sound understanding of the ecology and habitat requirements of many threatened species and they are well represented in the biodiversity action planning process from the national to the local scale.

1 Introduction

Instability is a feature of the populations and distribution of many butterfly and moth species in Great Britain and Ireland. Fluctuations are often severe and the extremes of extinction and colonisation are not uncommon. Indeed such instability may be a natural consequence for a group characterised by very short reproductive lives, a high dependence on the weather and often specialised niches.

There is little doubt that the high proportion of declines among butterflies, the best-studied group of *Lepidoptera*, has been a direct result of human activity. As is probably the case for many taxa, the massive decline and extinction of specialist species are being masked by the colonisations and expansions of generalists. Species that can tolerate degraded habitats, and eat exotic crops and garden plants, or that have benefited from the changing climate are spreading, while the majority succumb to habitat loss and changes in agricultural and forestry management.

2 State of knowledge

Despite all of the dramatic declines, expansions, extinctions and colonisations of species over the past 25 years, by far the most significant changes relate to improvements in our state of knowledge. The framework in which John Heath prepared his submission to the Changing Flora and Fauna of Britain conference (Heath 1974) was radically different from that which prevails now. Many of the facts that lepidopteran researchers and conservationists now take for granted were then unknown or in their infancy.

Systematic national recording of distributions had only just begun in 1973, and there were few data on abundance. The ecological requirements of many species (even fundamental ones such as larval foodplants) were poorly understood, and the underlying causes of distribution change were often a matter of generalisation and speculation. The importance now attached to metapopulation theory, parasite–host cycles, microclimate requirements and specific weather conditions in explaining change has developed in the past 25 years.

Concurrent with these scientific advances has come a growing public interest in nature conservation. An army of new recorders has swelled the ranks of natural historians, and non-governmental organisations (NGOs) have undergone a huge expansion in membership, staff and influence. Butterfly Conservation, whose remit covers all Lepidoptera, has expanded from about 500 members in 1973 to the current total of 8500.

The national policy framework has developed out of all recognition. The Wildlife and Countryside Act (1981) and Wildlife (Northern Ireland) Order (1985) introduced legal protection, at least from commercial exploitation, for a considerable number of Lepidoptera. The Red Data Book (Shirt 1987) highlighted many of the most threatened species and the statutory conservation agencies have organised species recovery programmes. In recent years, the biodiversity and species action planning process has further revolutionised Lepidoptera conservation.

These major changes have dramatically improved our knowledge of the changing Lepidoptera of Great Britain and Ireland and provided the will and means to halt and reverse the numerous declines that are evident.

3 Data capture

The datasets of distribution and abundance information available today would have been almost unimaginable in 1973.

3.1 Distribution data

Butterfly recording has a long history in Britain (Harding et al. 1995), but it was not until 1984 that the first comprehensive atlas of butterflies was published. The Atlas of Butterflies in Britain and Ireland (Heath et al. 1984) marked the fruition of the Lepidoptera Recording Scheme run by the Biological Records Centre (BRC) from 1967 until 1982. The scheme also included macro-moths with the intention of publishing distribution maps for this group. This ambition has been only partially realised, through the (ongoing) *Moths and Butterflies of Great Britain and Ireland* series (Heath and Emmet 1976 onwards). The macro-moth element of the scheme was discontinued after 1982, while butterfly records continued to be collated at a national level in a joint project with Butterfly Conservation.

Some organised recording was carried out in Ireland during the early stages of the *Lepidoptera* Recording Scheme and by the Irish Biological Records Centre, which was in existence between 1971 and the late 1980s. An atlas of butterflies in Ireland was published in 1992 (Hickin 1992).

During much of the 1980s and early 1990s there was no national scheme for any macro-moths and returns of butterfly records to BRC gradually declined. However, much of the considerable interest generated by the *Lepidoptera* Recording Scheme became refocused on to regional and county-based projects. Local recording schemes proliferated and resulted in the publication of many county lists and atlases (a process that continues apace today). The success of such schemes is clear from the high quality of the publications, the analysis they contain and from the frequency with which the atlases are updated. Many counties in England now produce an annual atlas of butterflies.

A survey by BRC in 1988 identified over 200 local butterfly projects and provided the impetus for a fresh look at the co-ordination of recording at a national level (Harding *et al.* 1995). The report of a working party drawn from Butterfly Conservation and BRC to examine this issue marked the birth of a new national survey, called Butterflies for the New Millennium (BNM).

Launched in 1995, the BNM project aims to be the largest and most comprehensive survey of butterflies ever undertaken in Great Britain and Ireland (Asher 1995; Fox 1998). It is a collaborative project involving local partnerships of Butterfly Conservation branches, county wildlife trusts, local records centres and a wide range of other organisations involved with biological recording, nature conservation and land management. An initial, intensive five-year survey (1995–9) will culminate in the publication of a new atlas in 2000 (Asher *et al.* in press). At the end of 1999, over 1.6 million records had already been amassed, representing 99% of the 10-km squares in Britain. There seems little doubt that the atlas and the dataset that underpins it will be comprehensive and authoritative. The Republic of Ireland joined the BNM initiative in 1998 with co-ordination being provided by the Dublin Naturalists' Field Club.

National recording of the rarer moths has also been revived. Since 1992 Butterfly Conservation has operated a national recording network for the rarer British macro-moths (Waring 1998a) with funding from the Joint Nature Conservation Committee (JNCC). The principal objective is to produce up-to-date distribution maps to aid conservation planning for the 200 or so species that fall into the *Red Data Book* and nationally scarce categories.

There are also several national recording schemes covering some micro-moth families, but space precludes a full discussion of those here.

3.2 Abundance data

Two long-running national schemes recording the abundance of *Lepidoptera* have operated throughout much of the period. The Rothamsted Insect Survey's network of light-traps has operated nightly since the early 1960s and has produced an enormous dataset of the catch numbers of nearly 700 species of macro-*Lepidoptera* from over 350 sites across Britain (Woiwod and Harrington 1994). The project is run by the Institute of Arable Crop Research, but volunteers operate most traps. The Butterfly Monitoring Scheme, which uses a standardised transect methodology to collect relative abundance data, was launched by Institute of Terrestrial Ecology and Nature Conservancy Council in 1976 and continues today (Pollard and Yates 1993a). These two national networks currently embrace about 90 and 115 sites respectively.

In addition to these nationally collated schemes, there has been a substantial growth in the number of regularly recorded light-traps and butterfly transects. For example, some 300 butterfly transects are operated on sites outside the 'official' Butterfly Monitoring Scheme, many walked by Butterfly Conservation volunteers. At the moment this wealth of data is generally only collated at a county level, but the benefits in seeing it drawn together at a national level are clear.

4 Workforce

It is difficult to gauge the number of lepidopterists in Britain and Ireland, but an estimate may be gained from the number of individuals who have contributed to the various recording schemes. The vast majority of these are amateur naturalists and some may not even consider *Lepidoptera* as their primary interest. Some 2000 individuals contributed butterfly records to the Heath *et al. Atlas* (1984) and at least 10 000 are expected to take part in Butterflies for the New Millennium. It has been estimated that several thousand contributors sent in moth records during the period of the *Lepidoptera* Recording Scheme and many of these will still be active recorders (Waring 1998a). Furthermore, there has been a clear growth in the number and size of moth groups and several thousands more people have probably taken up recording since the early 1980s. This is in no small part due to the publication of an excellent moth identification guide (Skinner 1984).

There is some overlap between butterfly and moth recorders, but an overall and probably very conservative estimate of 10 000 *Lepidoptera* recorders would seem fair. It is worth noting, as mentioned before, that 8500 people consider the fate of *Lepidoptera* to be important enough to join Butterfly Conservation.

5 Legislative and policy status

Lepidoptera, and in particular butterflies, have been comparatively well represented in legislation, treaties and policy documents during the last 25 years. Some of these are summarised in Table 18.1. Nevertheless, no *Lepidoptera* are legally protected in the Republic of Ireland or the Channel Islands, no butterflies in the Isle of Man and no moths in Northern Ireland.

The *Red Data Book* (Shirt 1987) and legislation have undoubtedly influenced the conservation of threatened *Lepidoptera* in Britain (for example through guiding the notification of protected areas), but the current impetus is clearly focused around the UK Biodiversity Action Plan (BAP; English Nature 1998). The BAP categories of Priority Species and Species of Conservation Concern contain 25 butterfly and 122 moth

Table 18.1 Numbers of *Lepidoptera* included in major legislative and policy documents.

	UK BAP[1] priority species	UK BAP[1] species of conservation concern	Wildlife and Countryside Act[2] Schedule 5 (full protection)	Wildlife (Northern Ireland) Order[3]	Red Data Book[4] categories 1–3
Butterflies	11	14	6	7	7
Moths	53	69	8	n/a	97
Lepidoptera	64	83	14	7	104

[1] UK BAP refers to the UK Biodiversity Action Plan (English Nature 1998).

[2] The Wildlife and Countryside Act 1981 (as amended) affords legal protection to *Lepidoptera* in England, Scotland and Wales. In addition to the species listed as receiving full protection under the Act, a further 18 species of butterfly are afforded a lower level of protection which covers commercial trade.

[3] The Wildlife (Northern Ireland) Order 1985 affords legal protection, broadly equivalent to the full protection category in Britain, to species in Northern Ireland.

[4] Shirt (1987).

species, and the drafting and implementation of action plans for the most threatened of these is at the forefront of *Lepidoptera* conservation.

6 Changes 1973–98

Many of the changes noted in 1973 (Heath 1974), were later demonstrated unequivocally with national distribution data. For butterflies, the 1984 *Atlas* (Heath *et al.* 1984) presented trends based on the distribution change over 150 years. This information was analysed by Warren *et al.* (1997) to produce a new 'red list' of British species based on the IUCN (World Conservation Union) criteria, which include rates of decline as well as rarity. The Butterfly Monitoring Scheme dataset continues to be widely used for research and produces annual indices of relative abundance for many species, allowing trends to be measured over time (Pollard and Yates 1993a).

The data from the Butterflies for the New Millennium project will thoroughly revise our knowledge of distributional change over the period. These results will not be available until 2001, but many important patterns have already emerged and some examples will be presented.

The distribution and abundance data for moths have not been analysed to the same extent as those for butterflies. *The Moths and Butterflies of Great Britain and Ireland* volumes that have been published (Heath and Emmet 1976 onwards) include discussion of distribution change, as do reports from the scarce macro-moth recording scheme. However, the distribution changes of many moths, particularly the common species, have not been assessed. Similarly, the data gathered by the Rothamsted Insect Survey have not been analysed to their full potential and neither abundance indices nor trends are produced as a matter of routine.

In this account examples will be presented to highlight recently detected change. Unfortunately, the lower level of recording and monitoring that has taken place in Ireland makes the assessment of change there extremely difficult. Reluctantly, therefore, the examples chosen have been selected only from Britain.

6.1 Losses and declines

Table 18.2 shows eight species that have become extinct in recent decades with the approximate date of the last confirmed sighting. In the case of the Large Tortoiseshell (*Nymphalis polychloros*), extinction is assumed to have occurred during the 1980s. Although sightings of Large Tortoiseshells continue, there is no good evidence of a wild breeding colony.

Two macro-moths that were believed extinct during the past 50 years have undergone recent changes in fortune. Only one colony of Blair's Wainscot (*Sedina buettneri*) was ever known in Britain and this disappeared around 1950. Single specimens were recorded on the south-east coast of England during the following decades, but these were thought to be immigrants. Then, in 1997, breeding was confirmed in Dorset, and subsequent searches revealed further sites (Davey 1998). Whether the species has recolonised Britain from mainland Europe or simply been discovered at previously overlooked colonies is unclear. Recolonisation, however, seems to account for the first records of Small Ranunculus (*Hecatera dysodea*) for nearly 60 years, which were also noted in 1997. Although only three individuals were seen, searches in 1998 revealed larvae on wild lettuces (*Lactuca* spp.) at a number of sites and the species has now been recorded from five 10-km squares (Agassiz and Spice 1998). Large Blue (*Maculinea arion*) butterflies are also flying once again in southern England, thanks to human-assisted reintroduction (Thomas 1995).

In addition to these extinctions, a huge number of species have declined over recent decades, many very rapidly. Table 18.3 shows the 32 butterfly species (more than half of the resident fauna) that declined by more than 10% in 10-km square distribution between the periods 1940–69 and 1970–82 (Warren *et al.* 1997). The species are arranged in order of severity of decline. The list is topped, not surprisingly, by the two species that have become extinct.

When the Butterflies for the New Millennium Atlas is published, new analysis of distribution change will be available. In the absence of these results, an indication of the overall post-1982 trend has been added to Table 18.3, based on the interim findings of the current survey. Declines have continued for many species including the Chequered

Table 18.2 Some recent extinctions of *Lepidoptera*.

Species	Approximate date of extinction
Essex Emerald (*Thetidia smaragdaria*)	*c.* 1991
Large Tortoiseshell (*Nymphalis polychloros*)	1980s
Large Blue (*Maculinea arion*)	1979
The Cudweed (*Cucullia gnaphalii*)	*c.* 1979
Viper's Bugloss (*Hadena irregularis*)	*c.* 1977
Lesser Belle (*Colobochyla salicalis*)	*c.* 1977
Feathered Ear (*Pachetra sagittigera*)	*c.* 1963
Spotted Sulphur (*Emmelia trabealis*)	*c.* 1960

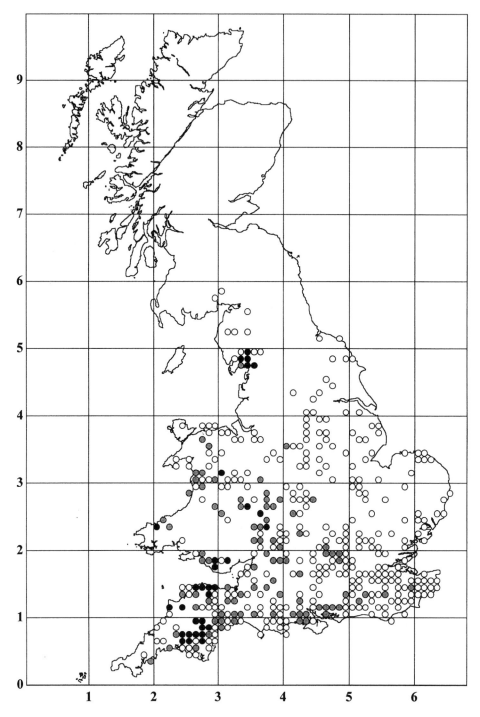

Figure 18.1 Map showing the recent decline of the High Brown Fritillary (*Argynnis adippe*).
Key: O pre-1970 (Biological Records Centre); ● 1970–82 (Biological Records Centre) and ● 1990–97 (Butterfly Conservation)

Skipper (*Carterocephalus palaemon*), Grizzled Skipper (*Pyrgus malvae*), Wood White (*Leptidea sinapis*), Brown Hairstreak (*Thecla betulae*), High Brown Fritillary (*Argynnis adippe*), Pearl-bordered Fritillary (*Boloria euphrosyne*), Small Pearl-bordered Fritillary (*Boloria selene*), Marsh Fritillary (*Eurodryas aurinia*) and Large Heath (*Coenonympha tullia*). A number of other species have continued to decline in some areas while showing stable or expanding distribution in other parts of their range. Only three of the 32 species in Table 18.3 have shown a significant re-expansion of distribution in recent years: the Brown Argus (*Aricia agestis*), White Admiral (*Ladoga camilla*) and Marbled White (*Melanargia galathea*).

The decline of the High Brown Fritillary (Fig. 18.1) has been one of the most sudden and severe of all British butterflies. It was once widespread across England and Wales in woodland clearings and bracken (*Pteridium aquilinum*)-dominated habitats where violets (*Viola* spp.), the larval foodplants, grew in abundance. Since the 1950s, the High Brown Fritillary has declined sharply, particularly in the east, and it is now reduced to about 50 definite colonies, with strongholds on Exmoor, Dartmoor and Morecambe Bay (Barnett and Warren 1995a). Over the period for which records are available this represents a decline (measured by 10-km square) of over 90%.

The speed of many declines is of particular concern and highlights the need for continuous survey and monitoring. Several high-profile species have declined to local extinction in recent years despite the best efforts of site managers and volunteers. The Chequered Skipper colonies in England were well known in the 1960s and many enjoyed high levels of site protection. Nevertheless, in under a decade the species was declared extinct in England, in 1975. Recent surveys of the Pearl-bordered Fritillary, organised by Butterfly Conservation, have suggested that the species has suffered a contraction of range of over 50% (of 10-km squares) in the past 15 years across Britain as a whole, with much greater losses in England and Wales (Brereton and Warren 1999). Again, many losses (particularly in southern England) have occurred at sites that have a high level of conservation effort.

In addition to the species in Table 18.3, the Wall Brown (*Lasiommata megera*) has recently shown signs of a significant decline. This butterfly's range includes much of England, Ireland, Wales and the Isle of Man and coastal areas of southern Scotland. The 1984 *Atlas* (Heath *et al.* 1984) showed little change in distribution since 1940, although the authors noted that many breeding areas had been lost to intensive agriculture. Collation of recent records is starting to give a clear picture of a major decline centred on the inland (and well recorded) areas of south-central and south-east England and the west Midlands. In 1997 the Butterfly Monitoring Scheme recorded the lowest population index of the Wall Brown since monitoring began (Greatorex-Davies and Pollard 1997). At individual sites within the scheme the decline has been most pronounced in southern England. These changes may be part of a short-term fluctuation, for which the species is renowned, or may signal the permanent and sudden collapse of a 'common' butterfly.

Rates of decline determined from a comparison of distribution maps from different periods may contain significant bias connected with the scale of measurement. In order to assess distribution at the national level, large recording units (for example 10-km squares) have traditionally been used. If declines are only detected when species are lost from an entire grid square, large-scale reductions in the number of colonies within a square will remain unnoticed. There is evidence to suggest that the rates of decline of a

Table 18.3 Declines of some butterfly species between 1940–69 and 1970–82, and recent trends.

Species	% decline in range 1940–69 to 1970–82[1]	Recent trend
Large Tortoiseshell	−81%	Probably extinct
Large Blue	−80%	Extinct, but reintroduced
Silver-spotted Skipper	−64%	Some recovery
High Brown Fritillary	−63%	Continued rapid decline
Heath Fritillary	−48%	Stable
Silver-studded Blue	−45%	Some local declines
Marsh Fritillary	−44%	Continued decline
Adonis Blue	−42%	Some recovery
Chequered Skipper	−38%	Continued decline
Pearl-bordered Fritillary	−38%	Continued rapid decline
Purple Emperor	−38%	Stable
White Admiral	−34%	Expansion
Brown Hairstreak	−31%	Continued decline
Grizzled Skipper	−31%	Continued decline
Silver-washed Fritillary	−30%	Some local declines and expansions
Wood White	−30%	Continued decline
Dark Green Fritillary	−29%	Some local declines and expansions
Brown Argus	−27%	Rapid expansion
Mountain Ringlet	−27%	Stable
Dingy Skipper	−26%	Some local declines
Duke of Burgundy	−26%	Some local declines
Grayling	−26%	Some local declines
Black Hairstreak	−25%	Stable
Green Hairstreak	−25%	Some local declines and expansions
Small Blue	−25%	Some local declines
Large Heath	−22%	Continued decline
White-letter Hairstreak	−22%	Some local declines and expansions
Small Pearl-bordered Fritillary	−21%	Continued rapid decline
Marbled White	−20%	Expansion
Lulworth Skipper	−19%	Some local expansions
Chalkhill Blue	−17%	Some local declines
Northern Brown Argus	−12%	Some local declines

[1] Source Warren *et al.* (1997).

number of butterflies of conservation concern have been significantly under-estimated because of this scale effect (Thomas and Abery 1995). The problems will be even greater for widespread species. The collection of survey data on smaller-scale grids and the inclusion of abundance measures will ensure that this bias is limited when the Butterflies for the New Millennium dataset is analysed.

Abundance measures generated by the Butterfly Monitoring Scheme may also not reflect widespread declines because of bias; the number of sites monitored is relatively low and almost all are nature reserves. This has several implications for the application of the results. Many of the rarer species do not occur on sufficient monitored sites to enable the calculation of annual indices. In addition, the sites used are not representative of the habitat conditions experienced by many of the more common species in the wider countryside. It is possible that populations might remain healthy on a monitored nature reserve, while undergoing severe decline as a result of habitat loss or deterioration in the surrounding area (and vice versa).

Many moths have also declined during recent decades (see Table 18.4 for examples), but assessment is more difficult because of lower recording coverage. Apparent declines and recoveries may result simply from changes in recording effort. The collation of distribution data across Great Britain and Ireland is limited to scarce moths and this accounts for the high number of UK BAP species that appear in the table. Nevertheless, evidence from regularly operated light-traps, particularly those linked to the Rothamsted Insect Survey, has suggested that a number of formerly common and widespread species may be in serious decline. Examples include the Garden Tiger (*Arctia caja*) and Stout Dart (*Spaelotis ravida*) (I. Woiwod pers. comm.). However, no trends are available to identify species that might be suffering declines in abundance and range.

Figs 18.2 and 18.3 illustrate the declining distributions of two UK BAP moths, the Brighton Wainscot (*Oria musculosa*) that has always had a restricted recorded range,

Table 18.4 Some examples of declining moths.

Species	UK BAP status[1]
Stout Dart (*Spaelotis ravida*)	
Garden Tiger (*Arctia caja*)	
Broom-tip (*Chesias rufata*)	Species of conservation concern
Four-spotted (*Tyta luctuosa*)	Priority species
Narrow-bordered Bee Hawk-moth (*Hemaris tityus*)	Priority species
Drab Looper (*Minoa murinata*)	Priority species
Waved Carpet (*Hydrelia sylvata*)	Priority species
Sword Grass (*Xylena exsoleta*)	Priority species
White-line Snout (*Schrankia taenialis*)	Priority species
Clay Fan-foot (*Paracolax tristalis*)	Priority species
Brighton Wainscot (*Oria musculosa*)	Priority species
Common Fan-foot (*Pechipogo strigilata*)	Priority species
Orange Upperwing (*Jodia croceago*)	Priority species
Heart (*Dicycla oo*)	Priority species
Dark Crimson Underwing (*Catocala sponsa*)	Priority species
Light Crimson Underwing (*Catocala promissa*)	Priority species

[1] UK BAP refers to the UK Biodiversity Action Plan (English Nature 1998).

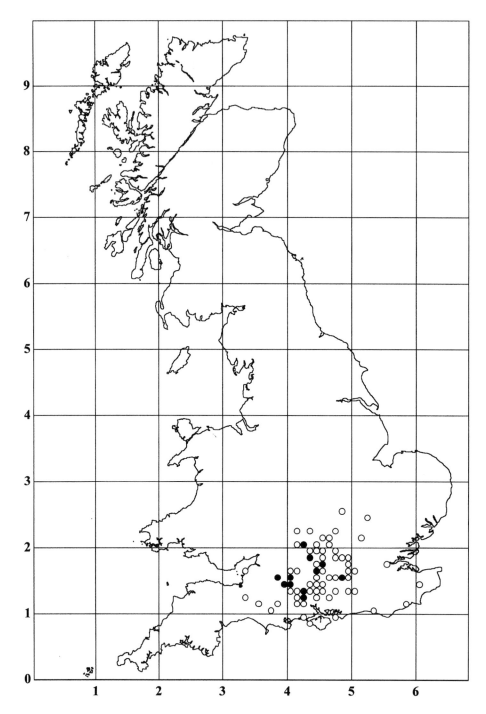

Figure 18.2 Map showing the recent decline of the Brighton Wainscot (*Oria musculosa*).
Key: O pre-1980; ● 1980–97.

National recording network for the rarer British macro-moths. Data courtesy of P. Waring and Joint Nature
Conservancy Committee.

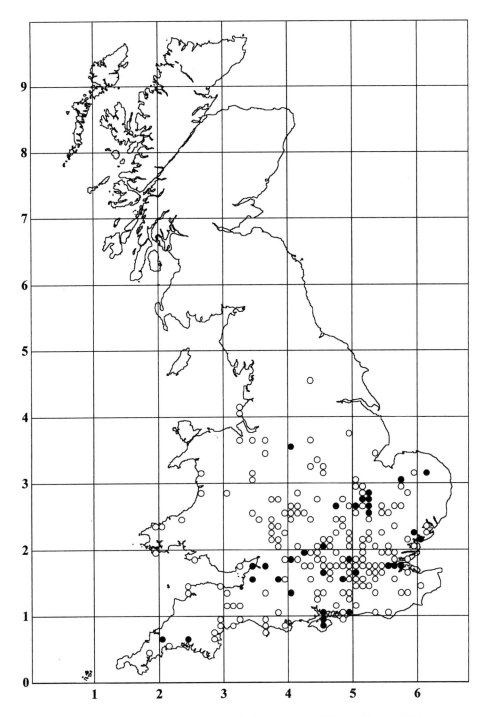

Figure 18.3 Map showing the recent decline of the White-spotted Pinion (*Cosmia diffinis*).
Key: O pre-1980; ● 1980–96.

National recording network for the rarer British macro-moths. Data courtesy of P. Waring and Joint Nature
Conservancy Committee.

and the White-spotted Pinion (*Cosmia diffinis*), a formerly widespread species. Many other examples could have been chosen. The White-spotted Pinion is an elm-feeding species (*Ulmus* spp.) and its decline following Dutch Elm Disease has been particularly rapid. Prior to the mid-1970s it was widespread in southern Britain, but there are few records since 1990. There are some recent signs of a recovery in Cambridgeshire, one of its few remaining strongholds (Waring 1996).

The species that have suffered severe declines are, almost without exception, dependent upon specific habitats that have themselves declined and become fragmented. Species of wetlands and heaths were among the first to suffer (Heath 1974) and Lepidoptera dependent on woodland clearings and chalk grasslands have followed quickly behind. It is increasingly accepted that even common and widespread species must also have declined. The rationale for this is simply the enormous loss of semi-natural habitat in the landscape. Species such as the Common Blue (*Polyommatus icarus*) and Small Copper (*Lycaena phlaeas*) will have suffered from the widespread reseeding and fertilisation of unimproved grasslands where their foodplants formerly grew. Even grass-feeding butterflies such as the Meadow Brown (*Maniola jurtina*) and Small Heath (*Coenonympha pamphilus*) may struggle to find suitable habitat on modern, agricultural pasture. These species are now often restricted to marginal areas where the grass is left to grow long (e.g. verges, wasteground and along hedgerows). Sufficient data are not yet available, either from distribution or transect recording, to permit analysis across Britain and Ireland that might support these unsubstantiated but widely held views.

6.2 Additions and increases

In contrast to the large number of declines, a smaller but still substantial group of species have fared reasonably well since 1973. Table 18.5 illustrates examples of species that are thought to have shown real expansion of range. Most are species of the wider countryside, adapted to habitats and foodplants that have suffered least (or even benefited) from the massive environmental changes wrought by human activity. However, the context of these expansions varies greatly from species to species.

In many cases, species are simply recolonising their historical range after periods of contraction. Climatic factors are thought to be responsible for most of these changes. The re-expansions of some butterflies began many decades ago and were noted in the 1984 *Atlas* (Heath *et al.* 1984). Examples included the Orange Tip (*Anthocharis cardamines*), White Admiral, Comma (*Polygonia c-album*) and Speckled Wood (*Pararge aegeria*). All of these species continue to expand and now occupy territory approximately equal to their historical ranges. The Comma has been expanding for much of the twentieth century, but is currently undergoing a particularly rapid phase of recolonisation (Fig. 18.4). Its range, as measured by 10-km square has expanded by over 60% in the past 16 years. Vagrant individuals have recently been recorded in Scotland, raising hopes of recolonisation after an absence of over a century.

Climatic factors are believed to have played a major part in the Comma's decline and recovery (Pratt 1987). It is, perhaps, not surprising that the Butterfly Monitoring Scheme figures for the Comma show a significant increase in abundance (Fig. 18.5). Individual sites with significant upward trends are all in the east and north of the butterfly's range, where most distribution change has also occurred. In addition to these

Table 18.5 Examples of *Lepidoptera* which are expanding or recovering their range.

Species	Area of main expansion/recolonisation
Small Skipper (*Thymelicus sylvestris*)	Northern England
Essex Skipper (*Thymelicus lineola*)	East Anglia and Midlands
Large Skipper (*Ochiodes venata*)	North-east England
Brimstone (*Gonepteryx rhamni*)	West Midlands
Orange Tip (*Anthocaris cardamines*)	Central Scotland
Purple Hairstreak (*Quercusia quercus*)	Midlands
Brown Argus (*Aricia agestis*)	East Anglia and Midlands
Holly Blue (*Celastrina argiolus*)	Midlands and Yorkshire
White Admiral (*Ladoga camilla*)	East Anglia, Midlands and Home Counties
Comma (*Polygonia c-album*)	Northern England, Midlands and East Anglia
Silver-washed Fritillary (*Argynnis paphia*)	South-east England
Speckled Wood (*Parage aegeria*)	Northern England, Midlands and East Anglia
Marbled White (*Melanargia galathea*)	South-east England and Yorkshire
Gatekeeper (*Pyronia tithonus*)	Midlands and Northern England
Ringlet (*Aphantopus hyperantus*)	Midlands and Northern England
Small Eggar (*Eriogaster lanestris*)	South-west England
Juniper Carpet (*Thera juniperata*)	South-east and Northern England and Midlands
Maple Prominent (*Ptilodonella cucullina*)	Eastern England
Varied Coronet (*Hadena compta*)	South-central and Northern England
Blair's Shoulder-knot (*Lithophane leautieri*)	Midlands and Northern England

increases, a number of Butterfly Monitoring Scheme sites have been colonised during the monitoring period. These are also mostly in eastern and northern England.

Other species have had much more recent upturns in fortune. The Marbled White has recolonised much of its historical range in recent years and the Silver-washed Fritillary (*Argynnis paphia*) and Small Eggar (*Eriogaster lanestris*) have shown minor but very welcome improvements in distribution. In the case of the Small Eggar (Fig. 18.6) some of the 'new' 10-km squares may result from improved recording, but others are thought to represent genuine expansion (P. Waring pers. comm.).

Several further species have shown dramatic expansions of range into areas where they were not formerly recorded. A few are relatively recent colonists of Great Britain and Ireland and their subsequent expansion is unsurprising. Blair's Shoulder-knot (*Lithophane leautieri*) is a well-documented example (Agassiz 1996; Waring 1998b). First recorded in 1951, Heath (1974) gave its distribution as 'from Sussex to Devon and Somerset and north to Buckinghamshire'. The moth, whose larvae feed on planted

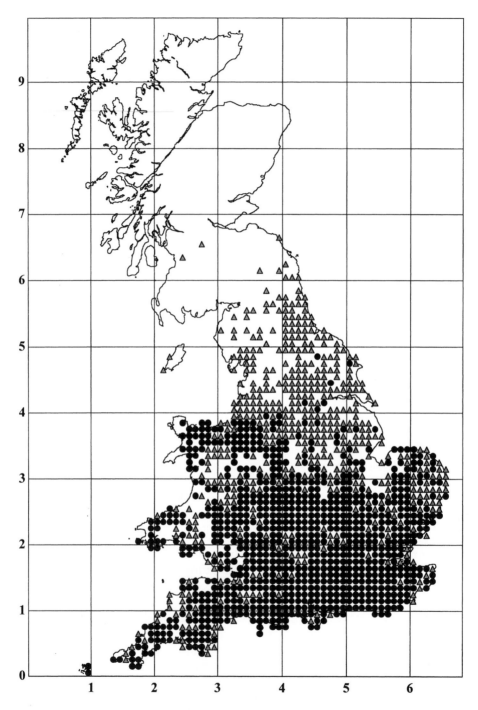

Figure 18.4 Map showing the recent expansion of the Comma (*Polygonia c-album*).
Key: ● 10 km squares recorded 1970–82 (Biological Records Centre); ▲ additional
10-km squares recorded 1995–97 (Butterflies for the New Millennium).

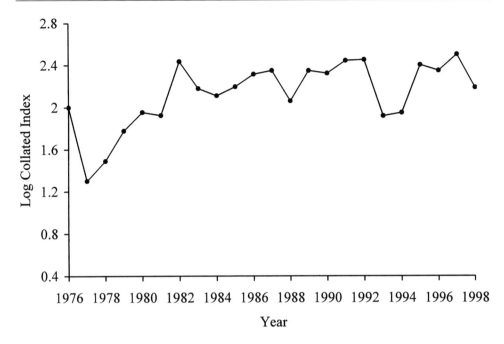

Figure 18.5 Butterfly Monitoring Scheme log collated index of the Comma (*Polygonia c-album*).

Data reproduced courtesy of the Centre for Ecology and Hydrology.

cypress trees (*Cupressus* spp.), has gradually colonised much of southern England. By 1998 it had been recorded from as far afield as Gwynedd, Cumbria and the Isle of Man.

More intriguing are the expansions of resident species that had previously limited distributions. These expansions can sometimes be related to a change in the ecology of the species concerned. For example, the range of the Juniper Carpet (*Thera juniperata*) was formerly limited by the distribution of its larval foodplant, Juniper (*Juniperus communis*). The moth was considered to be nationally scarce, but has become much more widespread by breeding on cultivated junipers in gardens (Waring 1992a).

The Brown Argus has achieved something similar. In Britain, the butterfly has traditionally been associated with habitats supporting the larval foodplants Common Rock-rose (*Helianthemum nummularium*), Common Stork's-bill (*Erodium cicutarium*) and, occasionally, Dove's-foot Crane's-bill (*Geranium molle*). In south-east England, for example, the Brown Argus was largely restricted to chalk downlands. In recent years, many new colonies have been discovered, particularly in eastern England and the Midlands, at sites where neither of the main 'traditional foodplants' grow. Although the species is known to use other foodplants in continental Europe, it seems not to have done so previously in Britain. Breeding has now been observed on Cut-leaved Crane's-bill (*G. dissectum*), Small-flowered Crane's-bill (*G. pusillum*) and Meadow Crane's-bill (*G. pratense*). It is likely that other factors have influenced this switch to previously under-utilised foodplants (Greatorex-Davies and Pollard 1997).

The influence of parasites on abundance and, perhaps, distribution fluctuations has been proposed for at least one of the butterflies in Table 18.5. Populations of the Holly Blue (*Celastrina argiolus*) can suffer extremely high levels of mortality due to a parasitic

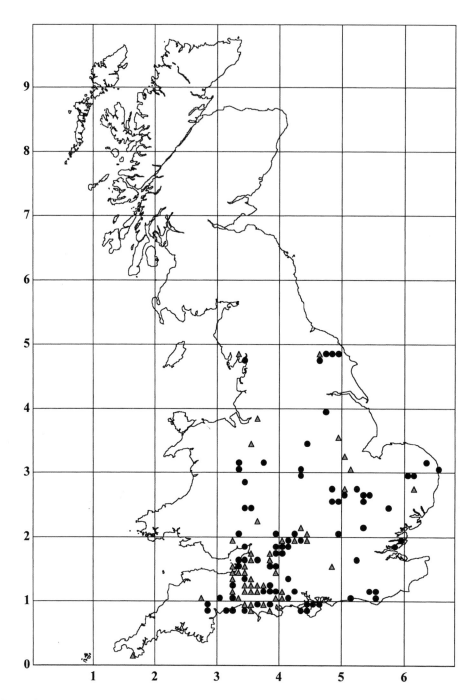

Figure 18.6 Map showing the recent expansion of the Small Eggar (*Eriogaster lanestris*).
Key: ● 1960 onwards records given in Emmet and Heath (1992); ▲ additional 10-km squares recorded 1980–96.

National recording network for the rarer British macro-moths. Data courtesy of P. Waring and Joint Nature Conservancy Committee.

wasp (*Listrodromus nycthemerus*). Once the butterfly is reduced to low levels of abundance, the parasite populations also collapse as many wasps fail to locate Holly Blue larvae in which to lay their eggs. In the subsequent period of lower parasite pressure, the reproductive success of the butterfly increases and populations build up rapidly. The populations of the two species may therefore track one another, forming a regular cycle of abundance. Transect data give some indication of such a cycle, but more work is required to confirm its existence and causes. The same data also suggest that weather factors influence the fluctuations and this may lessen the importance attached to parasites (Pollard and Yates 1993b). While the effect of the parasite–host cycle on the distribution of the Holly Blue has not been confirmed, it should be considered when analysing survey data. Many butterflies and moths suffer high levels of pre-adult mortality due to parasitic *Diptera* and *Hymenoptera*, and such effects may be much more widespread than has been recognised so far.

Altogether more mysterious are the expansions of residents, such as the Dotted Rustic (*Rhyacia simulans*) and the Essex Skipper (*Thymelicus lineola*). These have undergone large expansions since the 1970s, although the Dotted Rustic is now retracting again (Waring 1992b). The expansion of the Dotted Rustic is illustrated in Fig. 18.7. Despite the fact that the larval foodplant of the Dotted Rustic is unknown, it seems unlikely that either this species, or the Essex Skipper were previously limited by the availability of suitable habitat. It appears more likely that unknown climatic factors are influencing these species. Unfortunately, the Butterfly Monitoring Scheme does not produce an index for the Essex Skipper (because of identification problems) and no data are available for analysis against climatic variables.

The use of distribution data to determine trends relies on adequate recording coverage in the periods being compared. This is the case for most butterflies, but not for the majority of moth species in Great Britain and Ireland. Many apparent expansions in recent years reflect little more than increases in recording effort. A substantial number of species that were formerly listed as nationally scarce (i.e. occurring in up to a hundred 10-km squares) have been recorded at so many new locations that they are no longer regarded as such. Examples include the Obscure Wainscot (*Mythimna obsoleta*), Shaded Pug (*Eupithecia subumbrata*), Oak-tree Pug (*E. dodoneata*), Maple Pug (*E. inturbata*), Ochreous Pug (*E. indigata*), Sloe Pug (*Rhinoprora chloerata*), Lead-coloured Drab (*Orthosia populeti*) and Brown Scallop (*Philereme vetulata*). Other species that are still nationally scarce have nevertheless 'benefited' from the discovery of new colonies. The apparent expansions of the UK BAP species the Northern Dart (*Xestia alpicola alpina*), Rannoch Sprawler (*Brachionycha nubeculosa*), Dark Bordered Beauty (*Epione parallelaria*), Reed Leopard (*Phragmataecia castaneae*) and Silky Wave (*Idaea dilutaria*) are examples.

Table 18.6 lists some recent additions to the *Lepidoptera* of Great Britain and Ireland. Most of these moths are thought to be colonists, although two, The Silurian (*Eriopygodes imbecilla*) and possibly the Southern Chestnut (*Agrochola haematidea*), may be previously overlooked resident species. Two further species, Blair's Wainscot and Small Ranunculus, which were formerly thought extinct, have already been discussed in section 6.1. Agassiz (1996) presents a comprehensive discussion of colonisation by *Lepidoptera*.

The expansions of most resident species appear to be caused by climatic factors or, in the case of many moths, previous under-recording. A few species have been able to use novel or previously under-utilised foodplants that are more widely distributed than

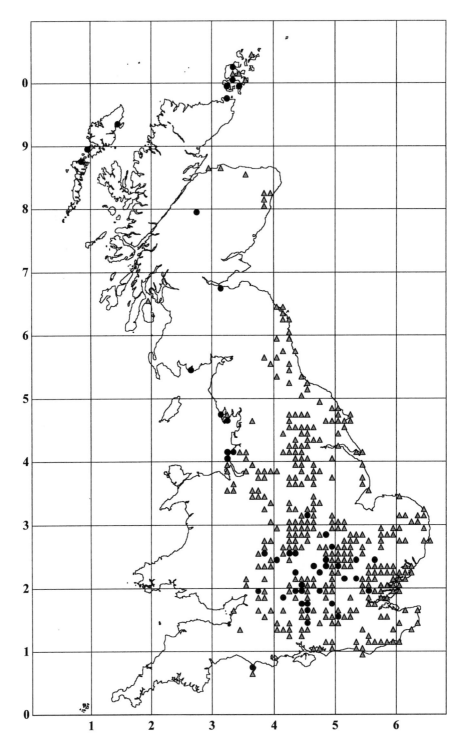

Figure 18.7 Map showing the recent expansion of the Dotted Rustic (*Rhyacia simulans*).
Key: ● 1960–78; ▲ additional 10-km squares 1980–91.

National recording network for the rarer British macro-moths. Data courtesy of P. Waring and Joint Nature
Conservancy Committee.

Table 18.6 Some recent 'additions' to the *Lepidoptera* fauna of Great Britain and Ireland.

Species	Year
Vitula biviella	1997
Cocylis molliculana	1993
Southern Chestnut (*Agrochola haematidea*)	1990
Tamarisk Pug (*Eupithecia ultimaria*)	1989
Cypress Carpet (*Thera cupressata*)	1984
Bloxworth Snout (*Hypena obsitalis*)	1982
Feathered Beauty (*Peribatodes secundaria*)	1981
Dioryctria schuezeella	1980
The Silurian (*Eriopygodes imbecilla*)	1972

their 'normal' species. However, climatic factors may also be influencing these changes. Most other rapidly expanding species are recent colonists. Habitat change (excluding the planting of exotics) appears to have had little or no positive effect on the vast majority of the *Lepidoptera*.

6.3 Aliens and introductions

The introduction or reintroduction of species, particularly butterflies, to sites is a regular occurrence. Sometimes this forms part of an agreed conservation strategy, but more often (re)introductions are carried out by individuals acting without the permission of the site owner or conservation bodies. It is clear that the vast majority of introductions in this category fail within a short space of time (Oates and Warren 1990).

The introduction of new or previously extinct species to Great Britain and Ireland is much less common and there are few documented examples from recent decades. The Geranium Bronze (*Cacyreus marshalli*) is a South African butterfly that has become an established pest of garden plants in southern Europe. In 1997 adult butterflies were observed in a Sussex garden, these being the first records of wild adults in Britain (Holloway 1998). Breeding behaviour was observed and there was one further record in 1998. It seems likely that this species arrived as an accidental introduction with imported plants in the horticultural trade.

The discovery of Gypsy Moth (*Lymantria dispar*) larvae in several Essex gardens featured prominently in the media in 1995. The species became extinct in Britain early in the twentieth century, although male individuals have been recorded along the south coast during recent decades, and media interest arose because the moth is a major forest pest in many parts of Europe and North America. The Forestry Authority implemented a control regime and the current status (if any) of the population is unknown. What seems more certain is that the population resulted from either a deliberate release or escape of stock that had been bred in captivity (Waring 1995).

A recently introduced species whose persistence seems more assured is the Cypress Tip (*Argyresthia cupressella*). This North American leaf-mining moth was first noticed in Suffolk in 1997. The species was subsequently found to be quite widespread in the county and was reported from Middlesex in 1998. As this species is new to Europe, it is very likely that introduction to Britain occurred directly from North America, presumably with imported foodplants (cypress, cedar and juniper species). The species is already too widespread for eradication to be attempted and further rapid spread seems likely (Agassiz and Tuck 1999).

7 Factors affecting change

The organised collection and analysis of unprecedented amounts of data in the past 25 years have reinforced what natural historians have known for years: the populations and distributions of many *Lepidoptera* are in a state of dramatic flux. Some of the major causes have also been recognised for many decades, for instance climate change and the destruction of habitat. However, recent years have witnessed many advances in our understanding of the factors affecting change. While most have been elucidated in relation to butterflies, they will undoubtedly be of importance to other *Lepidoptera*.

7.1 Habitat loss

Three main factors are now believed to have contributed to the many declines. The first and most obvious is the destruction of habitat. The intensification of agriculture and development of large-scale commercial forestry using exotic trees has led to the wholesale loss of semi-natural habitats and their associated faunas. The dramatic extent of these habitat losses has been well publicised (Nature Conservancy Council 1984). Fens, lowland heaths, flower-rich meadows, calcareous grasslands, peat bogs and ancient woodlands, among others, have been reduced to isolated remnants in many areas. Heath (1974) and others have given summaries of the effect of habitat destruction on *Lepidoptera*.

7.2 Management changes

Increasing importance has recently been attached to the loss of habitat resulting from changes in management, in particular, the abandonment of traditional techniques such as the coppicing of woodland and changes in grazing regimes (Warren 1992). Although not as immediate as habitat destruction, the repercussions of such changes are often severe. Detailed studies of butterfly and moth ecology have highlighted the importance of specific microclimate requirements. Changes in management can easily disrupt microclimatic conditions and this has helped to explain the loss of species from apparently suitable sites.

   The severe declines of a number of species can be attributed to the cessation of regular coppicing in woodlands. The species are adapted to early successional stages and colonies die out when the canopy closes and shade intensifies. In the past, rotational coppicing created a continuous supply of open glades that could be colonised by the *Lepidoptera*. The declines of the High Brown, Pearl-bordered, Small Pearl-bordered, Silver-washed and Heath (*Mellicta athalia*) Fritillaries, Wood White and Duke of Burgundy (*Hamearis lucina*) have all resulted, in part at least, from the abandonment of coppice woodland (Warren 1992). Many moths are also thought to have been detrimentally affected, including the UK BAP species Orange Upperwing (*Jodia croceago*),

Common Fan-foot (*Pechipogo strigilata*), Clay Fan-foot (*Paracolax tristalis*) and Drab Looper (*Minoa murinata*) (English Nature 1999).

In the west of Britain, several of the same species are more commonly associated with bracken-dominated habitats. Here too there have been declines, however, most notably by the High Brown and Pearl-bordered Fritillaries. Some of this is undoubtedly due to the destruction of habitat, but changes in grazing regimes are also significant. Too much livestock will destroy the bracken leaf-litter that provides the very warm microclimate necessary for larval development. If there are too few grazing animals, however, the leaf-litter will accumulate and smother the violets upon which the butterflies' caterpillars feed (Warren and Oates 1995).

Reduced grazing pressure on the chalk of southern Britain has contributed to the declines of a number of species. Fewer domestic stock and the effects of myxomatosis on rabbit populations have led to increased sward height and gradual invasion by scrub. Many of the *Lepidoptera* of chalk grassland are adapted to the hot microclimates that occur in short turf and these species have been lost as grazing intensity has declined. Examples include the Silver-spotted Skipper (*Hesperia comma*), Adonis Blue (*Lysandra bellargus*) and Silver-studded Blue (*Plebejus argus*) (Warren 1992). The Straw Belle (*Aspitates gilvaria*) and Black-veined Moth (*Siona lineata*) have also declined due to the succession (towards scrub) of their grassland habitats following changes in grazing pressure (English Nature 1999). All of these *Lepidoptera* are Priority Species in the UK BAP.

In the wet grasslands of western Britain, species such as the Marsh Fritillary, Narrow-bordered Bee Hawk-moth (*Hemaris tityus*) and Double-line (*Mythimna turca*) have all suffered as a result of loss of suitable habitat. Both habitat destruction (draining, ploughing, fertilising or reseeding) and reduced grazing (and over-grazing) have contributed to these declines (Barnett and Warren 1995b; English Nature 1999).

The succession of other habitats in response to reduced grazing has also had severe effects on associated *Lepidoptera*. For example, under-grazing caused the extinction of the Large Blue in thyme-rich (*Thymus praecox*) grasslands. Increasing sward height led to reduced ground temperatures and the decline of the ant species upon which the butterfly depends for its development (Thomas 1995). The encroachment of scrub on to heathland almost led to the extinction of one of our rarest moths, the Reddish Buff (*Acosmetia caliginosa*). This species is now recovering as a result of scrub clearance, captive breeding and reintroduction under the auspices of an English Nature Species Recovery Programme (Waring 1998c).

At the other end of the spectrum, over-grazing can also be a problem. Upland areas, in particular, have witnessed increases in grazing pressure as a result of the Common Agricultural Policy. Many species of butterfly and moth must have become much less common as a result. The Marsh Fritillary, for example, has declined or been lost from some sites because of increased grazing by sheep or a switch from cattle to sheep (Barnett and Warren 1995b).

These changes in traditional agricultural and woodland management have caused numerous declines. However, a few species have benefited, at least in the short term. The White Admiral, for example, is a truly woodland butterfly and has probably benefited from the cessation of coppicing (Pollard 1979). The same may also be true for the Speckled Wood. The increase in sward length on neglected grasslands has suited the Lulworth Skipper (*Thymelicus acteon*) and probably other more common species

(Thomas 1990). The expansions of two butterflies, the Marbled White and Essex Skipper, may be linked to the recent reduction in management of roadside and trackside verges and the subsequent increase in sward height.

7.3 Fragmentation and isolation

Habitat destruction and changes in management have created a fragmented landscape. Where habitat losses have been severe, for example in much of southern and eastern England, the remaining patches tend to be small and isolated. Recent studies of *Lepidoptera* populations distributed in this way suggest that there is a steady loss of species from small, isolated sites. Populations of butterflies occupying small sites are themselves likely to be small and thus more likely to become extinct as a result of chance events (such as fires, disease or unfavourable weather) and demographic effects (for instance difficulty in finding mates at low-population densities). Genetic effects such as inbreeding, which occur in small populations, may also play a significant role in local extinction as has been shown in the Glanville Fritillary (*Melitaea cinxia*) in Finland (Saccheri *et al.* 1998).

   Lepidoptera occupying fragmented habitats are, therefore, more susceptible to local extinction. If the available habitat patches are also isolated, the chance of colonisation is greatly reduced. These combined effects are believed to be responsible for the rapid loss of species from well-protected habitat patches. Many butterfly species in Great Britain and Ireland are now restricted to small, isolated islands of suitable habitat in an inhospitable countryside and the long-term consequences may be severe.

   A further dimension to this field of research is provided by the concept of metapopulations (Gilpin and Hanski 1991). This theory proposes that the continued persistence of species occupying a number of small habitat patches is dependent on occasional migration between the populations. Such migration allows the recolonisation of patches in which populations periodically become extinct, or may prevent (or delay) local extinction by boosting numbers and genetic diversity. At any one time the species may only occupy a proportion of the available habitat in an area, but is able to move to different patches as they become suitable. In effect, local colonies are linked and dependent upon each other for the long-term survival of the species.

   Many of the species that have declined are restricted to patchy or fragmented habitats and field studies have shown evidence of metapopulation structures for some butterflies and at least one moth, *Wheeleria spilodactylus* (Menendez and Thomas 1999). Examples include the Marsh Fritillary (Warren 1994), Silver-studded Blue (Lewis *et al.* 1997), Dingy Skipper (*Erynnis tages*) (Gutierrez *et al.* 1999) and Brown Argus (Wilson and Thomas 1999). The increasing evidence of metapopulations in many threatened *Lepidoptera* has implications for conservation. In particular, the survival of species may require the protection and suitable management of groups of habitat patches, even if some appear suboptimal or only hold temporary populations.

7.4 Climate

The relationships between *Lepidoptera* and climate are complex and remain unclear despite much research using the Butterfly Monitoring Scheme dataset (Pollard and

Yates 1993a). Most work has focused on the effects of short-term weather trends on the annual population indices of butterfly species.

The Butterfly Monitoring Scheme data show that many species benefit from warm, dry weather immediately preceding and during the flight period. Such conditions not only permit breeding behaviour, but also accelerate the development of pre-adult stages, reducing levels of predation and parasitism and raising the potential for an increased number of generations. Extra generations often lead to higher population sizes and may facilitate range extension. High temperatures may also make extra habitat available to species that are close to their climatic limits in Britain. Not all butterflies respond well to warm, dry summer weather, however; the Speckled Wood, for example, shows the opposite relationship (Pollard and Yates 1993a).

This type of analysis has provided many interesting results, but interpretation has remained difficult because of the complexity of the factors involved and the relatively short monitoring period. It has not yet been possible to separate the linked factors of temperature and precipitation, although it is certain that drought can have a drastic effect on populations, presumably through the dessication of larval foodplants.

It is clear, though, that weather conditions play a significant role in determining the population levels of many butterflies in Great Britain and Ireland and this is almost certain to have effects on distribution, at least at the local scale. Small, isolated populations may be driven to extinction by unfavourable weather conditions. However, most widespread declines during recent decades cannot be attributed to the changing climate: the cause has been habitat loss and change.

Climate may be more influential in the expansions of species that are already widespread. These appear not to have been previously limited by habitat or larval foodplant availability and climate has generally been invoked as the 'cause' of many recent expansions. The evidence, however, is circumstantial, relying on the comparison of periods of large-scale distribution change with the concurrent climatic characteristics.

The contraction and re-expansion of many species in Great Britain and Ireland and continental Europe appear to fit the general pattern of climatic change over the past two centuries. Species such as the Speckled Wood and Comma suffered major contractions of range during the nineteenth century, but re-expanded in the climatic amelioration that characterised the first half of this century (Heath *et al.* 1984). These expansions were temporarily halted during the 1950s and 1960s, when the climate again deteriorated. Several moths that had colonised southern England during the preceding period of favourable weather, now became extinct (e.g. Clifden Nonpareil, *Catocala fraxini*, and Lunar Double-stripe, *Minucia lunaris*) (Burton 1998).

More recently this deterioration has been reversed by anthropogenically induced climate change and many species appear to have responded by range expansion. The case for climate being the main causal factor behind many expansions (and perhaps some contractions) is strengthened by the simultaneous responses of many species across large areas of Europe (Burton 1998; Parmesan *et al.* 1999; Swaay and Warren 1999). It remains unclear whether species are benefiting directly, or via changes in habitat.

The future effects of climatic warming are the subject of great speculation. Many butterflies appear to reach their climatic limit in Great Britain and Ireland and summer temperature is believed to be the key restricting factor (Heath *et al.* 1984). Some may benefit from increased temperatures, and be able to expand their distributions, provided that suitable habitat remains. Other species may colonise from Europe. It is unclear whether the recent colonisations of moths represent the beginning of a response to climatic warming or simply a natural pattern of immigration, temporary occupation and extinction. No new butterfly species have yet become established in Great Britain and Ireland, although there was recent evidence suggesting a resident population of the Queen of Spain Fritillary (*Argynnis lathonia*) in Suffolk (Wilson 1998).

The significant climate change that is predicted will undoubtedly affect habitats. It is assumed that as the climate warms, habitats will shift northwards and to higher elevations. This immediately suggests that species dependent on habitat restricted to high altitude and northern latitude may be most at risk. A number of *Lepidoptera* fall into this category, for example the Mountain Ringlet (*Erebia epiphron*) and Northern Dart. The effects of global warming on species and habitats will probably be complex, though, and accurate predictions are not yet available.

Concern is not limited to species whose habitats are threatened by global warming. Many of the scarce *Lepidoptera* are restricted to habitat patches, separated by intensively managed landscape. If climate change makes their current habitat patches unsuitable, they may be unable to find or reach other, more favourable, sites.

7.5 Other causes of change

Collecting, atmospheric pollution and the use of insecticides have also been proposed as causes of decline (Warren 1992). In relation to the factors already outlined, these causes are insignificant in the vast majority of cases. Collecting may still pose a threat to a few very rare moths, such as the Fiery Clearwing (*Pyropteron chrysidiformis*) (English Nature 1999). Expansion in range of larval foodplants is another possible cause of change, but there is little evidence that this has been a significant factor, except in the case of exotic plants.

8 Prognosis for the next 25 years

Without widespread action to reverse the habitat change of recent decades the prognosis for a large proportion of our *Lepidoptera* is not good. The rapid distribution declines already recorded for some species may only mark the beginning of a much more widespread phenomenon. Many species are increasingly restricted to small, isolated habitat patches. Ecological theory predicts that such populations will tend to become extinct over time, particularly if neighbouring patches (which may form part of a metapopulation structure) are destroyed or become unsuitable. It is entirely possible that many species have already been reduced below this threshold for long-term survival, and that we are now sitting on a 'biodiversity time bomb'.

The potential effects of global warming will simply accentuate these problems for many species. Vegetation zones will probably shift, but without continuous habitat to move through, many species will be unable to follow and colonise new, suitable areas. Only species with high mobility and those that can breed in the current agricultural and urban landscape will be able to cope.

Thankfully, there are some encouraging signs that the scale of this problem has been realised. The UK BAP has brought a fresh and focused approach to species and habitat conservation. Butterfly Conservation, in acting as lead partner for *Lepidoptera* in this planning process, is highlighting the problems faced by individual species and achieving significant action on the ground. Detailed Species Action Plans have been written for 24 threatened butterflies in order to summarise ecological knowledge and identify the most urgent actions. However, sweeping changes are required to diffuse the 'biodiversity time bomb'. The solution will require radical alterations to the processes and policies that have driven the destruction of the countryside. Three priority areas must be tackled. First, all remaining areas of semi-natural habitat must be safeguarded. Secondly, appropriate management of such habitat must be maintained. This will often involve the reinstatement of traditional agricultural and forestry management practices. The final and perhaps most ambitious requirement will be the restoration of large areas of the wider countryside. This will be essential to allow species to expand, recolonise former areas and move with the tide of climate change.

ACKNOWLEDGEMENTS

Particular thanks are expressed to Martin Warren, Paul Waring, Nick Greatorex-Davies, Ian Woiwod and Mark Parsons for helpful comments on the manuscript and stimulating discussion. Many thanks are also due to Biological Records Centre, the Centre for Ecology and Hydrology, Paul Waring and the Joint Nature Conservation Committee for permission to reproduce data. Maps were produced using DMAP software written by Dr A. Morton.

REFERENCES

Agassiz, D.J.L. (1996) Invasions of Lepidoptera into the British Isles, in Emmet, A.M. (ed.) *The Moths and Butterflies of Great Britain and Ireland*, Vol. 3. Colchester: Harley Books, pp. 9–36.

Agassiz, D.J.L. and Spice, W.M. (1998) The return of the Small Ranunculus. *Entomologist's Record and Journal of Variation*, **110**, 229–232.

Agassiz, D.J.L. and Tuck, K.R. (1999) The Cypress Tip moth *Argyresthia cupressella* new to Britain. *Entomologist's Gazette*, **50**, 11–16.

Asher, J. (1995) Butterflies for the New Millennium. *Butterfly Conservation News*, **59**, 15.

Asher, J., Warren, M., Fox, R., Harding, P., Jeffcoate, G. and Jeffcoate, S. (in press) *The Millennium Atlas of Butterflies in Britain and Ireland*. Oxford: Oxford University Press.

Barnett, L.K. and Warren, M.S. (1995a) *High Brown Fritillary Species Action Plan*. unpublished report, Butterfly Conservation.

Barnett, L.K. and Warren, M.S. (1995b) *Marsh Fritillary Species Action Plan*. unpublished report, Butterfly Conservation.

Brereton, T.M. and Warren, M.S. (1999) Ecology of the Pearl-bordered Fritillary butterfly in Scotland and possible threats from bracken eradication measures in woodland grant schemes, in Taylor, J.A. (ed.) *Bracken Perceptions and Bracken Control in the British Uplands*, International Bracken Group Special Publication no. 2. Aberystwyth: International Bracken Group, pp. 62–73.

Burton, J.F. (1998) The apparent responses of European Lepidoptera to the climate changes of the past hundred years. *Atropos*, **5**, 24–30.

Davey, P. (1998) Further sites for Blair's Wainscot *Sedina buettneri* in Dorset, *Atropos*, **4**, 35.

English Nature (1998) *UK Biodiversity Steering Group, Tranche 2 Action Plans, Vol. 1.* Peterborough: English Nature.

English Nature (1999) *UK Biodiversity Steering Group, Tranche 2 Action Plans, Vol. 4. Invertebrates*, Peterborough: English Nature.

Fox, R. (1998) Butterflies for the New Millennium – a new atlas of butterflies in Britain and Ireland. *British Wildlife*, **9**, 176–179.

Gilpin, M. and Hanski, I. (eds) (1991) *Metapopulation Dynamics: Empirical and Theoretical Investigations*. London: Academic Press.

Greatorex-Davies, J.N. and Pollard, E. (1997) *Butterfly Monitoring Scheme: Report to Recorders 1997*. unpublished report, Institute of Terrestrial Ecology.

Gutiérrez, D., León-Cortés, J.L. and Thomas, C.D. (1999) Spatial distribution of a regionally rare butterfly in a fragmented landscape. Paper presented at the British Ecological Society Winter Meeting, Leicester, January 5–8.

Harding, P.T., Asher, J. and Yates, T.J. (1995) Butterfly monitoring 1 – recording the changes, in Pullin, A.S. (ed.) *Ecology and Conservation of Butterflies*. London: Chapman and Hall, pp. 3–22.

Heath, J. (1974) A century of change in the Lepidoptera, in Hawksworth, D.L. (ed.) *The Changing Flora and Fauna of Britain*. London: Academic Press, pp. 275–292.

Heath, J., Emmet, A.M. and others (eds) (1976 onwards) *The Moths and Butterflies of Great Britain and Ireland*, Vols 1–3, 7, 9–10 (others to be published). Colchester: Harley Books.

Heath, J., Pollard, E. and Thomas, J.A. (1984) *Atlas of Butterflies in Britain and Ireland*. Harmondsworth: Viking.

Hickin, N. (1992) *The Butterflies of Ireland*. Schull: Robert Rinehart Publishers.

Holloway, J. (1998) Geranium Bronze *Cacyreus marshalli*. *Atropos*, **4**, 3–6.

Lewis, O.T., Thomas, C.D., Hill, J.K., Brookes, M.I., Crane, T.P.R., Graneau, Y.A., Mallet, J.L.B. and Rose, O.C. (1997) Three ways of assessing metapopulation structure in the butterfly *Plebejus argus*. *Ecological Entomology*, **22**, 283–293.

Menendez, R. and Thomas, C.D. (1999) Spatial population structure in the patchily distributed Lepidoptera *Wheeleria spilodactylus*. Poster presented at the British Ecological Society Winter Meeting, Leicester, January 5–8.

Nature Conservancy Council (1984) *Nature Conservation in Great Britain*. Shrewsbury: Nature Conservancy Council.

Oates, M.R. and Warren, M.S. (1990) *A Review of Butterfly Introductions in Britain and Ireland*. Godalming: Worldwide Fund for Nature.

Parmesan, C., Ryrholm, N., Stefanescu, C., Hill, J.K., Thomas, C.D., Descimon, H., Huntley, B., Kaila, L., Kullberg, J., Tammaru, T., Tennant, J., Thomas, J.A., and Warren, M.S. (1999) Poleward shifts in geographical ranges of butterfly species associated with regional warming. *Nature*, **399**, 579–583.

Pollard, E. (1979) Population ecology and change in range of the White Admiral *Ladoga camilla* in England. *Ecological Entomology*, **4**, 61–74.

Pollard, E. and Yates, T.J. (1993a) *Monitoring Butterflies for Ecology and Conservation*. London: Chapman and Hall.

Pollard, E. and Yates, T.J. (1993b) Population fluctuations of the Holly Blue butterfly. *Entomologist's Gazette*, **44**, 3–9.

Pratt, C. (1987) A history and investigation into the fluctuations of *Polygonia c-album*: the Comma butterfly. *Entomologist's Record and Journal of Variation*, **99**, 69–80.

Saccheri, I., Kuussaari, M., Kankare, M., Vikman, P., Fortelius, W. and Hanski, I. (1998) Inbreeding and extinction in a butterfly metapopulation. *Nature*, **392**, 491–494.

Shirt, D.B. (ed.) (1987) *British Red Data Books, no. 2 Insects*. Peterborough: Nature Conservancy Council.

Skinner, B. (1984) *Colour Identification Guide to Moths of the British Isles*. London: Viking.

Swaay, C. van and Warren, M.S. (1999) *Red Data Book of European Butterflies (Rhopalocera)*, Nature and Environment, no. 99. Strasbourg: Council of Europe Publishing.

Thomas, C.D. and Abery, J.C.G. (1995) Estimating rates of butterfly decline from distribution maps: the effect of scale. *Biological Conservation*, **73**, 59–65.

Thomas, J.A. (1990) The conservation of Adonis Blue and Lulworth Skipper butterflies – two sides of the same coin, in Hillier, S.H., Walton, D.W.H. and Wells, D.A. (eds) *Calcareous Grasslands – Ecology and Management*. Huntingdon: Bluntisham Books, pp. 112–117.

Thomas, J.A. (1995) The ecology and conservation of *Maculinea arion* and other European species of large blue butterfly, in Pullin, A.S. (ed.) *Ecology and Conservation of Butterflies*. London: Chapman and Hall, pp. 180–197.

Waring, P. (1992a) On the current status of the Juniper Carpet moth. *Entomologist's Record and Journal of Variation*, **104**, 143–148.

Waring, P. (1992b) The spread of the Dotted Rustic *Rhyacia simulans* in Britain as reported by the national network for recording the rarer British macro-moths. *Entomologist's Record and Journal of Variation*, **104**, 311–314.

Waring, P. (1995) Wildlife reports: moths. *British Wildlife*, **6**, 393–395.

Waring, P. (1996) Wildlife reports: moths. *British Wildlife*, **7**, 324–326.

Waring, P. (1998a) A résumé of the history, operation and products of the work of the national recording network for the rarer British macro-moths and ideas on developments for the future. *National Moth Conservation Project News Bulletin*, **9**, 7–12.

Waring, P. (1998b) Wildlife reports: moths. *British Wildlife*, **10**, 54–56.

Waring, P. (1998c) Wildlife reports: moths. *British Wildlife*, **9**, 392–394.

Warren, M.S. (1992) The conservation of British butterflies, in Dennis, R.L.H. (ed.) *The Ecology of Butterflies in Britain*. Oxford: Oxford University Press, pp. 246–274.

Warren, M.S. (1994) The UK status and suspected metapopulation structure of a threatened European butterfly, the Marsh Fritillary. *Biological Conservation*, **67**, 239–249.

Warren, M.S. and Oates, M.R. (1995) The importance of bracken habitats to fritillary butterflies and their management for conservation, in Smith, R.T. and Taylor, J.A. (eds) *Bracken: An Environmental Issue*. Leeds: The International Bracken Group, pp. 178–181.

Warren, M.S., Barnett, L.K., Gibbons, D.W. and Avery, M.I. (1997) Assessing national conservation priorities: an improved red list of British butterflies. *Biological Conservation*, **82**, 317–328.

Wilson, R. (1998) The Queen of Spain Fritillary *Argynnis lathonia* – a new British breeding species? *Atropos*, **5**, 3–7.

Wilson, R.J. and Thomas, C.D. (1999) How landscape structure affects dispersal in a metapopulation of the Brown Argus butterfly, *Aricia agestis*. Paper presented at the British Ecological Society Winter Meeting, Leicester, January 5–8.

Woiwod, I.P. and Harrington, R. (1994) Flying in the face of change: the Rothamsted Insect Survey, in Leigh, R.A. and Johnston, A.E. (eds) *Long-term Experiments in Agricultural and Ecological Sciences*. Wallingford: CAB International, pp. 321–342.

Chapter 19

Grasshoppers, crickets and allied insects

Judith A. Marshall

ABSTRACT

By 1973, 42 species of *Orthoptera* were recorded as breeding in the British Isles, with a further three species restricted to the Channel Islands. Of these 45 species, 11 were established aliens normally breeding only in protected environments, and at least six of these had become established here during the preceding century. Since 1973, further introduced species have become established and, with the inclusion of *Dermaptera*, 58 species are now recognised of which 20 are established aliens.

In 1987 there were three *Orthoptera* species recognised in the British *Red Data Book* as endangered, and two as vulnerable. Two of the endangered species have since become the subject of Species Recovery Programmes, as has one of the vulnerable species.

The publication in recent years of several books on orthopteroid insects and, in 1997, the full *Atlas of Grasshoppers, Crickets and Allied Insects in Britain and Ireland* have greatly increased interest in and knowledge of this group.

1 State of knowledge

Orthoptera is here used in the very wide sense to include orthopteroid insects of both the *Orthoptera Saltatoria* and the *Orthoptera Cursoria*. The saltatorial *Orthoptera*, the true 'jumpers', are the grasshoppers, crickets, bush-crickets, camel-crickets and allied insects with well-developed hind legs. The cursorial *Orthoptera* include several distinct groups of insects now usually treated as separate orders: the *Blattodea* (cockroaches) and Mantodea (praying mantids), sometimes linked together as the *Dictyoptera*, the *Phasmida* (stick- and leaf-insects) and the *Dermaptera* (earwigs). The *Mantodea* are only represented in the UK by occasional adventive specimens.

When considering the history of the British *Orthoptera* from 1800 (Marshall 1974) the *Dermaptera* were not included, but they have been included in more recent works when discussing 'grasshoppers and allied insects' (Marshall and Haes 1988; Haes and Harding 1997).

The term 'native' used here in relation to orthopteroid insects, means that the species is considered to have arrived here independently of humans, after the last Ice Age and before the Neolithic period (Eversham and Arnold 1992). An alien is a species whose presence is a consequence of the activities of humans or of domesticated animals. Many species arrive as casual introductions, often with imported fruit, vegetables and other goods, but only if such introductions have been or are successfully breeding here are they regarded as 'established' species. The discovery of single imported specimens often arouses great interest but rarely presages a new resident.

2 Data capture and workforce

The collation of 10-km square distribution records of *Orthoptera* began in 1967 with the launch of the Insect Distribution Maps Scheme at the Biological Records Centre, and was originally run jointly with the *Odonata* scheme (Merritt *et al.* 1996). Preliminary distribution maps were produced in 1974 and the first *Provisional Atlas of Orthoptera* was published in 1978 (Skelton 1978). After a second *Provisional Atlas* (Haes 1979), the publication of the definitive *Atlas of Grasshoppers, Crickets and Allied Insects in Britain and Ireland* (Haes and Harding 1997) was the culmination of many years of effort, incorporating current, recent and historical data from over 1500 individual recorders. Haes and Harding summarise the history of recording in Britain and Ireland with full acknowledgements and a comprehensive list of the many people involved in the scheme since its inception.

3 Synopsis of changes 1800–1973

The earliest lists of orthopteroid insects contained only eight native species, six *Saltatoria* and two *Dermaptera* (Forster 1770). From 1800 to 1973 the story is one of gradual recognition and accurate identification of the native species present. The most recent addition, the lesser mottled grasshopper *Stenobothrus stigmaticus*, was first collected in 1962. By 1973, 35 native species of orthopteroid insects, including *Dermaptera*, had been recorded in Great Britain and Ireland, with a further three species restricted to the Channel Islands.

Forster had listed three established aliens, of which two cosmopolitan pest species, the common cockroach *Blatta orientalis* and the house-cricket *Acheta domesticus*, had probably been established here for many centuries (Marshall 1974). Other cockroaches followed, and the major cosmopolitan pest species had established breeding colonies here during the early part of the nineteenth century. Fifteen established aliens were listed by 1973, including the stick-insects that were accidentally introduced with plants from New Zealand in the early 1900s. Several species have become well established in gardens in the Scilly Isles, Cornwall and Devon. The status of these and other introduced and established species is discussed later (section 5.3 below).

4 Species in Red List categories, Schedule 5 and Species
Recovery Programmes

In 1987 there were three *Orthoptera* recognised in the British *Red Data Book* as endangered, two as vulnerable and one as rare. Two of the endangered species, the field cricket *Gryllus campestris* and the mole cricket *Gryllotalpa gryllotalpa*, have since become the subject of species recovery programmes. The third species listed in 1987 as endangered, the scaly cricket *Pseudomogoplistes squamiger*, was discovered at new localities in 1998 and has been shown to exist in larger numbers than was previously thought possible (see 5.2.1).

One of the species recognised as vulnerable, the wart-biter *Decticus verrucivorus*, has also become the subject of a species recovery programme, and the second of the two vulnerable species, the large marsh grasshopper *Stethophyma grossum*, while having no protected status, is at increasing risk in the UK and is the subject of a species action plan (see 5.2.3).

Gryllus campestris, *Gryllotalpa gryllotalpa* and *Decticus verrucivorus* were all listed as protected under Schedule 5 of the Wildlife and Countryside Act 1981 (see 5.2.3).

The heath grasshopper *Chorthippus vagans* was listed in 1987 as rare, and has a very restricted range in dry heathlands in Hampshire and Dorset. On Jersey it occupies a wider range of dry, often exposed grassy habitats, some however threatened by recreational or road-building activities.

5 Changes 1973–98

5.1 Losses and declines

In Britain the field cricket *Gryllus campestris* survives only in a very restricted habitat. Its requirements are short grass and a light sandy or chalk soil suitable for the nymphs to burrow into; crickets over-winter as penultimate instar nymphs and need the protection of burrows to survive. By 1973 the field cricket had become restricted to only two sites in West Sussex, and it died out at the smaller of these during 1988. It is probable that the poor summer meant that none of the nymphs in this small colony could grow large enough to survive the winter and complete the lifecycle.

The mole cricket *Gryllotalpa gryllotalpa* used to be a well-known insect of marshy areas, living in the warm, sandy areas along the edges of bogs and marshes (Sheppard 1998). It declined dramatically during the twentieth century on mainland Britain, and had been seen so rarely that it was considered possibly extinct. In 1994 it became the subject of a special project under English Nature's species recovery programme and in 1996 a survey was launched. Sadly this large and distinctive insect has proved extremely elusive, and although a mole cricket was found in a Cheshire garden in 1995 and a specimen was seen and heard calling in the same area in 1996, hopes of locating a colony in Cheshire were not fulfilled (Judd 1998). Some purported sightings resulting from the survey were potentially of mole crickets, but in the absence of well-substantiated recent records of the species, its status remains unclear.

The wart-biter *Decticus verrucivorus* was reported in 1974 as having a wider known distribution than at any time since 1800 (Marshall 1974). Nevertheless, it is known to survive at only three or four sites, and at very low densities (Brown 1997).

5.2 Additions and increases

There have been no additions to the known native orthopteroid fauna but there are increases to the known ranges, either by (1) the discovery of previously unknown sites; (2) by real expansion; or (3) by reintroduction of species to habitats from which they had previously disappeared.

The scaly cricket, previously referred to as *Pseudomogoplistes* (formerly *Mogoplistes*) *squamiger* but now known to be *P. vicentae* (Gorochov 1996) was first collected from Chesil Beach, Dorset in 1949, and had been regarded only as a probable native species (Marshall and Haes 1988). The occurrence of the species in low numbers in a very restricted habitat, beneath rocks and rubble on a shingle beach, gave rise to conjecture of possible introduction from the Mediterranean area, perhaps during the 1940s via the nearby Portland Naval Base. However, the discovery in 1992 of an unintentional pitfall trap (in the form of a discarded margarine tub) containing 90 dead and three live crickets (Timmins 1994), suggested that the population size could be relatively high. Further surveys were made on Chesil Beach and of potential sites on other shingle systems along the south coast of England (ibid. 1996). That no further specimens of *P. vicentae* were discovered at other sites by an experienced recorder of the species, who had trapped and released hundreds on Chesil Beach, gave further credence to the introduction theory. Traps were baited with grated cheese, white breadcrumbs, grated chocolate digestive biscuits and ripe banana – a proven attractive mixture! – and were placed only in shingle beaches bearing a close similarity to the Chesil Beach locations.

Events in the summer of 1998 dramatically changed the scaly cricket picture, as live specimens were discovered on the Channel Island of Sark and on the south coast of Devon at Branscombe (Sutton 1998). Baited pitfall traps, with the tasty addition of cat biscuits and mango chutney to the previous mixture, were deployed in the Branscombe area and a strong colony was reported there. Both of these new sites were discovered by entomologists, but by chance rather than as the result of a dedicated search for the species. The history of the scaly cricket in Britain from its first discovery in 1949 up to the end of 1998 has been fully described and reviewed (Sutton 1999a).

The scaly cricket was also discovered at a new locality on the north coast of France at Carolles, Manche (Morère and Livory 1999). Late in 1997 a pre-1951 record of the species at Granville on the Cherbourg Peninsula was noted, and duly reported in the newsletter of the Normandy Orthoptera Society (Stallegger 1998) in the hope that it would be searched for – and discovered – during 1998. Again by chance, the species was observed on a shingle beach at Carolles and, though the importance of the sighting was not initially appreciated, a female specimen was later collected (Beaufils 1999). A few days later a male was also discovered, and these two specimens from France were described as a new subspecies, *P. vicentae septentrionalis* (Morère and Livory 1999).

Further exciting discoveries were made in 1999, again on shingle beaches, on Guernsey in the Channel Islands and at St Brieux near Rosaires in Brittany (Sutton 1999b).

These new records may be considered to support theories of expansion as a result of global warming, or to be further evidence of a relict species surviving in isolated pockets. The true picture is currently obscured by the problems surrounding the correct identification of the species involved. This should be resolved by future studies on a wider range of material than currently available.

The scaly cricket in Dorset was originally recorded as *Mogoplistes squamiger*, a species with a wide recorded distribution in the Mediterranean area and on Madeira and the Canary Islands (Harz 1969). *M. squamiger* was later separated from other species of *Mogoplistes* in the newly described genus *Pseudomogoplistes* (Gorochov 1984). Gorochov went on to describe three further species of *Pseudomogoplistes* (1995, 1996), based mainly on the structure of the male genitalia. Two of these species, *P. byzantius* from Crimea (Ukraine) and Thasos (Greece) and *P. turcicus* from southern Turkey (Gorochov 1995), he described as fairly similar in appearance to *P. squamiger* from Croatia. Males of the third new species, *P. vicentae* from Morocco and Portugal, are described as distinctly different from males of the other three species in the shape of the supra-anal plate and internal genitalia (ibid. 1996).

Identification of the species within *Pseudomogoplistes* presents a considerable taxonomic problem as the original description of *squamiger* was based on a female, and both females and nymphs are 'indistinguishable … from other species of this genus' (Gorochov 1995). Male specimens from Chesil Beach and Sark have now been identified by Gorochov (pers. comm.) as *P. vicentae*, while specimens from Madeira held at The Natural History Museum, London, are similar to *P. byzantius* in the *P. squamiger* group.

The term 'scaly cricket' is applied to several closely related members of the *Mogoplistinae*, and it is inappropriate to apportion separate common names to the several species involved here when all are so very similar in appearance and habitat requirements. The subspecies recognised in France has been named as the 'northern scaly cricket', but should the populations from each isolated northern colony prove to be discrete species or subspecies, it may be useful in future to indicate their locale, for example as the 'Dorset northern scaly cricket' or the 'Sark northern scaly cricket'.

Since the majority of orthopteroid species are found in southern England, much of the work on recording has been concentrated in this region. In Surrey, 30 of the possible total of 35 species are known to be present and have been fully documented by David Baldock in the fifth in a series of county wildlife atlases; the completeness of the maps in this publication owes much to the use of a bat-detector for recording species distribution, a technique developed and fully utilised by Roger Hawkins (Baldock 1999).

Recording in Scotland, Wales, Ireland and the Channel Islands was summarised by Haes and Harding (1997). Additional distribution records have been noted for all areas during the twentieth century, although Ireland was known to be particularly under-recorded, and since 1974 four species have been added to the Irish list – three native bush-crickets and one established alien.

Orthopteroid success stories of the twentieth century are the extended ranges of three, possibly four, species of bush-cricket, partly as a result of an increase of macropterism in normally brachypterous species. The long-winged cone-head *Conocephalus discolor* has shown a dramatic increase in distribution area during the last 20 years, throughout western Europe (Kleukers *et al.* 1996). Until the 1970s this species was known in Britain as a rare and localised species in southern coastal areas, and had only been recognised as a native in the 1930s (Blair 1936). (The few previous records had been thought to be misidentifications of the short-winged cone-head *C. dorsalis*, possibly in its macropterous form.) New records of *C. discolor* from Hampshire were reported in the 1970s, and by the early 1980s it was found to be widely distributed in the New Forest and even common in quaking bogs. Haes (Kleukers *et al.* 1996) details the recorded spread of the species as it extended inland to Surrey and Oxfordshire and westwards to Cornwall during the 1990s. The distribution map for 1995, as published in 1997, presented a very different picture even from that of 1988. Although normally a long-winged species, the extra-macropterous form occurs more frequently in hotter summers, and single new records are often of this form. There is the possibility that continental specimens have contributed to the spread of this species here, particularly for the extreme south-western records in Cornwall and the Scilly Isles (J. Widgery pers. comm.).

The short-winged cone-head *C. dorsalis* has been more fully recorded in recent years, and has also extended its range in the Thames valley area, with the macropterous fully winged form '*burri*' Ebner found mainly during hot summers.

Roesel's bush-cricket *Metrioptera roeselii* has also shown a considerable increase in distribution area in recent years, particularly in the London area, taking advantage of urban wasteland and road and railway verges. Originally considered to be a strictly coastal or estuarine species, it has now extended its range from the Thames northwards through Suffolk and Bedfordshire, westwards through Oxfordshire and southwards through Sussex and Kent. Increased numbers of the fully-winged form *diluta* are observed in very hot summers and have probably contributed to the spread of the species.

The grey bush-cricket *Platycleis albopunctata* was known only from coastal habitats in southern Britain, until its discovery by M. Skelton in 1997 on a chalk ridge at least 2 km inland. In 1998 he found a second site on the same ridge in Dorset, and a further well-established colony 14 km inland on dry wasteground in Hampshire (Widgery 1999). This raises the possibility that the species may have reached other inland sites which have yet to be discovered.

The launch in 1991 by English Nature of their species recovery programme with the inclusion of the field cricket *Gryllus campestris* has enabled this species to be helped back from the brink of extinction in Great Britain. In 1991 the field cricket was only known to survive in the Coates Castle area of Sussex. The main field-cricket population centres within this area were ascertained so that the boundary of a Site of Special Scientific Interest (SSSI) could be established. This colony was so small that a captive-breeding programme was agreed upon, and established at the Invertebrate Conservation Centre at London Zoo. Captive-reared nymphs were released in 1992 at a site which had recently suffered extinction, and a small number of specimens survived to become adults the following year. Appropriate management of this site has continued to provide optimum conditions for the survival of the species in this area. Under the species recovery programme, autecological studies of the field cricket were undertaken in England in 1993 and in The Netherlands and Channel Islands during 1994 (Edwards 1994). These studies have shown that the species requires a cycle of habitat disturbance over 10–20 years for its survival, and that in an appropriate habitat it may disperse and colonise over a limited distance of about 200 m each year. The crickets prefer steep south-facing banks which are open to the sun throughout the year and have small areas of open ground; perhaps the name 'field cricket' is not totally appropriate as it is certainly not descriptive of their preferred habitat. A better understanding of the life history of the species has also resulted from these studies, including their surprisingly flexible use of burrows. New burrows are built regularly and crickets may change burrows almost on a daily basis; this is far from the previous ideas of territorial occupation (Edwards *et al.* 1996). The captive colony at London Zoo has bred successfully, with large outside over-wintering cages which are open to public view. As stocks of crickets become available they are released at suitable sites, appropriately managed to conform with the recently identified habitat requirements. The former site at Frensham, Surrey, is one of the areas thus recolonised, with press and television coverage of the event (Edwards 1998). All such releases of reared stock are either on former known sites, or areas considered appropriate for the survival of the species, within its known historical distribution area. Sites outside the known distribution area of the species are not considered as potential areas for introduction, no matter how suitable they may be.

The wart-biter *Decticus verrucivorus* is also the subject of a species recovery programme, and its ultimate survival in the country may well also depend upon appropriate site management. Autecological research on the species was conducted between 1987 and 1990, commissioned by the Nature Conservancy Council, to obtain detailed understanding of its habitat requirements. The studies showed that the vegetation structure and species composition of the habitat is of vital importance to the developing nymphs (Cherrill and Brown 1990a,b). The eggs hatch in early April, the young nymphs living in areas of short turf on warm south-facing slopes, enabling them to bask in the sun. After the fifth nymphal instar their behaviour changes, the sixth and seventh instar nymphs and adults preferring in live in dense grass tussocks. However, the adult females then move out again to areas of short, open turf to seek small patches of bare soil in which to lay their eggs. Thus the ideal habitat for the species would seem to be formed of a mosaic of short turf and grass tussocks (Cherrill 1993).

A captive-breeding programme was commenced at the Invertebrate Conservation Centre at London Zoo, and following the identification of suitable sites, some reintroductions of the species have been started. The long-term objective for the wart-biter is to secure its future in England, by the maintenance of existing colonies and the establishment of viable populations at new sites, within the known range of the species (Brown 1997).

The mole cricket *Gryllotalpa gryllotalpa* presents more of a problem in terms of conservation, since if there are no known breeding sites even excellent habitat management will make no difference to its status (Burton 1989). The continued presence of the species in the Channel Islands has permitted the collection of live material in order to study their lifecycle, habits and behaviour (Sheppard 1998). However the successful establishment of a captive-breeding population of material from Guernsey at London Zoo would not automatically result in the release of individuals in Great Britain.

It has been known for some years that there are several different species of mole cricket in Europe which have a very similar external appearance but may differ immensely in acoustic behaviour and chromosome number. Recent work has shown that cuticular hydrocarbon pattern may enable further separation of species, and there are currently 13 different named sibling species or races in the European and Mediterranean area (Broza *et al.* 1998).

Mole cricket chromosome numbers range from $2n = 12$ to $2n = 23$, but all specimens so far examined from northern Europe, including Great Britain, France and Guernsey, possess $2n = 12$ chromosomes. Although live material is required for the study of chromosomes, analysis of the cuticular hydrocarbons may be made from museum material with no detriment to the specimen, and continuing studies of northern European material show very similar patterns in specimens from Britain and Guernsey (Broza pers. comm.). If ample, healthy stocks of Guernsey mole crickets can be bred in captivity and can be shown to be extremely similar to material previously collected in Britain, the possibility of releasing live material in appropriate sites could be considered.

The large marsh grasshopper *Stethophyma grossum* has a scattered distribution in Great Britain on fragmented, boggy heathland and quaking bogs, and is at increasing risk through continued habitat depletion. The survival of this species in southern England may depend on the establishment of a successful conservation programme, through the instigation of appropriate habitat management and the development of a captive-breeding programme (Brown 1997).

5.3 Aliens and introductions

Since 1973 further introduced species have established breeding colonies here, and with the inclusion of the *Dermaptera*, 58 species are now recognised of which 20 are established aliens.

The status of a live alien orthopteroid insect captured in Britain may range from that of a well-established cosmopolitan pest species to a 'one-off' importation of a single tropical insect – and, depending on where the specimen is captured, the exact status may be extremely difficult to determine. Specimens captured in domestic environments are normally pest species, but may have been imported. Insects, particularly cockroaches, are easily transportable in a variety of ways and many live tropical cockroaches have been carried here with personal luggage, for example in suitcases, in the frame of a child's pushchair and in camping gear. Imported goods present even more opportunities, from plastic toys made and packaged abroad and carrying cosmotropical pests, to fruit and other foodstuffs carrying wild-living endemic species.

In recent years occasional specimens of a rare brachypterous South African cockroach *Temnopteryx* have been imported with grapes, but the main source of introduced non-pest cockroaches is with bananas, mostly neotropical in origin. Many of these are recently field-caught specimens in good health and may be sufficiently mobile to wander away from the bananas on to other fruit or vegetables, causing some concern about potential infestations.

Occasionally the casual introduction of a gravid female or a fertile ootheca may result in a temporary infestation. The temporary establishment and immediate eradication of two breeding colonies of the brown cockroach *Periplaneta brunnea* at London's Heathrow Airport in the 1960s was not considered to represent permanent establishment here (Marshall 1974). A briefly established and rapidly eradicated infestation of the cosmotropical harlequin cockroach *Neostylopyga rhombifolia* in a block of flats, was also not considered to represent permanent establishment. Persistence of a colony after repeated (though unsuccessful) attempts at eradication may well result in the permanent establishment and further distribution of alien species. The cinereous cockroach *Nauphoeta cinerea*, having been a casual immigrant for decades, has in recent years become established at Jersey Zoo (Haes and Harding 1997; M. Barclay pers. comm.). The species was originally kept in large-scale culture for use as food for reptiles, but many specimens eluded capture by the reptiles and the species now thrives in the cages though it is no longer reared specifically as a food.

Some tropical orthopteroid species have been introduced and have established breeding colonies in protected sites. These include cosmotropical pests such as the greenhouse camel-cricket *Tachycines asynamorus* and the Surinam cockroach *Pycnoscelus surinamensis*, both species that are likely to damage plants when feeding on them, and thus are usually eradicated as soon as they make their presence felt. The Surinam cockroach is parthenogenetic and a burrowing root-feeder, so a stray imported with plants may rapidly build up a thriving and potentially damaging colony.

The chance introduction in 1997 of a southern European cockroach, *Loboptera decipiens*, to a home in southern England resulted in a very healthy colony, to the consternation of the householders. The infestation was eradicated from the house and was thought to have been eliminated, but surprisingly reappeared in the garden during the summer of 1998. This is a small, dark-bodied, cream-edged species, the adult having tiny lobe-like wings, and is not normally a pest (Chinery 1986) – indeed many orthopterists would be delighted to welcome it to their garden, though perhaps not into their home. The survival of a breeding population for more than a year outdoors may merit inclusion on the established species list, perhaps as 'the lobe-winged cockroach', though without at this stage revealing the actual locality.

It should be noted that many species of cockroach are kept in culture both as food sources and as pets, and have on many occasions caused temporary infestations in homes and sheds, sometimes to the horror of the parents (or spouse) of the 'keeper' in question. Long-term wholesale infestations from this source are though unlikely. As well as cockroaches, crickets are also widely reared as a food source for reptiles and birds, and the use of the southern field cricket *Gryllus bimaculatus* may present a more serious problem. For several summers occasional specimens of this species have been seen and heard in various locations in southern England, and although breeding colonies may not become established, the presence of occasional specimens may present confusion with our native field cricket.

Locusts, both the desert locust *Schistocerca gregaria* and the migratory locust *Locusta migratoria*, are cultured commercially as a food source and for teaching purposes, so that stray specimens encountered out of doors may be introductions or have been released – deliberately or otherwise. Both these species may reach our shores under their own power, flying from Europe or even north Africa. After the 1987–88 plague of the desert locust in Africa, many specimens were carried across Europe with the easterly trade winds and even as far as the West Indies (Ritchie and Pedgley 1989). The large Egyptian grasshopper *Anacridium aegyptium* occasionally finds its way here as a stowaway with goods from the Mediterranean.

The New Zealand stick-insects and the long-cultured laboratory (or Indian) stick-insect *Carausius morosus* have had a well-established and documented history here for nearly a century (Brock 1991; Haes and Harding 1997). Although the New Zealand species are difficult to rear in captivity, they have long shown their ability to survive where appropriate food plants such as Japanese Cedar *Cryptomeria japonica*, cypress *Cupressus* spp. or *Chamaecyperis* spp., or various *Rosaceae* are available. Remains of cultures of both the prickly stick-insect *Acanthoxyla prasina geisovii* and the unarmed stick-insect *Acanthoxyla inermis* were discarded in 1985 at Greatwork in Cornwall, having apparently died out, but since 1988 both species have been in evidence in the garden on *Cupressus* species (A. James pers. comm.). The laboratory stick-insect is one of many tropical and subtropical species reared in culture here, often very successfully. For many years there have been transient populations of *C. morosus* resulting from cast-off cultures or discarded eggs, feeding mainly on privet, but this species probably does not over-winter without protection. Of the many other species in culture, at least two – the Corsican stick-insect *Bacillus rossius* and the pink-winged stick-insect *Sipyloidea sipylus* – have been reported as surviving for a number of years after being discarded (McNamara 1996).

Tropical species are unlikely to be able to survive for very long if released, even when ample stocks of food are apparently available. Eggs are occasionally discarded either by accident or design, and specimens have been known to be released (by a jaundiced parent) into an apparently suitable bramble-patch in a local wood (P.C. Barnard pers. comm.).

6 Factors affecting change

Changes in agricultural usage of land and continuing urbanisation may restrict the available habitat for many British species, but urban dereliction may aid the range extension of Roesel's bush-cricket *Metrioptera roeselii*, the long-winged cone-head *Conocephalus discolor* and the lesser marsh grasshopper *Chorthippus albomarginatus*. Disused railway-land seems particularly to suit the requirements of these species.

The effect of the cessation of grazing on marginal land, with the resultant spread of bracken, heathers, scrub and coarser grasses, has played a major part in the extinction of field-cricket colonies in the UK and also in The Netherlands and Jersey (Edwards *et al.* 1996).

Habitat depletion through changing land use may impose pressure on native species. It is probable that continued peat extraction will place the large marsh grasshopper at increasing risk in Great Britain (Brown and Cheesman 1997).

With the extensive and rapid removal of large areas of commercially exploited peatland, the vicinities of such large sites could well dry out over large areas well beyond the limits of the cleared peat. This could result in the serious reduction or even loss of populations of moisture dependent species, including the bog bush-cricket *Metrioptera brachyptera* the short-winged cone-head *Conocephalus dorsalis* and the mole cricket *Gryllotalpa gryllotalpa*. These species would be particularly at risk near the natural boundaries of their present range.

On the positive side, the recent expansions described earlier (5.2.2) may continue, particularly if climate warming continues.

7 Prognosis for the next 25 years

UK Biodiversity Species Action Plans have been established for four species recognised as endangered or vulnerable. The ultimate aim for all species involved is to reverse the decline of populations under threat of extinction, by ensuring favourable conservation status and habitat management at appropriate sites. Initial assessment of the needs of the insects and current management of their known sites will indicate options for future management.

On the credit side, the often urgent need for such assessments has attracted a core of both professional and amateur recorders, with many 'amateurs' having a highly professional knowledge of the group. All are aware of the key role which the most easily identified British orthopteroids provide as habitat indicators, comprising a limited number of species at the edge of their European ranges. Two notable veterans among the amateur recorders are Ian Menzies, who has contributed records of prime importance from the mid-1940s to the present day, and Michael Skelton who, since the 1970s, has provided a continuous supply of data of outstanding importance. Other front-line amateur recorders have provided specialist observations on specific groups, for example Paul Brock on *Phasmida*, John Paul on *Tetrigidae* and Chris Timmins on *Dermaptera*. Full acknowledgement of the many contributors and workers involved in the recording scheme over the years was made by Haes and Harding (1997).

The continued success of the *Orthoptera* recording scheme owes much to the inspiring enthusiasm of Chris Haes, voluntary organiser for nearly 20 years, and the continuing endeavours of John Widgery.

Mapping schemes continue to supply valuable information on the current distribution of species, and captive-rearing programmes enable the reintroduction of healthy stock to sites that are within the original natural range of the species. Computer generated mapping has enabled the rapid production of accurate and easily updated distribution maps. As a consequence, many local or county orthopteroid maps are now presented as 10-km square, tetrad and 1-km square series as a matter of course, an outstanding improvement both in time and publication cost over the Letraset or hand-dotted maps of the past.

REFERENCES

Baldock, D.W. (1999) *Grasshoppers and Crickets of Surrey*. Woking: Surrey Wildlife Trust.

Beaufils, M. (1999) Un heureux concours de circonstances. *L'Argiope*, **23**: 26–27.

Blair, K.G. (1936) *Conocephalus fuscus* Fab., a grasshopper new to Britain. *Entomologist's Monthly Magazine*, **72**: 273–274.

Brock, P.D. (1991) *Stick Insects of Britain, Europe and the Mediterranean*. London: Fitzgerald.

Brown, V.K. (1997) *Species Action Plan*, Decticus verrucivorus *(L.) (Orthoptera: Tettigoniidae)*, Report for English Nature. Peterborough.

Brown, V.K. and Cheesman, O.D. (1997) *Species Action Plan*, Stethophyma grossum *(L.) (*Mecostethus grossus *(L.)) (Orthoptera: Acrididae)*, Report for English Nature. Peterborough.

Broza, M., Blondheim, S. and Nevo, E. (1998) New species of mole crickets of the *Gryllotalpa gryllotalpa* group (Orthoptera: Gryllotalpidae) from Israel, based on morphology, song recordings, chromosomes and cuticular hydrocarbons, with comments on the distribution of the group in Europe and the Mediterranean region. *Systematic Entomology*, **23**: 125–135.

Burton, J.F. (1989) Close of play for mole-crickets. *Countryman*, **94**: 80–81.

Cherrill, A. and Brown, V.K. (1990a) The life cycle and distribution of *Decticus verrucivorus* (L.) within a chalk grassland in southern England. *Biological Conservation*, **53**: 124–143.

Cherrill, A. and Brown, V.K. (1990b) The habitat requirements of adults of the Wart-biter *Decticus verrucivorus* (L.) (Orthoptera: Tettigoniidae) in southern England. *Biological Conservation*, **53**: 145–157.

Cherrill, A. (1993) The Conservation of Britain's Wart-biter Bush-cricket. *British Wildlife*, **5**: 26–31.

Chinery, M. (1986) *Insects of Britain and Western Europe*. London: Collins.

Edwards, M. (1994) *Species Recovery Programme, Field Cricket*, Gryllus campestris. Project Report for 1994, Report for English Nature, Peterborough.

Edwards, M., Patmore, J.M. and Sheppard, D. (1996) The Field Cricket – preventing extinction. *British Wildlife*, **8**: 87–91.

Edwards, M. (1998) *English Nature Species Recovery Programme, Field Cricket*, Gryllus campestris, Project Report for 1998, Report for English Nature. Peterborough.

Eversham, B.C. and Arnold, H.R. (1992) Introductions and their place in British wildlife, in Harding, P.T. (ed.) *Biological Recording of Changes in British Wildlife* (Institute of Terrestrial Ecology symposium no.26). London: HMSO, pp. 44–59.

Forster, J.R. (1770) *A Catalogue of British Insects*. Warrington: Eyres.

Gorochov, A.V. (1984) A contribution to the taxonomy of modern Grylloidea (Orthoptera) with a description of new taxa [in Russian]. *Zoologicheskii Zhurnal*, **63**: 1641–1651.

Gorochov, A.V. (1995) Two new species of the genus *Pseudomogoplistes* Gorochov (Orthoptera: Mogoplistidae). *Zoosystematica Rossica*, **3** (1994): 249–250.

Gorochov, A.V. (1996) A new species of *Pseudomogoplistes* from Morocco and Portugal (Orthoptera: Mogoplistidae). *Zoosystematica Rossica*, **4** (1995): 292.

Haes, E.C.M. (ed.) (1979) *Provisional Atlas of the Insects of the British Isles. Part 6. Orthoptera*, 2nd edn. Huntingdon: Institute of Terrestrial Ecology.

Haes, E.C.M. and Harding, P.T. (1997) *Atlas of Grasshoppers, Crickets and Allied Insects in Britain and Ireland*. London: The Stationery Office.

Harz, K. (1969) Die Orthopteren Europas I. *Series Entomologica*, **5**: i–xx, 1–749.

Judd, S. (1998) *Mole Cricket,* Gryllotalpa gryllotalpa, *monitoring of possible Macclesfield colony and other associated regional work*. Report for English Nature, Peterborough.

Kleukers, R.M.J.C, Decleer, K., Haes, E.C.M., Kolshorn, P. And Thomas, B. (1996) The recent expansion of *Conocephalus discolor* (Thunberg)(Orthoptera: Tettigoniidae) in western Europe. *Entomologist's Gazette*, **47**: 37–49.

Marshall, J.A. (1974) The British Orthoptera Since 1800, in Hawksworth, D.L. (ed.) *The Changing Flora and Fauna of Britain*. London and New York: Academic Press, pp. 307–322.

Marshall, J.A. and Haes, E.C.M. (1988) *Grasshoppers and Allied Insects of Great Britain and Ireland*. Colchester: Harley Books.

McNamara, D. (1996) A note on *Bacillus rossius*, the Corsican stick insect. *Bulletin of the Amateur Entomologists' Society*, **55**, 31–32.

Merritt, R., Moore, N.W. and Eversham, B.C. (1996) *Atlas of the Dragonflies of Britain and Ireland*. London: The Stationery Office.

Morère, J.J. and Livory, A. (1999) Le grillon maritime de la Manche: une espèce nouvelle pour la France. *L'Argiope*, **23**: 29–37.

Ritchie, M. and Pedgley, D. (1989) Desert locusts cross the Atlantic. *Antenna*, **13**: 10–12.

Sheppard, D. (1998) The search for *Gryllotalpa*. *Biologist*, **45**: 155–157.

Skelton, M.J. (ed.) (1978) *Provisional Atlas of the Insects of the British Isles. Part 6. Orthoptera*. Huntingdon: Biological Records Centre.

Stallegger, P. (1998) Coordination Orthopteres Normandie. *Lettre de Liaison*, **5**: 1–16.

Sutton, P.G. (1998) The Scaly Cricket *Pseudomogoplistes squamiger:* new to Devonshire. *Bulletin of the Devon Invertebrate Forum*, **3**: 4–5.

Sutton, P.G. (1999a) The Scaly Cricket in Britain – A complete history from discovery to citizenship. *British Wildlife*, **10**: 145–151.

Sutton, P.G. (1999b) Distribution of the Scaly cricket *Pseudomogoplistes vicentae* in Europe: recent developments. Handout at the Orthopterists' Annual Meeting, 1999, The Natural History Museum, 17 November 1999.

Timmins, C.J. (1994) The population size of *Pseudomogoplistes squamiger* Fischer (Orth. Gryllidae). *Entomologist's Monthly Magazine*, **130**: 66.

Timmins, C.J. (1996) *Project to assess the distribution and status of* Pseudomogoplistes squamiger *(Fischer) in Britain*. Report for English Nature, Peterborough.

Widgery, J. (1999) Orthoptera Recording Scheme for Britain and Ireland. *Orthoptera Recording Scheme Newsletter*, **25**: 1–13.

Chapter 20

Dragonflies and damselflies

Stephen J. Brooks

ABSTRACT

The distribution and status of the 52 species of dragonfly recorded from the British Isles are among the best known of any invertebrate group. By 1990 the *Odonata* Mapping Scheme had received more than 160 000 UK records, from over 2000 recorders and achieved a coverage of about 87% of the British Isles.

Between 1945 and 1960 three species became extinct in Britain, while *Coenagrion lunulatum* (the Irish damselfly) was discovered in 1981 for the first time in the British Isles. Other species have suffered declines, although a few have increased their range.

The activities of extractive industries and the creation of many new garden ponds have helped to increase the range and abundance of several species. Additionally, in the last few years, perhaps in response to climate change, some species have expanded their ranges northwards and there appears to have been an increase in the abundance and numbers of species migrating to Britain, although none of these has established permanent breeding populations.

In the future, it seems likely that pressure will continue to grow on wetlands causing further contractions in the range of many dragonfly species. However, if the number of wetland reserves and quality of land management continue to improve this may help to slow the decline.

Global warming may cause an increase in the numbers of migrant species. Some of these may establish breeding populations, although this may be counter-balanced by declines in southern populations of cold-adapted native species.

1 Introduction

The abundance and diversity of dragonflies in Britain, like other members of the biota, have always been dynamic. Populations have fluctuated for thousands of years in response to climate change but, increasingly environmental change induced by the impact of humans has become more significant. Changes in land use during the last 50 years have perhaps caused the greatest changes in the fortunes of Britain's dragonfly fauna in the last 10 000 years and this has heightened awareness for the need of active conservation measures to maintain the fauna. Accounts on dragonfly conservation have been provided by Moore (1976) and Merritt *et al.* (1996) and this chapter has necessarily drawn heavily on this work and brings the assessment up to date.

2 State of knowledge

The distribution and status of dragonflies (*Odonata*) in Britain and Ireland are among the best known of any invertebrate group. A total of 52 species has been recorded, of which 38 have established breeding populations, one additional resident species has dubious taxonomic validity, three formerly resident species are now extinct, and 10 species are migrants of varying frequency and regularity, none of which has so far established viable breeding populations for more than a few years (Brooks and Lewington 1997). Dragonflies occur throughout Britain and Ireland but they are not evenly distributed. Most occur in the southern half of England, but a few are restricted to Scotland and Ireland (Table 20.1). Climate and habitat requirements play key roles in determining the distribution of the species. A few, like *Enallagma cyathigerum* (Common Blue Damselfly) and *Ischnura elegans* (Blue-tailed Damselfly), occur in a broad range of habitats and are widespread throughout the British Isles, but most species have more restricted distributions. For example, *Cordulegaster boltonii* (Golden-ringed Dragonfly) and *Orthetrum coerulescens* (Keeled Skimmer) occur throughout much of southern England and also along the west coast of Britain into northern Scotland. They appear to favour an oceanic climate of warm, wet summers and mild winters. These two species are further restricted by their habitat requirements since *C. boltonii* breeds in streams and *O. coerulescens* in streams, shallow bog-pools and seepages. Both these biotopes depend upon the wet climate of western Britain.

Some species, like *Orthetrum cancellartum* (Black-tailed Skimmer) and *Libellula fulva* (Scarce Chaser), apparently prefer the more continental climate of south-east England. *O. cancellatum* is further restricted to lakes in the early stages of succession, and *L. fulva* to slow-flowing rivers. Other species, such as *Aeshna caerulea* (Azure Hawker) and *Coenagrion hastulatum* (Northern Damselfly), occur only in Scotland and, while both *Ceriagrion tenellum* (Small Red Damselfly) and *Leucorrhinia dubia* (White-faced Darter) breed only in acid bog-pools, *C. tenellum* has a southern distribution whereas *L. dubia* occurs mainly in northern Britain.

Table 20.1 Regional distribution of *Odonata* species in the British Isles.

Region	Number of species
Scotland[1]	21
Ireland[2]	23
Northern England[3]	19
Wales	29
Southern England	34

[1] Includes two species restricted to Scotland.
[2] Includes one species restricted to Ireland.
[3] Species occurring in England north of the Humber estuary.

3 Distribution data

The history of dragonfly recording in Britain, summarised by Merritt *et al.* (1996), goes back to the earliest insect collections made by naturalists in the late seventeenth century. The collection of Leonard Plukenet (1642–1706) includes six species of dragonfly, collected in the Islington and Hampstead Heath areas of London (Hammond 1975). Despite the obvious changes in land use in the last 300 years, five of the six species known to Plukenet still occur in urban north London, holding on in parks and local nature reserves. Only one, *Brachytron pratense* (Hairy Dragonfly), a fenland species breeding in ditches and dykes with luxuriant aquatic vegetation, is no longer found there (Brooks 1989). Nevertheless, even this species has undergone a recent range expansion and has bred in the Lee Valley north of Waltham Abbey since 1995.

By the end of the nineteenth century all but four of the species now known to be resident in Britain and Ireland had been recorded. Monographs by Lucas on adult dragonflies (Lucas 1900) and larvae (Lucas 1930) kindled interest in the group and this was further fuelled by Cynthia Longfield's field guides (Longfield 1937, 1949) which included notes on the distributions of the species. However, perhaps because of difficulties in preserving the adult colours, dragonflies never became a popular group with insect collectors. Advances in macrophotography are partly responsible for the recent rise in public interest in dragonflies because they make such colourful and striking subjects and this has now largely replaced the need or desire to collect them.

The first published distribution maps of the British species, which appeared in Corbet *et al.* (1960), only provided a broad-brush impression although they were based on quite a large number of collated records. The current high level of knowledge about dragonfly distribution in Britain and Ireland is due to the success of the *Odonata* Recording Scheme (ORS). The ORS was inaugurated in 1968 by the Biological Records Centre under the direction of J. Heath. The first maps showing *Odonata* distribution based on 10-km squares were produced in 1974 (Skelton 1974) but the records were patchy, with only parts of southern England reasonably well recorded and relatively few recorders contributing to the scheme. In 1977 the ORS was galvanised by the appointment of D. Chelmick as organiser (followed in 1981 by R. Merritt) and by the publication of a new field guide by Hammond (1977) in which, for the first time, males and females of all the resident species were featured in detailed colour illustrations. This publication led to a large increase in the number of records being submitted to the ORS and by the time a provisional atlas was published in 1979 (Chelmick 1979) the overall distribution of most species was apparent.

In 1983 the rapidly growing interest in dragonfly recording became the catalyst for the formation of the British Dragonfly Society (BDS), which in turn introduced new dragonfly enthusiasts to dragonfly recording. Consequently, the number of records received by ORS reached a peak between 1983 and 1987. The most recently published atlas (Merritt *et al.* 1996) includes records up to the end of 1990, by which date the ORS had received more than 160 000 UK records and achieved a coverage of about 87% of all 10-km squares in the British Isles. The overall patterns of species distributions are now well known, although there are still significant gaps in the coverage of Ireland, and distribution maps for this country may still reflect recorder effort rather than genuine distribution patterns.

4 Who records the distribution of *Odonata*?

The majority of the 2000 dragonfly recorders in the British Isles are amateur enthusiasts. There are probably less than five biologists who work professionally on dragonflies in Britain and a similar number of PhD students at any one time. There are currently about 1300 members of BDS but, surprisingly perhaps, only about 30% of them submit records to ORS, even though it was integral in the formation of BDS and in 1996 BDS took over the running of ORS. Similarly, only about 40% of people who submit records to ORS are members of BDS. So most of the dragonfly records received by ORS come from amateur naturalists who have an interest in recording dragonflies but for whom dragonflies are not the main group of interest. In recent years there has been growing interest in dragonflies among bird-watchers, and dragonflies are living up to their name as 'the bird-watcher's insect' (Corbet *et al.* 1960). The attraction of dragonflies for bird-watchers may be partly because many bird-watchers spend much of their time at reservoirs and other large water-bodies, because dragonflies can be observed and identified through binoculars and telescopes, and because of interest in the annual influxes of migatory species. In 1982, Hammond's (1977) identification guide was the only book on dragonflies available in print. Since then, a plethora of regional and national field guides and books on dragonfly ecology have appeared. Interest in dragonflies has never been greater than it is now.

5 Synopsis of changes in distributions to 1973

The dragonfly fauna of Britain was probably exterminated during the last Ice Age. Most of Britain north of the Midlands was covered by an ice sheet, and the countryside to the south would have resembled arctic tundra, a habitat not noted for its diversity of dragonflies. However, following the retreat of the glaciers from about 14 000 years ago, the climate ameliorated and dragonflies began to colonise Britain. Indeed, remains of dragonflies have been found in lake sediments deposited in Britain during the late-Quaternary period (Coope 1995). Colonisation of these islands from Europe would have been impeded, first by the inundation of the Irish land bridge and then by the subsequent inundation of the land bridge between England and continental Europe; for these reasons the dragonfly fauna of Ireland is less diverse than that of Britain, and many species that occur in northern Europe, and which could presumably tolerate the cool British climate, do not occur here.

Clearance of woodland for early agriculture and the provision of ponds to water livestock would have had a beneficial impact on dragonfly faunas, and dragonflies have probably benefited from human activities throughout history. The desire to improve agricultural yields started to take its toll, however, from the seventeenth century onwards, with the progressive draining of the great expanses of meres, fens and levels in eastern England and Somerset. Agricultural intensification reached its peak in the two decades following the Second World War and this period coincided with the greatest contractions in dragonfly distributions.

Because there had been little systematic recording of dragonflies prior to 1968, and realistic coverage of the British Isles was not achieved before 1974, it is difficult to be precise about subtle changes in dragonfly distribution before this time. However, it is known that three species, each of which formerly had very restricted distributions in

England, became extinct between 1953 and 1963. *Coenagrion scitulum* (Dainty Damselfly) was first discovered in Benfleet, Essex in 1946. The main breeding site was restricted to just one pond but this was flooded by sea water in early 1953, and despite thorough searches the species has never been seen again in Britain. *Coenagrion armatum* (Norfolk Damselfly), first discovered in Britain in 1902, was known from Stalham, Sutton and Hickling Broads, Norfolk (Porritt 1912; Merritt *et al.* 1996). The population went into decline in the 1950s as the breeding sites began to dry up and became over-grown with reed, sallow and alder carr; this species was last seen in 1957 (Hammond 1977). In 1820, *Oxygastra curtisii* (Orange-spotted Emerald) was discovered on the Moors River, Dorset. It is the only species of dragonfly to have been originally described from Britain. Later it was also found on the neighbouring River Stour and, in 1946, it was discovered on the River Tamar, on the Devon–Cornwall border, but has never been seen there since. Unfortunately, by 1957, due to increasing shading by bankside vegetation, suitable breeding habitat had been reduced to a stretch of the West Moors River (Moore 1991a). The final cause of extinction was almost certainly sewage discharge into the river and the species was last recorded in 1963.

All three of these now-extinct species were never recorded from more than one or two localities and were thus very vulnerable to local catastrophic events. On the contrary, none of the dragonfly species currently breeding in Britain has such a restricted distribution and so do not appear to be threatened with immediate extinction. These extinctions emphasise the importance of maintaining several viable populations of a species over a relatively wide geographical area if long-term conservation is to be achieved.

Agricultural intensification in the 1950s and 1960s resulted in the drainage or pollution of many wetland biotopes. This was particularly severe in East Anglia where many, even formerly widespread, species became more local. Prominent among these were *Calopteryx virgo*, *Pyrrhosoma nymphula*, *Platycnemis pennipes*, *Coenagrion pulchellum*, *Ceriagrion tenellum*, *Brachytron pratense*, *Cordulia aenea*, *Libellula depressa* and *Sympterum danae*. Species dependent on acid-water habitats were particularly badly affected (Moore 1986). Another casualty of post-war agricultural practices was *Coenagrion mercuriale* which was lost from Cornwall and several sites in Devon. Peat extraction and the associated drainage resulted in the loss of *Ceriagrion tenellum* from the Somerset Levels and several of the English breeding sites of the rare *Leucorrhinia dubia*. In Scotland, extensive drainage and afforestation of the Highlands led to the destruction of many former breeding sites of *Aeshna caerulea* and *Somatochlora arctica*. Lowland heaths, especially in Dorset, were also reduced and fragmented by agricultural intensification, forestry and urban developments during the same period and this led to local reductions in the populations of species such as *Aeshna juncea*, *Ceriagrion tenellum* and *Orthetrum coerulescens*.

A few species appear to have extended their ranges in Britain during the twentieth century. At the beginning of the century both *Aeshna mixta* (Migrant Hawker) and *Orthetrum cancellatum* (Black-tailed Skimmer) were relatively uncommon and restricted to a few breeding colonies in south-east England. However, they have both been expanding westwards and northwards and by the mid-1970s occurred roughly to the south-east of a line drawn between the Humber and Severn Estuaries. This expansion has probably been assisted by the increasing number of flooded gravel, sand and clay pits since both species favour sites in the early stages of succession.

6 Species with formal conservation status and legal designations

Table 20.2 summarises the formal conservation status and legal designations applied to British *Odonata*. Nine species of dragonfly are currently listed under one of the *Red Data Book* categories (Shirt 1987) and two of these (*Oxygastra curtisii* and *Coenagrion mercuriale*) are listed in the European Union Habitats Directive and Bern Convention. *C. mercuriale* is also the subject of a Biodiversity Action Plan and *Aeshna isosceles* receives protection under Schedule 5 of the 1981 Wildlife and Countryside Act. A further nine species are listed as Nationally Scarce in the SSSI Guidelines (NCC 1989). The main focus of British and European legislation is to protect individuals of the species so designated from the activities of collectors. This form of legislation, while it may be essential for the conservation of mammals, birds and plants, is not appropriate for the conservation of most insects. The populations of even rare species of insects, including dragonflies, are usually large enough to withstand the activities of human collectors, just as they are able to survive the attentions of other predators. Legislation of this kind could even hinder the conservation of insects since it is sometimes necessary for those monitoring insect populations to collect voucher specimens to confirm identification. While collecting permits may be available on application, many amateur entomologists are discouraged by swingeing legislation, and yet these are the very people from whom much of the information about insect distributions has been obtained in the past. In order to conserve insects it is far more important to conserve their habitats. None of the existing legislation has so far achieved this adequately.

The Biodiversity Convention, signed by the British Government in Rio de Janeiro in 1992, led to the publication of the UK Biodiversity Action Plan (BAP) (Anon. 1994). The BAP has resulted in some useful conservation efforts. Short, medium and long lists of organisms were drawn up to prioritise the formulation of Species Action Plans (SAPs), each having specific objectives that would result in the conservation and consolidation of rare species. *Coenagrion mercuriale* was the only species of dragonfly to appear on the BAP short list. However, before any progress could be made towards achieving the objectives of the SAP, it was first necessary to find out some basic information about the biology and ecology of the species.

BDS and the Countryside Council for Wales (CCW) are co-ordinating the efforts of amateur and professional dragonfly enthusiasts to monitor populations of *C.*

Table 20.2 Numbers of *Odonata* species currently with conservation status or listed in EC or UK legislation.

	Number of species
EC Habitats Directive, Annexe II	2
EC Habitats Directive, Annexe IV	1
Bern Convention, Appendix II	2
Wildlife and Countryside Act, Schedule 5	2
Red Data Book	9
SSSI listing	9

mercuriale in Dorset and Wales, initially to establish the size of existing populations and to investigate dispersal and the movement of individuals between populations. This project has also successfully attracted Research Council funding for a PhD thesis.

7 Changes 1973–98

The most serious declines in dragonfly populations, resulting mainly from intensification of agriculture and forestry, occurred in the two decades following the Second World War when it was government policy for the UK to become self-sufficient in food production. No further species of dragonfly has become extinct in the British Isles since *Oxygastra curtisii* was last seen in 1963, and most species have not shown any major contractions since this period. However, many species have continued to suffer local extinction and some have suffered range contractions; despite these negative pressures, others have nevertheless expanded their ranges.

7.1 Losses and declines

In 1971 it appeared that *Lestes dryas* (Scarce Emerald Damselfly) had become extinct in England, although it still continued to occur in Ireland. Although one of the most widespread species of dragonfly in the Holarctic region, in England it seems always to have been restricted to the east where it occurred in fens and ponds from Sussex to south Yorkshire. The species typically breeds in shallow, heavily vegetated pools and ditches, which may dry out late in the summer. The over-wintering eggs are drought resistant, and the larvae are able to complete development in a few weeks before the pools begin to dry. The larvae are relatively large and active during the day so only survive in fish-free habitats, especially temporary pools. Such breeding sites are susceptible to drainage or can quickly become overgrown if not managed and many were lost to agricultural intensification during the 1950s and 1960s. By 1971 *L. dryas* could not be found in any of its last few remaining breeding localities in East Anglia and the marshes bordering the River Thames in Essex and Kent. The exact causes of the loss of 10 *L. dryas* populations have been described by Moore (1980). Then, in 1983, it was rediscovered in Essex (Benton and Payne 1983). With further searching it was found elsewhere along the Thames Marshes in Kent and Essex and also in Norfolk. The species seems to have been overlooked by recorders who may have assumed it was no longer worth closely examining specimens of the very similar Emerald Damselfly *L. sponsa*. However, the rediscovery of *L. dryas* encouraged recorders once more and the species now seems to be locally common, but nevertheless still vulnerable to changes in land use.

The Ruddy Darter *Sympetrum sanguineum* is largely restricted to south-east England, although there is recent evidence of a northward range expansion (see below). During the 1970s it was thought to have undergone a steep decline, but reassessment of pre-1960 records showed that many of these earlier records derived from mass immigrations and did not represent the occurrence of breeding populations (Merritt *et al.* 1996). This emphasises the importance of distinguishing between breeding and non-breeding records before a true understanding of dragonfly distribution can be obtained. This is a strong focus of the current dragonfly recording scheme. Other species with resident populations which also have periodic immigrations, or population movements within Britain, include *Aeshna mixta*, *Libellula depressa*, *L. quadrimaculata*, *Sympetrum striolatum* and *S. danae*.

Perhaps the most serious losses since 1973 have been suffered by the White-faced Darter *Leucorrhinia dubia*. The species is essentially boreal with strongholds in northern Scotland. Never common in England, it is largely restricted to acid-water pools on heaths and raised bogs in scattered locations from Surrey to the Scottish borders. Several important populations were lost due to drainage, afforestation and industrial-scale peat milling, notably at Thorne Moor in Yorkshire. The southernmost populations in Surrey now appear to have become extinct. The final colony at Thursley Common has been in decline for several years and no adults have been confirmed there since 1995. This is particularly disturbing since Thursley Common is a National Nature Reserve, and arguably Britain's premier dragonfly site. Consequently, dragonflies are a high priority in reserve management policy and *L. dubia* was a well-known speciality. The apparent failure of this challenge to conserve *L. dubia* emphasises how little is understood of the autecology of many of Britain's dragonfly species.

7.2 Additions and increases

In 1981, *Coenagrion lunulatum* was discovered in County Sligo, Ireland. This species had never before been recorded in the British Isles, although scattered populations occur throughout northern and eastern Europe. Following its discovery further populations were found and at present it is known from widely scattered localities in southern Northern Ireland and northern parts of the Irish Republic. Given that large areas of Ireland were under-recorded for *Odonata* in 1981, and that the species closely resembles several other coenagrionid dragonflies that are common in Ireland, *C. lunulatum* had almost certainly been overlooked until that time. Its scattered distribution and the observation that several colonies have died out since their discovery gives cause for concern about the continuing status of the species.

Since 1973 several species appear to have broadened their ranges in Britain and Ireland. It is however important to distinguish between genuine range expansion and the effects of change in recorder effort. For example, until recently it was a widely held view that *Ischnura pumilio* would only breed in shallow, acid-water seepages in bogs. The discovery of the species using seepages in chalk quarries (Cham 1991) led recorders to search for the species at other sites and it has now been found to be more widespread than had been previously suspected. It is now evident that *I. pumilio* requires warm, shallow, gently flowing water and that pH is largely irrelevant. There have been several new colonies of *Aeshna caerulea* discovered in the north-west Highlands of Scotland. This is almost certainly because dragonfly recorders have visited many new sites in this remote region and the species was previously under-recorded. Similarly, Scottish populations of *Somatochlora metallica* and *Cordulia aenea* were rediscovered in 1976 and 1978 respectively. The apparent absence of these species for many years almost certainly represented a hiatus in recording rather than local extinction and recolonisation since other colonies were remote from the Scottish sites.

Nevertheless, in south-east England *S. metallica* does seem to be expanding its range. This is one of the most closely monitored parts of the British Isles so the discovery of new breeding colonies here probably does represent a genuine range increase. Similarly, new breeding records of *Libellula fulva* in Kent and on the River Wey in Surrey and

Hampshire suggest that this species too has undergone a local expansion. Like *Orthetrum cancellatum* and *Aeshna mixta*, *L. depressa* has also consolidated its occurrence in south-east England because it favours sites in the early stages of succession, and so has benefited from the increasing numbers of flooded pits left after mineral extraction. Prince and Clarke (1993) postulate that the increase in abundance of dragonflies associated with flooded mineral extraction pits has resulted in a rise in the population of the hobby *Falco subbuteo*, a raptor that specialises in feeding on large insects such as *Odonata*. The new breeding colonies of the dragonflies discussed above have all been broadly within the existing range of each species. However, in recent years it has become apparent that there has been a significant northward (and westward) extension in range of a number of species, most notably *Erythromma najas*, *Aeshna cyanea*, *A. mixta*, *Anax imperator*, *Orthetrum cancellatum* and *Sympetrum sanguineum*. This may well be in response to recent climatic warming.

7.3 Aliens and introductions

Because of their powerful flight, dragonflies have excellent dispersal abilities and some have been recorded many hundreds, even thousands, of miles from their nearest known breeding sites. Even the more weakly flying damselflies can be carried great distances on wind currents. For this reason there are British records of several species that have not established long-term breeding populations. Some have only ever been recorded once, such as *Aeshna affinis* which was found in Kent in 1952. But others are more regular migrants which may turn up on an annual basis, sometimes in very large numbers. The most frequent migrants are species of darter dragonflies, in the genus *Sympetrum*. In most years there are records of *S. fonscolombii*, usually in south-west England following strong prevailing winds which carry individuals from their breeding grounds in south-west Europe. Sometimes breeding populations become established for short periods. There are currently several sites in England which have produced successive generations of the species over the last three or four years. *S. flaveolum* occurs less frequently but when it does appear, notably in 1926, 1945, 1955 and 1995, it may be present in huge numbers. *S. flaveolum* has a more easterly distribution in Europe than *S. fonscolombii* and so is usually first seen in eastern England following strong south-easterly winds. *S. vulgatum* has also occasionally been confirmed as a migrant, often in company with *S. flaveolum*. However, despite their frequent occurrences none of these species has successfully maintained breeding colonies for more than a few years. This may be because the British climate is too cool for them, although this seems unlikely in the case of *S. flaveolum* which maintains viable populations in southern Sweden. Alternatively, the year-to-year occurrence of these migrants may be too erratic for them to establish viable long-term populations that may be too small to survive without an annual top-up from newly arrived insects from abroad.

Another visitor which seems to be becoming more frequent is *Hemianax ephippiger*. Prior to the mid-1980s this species had been recorded only six times in Britain; since then however it has made almost annual appearances. This species breeds in north Africa and its arrival often coincides with the precipitation of desert dust. In the last few years there has been a sharp increase in the number of migrants recorded and *Anax parthenope* (recorded in 1996 and 1997), *Sympetrum pedemontanum* (1995), *Crocothemis erythraea* (1995) and *Anax junius* (1998) have all been recorded for the

Table 20.3 Exotic *Odonata* imported into Britain with tropical plants (after Brooks 1988).

South-east Asian species	South-east USA species
Ceriagrion cerinorulbellum	Argia fumipennis
Ischnura senegalensis	Enallagma signatum
Pseudagrion sp.	Ischnura posita
Anax gibbosulus	Erythemis simplicicoides
A. guttatus	
Crocothemis servilia	
Orthetrum sabina	
Rhodothemis rufa	
Tramea transmarina euryale	
Urothemis bisignata	

first time. This last species is a north American resident and when five individuals were discovered in Cornwall this was the first time that any dragonfly species had been confirmed crossing the Atlantic to Britain. The 1989 record for *Pantala flavescens* was only the fourth time this species had been recorded in Britain; although the species is well known for its migratory behaviour, it is usually restricted to tropical latitudes.

Several species of tropical dragonfly have been inadvertently imported into Britain with aquatic plants intended for the aquarists' trade. These dragonflies have probably arrived either as endophytically deposited eggs or as small larvae and have subsequently emerged as adults in this country. Brooks (1988) recorded adults of 14 imported species from a warehouse in north London: the majority originated in south-east Asia but some were from south-eastern USA (Table 20.3). It is unlikely that any of these species could maintain viable populations in the wild in Britain because of adverse climatic conditions, although there was some evidence of breeding of at least one of these species, *Ischnara posita*, within the warehouse. Nevertheless, it is possible that a widely distributed American species could become established if it were introduced into Britain in this way.

8 Factors affecting change

8.1 Agriculture

Most of the land surface of Britain is farmed: any changes in agricultural practice therefore have a profound ecological impact. Agricultural intensification in Britain over the last 50 years has had a major negative impact on dragonflies (Merritt *et al.* 1996). Populations of all species have been hardest hit in southern England and East Anglia, resulting in the national extinction of one species (*Coenagrion armatum*) and serious declines in several others. The main factors causing this decline are drainage of wetlands – which, for example, has resulted in the loss of 75% of ponds since 1888

(Pond Conservation Group 1993) and a 40% loss of wet grassland (RSPB 1998) – lowering of water-tables by over-abstraction, runoff of biocides from field sprays or sheep dips, and runoff of fertilisers causing eutrophication and promoting growth of algae, toxic cyanobacteria and deoxygenation.

In spite of these negative assaults by the farming industry there have been some positive steps which have helped to avert a catastrophic decline in dragonflies in the wider countryside. One of these was the formation of the Farming and Wildlife Advisory Group (FWAG) in 1976. This is an umbrella organisation bringing together farming, forestry and conservation groups with government departments. One positive result of this initiative is the creation and restoration of thousands of farm ponds which have helped to conserve dragonflies in the wider countryside (Moore 1991b). For example, between April and June in 1997 FWAG gave advice on the creation of 261 ponds, the restoration of 1524 and the management of 489 ponds (N.W. Moore pers. comm.).

8.2 Woodland management

For most species of dragonfly dense woodland is an anathema since they require breeding sites that are open to the sun. For this reason, dense coniferous afforestation has had a serious impact since this often results in drainage, acidification, shading of breeding sites and impoverishment of adult foraging sites by reducing prey densities. Nevertheless, the corduliids *Cordulia aenea* and *Somatochlora metallica* are virtually restricted to woodlands, at least in England, and these species may have been adversely affected by fragmentation of deciduous woodlands and the loss of woodland ponds. This may partly explain the rather patchy distribution of these species in Britain today.

8.3 Extraction industries

The efforts of the mineral extraction industries have resulted in a large increase in the number of flooded gravel, clay and chalk pits, especially in south-eastern England. Despite the initial habitat destruction when these pits were first dug, this activity has probably resulted in a net gain in dragonfly breeding sites. Rehabilitation of these sites is particularly valuable when nature conservation is borne in mind and they are restored as a mosaic of wetland features of differing surface area, volume and with extensive shallow margins (Andrews and Kinsman 1990).

Pristine peat bogs usually have few significant areas of open water. Therefore small-scale digging of peatlands, which produces a complex of small ditches and pools, has resulted in the creation of many valuable dragonfly habitats. However, the recent shift towards industrial-scale peat milling results in the wholesale removal of peat, exposing the mineral bedrock below. Now 98% of raised bogs and 90% of blanket bogs have been lost, mainly to peat extraction, forestry and agriculture (Foss 1997), and this has had a serious impact on the populations of acid-water specialists.

8.4 Water pollution

Declines in water quality due to industrial, agricultural and road runoff, together with pollution from sewage works, has had a serious impact on dragonfly numbers. Particularly badly effected are those species restricted to lowland rivers, and pollution has resulted in the national extinction of one species (*Oxygastra curtisii*). Another significant negative influence has been the management practices of water authorities and drainage boards which all too often result in the removal of bankside vegetation, the steepening and straightening of river banks, and canalisation to speed flow rates. Adult dragonflies require bankside and emergent vegetation for territorial perches, oviposition sites, roosting sites and feeding perches. Plentiful submerged aquatic vegetation is vital to larvae for concealment and for perches from which to hunt prey. They also require emergent vegetation for emergence supports during the transformation from larva to adult. However, in recent years there has been a trend towards more sensitive management practices and river restoration projects.

8.5 Climate change

In Britain, the 1990s have seen a series of record-breaking warm summers and mild winters and mean annual temperatures have risen by about 0.5° C since 1950. Despite this seemingly small increase there does appear to have been a response from dragonfly populations with a well-documented northward expansion of distributional range in several species. In addition, one essentially boreal species (*L. dubia*) has been lost from its last southern English localities, and there seems to have been a genuine increase in the abundance and diversity of non-breeding migratory species from continental Europe arriving in Britain, which cannot wholly be attributed to an increase in recorder effort.

8.6 Public interest

General public interest in nature conservation is currently high. Many people have become interested in wildlife gardening and this has resulted in large numbers of garden ponds being newly dug and managed sympathetically for wildlife. Many books are now available which describe how to create a wildlife pond and BDS have produced a booklet on pond creation with a view to encouraging dragonflies in particular (British Dragonfly Society 1992). Dragonflies have benefited from this activity and many species are now common and abundant in urban areas (Brooks 1989). Public interest in dragonflies has also increased enormously in the last two decades and this too has had a positive impact on dragonfly conservation. The British Dragonfly Society is burgeoning and dragonflies are receiving much more attention from other nature conservation societies such as the Royal Society for the Protection of Birds (RSPB). Thanks to this interest and the observations of amateur dragonfly enthusiasts we now know far more about dragonfly autecology and how to manage wetland sites in a way that will enhance dragonfly conservation.

8.7 National and international policies

The network of National Nature Reserves (NNR) and Sites of Special Scientific Interest (SSSI) has undoubtedly helped to conserve dragonflies in Britain, and helped to slow the decline in odonate populations. There are now 234 NNRs in Britain (Merritt *et al.* 1996) and by 1991, SSSIs covered 7% of the country (Moore 1991b). The NNR network is structured to reflect examples of different biotopes in Britain, based on plant communities. Despite the risk that this system might overlook important insect assemblages, it has served dragonflies well. All but one of the odonate species resident in Britain occur on one or more NNR. The exception is *Gomphus vulgatissimus* (Club-tailed Dragonfly) but this species occurs on several reserves managed by conservation non-governmental organisations (NGOs) (Merritt *et al.* 1996). Like NNRs, SSSIs are also selected for the most part on botanical or geological grounds. However, because the distribution of dragonflies is so well known, they are one of the few groups of invertebrates that can be used in SSSI selection. Knowledge of distribution is important since the selection criteria take into account the reduction in odonate diversity that occurs from south to north. An outstanding dragonfly assemblage in Orkney comprises seven species whereas one in central southern England must have at least 17 species to qualify.

While the SSSI system has undoubtedly protected many sites aspects of the system are flawed, and many sites are damaged and a few destroyed each year. Positive incentives that encourage nature conservation by landowners are inadequate. Landowners who are prevented from ploughing or draining a SSSI are compensated for profits foregone but do not receive financial encouragement to manage the site in a way that would enhance nature conservation. Therefore, SSSIs may decline in quality through neglect or lack of positive management as well as through outright destruction.

In addition to the statutory nature reserves, there are a large and expanding number of sites owned or managed by nature conservation NGOs. The largest landowner among these is the Royal Society for the Protection of Birds (RSPB). In recent years the RSPB has worked closely with BDS and has sought advice specifically to improve the management of its reserves for dragonflies. Not only has this conserved the populations of species already present but, with the construction of new dragonfly habitats, has encouraged additional species to colonise some reserves including several rarities (Pickess 1989). The positive impact of sympathetic management has been particularly well documented at the National Dragonfly Sanctuary in Ashton Wold. The Sanctuary is centred on a large lake which was originally quite turbid and poor in aquatic macrophytes due to the depredations of waterfowl and Chinese Water Deer; about five species of dragonfly bred in the lake. In order to encourage dragonflies, the lake was fenced to exclude the deer, aquatic plants were introduced and netted to prevent waterfowl reaching them, and emergent and bankside plants were encouraged. Within a few years 15 species had been recorded breeding at the site. This highlights how quickly dragonflies are able to respond to positive habitat management, because of their excellent dispersive abilities. Nevertheless, despite this network of nature reserves, without nature conservation measures also being focused on the wider countryside they will be insufficient to ensure the conservation of dragonflies in Britain in the future, especially in view of climate change, which may result in many of the nature conservation sites being 'in the wrong place'. Without hospitable areas of countryside linking the reserves they become isolated and, as the climate changes, dragonflies and other wildlife may be unable to track their optimum climatic envelopes as they have done in the past.

9 Prognosis for the next 25 years

The outlook for dragonfly conservation in the next 25 years is relatively bright. Piecemeal losses still occur, and these can be significant for species with restricted distributions, but the drastic declines of the 1950s and 1960s seem to have slowed and with the rise of public interest in dragonflies and our increasing knowledge of the appropriate conservation management of wetlands, it seems likely that this trend will continue. Nevertheless, demand for greenfield sites for development and roads continues to increase and even legal protection of sites of high nature-conservation value does not seem to take precedence over 'public interest'. Pressure from agricultural intensification continues to take a high toll and this will only be relieved with reform of the Common Agricultural Policy, leading to less emphasis on production and protecting markets, and more emphasis on extensification and management of the wider countryside with nature conservation in mind.

ACKNOWLEDGEMENTS

I am grateful to Dr Norman Moore and Mr David Winsland for their detailed comments on an earlier draft of this paper. Much of the information in this paper was drawn from the *Atlas of the Dragonflies of Britain and Ireland* by Merritt *et al.* (1996). I am grateful to these authors for making this information available and to all the dragonfly enthusiasts who submitted their records.

REFERENCES

Andrews, J. and Kinsman, D. (1990) *Gravel Pit Restoration for Wildlife: a Practical Manual.* Sandy: Royal Society for the Protection of Birds.

Anon. (1994) *Biodiversity. The UK Action Plan.* London: HMSO.

Benton, E. and Payne, R.G. (1983) On the rediscovery of *Lestes dryas* Kirby in Britain. *Journal of the British Dragonfly Society*, 1, 28–30.

British Dragonfly Society (1992) *Dig a Pond for Dragonflies.* Purley: British Dragonfly Society.

Brooks, S.J. (1988) Exotic dragonflies in north London. *Journal of the British Dragonfly Society*, 4, 9–12.

Brooks, S.J. (1989) The dragonflies (Odonata) of London: the current status. *London Naturalist*, 68, 109–131.

Brooks, S.J. and Lewington, R. (1997) *Field Guide to the Dragonflies and Damselflies of Great Britain and Ireland.* Hook: British Wildlife Publishing.

Cham, S.A. (1991) The scarce blue-tailed damselfly *Ischnura pumilio* (Charpentier): its habitat preferences in south-east England. *Journal of the British Dragonfly Society*, 7, 18–25.

Chelmick, D.G. (ed.) (1979) *Provisional Atlas of the Insects of the British Isles, Part 7, Odonata*, 2nd edn. Huntingdon: Institute of Terrestrial Ecology.

Coope, G.R. (1995) The effects of Quaternary climatic changes on insect populations: lessons from the past, in Harrington, R. and Stork, N.E. (eds) *Insects in a Changing Environment.* London: Academic Press, pp. 30–48.

Corbet, P.S., Longfield, C. and Moore, N.W. (1960) *Dragonflies.* London: Collins.

Foss, P. (1997) Ten years of the Save the Bogs Campaign, in Parkyn, L., Stoneman, R.E. and Ingram, H.A.P. (eds) *Conserving Peatlands.* Wallingford: CAB International, pp. 391–397.

Hammond, C.O. (1977) *The Dragonflies of Great Britain and Ireland.* London: Curwen.

Hammond, P.M. (1975) Seventeenth century British Coleoptera from the collection of Leonard Plukenet. *Entomologists' Gazette*, 26, 261–268.

Longfield, C. (1937) *The Dragonflies of the British Isles.* London: Warne.

Longfield, C. (1949) *The Dragonflies of the British Isles*, 2nd edn. London: Warne.

Lucas, W.J. (1900) *British Dragonflies* (Odonata). London: Upcott Gill.

Lucas, W.J. (1930) *The Aquatic (Naiad) Stages of the British Dragonflies (Paraneuroptera)*. London: Ray Society.

Merritt, R., Moore, N.W. and Eversham, B.C. (1996) *Atlas of the Dragonflies of Britain and Ireland*. London: HMSO.

Moore, N.W. (1976) The conservation of Odonata in Great Britain. *Odonatolagica*, 5, 37–44.

Moore, N.W. (1980) *Lestes dryas* Kirby – a declining species of dragonfly (Odonata) in need of conservation: notes on its status and habitat in England and Ireland. *Biological Conservation*, 17, 143–148.

Moore, N.W. (1986) Acid water dragonflies in Eastern England – their decline, isolation and conservation. *Odonatologica*, 15, 377–385.

Moore, N.W. (1991a) The last of *Oxygastra curtisii* (Dale) in England? *Journal of the British Dragonfly Society*, 7, 6–10.

Moore, N.W. (1991b) Recent developments in the conservation of Odonata in Great Britain. *Advances in Odonatology*, 5, 103–108.

Nature Conservancy Council (1989) *Guidelines for Selection of Biological SSSIs*. Peterborough: Nature Conservancy Council.

Pickess, B.P. (1989) The importance of RSPB reserves for dragonflies. *RSPB Conservation Review*, 5, 15–16

Pond Conservation Group (1993) *A future for Britain's ponds. An agenda for action*. Oxford: Pond Conservation Group.

Porritt, G.T. (1912) *Agrion armatum*, Charp. in the Norfolk Broads. *Entomologists' Monthly Magazine*, 48, 163.

Prince, P. and Clarke, R. (1993) The Hobby's breeding range in Britain – what factors have allowed it to expand? *British Wildlife*, 4, 341–346.

RSPB (1998) Conservation Action. *Birds*, 17, 54.

Shirt, D.B. (ed.) (1987) *British Red Data Books: 2. Insects*. Peterborough: Nature Conservancy Council.

Skelton, M.J.L. (ed.) (1974) *Insect Distribution Maps Scheme: Orthoptera, Dictyoptera and Odonata preliminary distribution maps*. Huntingdon: Biological Records Centre.

Chapter 21

Land slugs and snails

Robert A. D. Cameron and Ian J. Killeen

ABSTRACT

In general, changes in the land mollusc fauna of the British Isles over the last 25 years represent continuations of trends seen earlier. Limitations in the processes of data capture may conceal some declines, but it appears that a number of rare species, and a few commoner ones, have declined significantly. These declines are spread over a number of habitats. Very rare species, subject to statutory conservation policies, are now known from more sites than in 1973, but this is largely the result of improved surveying. Nearly all substantial increases in range have been in introduced species, some of which have spread very rapidly. These increases do not appear to relate to declines in native species. Climate change may play a part in some variations, but habitat destruction and modification is the most important factor in most declines. No native species have become extinct, and so the fauna has increased in size by the addition of new arrivals.

1 Introduction

In his review of changes in the land mollusc fauna of the British Isles, South (1974) was unlucky in writing just before the publication of the Conchological Society of Great Britain and Ireland's (CSGBI) first *Atlas* based on 10-km grid maps (Kerney 1976). He was, however, able to demonstrate both long-term changes (over the Holocene) and those over the previous century, and particularly since the first CSGBI vice-county census (Taylor and Roebuck 1885).

The 1976 CSGBI *Atlas* distinguished pre- and post-1950 records for some species, and the editor, M. P. Kerney, used the information obtained to indicate which species were declining or increasing on the British Isles distribution maps given in Kerney and Cameron (1979).

In this chapter, we have used data from all the above sources as a baseline for assessing changes since the 1970s. A second edition of the CSGBI *Atlas* was published after the initial preparation of this chapter (Kerney 1999), and we have taken account of it where possible. South (1974) demonstrated the difficulties in interpreting data collated on a vice-county basis; the same difficulties, albeit to a lesser extent, affect interpretation of successive grid-based censuses. We explore both pre- and post-1979 sources with critical attention to these difficulties.

2 Data capture and reliability

The incomplete state of our knowledge of the British Isles' fauna is illustrated by the discovery of 15 new species since 1973 (Table 21.1). Five of these additions are a consequence of segregating previously unrecognised sibling species. Among the remainder, those labelled as native are thought to be longstanding occupants which had been overlooked. *L. sempronii* was first collected in the 1930s, but more than 40 years elapsed before it was recognised and confirmed as a living species. These discoveries reflect the greater intensity and skill in surveying in recent years.

Since the publication of the 1976 *Atlas*, grid mapping has become the norm. Not only have more records been submitted to the CSGBI's recording scheme, but a number of more detailed studies of smaller areas have been undertaken, based on 2 × 2–km squares (tetrads). Lloyd-Evans (1975) pioneered this approach for a single 10 × 10–km square in Yorkshire, and two county atlases, for the Isle of Wight (Preece 1980) and Suffolk (Killeen 1992), have been published; others are in preparation.

In addition, the various national agencies responsible for conservation in England, Scotland and Wales have commissioned studies of particular species, especially those listed on the European Union Habitats and Species Directive (EUHSD) and the United Kingdom Biodiversity Action Plan (UKBAP). These new initiatives have increased the intensity of surveys in particular areas, and have also improved their quality. Many of them, particularly the commissioned surveys, have involved highly skilled recorders who have learnt to recognise the particular habitats of rare species, and to use sampling techniques which improve the chances of finding them.

Table 21.2 gives a striking demonstration of this for the four *Vertigo* species in priority categories of both EUHSD and UKBAP (see below). None of these species are

Table 21.1 Species added to the fauna of the British Isles since 1973, excluding greenhouse records.

Native	*Vertigo genesii*
	V. geyeri (in Great Britain, earlier in Ireland)
	V. arctica
	Lauria sempronii
	Perforatella rubiginosa
Uncertain status	*Tandonia rustica*
Introduced	*Helicodiscus singleyanus*
	Toltecia pusilla
	Lehmannia valentiana
	Cochlicella barbara
Segregates	*Arion distinctus*
	A. owenii
	A. flagellus
	Limax pseudoflavus
	Euconulus alderi

Table 21.2 Changes in the nominal UK distributions of the *Vertigo* species listed on the EUHSD and UKBAP.

Species	No. of 10-km square records		1999 Atlas
	1976 Atlas		
	pre-1950	post-1950	post-1965
Vertigo genesii	0	0	4
V. geyeri	0	0	16
V. angustior	4	1	9
V. moulinsiana	16	29	64

EUHSD: European Union Habitats and Species Directive; UKBPA: United Kingdom Biodiversity Action Plan.

believed to have increased their ranges or abundance since 1973, and have previously shown declines (Kerney 1976; Kerney and Cameron 1979).

This illustrates the general problem that the intensity and skill of surveying will affect the proportion of the actual distribution revealed. In the following sections, we attempt to discount this effect in order to identify real changes over the last 25 years.

In principle, there is a further source of data, free of this difficulty, in the mass of papers cataloguing the faunas of particular, precisely located sites, or giving precise locations in studies of single species. Resurveys, however, are rare; we give an example for one species below (p. 360). Regular monitoring of rare or declining species is increasing as the provisions of EUHSD and UKBAP are implemented.

Although most species may be correctly identified by recorders with a little experience, there are some difficult determinations, especially among slugs that require dissection of mature specimens. The CSGBI requires voucher specimens for all new vice-county records, and for any records which the recorder has reason to doubt. Nomenclature in this chapter follows Kerney *et al.* (1983), with the addition of *Arion flagellus*.

3 Changes prior to the late 1970s

Table 21.3 shows the number of species regarded as declining or increasing by Kerney (Kerney and Cameron 1979), categorised by status (native or introduced) and by rarity (number of 10 × 10 km squares occupied in Great Britain recorded in 1976). Declines are nearly all among native species, and rare species are more likely to be declining than common ones. This association of decline with rarity is partly an artifact of the surveying methods used: species which 'thin out' but remain within roughly the same number of grid squares or vice-counties will not show significant declines. Only introduced species show increases in range.

Table 21.4 lists the declining species by habitat. This is a somewhat arbitrary classification, since many species occur in more than one habitat. The wetland category is particularly heterogeneous, ranging from lowland fens to calcareous flushes at high altitudes; several species listed, for example *Vertigo moulinsiana* and *V. geyeri*, could never be found at the same site. The table shows, however, that declines were not confined to a single habitat type.

Table 21.3 Numbers and proportions of species declining and increasing in range as recorded by Kerney (Kerney and Cameron 1979), categorised by status (native or introduced) and by the number of 10-km grid squares recorded for each in Great Britain in the 1976 *Atlas* (Kerney 1976).

	No. of 10-km squares occupied			
	<10	*10–100*	*>100*	*Total*
'Native'	9	19	67	95
Declining (none increasing)	8 (89%)	9 (47%)	11 (16%)	28 (29%)
'Introduced'	9	1	13	23
Declining	1 (11%)	0	0	1 (4%)
Increasing	3 (33%)	0	5 (38%)	8 (35%)

Table 21.4 Declining species by habitat (Kerney 1979).

Wetlands	*Open, exposed*	*Woodland*
Catinella arenaria	*Truncatellina cylindrica*	*Acicula fusca*
Succinea oblonga	*Pomatias elegans**	*Leiostyla anglica*
Oxyloma sarsi	*Abida secale**	*Spermodea lamellata*
*Vertigo antivertigo**	*Pupilla muscorum**	*Ena montana*
V. substriata	*Helicella itala**	*Limax cinereoniger*
V. moulinsiana	*Monacha cartusiana**	*L. tenellus*
V. geyeri		*Zenobiella subrufescens*
*V. angustior**		*Helicodonta obvoluta*
V. lilljeborgi		

Others

Vertigo pusilla, V. alpestris (walls and rocks)

Balea biplicata (wet woodland, Thames valley)

*B. perversa** (rocks and tree trunks)

*Helicigona lapicida** (rocks and woods)

* Evidence for continued decline.

Table 21.5 Species increasing in the twentieth century (Kerney 1979) – all introductions.

Oxychilus draparnaudi	*Monacha cantiana*
Tandonia budapestensis	*Hygromia cinctella**
Boettgerilla pallens*	*H. limbata**
Deroceras caruanae	*Trichia striolata*

* Species showing continued substantial increases after 1976.

Table 21.5 lists the increasing species. With the exception of *Monacha cantiana*, all these species are common in gardens and similar anthropogenic habitats. The causes of these changes are considered below, in conjunction with those occurring since 1979.

4 Changes since the 1970s

Apart from the sources given above, the source for new records at vice-county level comes from the annual reports of the non-marine recorder for the CSGBI, published in the *Journal of Conchology*. Table 21.6 shows some of the species for which new vice-county records were accepted over the period 1980–98. In the first part of this table, the species listed are all either rare, or were recorded as declining in 1979, or both. Most have been the subject of intensive surveys, and there is no reliable evidence that either their distributions or densities have increased (cf. the particular case of the *Vertigo* species above).

The second part of Table 21.6 records increases in introduced species which are sufficiently large for there to be little doubt that they are authentic. The most dramatic increase is that of the small slug *Boettgerilla pallens*. First recorded in 1972, it was known from 47 vice-counties by 1990, and has spread across Europe in a manner reminiscent of that of the collared dove *Streptopelia decaocto* (cf. Williamson 1996) but unlike the bird, the slug has been transported passively. Many other species, not listed in Table 21.6, have also accumulated new vice-county records; these demonstrate the necessarily incomplete knowledge of ranges.

Despite the data in the new *Atlas* (Kerney 1999), some genuine declines are hard to identify, as they may be obscured by more intensive sampling. Wetland and woodland species appear to be controlled by the quantity and quality of their habitats. Destruction of ancient woodland, documented by Rackham (1976) has slowed over the last 25 years (ibid. 1986), and where such woodland survives, it appears to retain its molluscan fauna. The wetland situation is more complex. Wetlands known to contain rare or previously declining species are increasingly coming under close scrutiny, and are likely to be subject to management for conservation. Among the wetland species listed in Table 21.4, the relatively common *Vertigo antivertigo* has probably suffered the greatest continued loss of habitat. It is found in a wider range of wetland sites than the rarer species, and not all of this range receives the level of scrutiny and protection afforded to 'hotspots' with rare species.

Among the open-country species in Table 21.4, there is good evidence that several are continuing to decline in distribution and abundance. These declines can be seen in

Table 21.6 Selected species for which new vice-county records were accepted between 1980 and 1998, as given in the CSGBI non-marine recorder's reports in the *Journal of Conchology.* (A) Species which despite these new records are thought to be declining; (B) Species that are increasing substantially, all of which are introduced.

A

Species	Habitat	Result of
Vertigo moulinsiana	Wetland	Intensive surveys
V. angustior	Wetland	Intensive surveys
V. geyeri	Wetland	Intensive surveys
V. genesii	Wetland	Intensive surveys
Catinella arenaria	Wetland	Intensive surveys
Cochlodina laminata	S. woods	Confirmation of old northern records
Helicigona lapicida	Rocks and woods	?

B

Species	Current range	Native to
Boettgerilla pallens	Countrywide	East Europe
Theba pisana	Local, south-west	Mediterranean
Hygromia cinctella	Southern England	South-west France
Toltecia pusilla	Scattered	?Nearctic
Lehmannia valentiana	Scattered, south	Iberian peninsula

the published county atlases (Preece 1980; Killeen 1992) and in the new national *Atlas* (Kerney 1999). Many of these species are associated with short, sheep- and rabbit-grazed grassland on chalk and limestone, a habitat which has contracted in extent and deteriorated in quality as a result of changes in agricultural practice and of reductions in rabbit populations as a result of myxomatosis (Cameron *et al.* 1980a; Ingrouille 1995). The greatest decline has been in *Helicella itala*, which has disappeared from many of its inland sites. Fortunately *H. itala*, along with other affected species such as *Pupilla muscorum*, occurs at high densities in many sand-dune systems round the coast.

The changes in calcareous grassland, and particularly the shift to longer grasses and scrub, may be partly responsible for a considerable decline in another common species, *Cepaea nemoralis*, not featuring in the 1979 list. This species and its congener *C. hortensis* have been the subject of intensive studies of their colour and banding polymorphisms (Jones *et al.* 1977), which has resulted in records of vast numbers of accurately located populations. Some of these, first recorded in the 1950s and 1960s, have been resurveyed recently.

Cowie and Jones (1987) report a considerable decline in *Cepaea nemoralis* on the Marlborough Downs in Wiltshire; it is recorded as declining by both Preece (1980) and

Table 21.7 Change in *Cepaea* populations at Rickmansworth, Hertfordshire from 1965 to 1990.

	Habitat	1965 sites	Losses by 1990
Cepaea nemoralis	Hedges	21	13 (62%)
	Others	16	0
C. hortensis	Hedges	54	1 (2%)
	Others	28	1 (4%)

Killeen (1992), and Cameron (1992) gives data (Table 21.7) that show the same decline in a part of Hertfordshire. Yalden (pers. comm.) indicates the same for parts of Derbyshire, but anecdotal evidence suggests that this decline is not so evident further north. In some cases (Cowie and Jones 1987) *C. hortensis* has shown a corresponding increase in sites occupied but this is not universally reported. These changes are not confined to calcareous grassland, so factors other than those mentioned above may be involved.

Among the species listed for 'others' in Table 21.4, *Balea perversa* may be affected by atmospheric pollution (see below). *Helicigona lapicida* appears to have continued to decline, especially on the eastern side of Great Britain (Killeen 1992).

5 Conservation status and management

Table 21.8 lists species in the UK placed in priority categories under various conservation schedules and directories. These lists have been compiled using various criteria, including rarity (national and global), perceived threat to habitat types and evidence of decline. Thus *Geomalacus maculosus*, for example, is confined within the British Isles to a relatively small area of south-west Ireland, and globally to patches along the Atlantic coast of Europe. *Helix pomatia*, an introduced species to Britain, is on the list because of commercial exploitation for food across its range. Active monitoring and development of management plans already takes place for species on Annex II of the EUHSD and on the 'short-list' of the UKBAP; Drake (1997) is an example of this kind (in this case, for *Vertigo moulinsiana*).

6 The validity of comparisons over time

As shown above, an increase in the number of vice-county or grid-square records is not in itself sufficient evidence for an increase in range or abundance. Equally, an apparently static picture at this scale may conceal real declines, particularly in common species such as *Cepaea nemoralis*. In principle, a decrease in records may likewise not mean a real decline, if for any reason sampling is less intense or less focused, which might happen for common species in 'mundane' habitats like hedges. Changes in taxonomy and in the criteria required for reliable identification also have significant effects on the interpretability of data, especially at present for slugs. Some species are

Table 21.8 Conservation schedules for British terrestrial molluscs.

Species	EUHSD	RDB	BAP	WCA	Berne
Catinella arenaria		1	S	V	
Succinea oblonga		3	L		
Oxyloma sarsi		2	L		
Truncatellina cylindrica		2	L		
T. callicratis		3	L		
Vertigo moulinsiana	II	3	S		
V. arctica		1	L		
V. lilljeborgi		3	L		
V. geyeri	II	1	S		
V. genesii	II	1	S		
V. angustior	II	1	S		
Lauria sempronii		1	L		
Leiostyla anglica			L		
Ena montana		3	L		
Geomalacus maculosus	II				2
Tandonia rustica		4			
Limax tenellus			L		
Clausilia dubia			L		
Balea biplicata		3			
Monacha cartusiana		3	L		
Ashfordia granulata			L		
Perforatella rubiginosa		2			
Helicodonta obvoluta		3	L		
Helix pomatia	V		L		3

EUHSD   European Union Habitats and Species Directive Annexes.
RDB   British *Red Data Book* 3 (Bratton 1991).
BAP   UK Biodiversity Action Plan lists (L, long list; S, short list).
WCA   UK Wildlife and Countryside Act 1981 (1998) schedules.
Berne   Appendices to the Berne Convention.

notoriously hard to find, making interpretation of grid-square maps perilous (the sub-terranean slugs in the genus *Testacella* are the most obvious example).

While tetrad sampling provides a better basis for monitoring the details of change than coarser-scaled surveys, repeat sampling of accurately located sites, if possible with the same sampling methodology, gives the most reliable results. Excellent results of this approach, or one approaching it, can be seen in recent Swedish studies (Waldén 1992; von Proschwitz 1994). While possible for monitoring a small number of rare species, this is scarcely practical for all species on a nationwide basis. Good data sets do exist, however, for repeat surveys on some species or habitats. Where changes are substantial, and especially where they affect one part of the range more than another, the relatively coarse-grained data generally available can give reliable results.

7 Causes of change

7.1 Climate

South (1974) wrote towards the end of a period in which the British climate appeared to be cooling relative to the early part of the twentieth century. He suggested that various alien species had established themselves in that warmest period and had been restricted thereafter, suspecting that some native species apparently retreating southwards, for example *Pomatias elegans*, *Ena montana* and *Helicodonta obvoluta*, might be doing so under the influence of climatic change.

The latter species all had wider distributions in the mid-Holocene, the climatic optimum (Kerney 1968). Taking the longer view over the whole of the Holocene, there is a small suite of arctic-alpine and east European species that are now extinct or greatly restricted in the British Isles. These have undoubtedly been affected by climate variation and the associated change in available habitats (Kerney 1977; Kerney *et al.* 1980). The fauna of the climatic optimum, however, remains intact; there is no evidence for extinction of any native species in the last 5–6000 years. Some taxa including several *Vertigo* species, *Leiostyla anglica* and *Spermodea lamellata* appear to have retreated north-westwards, though not obviously in the last 25 years, while others, for example *Helicodonta obvoluta* and *Monacha cartusiana*, have retreated to the south and east. Relative to the effects of human disturbance, the effects of climate appear slight and ambiguous for native species in recent times.

With some qualifications, this appears also to be the case for introductions from more southerly regions. Some aliens, for example *Helix aspersa* and *Monacha cantiana*, have maintained large ranges, although anecdotal evidence suggests that the former suffers heavy mortality in hard winters; it has certainly increased greatly in numbers in many urban areas in recent years. Others have either become extinct (*Cernuella neglecta*, *Bradybaena fruticum*) or remained confined in very few populations (*Trochoidea elegans*, *Cochlicella barbara*). A few have shown significant range expansions over the last 20 years – *Hygromia cinctella* (Willing 1998), *H. limbata*, *Theba pisana* and *Lehmannia valentiana*. While these have no doubt largely been a result of passive dispersal, their survival in new locations may owe something to the mild winters of the period. *L. valentiana* in particular was a 'greenhouse alien', but it is now being found in normal semi-natural habitats.

7.2 Introductions and competition

In principle, successful invading species might compete with established natives, eliminating them or restricting their geographical or habitat ranges. While it is impossible to be sure of the status of every species, it seems likely that 25–30 species (20–25% of the fauna outside greenhouses) have been introduced successfully over a period extending from pre-Roman Iron Age times to the twentieth century; no native species is known to have become extinct in the British Isles over that period. While a few species are common in woods and hedgerows, most are at peak abundance in extremely anthropogenic habitats, especially gardens. There is a suggestion that the native, largely self-fertilising slug *Arion ater* may be retreating, while its closely related, largely out-crossing and possibly introduced congener *A. rufus* is advancing (Ellis 1969). Other-wise, there is no clear case where any decline in a native species can be associated with an increase in an introduced species, although this is based on ranges rather than local densities. Competition appears to be at best a weak force in structuring the land mollusc fauna (Boycott 1934).

7.3 Habitat change

Habitat change is undoubtedly the major cause of faunal changes, particularly declines, in distributions of native species. In the case of woodland and wetland species, the habitats themselves are either destroyed or subject to extreme modification by design; in the case of calcareous grasslands, the change has been more accidental, although conversion of grassland to arable has played a part (Cameron *et al.* 1980a).

More subtle changes in habitat may be missed. While coppicing appears to have rather little impact on woodland faunas, other management practices, for example the use of herbicides on hedge bottoms and verges, may have an effect (Cameron *et al.* 1980b). Holyoak (1978) provides good evidence that *Balea perversa* is adversely affected by atmospheric pollution, which destroys its food supply and acidifies its substrate. Other species, for example *Clausilia bidentata*, may be similarly affected, and a number of distributions in Kerney (1976, 1999) show gaps coincident with areas of heavy pollution. In Sweden, acid rain has contributed to significant density reductions which can be reversed by liming (Gärdenfors 1987). Some sources of pollution have declined in the last 25 years, but there is as yet no clear evidence of corresponding spreads in ranges by snails, which are slow dispersers.

8 Future trends

The native land mollusc fauna of the British Isles has proved durable in the face of drastic environmental change brought about by human activities. This contrasts with the fate of such faunas on many oceanic islands (Cameron and Cook 1996). Introduced species appear not to displace natives, and so the fauna of the region has become richer over time, and is likely to continue doing so.

While there have been no extinctions, there have been numerous reductions in range, and the ranges of many rare species are very fragmented. They survive in this state because minimum viable population areas are small (Soulé 1986) and very small areas may support huge populations (approximately 800 m^{-2} for *Vertigo angustior* in parts of one site; Sharland in prep.). Despite the protection and management given to an increasing number of sites containing rare species, increasing fragmentation increases the risk of local extinctions, and most species have poor powers of dispersal. The continued, if slower, erosion of ancient woodland and wetland sites, their small sizes and continued isolation from one another all increase the risks of local and eventually national extinctions. There is a case for experimental introductions into reconstituted or successional habitats within existing geographical ranges.

While evidence for the effects of short-term climatic change is slight, greater change sustained over several decades might be expected to have an effect. In this context, rainfall regimes and water table levels are likely to be particularly important. Several species such as, for example, *Arianta arbustorum*, have a much more restricted range of habitats in the drier south-east of England than elsewhere.

REFERENCES

Boycott, A.E. (1934) The habitats of land Mollusca in Britain. *Journal of Ecology*, 22, 1–38.

Bratton, J.H. (ed.) (1991) *British Red Data Books: 3. Invertebrates other than insects.* Peterborough: Joint Nature Conservation Committee.

Cameron, R.A.D. (1992) Change and stability in *Cepaea* populations over 25 years: a case of climatic selection. *Proceedings of the Royal Society of London, B*, 248, 181–187.

Cameron, R.A.D. and Cook, L.M. (1996) Diversity and durability: Responses of the Madeiran and Porto-Santan snail faunas to natural and human-induced environmental change. *American Malacological Bulletin*, 12, 3–12.

Cameron, R.A.D., Carter, M.A. and Palles-Clark, M.A. (1980a) *Cepaea* on Salisbury Plain: patterns of variation, landscape history and habitat stability. *Biological Journal of the Linnean Society of London*, 14, 335–358.

Cameron, R.A.D., Down, K. and Pannett, D.J. (1980b) Historical and environmental influences on hedgerow snail faunas. *Biological Journal of the Linnean Society*, 13, 75–87.

Cowie, R.H. and Jones, J.S. (1987) Ecological interactions between *Cepaea nemoralis* and *Cepaea hortensis*: competition, invasion but no niche displacement. *Functional Ecology*, 1, 91–97.

Drake, C.M. (ed.) (1997) *Vertigo moulinsiana* surveys and studies commissioned in 1995–96. *English Nature Research Reports*, 217, 1–68.

Ellis, A.E. (1969) *British Snails*, 2nd edn. Oxford: Oxford University Press.

Gärdenfors, U. (1987) *Impact of Airborne Pollution on Terrestrial Invertebrates*. Solna: National Swedish Environmental Protection Board.

Holyoak, D.T. (1978) Effects of atmospheric pollution on the distribution of *Balea perversa* (Linnaeus) (Pulmonata: Clausiliidae) in southern Britain. *Journal of Conchology*, 29, 319–324.

Ingrouille, M. (1995) *Historical Ecology of the British Flora*. London: Chapman and Hall.

Jones, J.S., Leith, B.H. and Rawlings, P. (1977) Polymorphism in *Cepaea*: A problem with too many solutions? *Annual Review of Ecology and Systematics*, 8, 109–143.

Kerney, M.P. (1968) Britain's fauna of land Mollusca and its relation to the post-glacial thermal optimum. *Symposia of the Zoological Society of London*, 22, 273–291.

Kerney, M.P. (ed.) (1976) *Atlas of the Non-Marine Mollusca of the British Isles*. Cambridge: Institute of Terrestrial Ecology.

Kerney, M.P. (1977) British Quaternary non-marine Mollusca: a brief review, in Shotton, F.W. (ed.) *British Quaternary Studies: Recent advances*. Oxford: Clarendon Press, pp. 31–42.

Kerney, M.P. (1999) *Atlas of the Land and Freshwater Molluscs of Britain and Ireland*. Colchester: Harley Books.

Kerney, M.P. and Cameron, R.A.D. (1979) *Field Guide to the Land Snails of Britain and North-West Europe*. London: Collins.

Kerney, M.P., Cameron, R.A.D. and Jungbluth, J.H. (1983) *Die Landschnecken Nord- und Mitteleuropas*. Hamburg and Berlin: Paul Parey.

Kerney, M.P., Preece, R.C. and Turner, C. (1980) Molluscan and plant biostratigraphy of some late Devensian and Flandrian deposits in Kent. *Philosophical Transactions of the Royal Society of London, B*, **291**, 1–43.

Killeen, I.J. (1992) *The Land and Freshwater Molluscs of Suffolk*. Ipswich: Suffolk Naturalists' Society.

Lloyd-Evans, L. (1975) The biogeography of snails in Yorkshire. *Naturalist*, **100**, 1–12.

Preece, R.C. (1980) *An Atlas of the Non-Marine Mollusca of the Isle of Wight*. Isle of Wight: IOW County Council.

von Proschwitz, T. (1994) Zoogeographical studies on the land Mollusca of the province of Dalsland (SW Sweden). *Acta Regiae Societatis Scientarum et Litterarum Gothoburgensis, Zoologica*, **15**, 1–152.

Rackham, O. (1976) *Trees and Woodland in the British Landscape*. London: Dent.

Rackham, O. (1986) *The History of the Countryside*. London: Dent.

Soulé, M.E. (ed.) (1986) *Conservation Biology*. Sunderland, MA: Sinauer Associates.

South, A. (1974) Changes in composition of the terrestrial mollusc fauna, in Hawksworth, D.L. (ed.) *The Changing Fauna and Flora of Britain*. London: Academic Press, pp. 255–274.

Taylor, J.W. and Roebuck, W.D. (1885) Census of the authenticated distribution of British land and freshwater Mollusca. *Journal of Conchology*, **4**, 319–336.

Waldén, H.W. (1992) Changes in a terrestrial mollusc fauna (Sweden: Göteborg Region) over 50 years by human impact and natural succession, in Gittenberger, E. and Goud, J. (eds) *Proceedings of the Ninth International Malacological Congress*. Leiden: Unitas Malacologica, pp. 387–402.

Williamson, M.H. (1996) *Biological Invasions*. London: Chapman and Hall.

Willing, M. (1998) Alien landsnails on the move: the spread of the girdled snail *Hygromia cinctella* and the sandhill snail *Theba pisana*. *Conchologists' Newsletter*, **9**, 108–110.

Chapter 22

Birds

David W. Gibbons and Mark I. Avery

ABSTRACT

A suite of annual monitoring schemes has tracked the fortunes of two-thirds of UK breeding bird species since the early 1970s. Many other species are censussed periodically and distributions of all are mapped every 20 years. Wintering waterfowl have been monitored since the late 1940s.

Two hundred and forty-two species of bird bred in the UK at some time between 1800 and 1997; 232 between 1970 and 1997. Nearly 40 more species bred in the last quarter of the twentieth century than during the first half of the nineteenth. From 1970 to 1997 there was a net increase of about four breeding species per decade, continuing a trend started in 1950.

Populations of introduced species, rare breeding species, most breeding seabirds and birds of prey and many non-breeding waterfowl increased during 1970–97. In contrast, populations of woodland species fell moderately, and farmland birds steeply, during the last quarter of the twentieth century.

Targeted conservation action, ranging from species reintroductions to customised agri-environment schemes, has benefited populations of many rarer species. Seabirds may be being maintained at artifically high levels by current fisheries practices. Populations of birds of prey increased as a result of the lessening of human persecution and the removal of organochlorine pesticides. The success of wildfowl is probably due to an increasing number of artifical water-bodies, a reduction in hunting pressure, the phasing out of lead, increased eutrophication and the management of wetlands for nature conservation. Profound changes in crop and grass production, brought about through agricultural intensification, have played a major part in the declines of many common farmland birds.

1 State of knowledge

Though imperfect, knowledge of the status of Britain's avifauna is probably greater than for any other taxonomic group. There are a number of reasons for this: there are relatively few species and their taxonomy is well known; most are readily identifiable with little specialist training; many are relatively visible, and those that are not are frequently audible. There are tens of thousands of amateur enthusiasts, and birds have a high public profile as a consequence. The information gained by these enthusiasts is captured by a small number of organisations with professional staff who synthesise it to determine the status of individual species and of Britain's avifauna as a whole.

The changing fortunes of Britain's breeding birds have been well documented. Most notable among these publications are those of Alexander and Lack (1944), Parslow (1973), Sharrock (1974a), Gibbons et al. (1996a) and Holloway (1996). Between them, these publications summarised the status of individual breeding species for the period 1800–1995. The more recent of these publications have had a wealth of information sources to call upon. The distributions, and changing distributions, of Britain's birds are well known from two breeding season atlases (Sharrock 1976; Gibbons et al. 1993) and one wintering atlas (Lack 1986). The status of Britain's two dozen breeding seabirds has been documented by Cramp et al. (1974) and Lloyd et al. (1991). The status of Britain's common breeding birds are particularly well known and has been documented in Marchant et al. (1990) and, more recently, in Crick et al. (1998). Other notable publications have been: Owen et al. (1986), which discussed the status of waterfowl in Britain; Batten et al. (1990), which summarised the status of Red Data birds in Britain; and Thom (1986) and Lovegrove et al. (1994), which documented the status of the Scottish and Welsh avifaunas respectively. Knowledge of the status and distribution of Britain's birds has recently been placed in a broader, continental context, with the publication of an atlas of European breeding bird distributions (Hagemeijer and Blair 1997) and an assessment of the conservation status of European birds (Tucker and Heath 1994). The information contained within all of these publications has, to a greater or lesser extent, been based on information collected by one of several bird monitoring schemes.

2 Monitoring

Britain's breeding and non-breeding birds are routinely and comprehensively monitored (Table 22.1). Most schemes cover the UK rather than Britain and some encompass all of Britain and Ireland. These schemes are most obviously divided into breeding and non-breeding season schemes, with the breeding season better monitored (Table 22.1). There are several reasons for this. First, because of their generally territorial nature, birds are more evenly and predictably distributed across the countryside in the breeding season, making them amenable to simple monitoring techniques. Second, a large number of species are resident in Britain, thus counting them in the non-breeding season is unlikely to yield much information on trends over and above that from the breeding season. Third, the most important non-breeding immigrants, the waterfowl, are all well covered by a single scheme (WeBS); there are thus few important groups of non-breeding birds that are not monitored (wintering thrushes being one exception). Though most of these schemes focus on monitoring population levels, others cover demographic parameters such as productivity and survival.

2.1 Monitoring of breeding bird populations

The longest running breeding-bird monitoring scheme is the Common Birds Census (CBC; Marchant et al. 1990). Since 1962, this annual scheme has monitored the population trends of common, terrestrial breeding birds using a method that involves observers mapping the numbers of bird territories on sample survey plots, most often in

Table 22.1 Bird monitoring schemes in Great Britain.

Scheme	Lead partner	Others	Season	Species covered	Scope	Distbn	Popn	Demog	Start	Freq	Sample or census	Vols (V) or profs (P)
Common Birds Census (CBC)	BTO	JNCC	breeding	common, terrestrial	UK		√		1962	1y	sample	V
Waterways Birds Survey (WBS)	BTO	JNCC	breeding	riparian	UK		√		1974	1y	sample	V
Breeding Bird Survey (BBS)	BTO	JNCC, RSPB	breeding	common, terrestrial	UK	(√)	√		1994	1y	sample	V
Wetland Birds Survey (WeBS)	WWT	BTO, JNCC, RSPB	non-breeding	waterfowl	UK	√	√		1993 (1947)	1y	census	V
Rare Breeding Birds Panel (RBBP)	—	BB, BTO, JNCC, RSPB	breeding	rare species	UK	(√)	√		1973	1y	census	V
Statutory Conservation Agencies/RSPB Annual Breeding Bird Scheme (SCARABBS)	RSPB	JNCC, EN, SNH, CCW, DoENI	breeding	Red Data	UK	(√)	√		1998 (1961)	1y	variable	V and P
Seabird 2000, Seabird Colony Register, Operation Seafarer	JNCC	Seabird Group, RSPB, SOTEAG	breeding	seabirds	B and I	√	√		1969	15y	census	V and P
Seabird Monitoring Programme (SMP)	JNCC	Seabird Group, RSPB, SOTEAG	breeding	seabirds	B and I	(√)	√	√	1986	1y	sample	V and P
Breeding atlases	BTO	IWC, SOC	breeding	all	B and I	√	(√)		1968	20y	census	V
Winter atlas	BTO	IWC	winter	all	B and I	√	(√)		1981	>20y	census	V

continued on next page

Table 22.1 Bird monitoring schemes in Great Britain. (cont.)

Scheme	Lead partner	Others	Season	Species covered	Scope	Distbn	Popn	Demog	Start	Freq	Sample or census	Vols (V) or profs (P)
Nest Record Scheme (NRS)	BTO	JNCC	breeding	most	UK			√	1939	1y	sample	V
Constant Effort Sites (CES) ringing	BTO	JNCC	breeding	various	UK		(√)	√	1983	1y	sample	V

Organisations

BTO   British Trust for Ornithology
WWT   Wildfowl and Wetlands Trust
RSPB   Royal Society for the Protection of Birds
JNCC   Joint Nature Conservation Committee
BB   British Birds
EN   English Nature
SNH   Scottish Natural Heritage
CCW   Countryside Council for Wales
DoENI   Department of the Environment for Northern Ireland
SOTEAG   Shetland Oil Terminal Environmental Advisory Group
IWC   Irish Wildbird Conservancy (now BirdWatch Ireland)
SOC   Scottish Ornithologists' Club

Scope

UK   United Kingdom
B and I   Britain and Ireland

Distbn   monitoring of distributions
Popn   monitoring of populations
Demog   monitoring of demographic populations;
     √ measured by scheme;
     (√) measured by scheme in part
Freq   frequency of scheme; e.g. 1y = annual, 20y = every 20 years

Vols (V) or profs (P); if V, majority of fieldwork is by volunteers; if P, majority is by professionals farmland or woodland. A sister survey, the Waterways Bird Survey (WBS), was introduced in 1974 to monitor riparian species. Results of the CBC and WBS are reported annually in *BTO News*, with long-term trends summarised in Marchant *et al.* (1990) and Crick *et al.* (1998). The CBC has been invaluable in revealing the population trends of Britain's breeding birds; despite this, however, it does have a number of limitations. Most survey plots are in south and east England and are therefore not representative of Britain as a whole; in addition observers choose their own plots so the areas sampled may be unrepresentative. Only a few habitats are covered, and only 200–300 plots are surveyed each year because of the time-consuming nature of the work, both in the field and office (Gregory *et al.* 1998).

Because of these limitations, the Breeding Bird Survey (BBS) was introduced in 1994, and will eventually replace the CBC. BBS fieldworkers record all birds seen or heard while walking two parallel 1-km-long transects within randomly allocated 1-km squares of the Ordnance Survey National Grid. In 1997, an astonishing 2169 BBS squares were covered. These survey plots are widely distributed across Britain, cover all the major habitats and because of this and the large number of plots, cover more species than the CBC. Because of the formal sampling design, BBS population trends will be representative of Britain as a whole. The BBS will monitor trends of 100 species with at least moderate precision; this is nearly half of all breeding species and is about half as many species again as covered by the CBC and WBS. The results of the BBS are reported annually (e.g. Gregory *et al.* 1998). Although the success of the BBS will be measured most in the long term, after only four years of fieldwork interesting trends had begun to emerge by 1997. A very similar scheme, the Countryside Bird Survey (CBS), was started in 1998 in the Republic of Ireland (Murphy and Coombe 1998). Work is currently under way to introduce a similar design and method into the WBS.

At the other end of the abundance spectrum, rare breeding birds, those with fewer than 300 pairs breeding in UK, are monitored by the Rare Breeding Birds Panel. Since 1973 this panel of experts has collated records of rare breeding species from observers, mainly via a network of county organisers (Spencer and RBBP 1992). Unlike the CBC or BBS, the RBBP aims to report what amounts to complete annual censuses of all rare breeding species. There is no sampling design – indeed it would be hard to see how there could be for most species – and no field method; it is simply a collation of records. For some species the RBBP reports the great majority of breeding pairs, particularly when the species in question is the focus of detailed work by conservation organisations or occurs in areas with many birdwatchers. Examples are stone curlew *Burhinus oedicnemus,* or Dartford warblers *Sylvia undata* in England, or red kite *Milvus milvus* in Wales. For other species, however, such as redwing *Turdus iliacus*, the proportion of the population reported may be unknown. Only when full national surveys of these species are undertaken (e.g. Gibbons and Wotton 1996; Underhill *et al.* 1998; Wotton *et al.* 1998) is this known. Despite this, the RBBP is one of the most cost-effective of monitoring schemes, covering more than a quarter of all breeding species for a modest financial investment. The great value of the RBBP is that it brings together records that would have been collected anyway, makes them available by publishing them and thus encourages further data collection.

The highly localised distribution of seabirds, with huge numbers breeding at a relatively small number of traditional locations, makes them amenable to census and the two dozen species that breed in Britain and Ireland are monitored by two closely related schemes. The first is a complete census of all breeding seabirds, undertaken every 15 years or so. There have been two of these, with the fieldwork for 'Operation Seafarer' (Cramp *et al*. 1974) and the 'Seabird Colony Register' (Lloyd *et al*. 1991) undertaken mainly during 1969–70 and 1985–87 respectively. Fieldwork for the third, 'Seabird 2000', will be carried out during 1999–2001. Though attempting to cover all seabird colonies, the survey concentrates most comprehensively on those on the coast. Thus coverage of an entirely coastal species, such as the gannet *Morus bassanus*, is near complete, while that of a species that also breeds inland, such as the common gull *Larus canus*, is less comprehensive. 'Seabird 2000' recognises that complete coverage may not be practical, so has designed a sampling regime that ensures that at least all the most important colonies are counted. To ensure that any major changes occurring in between these censuses are detected, the Seabird Monitoring Programme (SMP) was instigated in 1986 (see for example Thompson *et al*. 1998). As part of this programme, seabirds are counted at sample colonies throughout Britain and Ireland; for pragmatic reasons this sample is, unfortunately, non-random. Despite the apparently high level of monitoring of seabird species, there still remains substantial uncertainty over the population trends of the nocturnal, burrow-nesting Manx shearwater *Puffinus puffinus*, and both the storm (*Hydrobates pelagicus*) and Leach's (*Oceanodroma leucorhoa*) petrels. Great advances in developing monitoring techniques for these species were, however, made in the late 1990s (Gilbert *et al*. 1998; Ratcliffe *et al*. 1998).

A substantial proportion of the breeding species that are neither rare, nor common, nor seabirds but are listed as 'Red Data' birds by Batten *et al*. (1990) are monitored within the joint statutory conservation agencies/RSPB annual breeding bird scheme, known by the awkward acronym SCARABBS. Like the other schemes, this is generic in that it covers about 40 species, yet survey work is largely undertaken on a species-by-species basis. The reason for this is that such species are too rare to be covered by BBS, yet too common to be reliably monitored by the RBBP, or they occur in areas with few birdwatchers (such as north and west Scotland). Because of this it is necessary to undertake single-species surveys tailored to the distribution and natural history of each species. This is expensive, so these species are surveyed only intermittently, typically every five or 10 years depending on the conservation status of the species concerned. Although SCARABBS has only formally been in existence since 1998 it has been around since the early 1960s under a different guise. Thus, for example, there have been four Dartford warbler surveys, one in each decade from the 1960s to the 1990s (Bibby and Tubbs 1975; Robins and Bibby 1985; Gibbons and Wotton 1996). The fifth survey is scheduled for 2004. The results of surveys undertaken within this scheme are published in the scientific literature: for example, golden eagle *Aquila chrysaetos* (Green 1996); merlin *Falco columbarius* (Rebecca and Bainbridge 1998); peregrine *Falco peregrinus* (Crick and Ratcliffe 1995), red-throated diver *Gavia stellata* (Gibbons *et al*. 1997); greenshank *Tringa nebularia* (Hancock *et al*. 1997); common scoter *Melanitta nigra* (Underhill *et al*. 1998); black grouse *Tetrao tetrix* (Hancock *et al*. 1999); nightjar *Caprimulgus europaeus* (Morris *et al*. 1994); corncrake *Crex crex* (Hudson *et al*. 1990; Green 1995) and Cetti's warbler *Cettia cetti* (Wotton *et al*. 1998). Although not strictly within SCARABBS, other single (occasionally multi-) species surveys are undertaken. Possibly the best examples of these are the surveys of breeding waders (e.g. Smith 1983; Shrubb and Lack 1991; O'Brien and Smith 1992; Brindley *et al*. 1998).

2.2 Monitoring of non-breeding bird populations

Because of its geographic location along the North Atlantic flyway, its relatively mild climate and extensive areas of wetland, most notably estuaries, Britain is of outstanding international importance for wintering waterfowl (mainly ducks, geese, swans and waders – see Rose and Scott 1997 for a full definition). These species are monitored by the Wetland Birds Survey (WeBS), and this is the only large-scale bird population monitoring scheme in the non-breeding season. Although WeBS formally started in 1993, it was in practice a merger of two earlier schemes, one of which – the National Wildfowl Counts – started in 1947 (Cranswick *et al.* 1997a). The scheme purports to be a complete annual census of all waterfowl. In practice, however, the proportion of the British population that is actually counted each year varies from species to species, dependent upon their distribution. In general a greater proportion of a highly congregatory species (for instance pink-footed goose *Anser brachyrhynchus*) is counted than of a widespread species (such as mallard *Anas platyrhynchos*). Synchronised counts are conducted monthly, mainly from September to March, though with optional year-round counts at some sites. At coastal sites, all counts are undertaken around high tide as waterfowl move closer inland at this time, often roosting in large numbers thus facilitating counting. In order to determine the feeding grounds of estuarine birds, a programme of low-tide counts was introduced in the early 1990s. Results of WeBS are reported annually (e.g. Cranswick *et al.* 1997b). A number of other surveys are undertaken within WeBS, the most notable of which are surveys of waterfowl on non-estuarine coasts (Moser and Summers 1986; Browne *et al.* 1996). The WeBS model has recently been adopted in the Republic of Ireland, and the winter of 1994–95 saw the first winter of fieldwork for the Irish Wetland Birds Survey (I-WeBS; Delany 1996).

There can be little doubt that it is the vagrants and scarce migrants that cause most excitement among British and Irish birdwatchers. Although no formal monitoring scheme exists for these species, records are comprehensively reported in county bird reports, *Birding World* and *British Birds*. These records are occasionally summarised, thus performing a monitoring function (e.g. Dymond *et al.* 1989). The tradition for collating records of scarce migrants, species that pass through Britain annually on migration but only in small numbers, was set by Sharrock (1974b), but has been more recently developed by Fraser *et al.* (1997, 1999). It is not uncommon to see a build-up of records of a particular species prior to it colonising as a breeding species. Probably the best recent example is the little egret, *Egretta garzetta*, which was increasingly reported during the late 1980s and 1990s with breeding first taking place in 1996 (Lock and Cook 1998). The spoonbill followed a similar pattern during the 1990s, though breeding has not yet occurred (Ogilvie *et al.* 1999).

2.3 Monitoring of distributions

Because they approximate complete censuses, some population monitoring schemes – most notably WeBS, RBBP, the 15-yearly seabird censuses and some single-species surveys – also monitor distribution on a variety of timescales. This is not the case, however, for most species whose distributions are only monitored intermittently by atlases. The 10-km square distributions of all species that breed in Britain and Ireland have been determined twice, in 1968–72 (Sharrock 1976) and 1988–91 (Gibbons *et al.* 1993). The next breeding atlas is scheduled for 2008. In principle, these surveys allow unbiased estimates of change in distribution of all species over a 20-year timescale. In practice, however, the 1968–72 atlas incorporated no measures of fieldwork effort. Because of this, for some species it was hard to be certain whether or not the changes observed between the late 1960s and late 1980s were real or the result of changes in local enthusiasm for such survey work. Such problems have beset much atlas work, not just for birds (Rich 1997). The 1988–91 breeding atlas attempted to resolve this problem for the future by spending a set amount of time within a number of smaller grid squares (tetrads) in each 10-km square. Not only will this allow more precise estimates of change when the third breeding atlas is undertaken, it has also allowed maps of abundance to be produced for the first time; these were based on the proportion of tetrads in a 10-km square in which each species was recorded. Similar methods have been adopted elsewhere (Harrison *et al.* 1997; Schmid *et al.* 1998). Over the last two to three decades, a plethora of local, generally county-wide, breeding bird atlases have been published (e.g. Sitters 1988); some have even been repeated (e.g. Smith *et al.* 1993; Dazley and Trodd 1994).

There has only been a single atlas of wintering bird distributions (Lack 1986), based on data collected between 1981 and 1984. This is likely to be repeated, though on what timescale in uncertain – perhaps 40 years. The winter atlas was quantitative in that most counts were timed and corrected to a standard six-hour period.

2.4 Monitoring demography

Two main schemes monitor demographic parameters, most notably breeding success and survival. Monitoring demography can provide a forewarning of adverse population trends to come, as well as retrospectively helping to unravel the causes of population declines. The Constant Effort Sites (CES) scheme, which started in 1983, uses changes in the number of birds caught in mist nets to monitor changes in population levels of common passerines of scrub and wetland. By calculating the proportion of young birds in the total catch, changes in breeding success are also estimated (Peach *et al.* 1996). Year-to-year comparisons are only possible because the same number of nets is used each year, in the same positions, for the same period of time on similar dates. Between-year recaptures of ringed birds are also used to calculate annual survival rates (Peach 1993). In 1996 there were 122 CES sites in the UK (Crick *et al.* 1998).

The Nest Record Scheme (NRS), which monitors annual variation in breeding success, has been running for nearly half a century and more than a million records have been collected. Observers provide information from each nest they find from which date of laying, clutch and brood size and daily nest failure rates are calculated (Crick *et al.* 1998). The Seabird Monitoring Programme (SMP) monitors annual variation in seabird breeding success, and Raptor Study Groups monitor raptor breeding success, in a similar manner.

2.5 Partnership

All of these schemes are organised as partnerships at two separate levels. First, they are partnerships between organisations; Table 22.1 lists not only the lead organisation responsible for running each scheme, but also the other partners involved. While the

British Trust for Ornithology (BTO) most often takes the lead, several others crop up time and again as equal partners. Most notable among these are the Royal Society for the Protection of Birds (RSPB), the Wildfowl and Wetlands Trust (WWT) and the Joint Nature Conservation Committee (JNCC), itself acting on behalf of the individual country statutory conservation agencies. These partnerships between governmental and non-governmental organisations allow both knowledge and the financial burden to be shared and ensure that all stakeholders in bird monitoring have access to and use the same data.

The second level of partnership is between volunteers and professionals. With the exception of some seabird monitoring and much of the survey work undertaken within SCARABBS, most fieldwork is undertaken by volunteers, themselves often members of BTO, WWT or RSPB. These fieldworkers are guided, and the resulting data analysed, by a small number of professional staff within these organisations often working through a network of regional volunteer organisers.

The level of volunteer input into fieldwork is staggering. For example: about 250 volunteers cover the 200–300 CBC plots; more than 100 volunteers survey the 120 WBS plots; more than 1500 fieldworkers are involved in the BBS; each of the 120 or so CES sites will be manned by a small number of bird ringers; several thousand volunteers participate annually in WeBS and 5000 or more birdwatchers helped collect data for each of the atlases. On top of this are the volunteer networks such as the BTO's Regional Representatives network and the County Bird Recorders. Particular species or groups of species have dedicated enthusiast groups, the most notable being the Raptor Study Groups, but there are also study groups for black-necked grebe *Podiceps nigricollis*, black grouse, ring ouzel *Turdus torquatus* and golden oriole *Oriolus oriolus*, to name but a few. Membership of the BTO and WWT are respectively at 10 500 and 70 000 while membership of the RSPB topped 1 million in 1998. Support for birds has never been greater.

3 Synopsis of changes to 1970

Sharrock's (1974a) review showed that the number of species breeding in Britain and Ireland was fairly static from 1800 to 1949, but increased markedly from 1950 to the early 1970s. This increase was equivalent to a net gain of about five species per decade. Sharrock also showed that, on balance, successful species (those increasing in numbers or range) outweighed failing species (decreasing in numbers or contracting in range).

3.1 Changes in status of Britain's breeding birds since 1970

The appendix to this chapter (see pp. 392–398), adapted from Gibbons *et al.* (1996a), lists all species that have bred in the UK since 1800 and highlights those that bred at some time during the period 1970–97 and those that bred for the first time during 1970–97 (17 species in total). Table 22.2, again adapted from Gibbons *et al.* (1996a), summarises the total number of breeding species, including and excluding those of introduced origin, that bred in the UK in five separate time periods. The results are only marginally affected by increasing or decreasing the geographical scale, to Britain or Britain and Ireland, respectively.

A total of 242 species of bird have bred in the UK at some time since 1800, although

Table 22.2 The changing species richness of the UK avifauna.

Category of breeding species	1800–49	1850–99	1900–39	1940–69	1970–97	1800–1997
No. native species	190	188	193	207	218	228
No. introduced species	4	6	9	10	14	14
No. all species combined	194	194	202	217	232	242

14 of these are now naturalised species with self-sustaining populations. The number of species breeding in the UK has increased by 20% since 1800 with nearly 40 more species breeding in the last quarter of the twentieth century than in the first half of the nineteenth. While nearly one-third of this increase was as a consequence of introductions, intentional or otherwise (Gibbons *et al.* 1996a), the rest was apparently due to the balance between colonisation and extinction being in favour of colonisation. Even within the period 1970–97, a steady increase in numbers of breeding species is apparent from the reports of the RBBP (Fig. 22.1) with a net increase of about four species per decade, very little of which is accounted for by introduced species. Only a few formerly regular breeding species have been lost from the UK since 1800: the great auk *Pinguinus impennis* is now globally extinct; the great bustard *Otis tarda* has been extinct in the UK for 170 years and the black tern *Chlidonias niger* and Kentish plover *Charadrius alexandrinus* disappeared during the last half of the twentieth century. Although the hoopoe *Upupa epops*, wryneck *Jynx torquilla* and red-backed shrike *Lanius collurio* were all in apparently terminal long-term decline during the last half of the twentieth century, there are still a few records of these species in most years.

The increase in species richness between 1970 and 1997 continues the trend, and at about the same rate, documented by Sharrock (1974a). It is likely that some of this

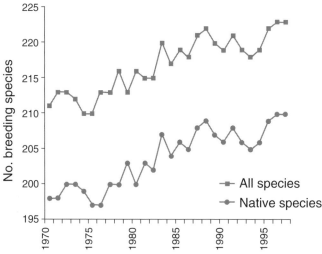

Figure 22.1 The annual number of species of bird that bred in Britain between 1970 and 1997. Trends for native species and all species (i.e. native plus feral and introduced species) are shown separately.

increase is real, but some artefactual. Sharrock suggested that much of the earlier increase was a consequence of improved site and species protection. Given that the extent of both these activities has increased yet further since 1970, it is likely that much of the recent increase can be attributed to these factors. During the nineteenth and first half of the twentieth century the eggs of a rare, potential colonist would have been much sought after by egg-collectors. Fortunately this pastime is now much less widespread (Bibby et al. 1990).

There are two potential artefactual reasons. First, part of the trend must be simply due to the increasing number of observers. Although a rare colonist might have passed unnoticed 50 years ago, this would be much less likely to be the case now. Second, the manner in which the number of breeding species has been calculated might artificially inflate some of the figures. Obtaining evidence of breeding, rather than simple presence in suitable habitat during the breeding season, can be difficult (and in some cases illegal). Species have only been admitted to the list of breeding species once breeding has been proven, but have been kept on the list in subsequent years if reported by RBBP even with no proof of breeding. Removing near-extinct species from the list is similarly difficult, as species may still be recorded for several years without any clear evidence of breeding. Despite this, there can be little doubt that species such as the red-necked grebe *Podiceps grisegena*, little egret, goldeneye *Bucephala clangula*, purple sandpiper *Calidris maritima* and Cetti's warbler have colonised very successfully since 1970 and none of this success can be argued away as recording artefact.

3.2 The changing fortunes of individual species and species' groupings

Because of the comprehensive monitoring of bird species in the UK, it is possible to quantify the change in status of most species of breeding birds and non-breeding waterfowl (at least). Fig. 22.2 documents, in highly summarised form, the change in status of various groupings of breeding species.

The manner in which the data for Fig. 22.2 were calculated will be published in detail elsewhere, though a synopsis of the method is given here (see also DETR 1998 and Gregory *et al.* 1999). The figure is based on the 'annual' population indices that are available for 186 of the 232 species that bred in the UK during 1970–97. Many of the species for which annual indices could not be calculated were extremely rare (populations not exceeding 20 pairs). Of the 186, 172 were native wild birds, the remaining 14 being of introduced or feral origin. In about a third of all cases the annual indices have had to be interpolated from surveys undertaken several years apart by assuming a constant annual rate of change between surveys. An annual index for each group was constructed by aggregating the individual species' population indices, each based on 1970 = 100. Each species was given equal weight in the index. Each group's annual index value was constructed using a logarithmic transformation of each species' series and then taking the exponential of the average to form the overall index. This transformation was necessary because of the skewed nature of the distributions, in other words, because population increases can be infinite, but population declines can only be 100%. The value for each grouping plotted in Fig. 22.2 represents the mean population index across all species in that group for the five-year period 1992–97. Species in a group whose mean index is 200 or 50 will, on average, have doubled or halved,

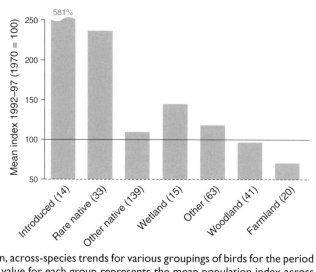

Figure 22.2 Mean, across-species trends for various groupings of birds for the period 1970 to 1992–97. The value for each group represents the mean population index across all species in that group for the five-year period 1992–97 compared to that in 1970 (arbitrarily set to 100). Species in a group whose mean index is 200 or 50 will, on average, have doubled or halved, respectively in population between 1970 and 1992–97. More complete details are provided in the text. The classification of species by habitat follows Gibbons *et al.* (1993). Note that 'wetland', 'other', 'woodland' and 'farmland' are subsets of 'other native'.

respectively in population between 1970 and 1992–97. The classification of species by habitat follows Gibbons *et al.* (1993).

The success of the 14 introduced species is apparent (Fig. 22.2), with populations of this group having increased more than five-fold since 1970. Most of these species are waterfowl or game birds, introduced either as additions to waterfowl collections from where they have escaped into the wild (such as the Mandarin duck *Aix galericulata*) or for sport (the red-legged partridge *Alectoris rufa*). The rapid increase in numbers of introduced waterfowl has occasionally created serious problems. The Canada goose, *Branta canadensis*, which now numbers more than 60 000 birds in Britain alone (Stone *et al.* 1997) has become a nuisance on many waterside amenity grasslands. Considerably more worrying from a conservation perspective, though, is the exponential increase in numbers of the ruddy duck *Oxyura jamaicensis*, unintentionally introduced in the mid-1950s. Probable descendants of the early feral stock have been recorded in increasing numbers in continental Europe, and it is now known that this species is hybridising with the much rarer, globally threatened white-headed duck *Oxyura leucocephala* in Spain (Hughes 1996).

Several other non-native species do breed in Britain and Ireland, although the extent to which their populations are viable in the wild is not always clear. Thus, for example, although it is possible that there may be a few self-sustaining populations of muscovy duck *Cairina moschata* (Gibbons *et al.* 1993), it is unlikely that those of the Monk parakeet *Myiopsitta monachus* are.

Across the 172 native species used in the index, the mean population trend has been slightly upwards, with the average species' population increasing by about 25%, thus apparently continuing the general improving trend in species status documented by Sharrock (1974a). As Fig. 22.2 shows, however, this rising trend is strongly influenced by the marked success of rarer native species. Populations of these 33 species, which numbered fewer than 500 breeding pairs in the UK in the mid-1990s (Stone *et al.* 1997), more than doubled in size between 1970 and 1992–97. The reasons for this marked success are similar to those for the apparent increase in species richness, part real – increased site and species protection, targeted species recovery programmes and species-specific site management – and part artefact –improved counting in more recent years. Once again, however, it is hard to argue away the strong upward trends in populations of goldeneye, marsh harrier *Circus aeruginosus*, osprey *Pandion haliateus*, hobby *Falco subbuteo*, avocet *Recurvirostra avosetta*, Mediterranean gull *Larus melanocephalus* and Cetti's warbler as simple recording artefacts.

Across the 139 more common native species in the index (those with more than 500 breeding pairs in the UK; Stone *et al.* 1997), there has still been a very slight rise in mean population trend, amounting to about 8% since 1970. Closer examination of the data, however, shows that most of this increase had occurred by the mid-1970s, since when the trend has been very stable (see Figure in DETR 1998). It is probable that a major part of this early rise was simply the recovery of populations of resident species from the extremely cold winters of 1962–63 and 1963–64. The impact of these winters has been well documented (e.g. Greenwood and Baillie 1991).

By exploding the more common species trend into individual habitat groupings, interesting patterns are revealed (Fig. 22.2); specifically, that species associated with wetlands have fared rather well, while species on farmland have fared extremely badly. The improving status of species of wetland habitats was also documented by Sharrock (1974a), and is driven by rising populations of such species as great-crested grebe *Podiceps cristatus*, grey heron *Ardea cinerea*, mute swan *Cygnus olor*, mallard, tufted duck *Aythya ferina*, coot *Fulica atra* and little ringed plover *Charadrius dubius*. Though only about 10% of breeding species are associated with farmland, they are mostly still common and widespread and thus even a small percentage decline means a great numerical loss of individual birds. Population levels of most farmland specialists

Table 22.3 Population trends of farmland specialists over the 25-year period 1972–96.

Species	Scientific name	% change
Grey partridge	*Perdix perdix*	– 78
Lapwing	*Vanellus vanellus*	– 42
Turtle dove	*Streptopelia turtur*	– 62
Skylark	*Alauda arvensis*	– 60
Starling	*Sturnus vulgaris*	– 45
Tree sparrow	*Passer montanus*	– 87
Linnet	*Carduelis cannabina*	– 41
Yellowhammer	*Emberiza citrinella*	– 60
Corn bunting	*Miliaria calandra*	– 74

Data are taken from the BTO/JNCC CBC as presented in Crick *et al.* (1998) and are for all CBC habitats. The definition of 'specialist' follows Siriwardena *et al.* (1998).

have declined by half, on average, since the mid-1970s (Fuller *et al.* 1995; Crick *et al.* 1998; Siriwardena *et al.* 1998; Table 22.3).

Although woodland species have fared reasonably well compared to those on farm-land (Fuller *et al.* 1995; Fig. 22.2), with only a drop of about 3% in mean trend since 1970, they have declined somewhat more since the mid- to late 1970s (amounting to a 10–15% decline, see DETR 1998). Within this general shallow declining trend, some woodland species have declined markedly since the early 1970s. Most notable among these are the tree pipit *Anthus trivialis*, the dunnock *Prunella modularis*, the three

Table 22.4 Population trends of breeding seabirds on British and Irish coasts, 1969–87. Adapted from Table 1.1 in Thompson *et al.* (1998). See text for information on more recent trends.

Species	Scientific name	% change
Fulmar	*Fulmarus glacialis*	+85
Manx shearwater	*Puffinus puffinus*	?
Storm petrel	*Hydrobates pelagicus*	?
Leach's petrel	*Oceanodroma leucorhoa*	?
Gannet	*Morus bassanus*	+36
Cormorant	*Phalacrocorax carbo*	+30
Shag	*P. aristotelis*	+40
Arctic skua[a]	*Stercorarius parasiticus*	+220
Great skua[a]	*Catharacta skua*	+150
Black-headed gull	*Larus ridibundus*	+13
Common gull	*L. canus*	+21
Lesser black-backed gull	*L. fuscus*	+31
Herring gull	*L. argentatus*	−43
Great black-backed gull	*L. marinus*	+3
Kittiwake	*Rissa tridactyla*	+22
Sandwich tern	*Sterna sandvicensis*	+53
Roseate tern[b]	*S. dougallii*	−80
Common tern	*S. hirundo*	−1
Arctic tern[c]	*S. paradisaea*	−14
Little tern	*S. albifrons*	+40
Guillemot	*Uria aalge*	+118
Razorbill	*Alca torda*	probably +
Black guillemot	*Cepphus grylle*	probably +
Puffin	*Fratercula arctica*	slightly +?

a, b and c, incorporating data from 1992, 1993 and 1989 respectively.

common species of thrush, blackbird *Turdus merula*, song thrush *T. philomelos* and mistle thrush *T. viscivorus*, and two of the rarer species of tit, marsh and willow, *Parus palustris* and *P. montanus*. Between them populations of these species fell by 31 to 56% across all CBC plots during 1972–96 (Crick *et al.* 1998). More alarmingly, the BBS has documented significant declines of all three species of thrush even over the very short period of 1994–97 (Gregory *et al.* 1998).

The population trends of coastal breeding seabirds over the period 1969–87 are listed in Table 22.4 and form part of the 'other' category in Fig. 22.2. Trends of most species, with the exception of some of the terns, most notably roseate *Sterna dougallii*, rose between 1970 and the late 1980s. Although the SMP has shown that some species, for example fulmar *Fulmarus glacialis*, gannet *Morus bassanus* and guillemot *Uria aalge* have continued to increase in numbers, more recent data and analyses have suggested that population levels of several of the terns are now falling. Little tern *Sterna albifrons* numbers fell by 39% between 1975 and 1998 (Ratcliffe *et al.* 2000), while the number of arctic terns, *Sterna paradisaea*, on Orkney and Shetland (which hold 85% of the British and Irish population) fell by more than 60% between 1980 and 1994 (Brindley *et al.* 2000).

Among the wintering wildfowl it is a struggle to find a species that has declined in Britain since 1970; most have either increased or remained at a similar level (Cranswick *et al.* 1997b). Although the average WeBS index value for the subspecific European white-fronted goose, *Anser albifrons albifrons*, during the 1990s was the lowest of any five-year period since the 1960s, the north-west European population has increased by an order of magnitude over the same period (Cranswick *et al.* 1997b). The WeBS trends do, however, document modest declines for three species of wintering duck, mallard, pintail *Anas acuta* and pochard *Aythya ferina*, for variable time periods since 1970. The indices for winter waders in the UK (Table 22.5) again show that populations of these species are reasonably buoyant.

Table 22.5 Five-year mean WeBS indices for winter wader numbers in the UK. Data summarised from Cranswick *et al.* (1997). 71–75 value refers to the five-year mean of the annual indices for the winters 1971–72 to 1975–76, etc.

Species	Scientific name	71–75	76–80	81–85	86–90	91–95
Oystercatcher	*Heamatopus ostralegus*	110	127	137	156	139
Ringed plover	*Charadrius hiaticula*	108	101	96	119	108
Grey plover	*Pluvialis squatarola*	128	176	249	416	531
Knot	*Calidris canutus*	96	68	77	86	86
Sanderling	*C. alba*	120	111	95	101	94
Dunlin	*C. alpina*	107	95	76	82	91
Black-tailed godwit	*Limosa limosa*	86	91	108	148	220
Bar-tailed godwit	*L. lapponica*	105	117	127	120	104
Curlew	*Numenius arquata*	115	105	109	129	145
Redshank	*Tringa totanus*	105	94	85	109	106
Turnstone	*Arenaria interpres*	107	114	120	143	129

4 Setting species priorities

The introduction of national and international *Red Data Books* and lists has been a notable feature of the last three decades. These document species that are of conservation concern for a wide variety of taxa across many geographical areas (e.g. Shirt 1987; Collar *et al.* 1994). *Red Data Birds in Britain*, the first such list for British birds, was produced in 1990 (Batten *et al.* 1990) and, in common with other Red Data lists, species were admitted on the basis of quantitative scientific criteria, such as rarity, population decline and localisation. In addition, however, species were also admitted on the basis of the importance of Britain to the species in question, specifically the proportion of the north-west European population that occurred in Britain. By 1996 such a wealth of new data on bird population levels and trends in the UK (e.g. Marchant *et al.* 1990; Gibbons *et al.* 1993) and Europe (Tucker and Heath 1994) had become available that the list was updated and refined. The new list, produced by non-governmental bird conservation organisations and entitled *Birds of Conservation Concern*, divided all species regularly occurring in Britain into three categories: red, amber and green (Avery *et al.* 1994; Gibbons *et al.* 1996b). Red-listed species (of which there were 36) were globally threatened or in rapid decline, either over the previous 25 years or historically. Amber-listed species (110) were in moderate decline, rare, localised, internationally important or of an unfavourable conservation status in Europe. All remaining species were green-listed. A very similar governmental listing was also produced (JNCC 1996). The biggest shift in composition between the 1990 and 1996 listings was the inclusion of a large number of common and widespread species that had undergone a substantial decline by 1996; many of these were species of the farmed environment.

The list of birds of conservation concern and the *Red Data Book* before it have been extremely influential in determining conservation priorities, from deciding which areas of governmental policy need to be changed to deciding on nature reserve acquisition strategies. A substantial number of red-listed species are currently subject to species recovery programmes and most are listed in the UK Government's Biodiversity Action Plan (DoE 1994, 1995).

5 The causes of change

The causes of population change are frequently harder to determine than the changes themselves.

Targeted conservation action has undoubtedly benefited populations of several rarer species whose earlier declines have been reversed in recent decades. The effects of this action are clear in Fig. 22.2. Such targeted action has taken a variety of forms such as species and nest protection, species reintroduction, acquisition of land for nature reserves, species-specific site management and targeted agri-environment measures. Some of these approaches may be given added weight if supported by legislation such as the Wildlife and Countryside Act 1981, and all are now conducted within an action planning process, either those of the government (DoE 1994, 1995) or individual conservation organisations (the RSPB, for example). There are several good examples of such targeted conservation action; all were introduced following periods of research that elucidated the causes of the species' decline. Stone curlew numbers rose during the 1990s as fieldworkers were employed to locate and protect nests from agricultural operations (Green 1988). The population of red kite, which declined to near extinction at the turn of the twentieth century, is now flourishing in part due to reduced human persecution following tougher legislation but also as a result of its reintroduction into parts of its former range in England and Scotland. Both corncrake and cirl bunting *Emberiza cirlus* numbers rose during the 1990s through closely targeted farmland management schemes, one introduced specifically for corncrakes, the other an existing scheme (Countryside Stewardship) tweaked specifically with cirl buntings in mind. The marked rise in numbers of avocet since 1970 has largely been a consequence of the acquisition of land for this species, followed by management to create suitable nesting habitats and the hydrological conditions necessary for its favoured prey. More recently, recovery programmes have been initiated for a number of other species such as bittern *Botaurus stellaris*, capercaillie *Tetrao urogallus* and black grouse, although success for these species remains elusive. Such targeted species-specific measures, while demonstrably successful, are expensive and often attempt on a small scale to address large-scale environmental changes that also affect large numbers of more widespread species. Seeking to change national (or international) legislation and policy over a broader scale is, if successful, a more sustainable long-term approach, though will commonly be less direct and may be geographically too imprecise for rarer species.

Since 1970, seabird populations have been fairly buoyant although this may, in part, be due to current fisheries practices. Several species of seabird, especially gannet, great skua *Catharacta skua*, and the larger gulls, obtain a significant amount of their diet from trawler discards – these are fish that are caught but which are too small, surplus to quota, or of a non-commercial species and are thrown back. Fulmar and kittiwake *Rissa tridactyla*, also take large quantities of discarded fish offal. It is possible that the populations of these species have been maintained at artificially high levels by these discarding practices and it is estimated that discards account for nearly a third of the 600 000 tons of food eaten by North Sea seabirds. To avoid over-exploitation of fish stocks the European Union plans to reduce both fishing effort (by decommissioning the fishing fleet) and the discarded bycatch by improved fishing technology (introducing net mesh sizes that only catch larger fish). The introduction of such a laudable sustainable fishery may, unfortunately, lead to declines in populations of those species that have come to rely on discards. There may also be knock-on effects for other species, too, as, for example, great skuas switch from discards to preying on terns (Owen and Dunn 1996). The very close link between fish stocks, seabird breeding success and ultimately seabird populations was highlighted by the sandeel (*Hyperophus lanceolatus* and *Ammodytes tobianus*) crisis of the late 1980s. Between 1984 and 1990 arctic terns on Shetland failed to breed successfully because of a lack of sandeels on which they depended (Monaghan *et al.* 1989; Avery 1990, 1993). In 1990 there was considerable debate on Shetland as to whether a single arctic tern fledged from any of Shetland's 500 or so colonies. As a precautionary measure to protect its stock, the Shetland sandeel fishery was closed in 1991. Even though the fishery has recovered somewhat, arctic tern populations still declined during the early 1990s (Brindley *et al.* 1999).

Populations of many birds of prey were low during the first half of the twentieth century, initially because of persecution by humans – because of the perceived threat they posed to game interests – and during the 1950s and 1960s through organochlorine contamination from pesticides. UK-wide legal protection for birds of prey was introduced in the 1950s and 1960s and organochlorine pesticides phased out soon after. As a consequence, bird of prey numbers have recovered, in some cases – for example pere-grine (Crick and Ratcliffe 1995) and marsh harrier (Underhill-Day 1998) – very strongly. Despite this only a single species, the hobby, a migratory species, has shown a true population increase rather than a recovery (Gibbons et al. 1993; Parr 1994; UK Raptor Working Group 1998). Populations of several birds of prey – most notably the hen harrier *Circus cyaneus* – are still maintained at low levels by illegal human persecu-tion (Etheridge et al. 1997). The kestrel *Falco tinnunculus*, a widespread species of farmland, is declining in the UK (Crick et al. 1998) as well as across much of Europe (Tucker and Heath 1994), presumably because of agricultural change.

Populations of many species associated with wetlands have increased, particularly breeding and non-breeding wildfowl. The almost universal rise in numbers of most duck species in the non-breeding season has been attributed to an increase in the number of water-bodies, particularly gravel-pit complexes, and an understandable pre-dilection to acquire this habitat for nature reserve management (Owen et al. 1986; Kirby et al. 1995). Reductions in hunting pressure, the establishment of refuges from hunting, the phasing out of lead for angling and gunshot and increased eutrophication may also have played a part. Although species of open water have fared well, the same cannot be said to be true for those associated with the wetland fringes or low-lying damp ground. Although the era of mass land drainage may now be passed, agricultural intensification has still led to a loss of such habitat, with a concomitant decline in its associated species, most notably breeding waders such as lapwing *Vanellus vanellus*, snipe *Gallinago gallinago* and redshank *Tringa totanus* (O'Brien and Self 1994; Langston et al. 1998). The introduction of the Environmentally Sensitive Areas (ESA) scheme under European Union agri-environment regulations may help. Currently, however, outside of nature reserves there has been little uptake of the higher tiers of this scheme that can deliver suitable conditions for breeding waders (e.g. Weaver 1995).

The biggest environmental change over the last 30 years, however, has undoubtedly been agricultural intensification (O'Connor and Shrubb 1986; Pain and Pienkowski 1997) which has mostly manifested itself as profound changes in techniques of crop and grass production (Fuller et al. 1995). There is strong evidence that these recent trends have played a major part in the population declines of many farmland species (e.g. Evans et al. 1995; Donald and Aebischer 1997).

Spring sowing of cereals, especially wheat, became far less frequent during the 1970s such that by the early 1980s autumn sowing was the dominant practice in lowland Brit-ain. This change had several consequences. First, winter stubbles, an important source of over-winter food for many seed-eating birds such as the corn bunting *Miliaria calandra* and cirl bunting, became much reduced (Donald and Evans 1994; Evans and Smith 1994). Second, species such as the lapwing and skylark *Alauda arvensis*, which traditionally nested in spring-sown crops, found themselves unable to breed success-fully in the earlier-maturing winter-sown varieties (Schlapfer 1988; Shrubb and Lack 1991; Wilson et al. 1997). Third, the nests of late-nesting species, such as the corn bunt-ing, were destroyed by increasingly earlier harvesting (Crick et al. 1994). Agriculture has become geographically polarised since the 1970s, with arable dominating in east-ern Britain and pasture in the west (Grigg 1989); this has had inevitable knock-on effects on species requiring a diversity of habitats, such as the lapwing (Shrubb and Lack 1991). Chemical pesticide usage (for example, insecticide, herbicide and fungi-cide) has increased dramatically since 1970 and although there is little evidence of any direct effect, indirectly it has led to a reduction in food (insects and weed seeds) available to chicks and adults (e.g. Potts 1986; Campbell *et al.* 1997). Although less is known about the impact of changes in pastoralism on bird populations, changes in the management of mown grassland have been a major cause of decline of the corncrake in Britain and Ireland (Green and Stowe 1993; Stowe *et al.* 1993). In the uplands of Britain, there has been a marked increase in sheep-stocking densities. There is strong circumstantial evidence that these higher densities lead to poorer breeding success of some upland species, such as the black grouse (Baines 1996), as over-grazing removes the insect and plant food necessary for their survival.

6 The future

Crystal-ball gazing is always a thankless task, nevertheless it is a fairly certain bet that bird populations will change as much in the next three decades as they have in the past three. It also seems certain that most of these changes will be caused by, or heavily influenced by, human activity and therefore the future for wildlife is to a large extent in our own hands. The following areas are likely to be of profound importance over the next few decades: agricultural intensification, climate change, fisheries policy and development pressure.

At the end of the twentieth century the declines in populations of common farmland birds appeared to be the most pressing conservation issue. Other wildlife has also declined on farmland but the bird data are of such greater quality that they act as a rallying point for concerns about farmland wildlife in general. The Agenda 2000 review of the Common Agricultural Policy (CAP) in 1999 failed to deliver the radical change needed to reverse the trends in farmland bird populations seen since 1970. Possibly the only noticeable change that might benefit bird populations was the switch from headage payments to area payments for stock (that is, paying farmers for the area of land holding stock, rather than the number of stock); this could well help to reduce over-grazing in upland areas.

So what is the future for farmland birds? The current growth in consumer demand for organically farmed products offers one potential brake against the increasing intensification of agricultural production. However, it will take a significant switch of money from production subsidies into payments for environmentally friendly farming before we are likely to see recoveries in farmland bird populations. The future of some of the UK's most widespread species, such as the skylark, remains uncertain and this may well be magnified by the introduction of genetically modified crops that are likely to accelerate the rate of agricultural change. We may face continuing declines of wildlife in the countryside and an increased reliance on nature reserves to support populations of once widespread and common species.

Even though it is frequently claimed that the populations and ranges of bird species in the UK have been affected by climate change, the evidence for this is still limited. In fact the best evidence (both in the UK and possibly globally) for the impact of a warming environment on wildlife is a phenological, rather than population, change. Crick *et al.* (1997) have shown that a wide variety of British bird species have begun to lay increasingly early since 1970 and this is attributed to rising temperatures. Species whose populations are held in check by low temperature-induced over-winter mortality (Greenwood and Baillie 1991) are those whose populations are most likely to rise as a consequence of climate change. The Dartford warbler is potentially one such species (Bale *et al.* 1996), as it is an insectivorous resident at the northern edge of its global range in Britain and its population is always knocked back after hard winters; its numbers have, however, risen dramatically since 1960 (Gibbons and Wotton 1996). Recent evidence suggests that the ranges of southerly-distributed British birds may have shifted northwards by nearly 20 km between 1970 and 1990 (Thomas and Lennon 1999). If this were to happen more generally, then a suitable patchwork of habitats would need to be created to accommodate these range changes. Britain may also receive a modest influx of species from continental Europe as a consequence of climate change; two possible candidates are the fan-tailed warbler *Cisticola juncidus* and crested lark *Galerida cristata* (Hagemeijer and Blair 1997). It is also possible that climate change will have more of an impact on migrants on their wintering grounds and on migration than on their breeding grounds in Britain.

Sea-level rise, due in part to thermal expansion of the world's oceans induced by climate change, will threaten many coastal sites of international importance for birds. Such sites need protection, either through 'hard' defences (walls to prevent sea encroachment) or through habitat recreation to replace sites that are lost.

Ironically, a move towards renewable energy production, which will be absolutely essential if the UK is to meet targets for carbon dioxide emission reductions, may also threaten birds. Terrestrial and marine windfarms, large-scale biomass production, tidal barrages and hydroelectric schemes can all have serious impacts on birds. Of these, marine windfarms will perhaps have most to recommend them when biodiversity, landscape, economic and power production concerns are integrated.

Fishery policy is likely to influence seabird numbers greatly. The dependence of many species on discards means that these species are vulnerable to changes in fishery policy and practice. Some feel that reductions in seabird numbers may be the price we have to pay for a more sustainable management regime for fisheries which would, it would be hoped, prove beneficial to marine biodiversity as a whole.

It is ironic that despite a broadly stable human population in the UK, the pressures from the demands of that population continue to grow. There is a projected demand for around 4 million new homes in the UK by 2020; these homes require roads, water, energy and other infrastructure. This continuing development will put pressures on the countryside as a whole including nominally 'protected' wildlife sites; a planning system that takes wildlife protection fully into account is undoubtedly part of the answer. Lifestyle changes, such as greener travel and lower water consumption, would help reduce our ecological impact but there is no certainty that these will occur quickly enough to have much of an impact over the next few decades. Only a more ambitious programme of habitat re-creation coupled with stringent controls for the very best wildlife sites will be able to maintain what are already sadly depleted wildlife resources.

Conservationists are often accused of scare-mongering and it is salutary to look back over recent decades and see that the losses of birds have been more or less balanced by gains. However, the emerging picture is that birds and other wildlife are under great pressure and the tendency is for them to be lost from the wider countryside and become more and more concentrated into smaller areas. It is difficult to escape the conclusion that unless we consciously expend effort and money on improving the conservation status of the UK's birds then the future looks bleaker than the past; there are, however, many reasons for not despairing. Public opinion in the UK is strongly supportive of wildlife conservation. As part of the UK Biodiversity Action Plan, the government has set challenging and specific targets for the recreation of lost habitats and for the restoration of population levels of declining species. In 1999 the UK Government published the first of an annual series of Sustainability Indicators (DETR 1998), including a wildlife indicator based on population levels of breeding birds. Such specific commitment to reporting on the state of wildlife and to its maintenance and recovery marks a new chapter of government involvement in the future of birds and wider biodiversity in the UK.

REFERENCES

Alexander, W. B. and Lack, D. (1944) Changes in status among British breeding birds. *British Birds*, **38**, 42–45, 62–69, 82–88.

Avery, M. I. (1990) Seabirds, fisheries and politics. *RSPB Conservation Review*, **4**, 36–39.

Avery, M. I. (1993) Arctic tern, in Gibbons, D. W., Reid, J. B. and Chapman, R. A. (eds) *The New Atlas of Breeding Birds in Britain and Ireland: 1988–1991*. London: T. and A. D. Poyser, pp. 220–221.

Avery, M. I., Gibbons, D. W., Porter, R., Tew, T., Tucker, G. and Williams, G. (1994) Revising the British Red Data List for birds: the biological basis of U.K. conservation priorities. *Ibis*, **137**, S232-S239.

Baines, D. (1996) The implications of grazing and predator management on the habitats and breeding success of Black Grouse *Tetrao tetrix*. *Journal of Applied Ecology*, **33**, 54–62.

Bale, J. S., Jones, T. H. and Gibbons, D. (1996) Impacts of climate change. *Journal of the Zoological Society of London*, **240**, 593–597.

Batten, L. A., Bibby, C. J., Clement, P., Elliott, G. D., and Porter, R. F. (1990) *Red Data Birds in Britain*. London: T. and A. D. Poyser.

Bibby, C. J. and Tubbs, C. R. (1975) Status, habitats and conservation of the Dartford Warbler in England. *British Birds*, **68**, 177–195.

Bibby, C. J., Robinson, P. and Bland, E. (1990) The impact of egg-collecting on scarce breeding birds. *RSPB Conservation Review*, **4**, 22–35.

Brindley, E., Norris, K., Cook, A., Babbs, S., Foster-Brown, C., Massey, P., Thompson, R. and Yaxley, R. (1998) The abundance and conservation status of redshank *Tringa totanus* nesting on saltmarshes in Great Britain. *Biological Conservation*, **86**, 289–297.

Brindley, E., Mudge, G., Dymond, N., Lodge, C., Ribbands, B., Steele, D., Ellis, P., Meek, E., Suddaby, D. and Ratcliffe, N. (1999) The status of Arctic Terns on Orkney and Shetland in 1994. *Atlantic Seabirds*, **1**, 135–143.

Browne, S. J., Austin, G. E. and Rehfisch, M. M. (1996) Evidence of decline in the United Kingdom's non-estuarine coastal waders. *Wader Study Group Bulletin*, **80**, 25–27.

Campbell, L. H., Avery, M. I., Donald, P., Evans, A. D., Green, R. E. and Wilson, J. D. (1997) *A review of the indirect effects of pesticides on birds*. Peterborough: Joint Nature Conservation Committee.

Collar, N. J., Crosby, M. J. and Stattersfield, A. J. (1994) *Birds to Watch 2: the world check-list of threatened birds*. Cambridge: BirdLife International.

Cramp, S., Bourne, W. R. P. and Saunders, D. (1974) *The Seabirds of Britain and Ireland*. London: Collins.

Cranswick, P. A., Kirby, J. S., Salmon, D. G., Atkinson-Willes, G. L., Pollitt, M. S. and Owen, M. (1997a) A history of wildfowl counts by WWT. *Wildfowl*, **47**, 217–230.

Cranswick, P. A., Waters, R. J., Musgrove, A. J. and Pollitt, M. S. (1997b) *The Wetland Bird Survey 1995–96*. Slimbridge: BTO/WWT/RSPB/JNCC.

Crick, H. Q. P., Dudley, C., Evans, A. D. and Smith, K. W. (1994) Causes of nest failure among buntings in the UK. *Bird Study*, 41, 88–94.

Crick, H. Q. P. and Ratcliffe, D. A. (1995) The Peregrine *Falco peregrinus* population of the United Kingdom in 1991. *Bird Study*, 42, 1–19.

Crick, H. Q. P., Dudley, C., Glue, D. E. and Thompson, D. L. (1997) UK birds are laying eggs earlier. *Nature*, 388, 256.

Crick, H. Q. P., Baillie, S. R., Balmer, D. E., Bashford, R. I., Beaven, L. P., Dudley, C., Glue, D., Gregory, R. D., Marchant, J. H., Peach, W. J. and Wilson, A. M. (1998) *Breeding birds in the wider countryside: their conservation status (1972–1996)*. Thetford: British Trust for Ornithology.

Dazley, R. A. and Trodd, P. (1994) *An Atlas of Breeding Birds in Bedfordshire, 1988–92*. Bedfordshire Natural History Society.

Delany, S. N. (1996) *Irish Wetland Bird Survey, 1994/95*. Dublin: BirdWatch Ireland.

Department of the Environment, Transport and the Regions (1998) *Sustainability counts: consultation paper on a set of 'headline' indicators of sustainable development*. London: The Stationery Office.

DoE (1994) *Biodiversity: the UK Action Plan*. London: HMSO.

DoE (1995) *Biodiversity: the UK Biodiversity Steering Group Report*, 2 vols. London: HMSO.

Donald, P. F. and Evans, A. D. (1994) Habitat selection by corn buntings *Miliaria calandra* in winter. *Bird Study*, 41, 199–210.

Donald, P. F. and Aebischer, N. J. (eds) (1997) *The Ecology and Conservation of Corn Buntings Miliaria Calandra* [UK Nature Conservation no. 13]. Peterborough: Joint Nature Conservation Committee.

Dymond, J. N., Fraser, P. A. and Gantlett, S. J. M. (1989) *Rare birds in Britain and Ireland*. Calton: T. and A. D. Poyser.

Etheridge, B., Summers, R. W. and Green, R. E. (1997) The effects of illegal killing and destruction of nests by humans on the population dynamics of the hen harrier *Circus cyaneus* in Scotland. *Journal of Applied Ecology*, 34, 1081–1105.

Evans, A., Appleby, M., Dixon, J., Newberry, P. and Swales, V. (1995) What future for lowland farmland birds in the UK? *RSPB Conservation Review*, 9, 32–40.

Evans, A. D. and Smith, K. W. (1994) Habitat selection of cirl buntings *Emberiza cirlus* wintering in Britain. *Bird Study*, 41, 81–87.

Fraser, P. A., Lansdown, P. G. and Rogers, M. J. (1997) Report on scarce migrant birds in Britain in 1995. *British Birds*, 90, 413–439.

Fraser, P. A., Lansdown, P. G. and Rogers, M. J. (1999) Report on scarce migrant birds in Britain in 1996. *British Birds*, 92, 3–35.

Fuller, R. J., Gregory, R. D., Gibbons, D. W., Marchant, J. H., Wilson, J. D., Baillie, S. R., and Carter, N. (1995) Population declines and range contractions among lowland farmland birds in Britain. *Conservation Biology*, 9, 1425–1441.

Gibbons, D. W., Reid, J. R. and Chapman, R. A. (1993) *The New Atlas of Breeding Birds in Britain and Ireland: 1988–1991*. London: T. and A. D. Poyser.

Gibbons, D. W. and Wotton S. (1996) The Dartford Warbler in the United Kingdom in 1994. *British Birds*, 89, 203–212.

Gibbons, D. W., Avery, M. I. and Brown, A. F. (1996a) 'Population trends of breeding birds in the United Kingdom since 1800'. *British Birds*, 89, 291–305.

Gibbons, D. W., Avery, M., Baillie, S., Gregory, R., Kirby, J., Porter, R., Tucker, G. and Williams, G. (1996b) Bird species of conservation concern in the United Kingdom, Channel Island and Isle of Man: revising the Red Data list. *RSPB Conservation Review*, 10, 7–18.

Gibbons, D. W., Bainbridge, I. P., Mudge, G. P., Tharme, A. P. and Ellis, P. M. (1997) The status and distribution of the Red-throated Diver *Gavia stellata* in Britain in 1994. *Bird Study*, 44, 194–205.

Gilbert, G., Gibbons, D. W. and Evans, J. (1998) *Bird monitoring methods*. Sandy: RSPB.

Green, R. E. (1988) Stone curlew conservation. *RSPB Conservation Review*, **2**, 30–33.

Green, R. E. and Stowe, T. J. (1993) The decline of the corncrake *Crex crex* in Britain and Ireland in relation to habitat change. *Journal of Applied Ecology*, **30**, 689–695.

Green, R. E. (1995) The decline of the Corncrake *Crex crex* in Britain continues. *Bird Study*, **42**, 66–75.

Green, R. E. (1996) The status of the Golden Eagle in Britain in 1992. *Bird Study*, **43**, 20–27.

Greenwood, J. J. D. and Baillie, S. R. (1991) Effects of density-dependence and weather on population changes of English passerines using a non-experimental paradigm. *Ibis*, **133**, (suppl. 1), 121–133.

Gregory, R. D., Bashford, R. I., Beaven, L. P., Marchant, J. H., Wilson, A. M. and Baillie, S. R. (1998) *The Breeding Bird Survey 1996–1997* [BTO Research Report no. 203]. Thetford: BTO.

Gregory, R. D., Gibbons, D. W., Impey, A. and Marchant, J. H. (1999) *Generation of the headline indicator of wild bird populations* [BTO Research Report no. 221]. Thetford and Sandy: BTO and RSPB.

Grigg, D. (1989) *English Agriculture: an historical perspective*. Oxford: Basil Blackwell.

Hagemeijer, W. J. M. and Blair, M. J. (1997) *The EBCC Atlas of European Breeding Birds*. London: T. and A. D. Poyser.

Hancock, M. H., Gibbons, D. W. and Thompson, P. S. (1997) The status of breeding greenshank *Tringa nebularia* in the United Kingdom in 1995. *Bird Study*, **44**, 290–302.

Hancock, M. H., Baines, D., Gibbons, D., Etheridge, B. and Shepherd, M. (1999) The status of male black grouse *Tetrao tetrix* in Britain in 1995–96. *Bird Study*, **46**, 1–15.

Harrison, J. A., Allan, D. G., Underhill, L. G., Herremans, M., Tree, A. J., Parker, V. and Brown, C. J. (1997) *The Atlas of Southern African Birds*, Vol. 1: *Non-passerines*. Johannesburg: BirdLife South Africa.

Holloway, S. (1996) *The Historical Atlas of Breeding Birds in Britain and Ireland: 1875–1900*. London: T. and A. D. Poyser.

Hudson, A. V., Stowe, T. J. and Aspinall, S. J. (1990) Status and distribution of Corncrakes in Britain in 1988. *British Birds*, **83**, 173–187.

Hughes, B. (1996) *The feasibility of control measures for North American Ruddy Ducks* Oxyura jamaicensis *in the United Kingdom*. London: HMSO.

Joint Nature Conservation Committee (1996) *Birds of Conservation Importance*. Peterborough: JNCC.

Kirby, J. S., Salmon, D. G., Atkinson-Willes, G. L. and Cranswick, P. A. (1995) Index numbers for waterbird populations III: long-term trends in the abundance of wintering wildfowl in Great Britain, 1966/67–1991/92. *Journal of Applied Ecology*, **32**, 536–551.

Lack, P. (1986) *The Atlas of Wintering Birds in Britain and Ireland*. Calton: T. and A. D. Poyser.

Langston, R., Burston, P., Wilkinson, N., Renwick, N. and Reed, S. (1998) The performance of site safeguard in the UK. *RSPB Conservation Review*, **12**, 9–17.

Lloyd, C., Tasker, M. L., and Partridge, K. (1991) *The Status of Seabirds in Britain and Ireland*. London: T. and A. D. Poyser.

Lock, L. and Cook, K. (1998) The Little Egret in Britain: a successful colonist. *British Birds*, **91**, 173–280.

Lovegrove, R., Williams, G. and Williams, I. (1994) *Birds in Wales*. London: T and A. D. Poyser.

Marchant, J. H., Hudson, R., Carter, S. P. and Whittington, P. A. (1990) *Population Trends in British Breeding Birds*. Tring: BTO.

Monaghan, P. E., Uttley, J. D., Burns, M. D., Thaine, C. and Blackwood, J. (1989) The relationship between food supply, reproductive effort and breeding success in Arctic Terns *Sterna paradisaea*. *Journal of Animal Ecology*, **58**, 261–274.

Morris, A., Burgess, D., Fuller, R. J., Evans, A. D. and Smith, K. W. (1994) The status and distribution of nightjars (*Caprimulgus europaeus*) in Britain in 1992. *Bird Study*, **41**, 181–191.

Moser, M. E. and Summers, R. W. (1986) Wader populations on the non-estuarine coasts of Britain and Northern Ireland: results of the 1984–85 Winter Shorebird Count. *Bird Study*, 34, 71–81.

Murphy, J. and Coombe, R. (1998) *Countryside Bird Survey: Interim Report*. Dublin: BirdWatch Ireland and Heritage Service, National Parks and Wildlife.

O'Brien, M. and Smith, K. W. (1992) Changes in the status of waders breeding on wet lowland grassland in England and Wales between 1982 and 1989. *Bird Study*, 39, 165–176.

O'Brien, M. and Self, M. (1994) Changes in the numbers of breeding waders on lowland wet grassland in the UK. *RSPB Conservation Review*, 8, 38–44.

O'Connor, R. J. and Shrubb, M. (1986) *Farming and birds*. Cambridge: Cambridge University Press.

Ogilvie, M. and The Rare Breeding Birds Panel (1999) Rare breeding birds in the United Kingdom in 1996. *British Birds*, 92, 120–154.

Owen, D. and Dunn, E. (1996) Recent developments in marine conservation. *RSPB Conservation Review*, 10, 25–31.

Owen, M., Atkinson-Willes, G. L. and Salmon, D. G. (1986) *Wildfowl in Great Britain*. 2nd edn, Cambridge: Cambridge University Press.

Pain, D. and Pienkwoski, M. W. (eds) (1997) *Farming and birds in Europe*. Academic Press: London.

Parr, S. J. (1994) Population changes of breeding Hobbies Falco subbuteo in Britain. *Bird Study*, 41, 131–135.

Parslow, J. L. F. (1973) *Breeding Birds of Britain and Ireland*, Berkhamsted: T. and A. D. Poyser.

Peach, W. (1993) Combining mark-recapture data sets for small passerines, in Lebreton, J.-D. and North, P. M. (eds) *Marked Individuals in the Study of Bird Populations*. Basel: Birkhauser Verlag.

Peach, W. J., Buckland, S. T. and Baillie, S. R. (1996) The use of constant effort mist-netting to measure between-year changes in the abundance and productivity of common passerines. *Bird Study*, 43, 142–156.

Potts, G. R. (1986) *The Partridge: Pesticides, Predation and Conservation*. London: Collins.

Ratcliffe, N., Vaughan, D., Whyte, C. and Shepherd, M. (1998) Development of playback census methods for Storm Petrels *Hydrobates pelagicus*. *Bird Study*, 45, 302–312.

Ratcliffe, N., Pickerell, G. and Brindley, E. (2000) Population trends of Little and Sandwich Terns in Britain and Ireland during a 30-year study. *Atlantic Seabirds*, 1.

Rebecca, G. W. and Bainbridge, I. P. (1998) The breeding status of the Merlin *Falco columbarius* in Britain in 1993–94. *Bird Study*, 45, 172–187.

Rich, T. (1997) Squaring the circles – bias in distribution maps. *British Wildlife*, 9, 213–220.

Robins, M. and Bibby, C. J. (1985) Dartford warblers in 1984 in Britain. *British Birds*, 78, 269–280.

Rose, P. M., and Scott, D. A. (1997) *Waterfowl Population Estimates*, 2nd edn. Wageningen: Wetlands International.

Schlapfer, A. (1988) Populationsokologie der Feldlerche *Alauda arvensis* in der intensiv genutzen Agrarlandschaft. *Ornithologische Beobachter*, 85, 309–371.

Schmid, H., Luder, R., Naef-Danzer, B., Graf, R. and Zbinden, N. (1998) *Atlas des Oiseaux Nicheurs de Suisse*. Sempach: Station Ornithologique Suisse.

Sharrock, J. T. R. (1974a) The changing status of breeding birds in Britain and Ireland, in Hawksworth, D. L. (ed.) *The Changing Flora and Fauna of Britain*. London and New York: Academic Press, pp. 203–220.

Sharrock, J. T. R. (1974b) *Scarce Migrant Birds in Britain and Ireland*, Berkhamsted: T. and A. D. Poyser.

Sharrock, J. T. R. (1976) *The Atlas of Breeding Birds in Britain and Ireland*. Berkhamsted, T. and A. D. Poyser.

Shirt, D. B. (ed.) (1987) *British Red Data Books 2: Insects*. Peterborough: Nature Conservancy Council.

Shrubb, M. and Lack, P. C. (1991) The numbers and distribution of Lapwings *V. vanellus* nesting in England and Wales in 1987. *Bird Study*, 38, 20–37.

Siriwardena, G. M., Baillie, S. R., Buckland, S. T., Fewster, R. M., Marchant, J. H. and Wilson, J. D. (1998) Trends in the abundance of farmland birds: a quantitative comparison of smoothed Common Birds Census indices. *Journal of Applied Ecology*, 35, 24–43.

Sitters, H. P. (1988) *Tetrad Atlas of the Breeding Birds of Devon*. Yelverton: Devon Bird-watching and Preservation Society.

Smith, K. W. (1983) The status and distribution of waders breeding on wet lowland grasslands in England and Wales. *Bird Study*, 30, 177–192.

Smith, K. W., Dee, C. W., Fearnside, J. W., Fletcher, E. W. and Smith, R. N. (1993) *The Breeding Birds of Hertfordshire*. Hertfordshire Natural History Society.

Spencer, R. and RBBP (1992) The Rare Breeding Birds Panel. *British Birds*, 85, 117–122.

Stone, B. H., Sears, J., Cranswick, P. A., Gregory, R. D., Gibbons, D. W., Rehfisch, M. M., Aebischer, N. J. and Reid, J. B. (1997) Population estimates of birds in Britain and in the United Kingdom. *British Birds*, 90, 1–22.

Stowe, T. J., Newton, A. V., Green, R. E., and Mayes, E. (1993) The decline of the corncrake *Crex crex* in Britain and Ireland in relation to habitat. *Journal of Applied Ecology*, 30, 53–62.

Thom, V. M. (1986) *Birds in Scotland*. Calton: T. and A. D. Poyser.

Thomas, C. D. and Lennon, J. J. (1999) Birds extend their ranges northwards. *Nature*, 399, 213.

Thompson, K. R., Brindley, E. and Heubeck, M. (1998) *Seabird Numbers and Breeding Success in Britain and Ireland, 1997*. Peterborough: Joint Nature Conservation Committee.

Tucker, G. M. and Heath, M. F. (1994) *Birds in Europe: Their Conservation Status*. Cambridge: BirdLife International.

Underhill, M. C., Gittings, T., Callaghan, D. A., Kirby, J. S., Hughes, B. and Delaney, S. (1998) Pre-breeding status and distribution of the Common Scoter *Melanitta nigra* in Britain and Ireland in 1995. *Bird Study*, 45, 146–156.

Underhill-Day, J. (1998) Breeding Marsh Harriers in the United Kingdom, 1983–95. *British Birds*, 91, 210–218.

UK Raptor Working Group. (1998) *DETR/JNCC Raptor Working Group's Progress Report to Ministers*. London: DETR/JNCC.

Weaver, D. J. (1995) *Broads ESA Breeding Wader Survey 1995*. Sandy: RSPB.

Wilson, J. D., Evans, J., Browne, S. J. and King, J. R. (1997) Territory distribution and breeding success of skylarks *Alauda arvensis* on organic and intensive farmland in Southern England. *Journal of Applied Ecology*, 34, 1462–1478.

Wotton, S., Gibbons, D. W., Dilger, M. and Grice, P. V. (1998) Cetti's warblers in the United Kingdom and the Channel Islands in 1996. *British Birds*, 91, 77–89.

Appendix The breeding bird species of Britain.

Species	Scientific name	Bred during 1970–97	First bred 1970–97
Red-throated diver	*Gavia stellata*	√	
Black-throated diver	*G. arctica*	√	
Great northern diver	*G. immer*		
Little grebe	*Tachybaptus ruficollis*	√	
Great crested grebe	*Podiceps cristatus*	√	
Red-necked grebe	*P. grisegena*	√	√
Slavonian grebe	*P. auritus*	√	
Black-necked grebe	*P. nigricollis*	√	
Fulmar	*Fulmarus glacialis*	√	
Manx shearwater	*Puffinus puffinus*	√	
Storm petrel	*Hydrobates pelagicus*	√	
Leach's petrel	*Oceanodroma leucorhoa*	√	
Gannet	*Morus bassanus*	√	
Cormorant	*Phalacrocorax carbo*	√	
Shag	*P. aristotelis*	√	
Bittern	*Botaurus stellaris*	√	
Little bittern	*Ixobrychus minutus*	√	√
Little egret	*Egretta garzetta*	√	√
Grey heron	*Ardea cinerea*	√	
Mute swan	*Cygnus olor*	√	
Whooper swan	*C. cygnus*	√	
Greylag goose	*Anser anser*	√	
Canada goose*	*Branta canadensis*	√	
Barnacle goose*	*B. leucopsis*	√	
Egyptian goose*	*Alopochen aegyptiacus*	√	
Shelduck	*Tadorna tadorna*	√	
Wood duck*	*Aix sponsa*	√	
Mandarin duck*	*A. galericulata*	√	
Wigeon	*Anas penelope*	√	
Gadwall	*A. strepera*	√	
Teal	*A. crecca*	√	
Mallard	*A. platyrhynchos*	√	
Pintail	*A. acuta*	√	
Garganey	*A. querquedula*	√	
Shoveler	*A. clypeata*	√	
Red-crested pochard*	*Netta rufina*	√	
Pochard	*Aythya ferina*	√	
Tufted duck	*A. fuligula*	√	
Scaup	*A. marila*	√	

continued on next page

Appendix The breeding bird species of Britain. (cont.)

Species	Scientific name	Bred during 1970–97	First bred 1970–97
Eider	*Somateria mollissima*	√	
Long-tailed duck	*Clangula hyemalis*		
Common scoter	*Melanitta nigra*	√	
Goldeneye	*Bucephala clangula*	√	√
Red-breasted merganser	*Mergus serrator*	√	
Goosander	*M. merganser*	√	
Ruddy duck*	*Oxyura jamaicensis*	√	
Honey buzzard	*Pernis apivorus*	√	
Red kite	*Milvus milvus*	√	
White-tailed eagle	*Haliaeetus albicilla*	√	
Marsh harrier	*Circus aeruginosus*	√	
Hen harrier	*C. cyaneus*	√	
Montagu's harrier	*C. pygargus*	√	
Goshawk	*Accipiter gentilis*	√	
Sparrowhawk	*A. nisus*	√	
Buzzard	*Buteo buteo*	√	
Golden eagle	*Aquila chrysaetos*	√	
Osprey	*Pandion haliaetus*	√	
Kestrel	*Falco tinnunculus*	√	
Merlin	*F. columbarius*	√	
Hobby	*F. subbuteo*	√	
Peregrine	*F. peregrinus*	√	
Red grouse	*Lagopus lagopus*	√	
Ptarmigan	*L. mutus*	√	
Black grouse	*Tetrao tetrix*	√	
Capercaillie	*T. urogallus*	√	
Red-legged partridge*	*Alectoris rufa*	√	
Grey partridge	*Perdix perdix*	√	
Quail	*Coturnix coturnix*	√	
Pheasant*	*Phasianus colchicus*	√	
Golden pheasant*	*Chrysolophus pictus*	√	
Lady Amherst's pheasant*	*C. amherstiae*	√	
Water rail	*Rallus aquaticus*	√	
Spotted crake	*Porzana porzana*	√	
Baillon's crake	*P. pusilla*		
Corncrake	*Crex crex*	√	
Moorhen	*Gallinula chloropus*	√	
Coot	*Fulica atra*	√	
Crane	*Grus grus*	√	√

continued on next page

Appendix   The breeding bird species of Britain. (cont.)

Species	Scientific name	Bred during 1970–97	First bred 1970–97
Great bustard	*Otis tarda*		
Oystercatcher	*Haematopus ostralegus*	√	
Black-winged stilt	*Himantopus himantopus*	√	
Avocet	*Recurvirostra avosetta*	√	
Stone-curlew	*Burhinus oedicnemus*	√	
Little ringed plover	*Charadrius dubius*	√	
Ringed plover	*C. hiaticula*	√	
Kentish plover	*C. alexandrinus*		
Dotterel	*C. morinellus*	√	
Golden plover	*Pluvialis apricaria*	√	
Lapwing	*Vanellus vanellus*	√	
Temminck's stint	*Calidris temminckii*	√	√
Purple sandpiper	*C. maritima*	√	√
Dunlin	*C. alpina*	√	
Ruff	*Philomachus pugnax*	√	
Snipe	*Gallinago gallinago*	√	
Woodcock	*Scolopax rusticola*	√	
Black-tailed godwit	*Limosa limosa*	√	
Whimbrel	*Numenius phaeopus*	√	
Curlew	*N. arquata*	√	
Redshank	*Tringa totanus*	√	
Greenshank	*T. nebularia*	√	
Green sandpiper	*T. ochropus*		
Wood sandpiper	*T. glareola*	√	
Common sandpiper	*Actitis hypoleucos*	√	
Spotted sandpiper	*A. macularia*	√	√
Red-necked phalarope	*Phalaropus lobatus*	√	
Arctic skua	*Stercorarius parasiticus*	√	
Great skua	*Catharacta skua*	√	
Mediterranean gull	*Larus melanocephalus*	√	
Little gull	*L. minutus*	√	√
Black-headed gull	*L. ridibundus*	√	
Common gull	*L. canus*	√	
Lesser black-backed gull	*L. fuscus*	√	
Herring gull	*L. argentatus*	√	
Great black-backed gull	*L. marinus*	√	
Kittiwake	*Rissa tridactyla*	√	
Gull-billed tern	*Gelochelidon nilotica*		
Sandwich tern	*Sterna sandvicensis*	√	

continued on next page

Appendix The breeding bird species of Britain. (cont.)

Species	Scientific name	Bred during 1970–97	First bred 1970–97
Roseate tern	S. dougallii	√	
Common tern	S. hirundo	√	
Arctic tern	S. paradisaea	√	
Little tern	S. albifrons	√	
Black tern	Chlidonias niger		
Great auk	Pinguinus impennis		
Guillemot	Uria aalge	√	
Razorbill	Alca torda	√	
Black guillemot	Cepphus grylle	√	
Puffin	Fratercula arctica	√	
Pallas's sandgrouse	Syrrhaptes paradoxus		
Rock dove/Feral pigeon*	Columba livia	√	
Stock dove	C. oenas	√	
Woodpigeon	C. palumbus	√	
Collared dove	Streptopelia decaocto	√	
Turtle dove	S. turtur	√	
Rose-ringed parakeet*	Psittacula krameri	√	√
Cuckoo	Cuculus canorus	√	
Barn owl	Tyto alba	√	
Snowy owl	Nyctea scandiaca	√	
Little owl*	Athene noctua	√	
Tawny owl	Strix aluco	√	
Long-eared owl	Asio otus	√	
Short-eared owl	A. flammeus	√	
Nightjar	Caprimulgus europaeus	√	
Swift	Apus apus	√	
Bee-eater	Merops apiaster		
Kingfisher	Alcedo atthis	√	
Hoopoe	Upupa epops	√	
Wryneck	Jynx torquilla	√	
Green woodpecker	Picus viridis	√	
Great spotted woodpecker	Dendrocopos major	√	
Lesser spotted woodpecker	D. minor	√	
Woodlark	Lullula arborea	√	
Skylark	Alauda arvensis	√	
Shorelark	Eremophila alpestris	√	√
Sand martin	Riparia riparia	√	
Swallow	Hirundo rustica	√	
House martin	Delichon urbica	√	

continued on next page

Appendix   The breeding bird species of Britain. (cont.)

Species	Scientific name	Bred during 1970–97	First bred 1970–97
Tree pipit	*Anthus trivialis*	√	
Meadow pipit	*A. pratensis*	√	
Rock pipit	*A. petrosus*	√	
Yellow wagtail	*Motacilla flava*	√	
Grey wagtail	*M. cinerea*	√	
Pied wagtail	*M. alba*	√	
Dipper	*Cinclus cinclus*	√	
Wren	*Troglodytes troglodytes*	√	
Dunnock	*Prunella modularis*	√	
Robin	*Erithacus rubecula*	√	
Nightingale	*Luscinia megarhynchos*	√	
Bluethroat	*L. svecica*	√	√
Black redstart	*Phoenicurus ochruros*	√	
Redstart	*P. phoenicurus*	√	
Whinchat	*Saxicola rubetra*	√	
Stonechat	*S. torquata*	√	
Wheatear	*Oenanthe oenanthe*	√	
Ring ouzel	*Turdus torquatus*	√	
Blackbird	*T. merula*	√	
Fieldfare	*T. pilaris*	√	
Song thrush	*T. philomelos*	√	
Redwing	*T. iliacus*	√	
Mistle thrush	*T. viscivorus*	√	
Cetti's warbler	*Cettia cetti*	√	√
Grasshopper warbler	*Locustella naevia*	√	
Savi's warbler	*L. luscinioides*	√	
Moustached warbler	*Acrocephalus melanopogon*		
Sedge warbler	*A. schoenobaenus*	√	
Marsh warbler	*A. palustris*	√	
Reed warbler	*A. scirpaceus*	√	
Icterine warbler	*Hippolais pallida*	√	√
Dartford warbler	*Sylvia undata*	√	
Lesser whitethroat	*S. curruca*	√	
Whitethroat	*S. communis*	√	
Garden warbler	*S. borin*	√	
Blackcap	*S. atricapilla*	√	
Wood warbler	*Phylloscopus sibilatrix*	√	
Chiffchaff	*P. collybita*	√	
Willow warbler	*P. trochilus*	√	

continued on next page

Appendix The breeding bird species of Britain. (cont.)

Species	Scientific name	Bred during 1970–97	First bred 1970–97
Goldcrest	*Regulus regulus*	√	
Firecrest	*R. ignicapillus*	√	
Spotted flycatcher	*Musciapa striata*	√	
Pied flycatcher	*Ficedula hypoleuca*	√	
Bearded tit	*Panurus biarmicus*	√	
Long-tailed tit	*Aegithalos caudatus*	√	
Marsh tit	*Parus palustris*	√	
Willow tit	*P. montanus*	√	
Crested tit	*P. cristatus*	√	
Coal tit	*P. ater*	√	
Blue tit	*P. caeruleus*	√	
Great tit	*P. major*	√	
Nuthatch	*Sitta europaea*	√	
Treecreeper	*Certhia familiaris*	√	
Golden oriole	*Oriolus oriolus*	√	
Red-backed shrike	*Lanius collurio*	√	
Jay	*Garrulus glandarius*	√	
Magpie	*Pica pica*	√	
Chough	*Pyrrhocorax pyrrhocorax*	√	
Jackdaw	*Corvus monedula*	√	
Rook	*C. frugilegus*	√	
Carrion crow	*C. corone*	√	
Raven	*C. corax*	√	
Starling	*Sturnus vulgaris*	√	
House sparrow	*Passer domesticus*	√	
Tree sparrow	*P. montanus*	√	
Chaffinch	*Fringilla coelebs*	√	
Brambling	*F. montifringilla*	√	
Serin	*Serinus serinus*	√	
Greenfinch	*Carduelis chloris*	√	
Goldfinch	*C. carduelis*	√	
Siskin	*C. spinus*	√	
Linnet	*C. cannabina*	√	
Twite	*C. flavirostris*	√	
Redpoll	*C. flammea*	√	
Common crossbill	*Loxia curvirostra*	√	
Scottish crossbill	*L. scotica*	√	
Parrot crossbill	*L. pytyopsittacus*	√	√
Scarlet rosefinch	*Carpodacus erythrinus*	√	√

continued on next page

Appendix   The breeding bird species of Britain. (cont.)

Species	Scientific name	Bred during 1970–97	First bred 1970–97
Bullfinch	*Pyrrhula pyrrhula*	√	
Hawfinch	*Coccothraustes coccothraustes*	√	
Lapland bunting	*Calcarius lapponicus*	√	√
Snow bunting	*Plectrophenax nivalis*	√	
Yellowhammer	*Emberiza citrinella*	√	
Cirl bunting	*E. cirlus*	√	
Reed bunting	*E. schoeniclus*	√	
Corn bunting	*Miliaria calandra*	√	

* species of introduced or feral origin.
The minimum requirement to qualify as a breeding species was arbitrarily taken as a record of a single pair with eggs.

This table lists all species that have bred in Britain since 1800; the listing would be very similar for UK or Britain and Ireland combined. Those species that bred during 1970–97 and, specifically, those that bred for the first time in 1970–97 are tick marked. By default, species with no ticks in the 'Bred during 1970–97' column have bred in Britain but not during 1970–97. Note: Kentish plover did breed in the Channel Islands post-1970; black tern did breed in Northern Ireland in 1975, but not in Britain; barnacle goose bred in Northern Ireland before 1970, but not in Britain.

Chapter 23

Mammals

Gordon B. Corbet and D. W. Yalden

ABSTRACT

The past 25 years have seen considerable progress in understanding the distribution of British mammals, both from re-evaluation of historical records and from specific surveys of several of the species regarded as important for legal and conservation reasons. Otters, brown hares, badgers and water voles have been studied by systematic structured surveys. Most of the carnivores, and the deer, are continuing to recover ranges that had been lost over previous decades or centuries. The greatest declines at present seem to be two rodents threatened by introduced species: the red squirrel's range continues to shrink as that of the grey squirrel spreads, and the water vole seems to be declining as a consequence of predation by American mink combined with a reduction in habitat caused by agricultural change.

1 Introduction

Since the 1973 account (Corbet 1974), knowledge of mammals in Britain has advanced on all fronts; greater archaeological knowledge and assessment have improved understanding of the historical record, better recording and analysis have changed our perception of the current fauna, and better appreciation of biodiversity is altering the legal and conservationist response to the state of the mammal fauna. As well as a series of distribution atlases (Arnold 1978, 1984, 1993) and two further editions of the *Handbook of British Mammals* (Corbet and Southern 1977; Corbet and Harris 1991), there has been a symposium on changes in selected species (Harris 1989) and an attempt has been made to estimate the numbers of each British mammal (Harris *et al.* 1995). A number of specific surveys and resurveys have been undertaken (e.g. of badgers, otters, brown hares, water voles and feeding bats; see below), and a general summary has been presented by Yalden (1999). About 1700 Mammal Society members and 2000 Bat Conservation Trust members contribute regular information on distribution and participate in surveys, along with professional members of the Deer Commission, Sea Mammals Research Unit, Game Conservancy, English Nature, Scottish Natural Heritage, Countryside Commission for Wales and the Irish National Parks and Wildlife Service, as well as gamekeepers and foresters. The 60 or so species present in Great Britain (41 native and 19 introduced), like the 30 in Ireland (perhaps 21 of them native) are at least being recorded well, and the decline of the water vole has been recorded as it happens (Strachan and Jefferies 1993), in contrast to the decline of the otter which was only recognised 20 years later (Chanin and Jefferies 1978).

2 The legal framework

Deer and seals have been protected by legislation limiting the seasons in which they can be culled and the weapons that can be used since the Deer Act of 1963 and the Seals Act of 1970. The 1981 Wildlife and Countryside Act gave protection to a number of British mammals, specified on Schedule 5 of that Act, from deliberate killing and sale or possession. The species listed included all the bats, pine marten *Martes martes*, wildcat *Felis silvestris*, red squirrel *Sciurus vulgaris*, common dormouse *Muscardinus avellanarius* and walrus *Odobenus rosmarus*. Bats are also protected from disturbance of their roost sites. Species may be added to the schedule, so the otter *Lutra lutra* and, most recently, the water vole *Arvicola terrestris* have been added. The badger *Meles meles* has meantime been the subject of a series of laws designed to protect both the animal itself and its setts from disturbance, particularly from illegal badger digging. Release of alien species into the wild is prohibited by the 1981 Act, and most of the recently established aliens have been specifically listed in Schedule 9 (coypu *Myocastor coypus*, edible dormouse *Glis glis*, mink *Mustela vison*, grey squirrel *Sciurus carolinensis*, crested porcupines *Hystrix cristata* and *H. hodgsoni* (now *H. brachyura*) red-necked wallaby *Macropus rufogriseus*; the Chinese muntjac *Muntiacus reevesi* has been recently added to this number).

3 Changes in the historical record

A better analysis of the archaeological record suggests a number of changes to the historical outline offered previously (Corbet 1974), and one important confirmation. The notion that the reindeer *Rangifer tarandus* might have survived into historical times was then doubted; both a reanalysis of the archaeological and historical evidence (Clutton-Brock and MacGregor 1988) and radiocarbon dating of significant Scottish specimens (Kitchener and Bonsall 1997) support the view that the reindeer died out long before historical times, probably about 8300 years ago. Other species whose survival was thought, largely on the arguments of Harting (1880), to extend variously into post-Roman and mediaeval Britain also seem likely to have died out earlier than he thought. The brown bear *Ursus arctos* survived into Roman times, but evidence for its survival as late as the tenth century is doubtful. The wild boar *Sus scrofa* probably died out at the end of the thirteenth rather than the seventeenth century, the later records cited by Harting being evidence of reintroductions for hunting purposes. Aurochs *Bos primigenius* and elk *Alces alces* seem to have died out in Bronze Age times, about 3500 years ago. The notion that 'wild' park cattle, such as those of Chillingham, might be surviving aurochs, an idea that Harting at least suggested, is not supportable. The beaver *Castor fiber* certainly did survive into Saxon times, on archaeological and other evidence, and Harting's discussion of its survival into mediaeval times is among the best supported of his late survivals. Wolves *Canis lupus* certainly survived in Scotland to about 1700, and slightly later in Ireland, as Harting reported, but the later records become increasingly heroic and perhaps mythical. One animal omitted by both Harting and Corbet turns out to have survived much more recently than heretofore suspected; a lynx *Felis lynx* from northern Scotland has been radiocarbon dated to 1770 years ago, that is, equivalent to Roman times in southern Britain (Kitchener and Bonsall 1997). Another survivor from the late glacial period was the northern vole *Microtus oeconomus*, which apparently lingered on the Scilly Isles to Bronze Age times. Among marine mammals, the grey whale *Eschrichtius robustus,* a species restricted to coastal waters, was exterminated in the Atlantic about 400 years ago, the latest records being from Devon and Cornwall.

The historical record from Ireland has also been markedly improved by the judicious application of radiocarbon dating to archaeological material (Woodman *et al.* 1997). The survival of some species through from late glacial times into the warmer post-glacial period has been confirmed. Although previously only reindeer and Irish elk (giant deer, *Megaceros giganteus*) were known to occur in Ireland in the late glacial era, stoat *Mustela erminea*, Irish hare *Lepus timidus hibernicus*, wolf and brown bear were present with them, still present in the post-glacial period, and may be regarded as survivors. Wild boar and lynx arrived in Ireland early in the post-glacial age, but the red deer *Cervus elaphus*, long regarded as a native mammal, is not so far an attested presence before about 4200 years ago, and may have been introduced by humans or swum over from Scotland late in the post-glacial period.

Knowledge of introductions to Great Britain is also somewhat improved by a better archaeological record and interpretation of the historical accounts. There are now several Iron Age records of the house mouse *Mus domesticus* in Britain, as well as enough Roman records to be certain that it was well established here then. The ship rat *Rattus rattus*, always regarded with confidence as an animal that accompanied returning Crusaders in the twelfth century, now turns out to have been introduced originally by the Romans, perhaps died out in Saxon times, but was certainly present again in Anglo-Scandinavian York and continuously present thereafter (O'Connor 1992). It is now much more certain that the Normans introduced both fallow deer *Dama dama* and rabbits *Oryctolagus cuniculus*, the former about a century before the latter. Conversely, some species that have always been regarded as native, notably the harvest mouse *Micromys minutus* and brown hare *Lepus europaeus*, may, because of their absence from early archaeological records, have to be reclassified as introductions (Yalden 1999).

4 Changes to the mammal fauna since 1973

Two species considered British in 1974 have since become extinct. The mouse-eared bat *Myotis myotis* never had more than a toehold in Britain, although it was once widespread and numerous in mainland Europe. A small colony of about 10–12 individuals was discovered in Dorset in 1956, but had died out by 1980; it is believed that excessive disturbance and human persecution was the cause. Meanwhile, another small colony of perhaps 50 bats had been discovered hibernating in Sussex in 1969, and their locality (unlike the Dorset site) was not publicised. The breeding site was never discovered, but was believed to have been destroyed in some accident in 1974, because only males returned to the hibernaculum in that and subsequent years, in steadily declining numbers. A single male survived from 1985 to 1990. The coypu *Myocastor coypus* has been successfully eliminated by a determined programme of trapping organised by the Ministry of Agriculture, Fisheries and Food; it therefore joins the muskrat *Ondatra zibethica* as one of the two examples of a totally successful pest-control programme. The population of coypu was estimated to be around 6000 in 1981 when the final trapping began; this required 24 trappers, 216 000 trap-nights and cost £2.5 million (Gosling and Baker 1988, 1989).

Conversely, at least two species have been added to the mammal fauna in this period. Nathusius' pipistrelle *Pipistrelllus nathusii* has been apparently spreading westwards in Europe, and becoming more frequently recorded as a migrant or vagrant in the British Isles. Courtship behaviour ('singing') was reported from Somerset in 1995 (Barlow and Jones 1996), and breeding roosts were reported from both Northern Ireland and Lincolnshire in 1997. Meanwhile, the study of the calls of common pipistrelle *Pipistrellus pipistrellus* that led to the discovery of Nathusius' pipistrelle in Somerset was part of a co-operative study which has shown that two cryptic species have been masquerading under that name. One produces most energy in its echolocation calls at 55 kHz, and is truly *Pipistrellus pipistrellus*, while the other produces most energy at 45 kHz, and ought to be called *P. pygmaeus* (Barrett *et al.* 1997; Jones and Barratt 1999).

It is less certain whether any further introductions should be added to the list of British mammals, but wild boar, escaped from specialist venison farms, were reported to be established in various parts of southern England in 1997. Other possible introductions have been discussed by Baker (1990). Various rodents, ungulates and even carnivores have survived in the wild for months or years, but so far without, apparently, establishing breeding colonies.

5 Population changes in British mammals

Most of the information that has accumulated since the previous account serves to highlight current changes in the abundance and/or range of many of the most conspicuous mammals in the British countryside. It is worthwhile taking each order in turn. A summary of current status is provided by Harris *et al.* (1995) and Yalden (1999); detailed references for this section are given by them.

5.1 Insectivores

The mole *Talpa europaea*, hedgehog *Erinaceus europaeus* and three shrews (*Sorex araneus, S. minutus, Neomys fodiens*) remain both widespread and common, with little evidence of significant changes in range or abundance for four of them: for the hedgehog, some evidence of a long-term decline is indicated by the declining bag obtained during game protection. No other evidence of population change is available, and it is unclear whether this decline, from about 3 per km^2 in 1960 to about 0.8 per km^2 in 1989, reflects a genuine decline in hedgehogs or a declining interest in persecuting them. Some estates no longer report them, since they were given partial legal protection in 1981 (Tapper 1992). If it is a genuine decline, it might reflect the increasing volume of road traffic, or it might indicate increasing predation by the undoubtedly increasing badger population. The hedgehog also shows some significant increases in range, as a consequence of introduction to both Harris in the Outer Hebrides and to other islands; the threat it poses to ground-nesting waders is a problem. The isolated Isles of Scilly population of *Crocidura suaveolens*, presumably introduced but before the Bronze

Age, is now known to be more numerous than previously suggested (cf. Harris *et al.* 1995; Temple and Morris 1998).

5.2 Chiropterans

These have already been mentioned as losing one species and gaining two. Other species appear to have become scarcer, perhaps in response to poorer summer weather, or to such human-induced changes as loss of food through pesticide use, habitat loss and direct persecution. Both greater horseshoe bats *Rhinolophus ferrumequinum* and house-dwelling bats, mostly *Pipistrellus* sp., have declined in abundance during the 1980s, as documented by direct counts at roost sites, and the two horseshoe bats remain absent from parts of their former range, especially in south-eastern England. On the other hand, the increased activities of bat recorders, and the application of new methods of study, including radio-tracking and netting at mating sites (cave entrances and the like) in autumn, have increased our knowledge of the range, abundance and habitat requirements of many of the species, including the rarer ones. The first known breeding site of Bechstein's bat *Myotis bechsteini*, perhaps the rarest British bat, was only discovered in 1997.

5.3 Lagomorphs

The lagomorphs have had mixed fortunes. The rabbit *Oryctolagus cuniculus* continues to make a slow but steady recovery from the devastating epidemic of myxomatosis in 1953–54. In 1995, it was considered to be back to about 40% of its pre-myxomatosis population, but it is more patchily distributed than before, and certainly less abundant in Wales. The brown hare *Lepus europaeus* seems to have been in decline throughout the twentieth century, and this is documented both by the game bag (Tapper 1992) and by a specific survey undertaken in 1991–93 (Hutchings and Harris 1996). It is unclear how much of the decline is the result of increased predation, to which they are certainly susceptible, and how much is due to the impoverished habitat provided by the polarised agricultural landscape of the 1980s and 1990s. Because the species is especially one of farmland, and the best populations are on sporting estates where cover and food are retained, the suspicion lies heavily on agricultural change as the agent of destruction. Mountain hares *Lepus timidus* are largely free from that particular threat, though often dependent on the maintenance of heather moorland for grouse shooting. Introduced populations in North Wales and the Lake District have died out, but those in the Peak District and Southern Uplands remain viable. In Highland Scotland the species remains common and widespread, though habitat loss (through afforestation) must have reduced numbers. In Ireland, the introduced brown hare remains very scarce, and the Irish hare *L. timidus hibernicus* is widespread on farmland as well as moorland.

5.4 The carnivores

The carnivores show some of the most noticeable changes in population and distribution since the 1973 review. The wildcat *Felis silvestris* has been the subject of study because of its restricted range and its history of persecution during the nineteenth century (Langley and Yalden 1977). It apparently recovered its range in Highland Scotland by the 1960s, and has shown little subsequent expansion; evidently the industrialised Central Lowlands and the farmed eastern fringe are not suitable habitat (Easterbee *et al.* 1991). A major feature of that unsuitability seems, however, to be the presence of domestic or feral cats, with which the wildcat has hybridised. It is currently uncertain just how much of a pure wildcat population there may be, and this has serious implications for conservation. The wildcat is protected from persecution under the 1981 Wildlife and Countryside Act, but feral cats are not, nor should they be. Distinguishing the two is decidedly problematical, however, and neither a gamekeeper nor a naturalist could be expected always to make a reliable field identification (Daniels *et al.* 1998).

The fox *Vulpes vulpes* was and is widespread and common, but in particular has now recovered its abundance in East Anglia, where it had been scarce (Tapper 1992). It seems, from vermin bags, to have become more and more abundant throughout the period. This also applies to the badger *Meles meles*, which has been the subject of two specific surveys, in 1985–88 (Cresswell *et al.* 1990) and in 1994–97 (Wilson *et al.* 1997). These were structured surveys, examining around 2300 1-km squares distributed across the range of available habitats in Great Britain. They suggest an increase from 43 000 to 50 000 social groups of badgers in the nine-year period. This appears to result from reduced mortality as legislation to protect both badgers and their setts from illegal persecution ('badger digging') took effect during the 1990s.

The otter *Lutra lutra* has been the subject of a similar series of national surveys. Early surveys, in 1977–82, showed that it was present in only 6% of its possible range in England, 20% in Wales, but 73% in Scotland and 92% in Ireland. Repeat surveys in England have shown a recovery to 10% in 1984–86 and 22% in 1991–94; similarly in Wales, it recovered to 39% in the mid-1980s and 52% in 1991. In Scotland, too, parts of the Central Lowlands that had been vacant were reoccupied during the 1980s. In parts of eastern England, reintroductions of captive-bred otters have been used to bolster the recovery. Even so, it is calculated that it will take another 100 years for a full recovery in England (Strachan and Jefferies 1996).

Two other mustelids that were severely restricted by nineteenth-century persecution, the polecat *Mustela putorius* and the pine marten *Martes martes*, have also shown partial but slow recovery of their former ranges (Langley and Yalden 1977). The polecat was restricted to western Wales by 1915; it had spread slowly back through Wales and the Marches by 1968 (Walton 1968) and has since spread more rapidly into the English Midlands (Birks and Kitchener 1999). Surreptitious releases of captive-bred animals into the Lake District and Argyll have also seen them re-established there. The pine marten has had a less certain recovery. In Scotland it survived in the extreme north-west in 1915 – but only just – and took until 1946 to spread through north-western Scotland as far as the Great Glen. By the 1960s it had spread across the Glen into the Grampian Mountains, and it is now widespread in the Highlands. Meanwhile, an introduction of wild-caught animals into the Southern Uplands (to Galloway) in 1980–81 has seen a small population established there. The pine marten had also survived in 1915 in England, in the Lake District, and in Snowdonia in Wales; these populations persisted through to the 1960s, apparently without increasing in abundance or range, and it is rather doubtful whether they have survived. This seems odd, in view both of the evident recovery in Scotland, and the similar operation of favourable factors (reduced numbers of gamekeepers, increased afforestation) which are believed to have been responsible there. It also contrasts with the spread of the polecat: perhaps some poorly understood difference in their ecologies has produced these different responses. One possible factor is the sharp increase in sheep numbers in both Wales and the Lake District, which might have adversely affected the upland prey base for the pine marten populations. Polecats tend to favour the lowlands, and prey heavily on common rats and rabbits around farms.

The two widespread and abundant mustelids, the weasel *Mustela nivalis* and stoat *M. erminea*, have remained so throughout the period. However, their numbers, as reflected in game bags, reveal somewhat different fortunes over the last 25 years (Tapper 1992). Stoats, which prey heavily on rabbits, have been recovering as and where rabbit numbers recover; whereas weasels, which prey heavily on field voles, have declined steadily because, perhaps, the longer grassland favoured by field voles has been lost to rabbits, agricultural changes and more sheep. One other mustelid has continued to increase and spread thoughout the period: the introduced American mink *M. vison* was still confined to western Britain in 1974, but has since spread thoughout both main islands and to several of the smaller ones as well. While initial concerns that it was having a serious impact on native wildlife were apparently exaggerated, serious impacts on water voles and on seabird colonies have now been documented (Craik 1997; Strachan *et al.* 1998). It is, however, possible that the recovering polecat and otter populations are having an effect in repressing the mink population (Strachan and Jefferies 1993).

The two species of seal have had mixed fortunes over the 25 years. The harbour (common) seal *Phoca vitulina* is actually the rarer of the two British residents, concentrated on more sheltered coasts. It was subject to appreciable commercial hunting during the 1960s, leading to legal protection in 1970. Resurveys of some of the hunted populations suggest that they had still not fully recovered by the mid-1980s. Then in 1988 a 'new' virus, phocine distemper virus, appeared among common seals in the seas between Denmark and Norway, spread west through the Wadden Sea, and reached the Wash in late summer. It is thought to have killed about 50% of the Wash population, and 10–20% of that up the east coast of Scotland, but barely affected the common seals around Orkney, Shetland, western Scotland or Ireland. The grey seal *Halichoerus grypus* was barely affected by the virus, and its numbers continue to increase as they did throughout the twentieth century. Most major colonies are surveyed annually, and this is one species for which both the range and population are well established. Commercial hunting in Orkney during the 1960s and culls on the Farne Islands in 1963–65, 1972 and 1975 did limit the rate of increase, but public pressure led to the abandonment of culls in later years. The population is increasing at about 6% per year.

5.5 Rodents

The biggest concern among the rodents is the continuing replacement of the red squirrel *Sciurus vulgaris* by its grey American cousin *S. carolinensis*. In 1971, the red squirrel still occurred quite widely in East Anglia and Wales, as well as in various smaller pockets including Cannock Chase, the Peak District and parts of Lincolnshire and Cornwall (Lloyd 1983). Since then, it has continued to retreat, even in former strongholds such as the Lake District. The grey squirrel has continued to flourish, and there is now no doubt that its spread is the main cause of decline of the red squirrel. While the grey squirrel had been slow to spread into East Anglia, and slow also to spread into the

Southern Uplands in Scotland, it did so rapidly during the 1980s and 1990s, and recent evidence from Ireland suggests that a similar though delayed replacement is under way there. It remains uncertain whether the red squirrel can sustain itself in the Caledonian pine forests of Scotland, and in the new conifer plantations elsewhere; if not, island populations, as on the Isle of Wight, Brownsea, Anglesey and Skye, may be the only secure ones.

No less worrying is the evident decline of the water vole *Arvicola terrestris*, largely as a result of predation by the introduced mink, but exacerbated by the destruction of bankside vegetation through over-grazing, cultivation and insensitive drainage operations (Strachan and Jefferies 1993). This decline only became apparent during the specific survey of water vole distribution, and it highlights similar concerns about the status of the field vole *Microtus agrestis*, another grassland specialist which should be similarly vulnerable to the loss of grasslands. This is still common and widespread – indeed the most common wild mammal in Britain according to Harris *et al.* (1995) – but the evidence of various studies of its predators is that it has become much less common during the 1980s and 1990s. It is less dominant among the prey of barn owls *Tyto alba* than it was in the 1970s, and also less available, apparently, to weasels: the converse change is that wood mice *Apodemus sylvaticus* appear to be either more numerous, or at least more readily available, particularly in the arable countryside of eastern Britain. Specific surveys of two rarer rodents, the harvest mouse *Micromys minutus* and dormouse *Muscardinus avellanarius,* confirm that they are less widespread than they once were. For the dormouse, this is a change compared with Victorian Britain, and seems to reflect the decline in coppicing and hedgerow management since then (Bright *et al.* 1996). For the harvest mouse, a Mammal Society survey in the 1970s actually found it to be as widespread as in Victorian times, even confirming the doubtful records from southern Scotland (Harris 1979), but subsequent agricultural changes have produced a serious decline in abundance, less definitely in range as well, since then. The common rat *Rattus norvegicus* and house mouse *Mus domesticus* remain widespread as commensal rodents, but appear to be much less common than formerly, perhaps in response to better hygiene and food storage. Meanwhile the ship rat *Rattus rattus*, formerly common and feared as the carrier of plague, is now reduced to apparently transient occurrences in a few ports and small populations on three small islands: Lundy, Garbh Eilean (Shiant Islands, Outer Hebrides) and Inchcolm, Firth of Forth. In Ireland, the bank vole *Clethrionomys glareolus*, apparently introduced about 1950, continues to spread at about 3 km per year (Smal and Fairley 1984).

5.6 Artiodactyls

The artiodactyls continue to flourish. Roe deer *Capreolus capreolus* have continued to spread across southern England after their reintroduction to Dorset in 1800, rather slowly from the East Anglian reintroduction in 1884, and southwards from the Scottish population. Roe deer are the most abundant and widespread deer now in Great Britain (they have never occurred in Ireland). Red deer *Cervus elaphus* are also more abundant now than they were 25 years ago, but the bulk of the population, perhaps 347 000, remains in Scotland and there are only 12 500 in England. The increasing Scottish population continues to trouble forestry and conservation interests because of its impact on both plantations and native vegetation. Intensive research has been carried out on the ecology of red deer, especially on the island of Rum (Clutton-Brock *et al.* 1982, 1989). The biggest range changes concern two of the introduced species. The muntjac *Muntiacus reevesi* continues to spread from the centre of introduction at Woburn Park, Bedfordshire, although deliberate releases well beyond that original focus have accelerated its spread (Chapman *et al.* 1994). The Sika *Cervus nippon* has been expanding its Scottish range in particular, apparently finding the dense thicket stages of coniferous plantations especially suitable. Unfortunately it has also managed to inter-breed very successfully with the native red deer, and the result is an increasingly hybrid population that is neither species. The feeling of the geneticists involved is that introgression is already too far advanced to be reversed, and that on mainland Scotland the native deer is essentially lost, as it is already in the Wicklow Mountains of Ireland; if true, it is important that Sika or hybrids are not allowed to reach the islands. Meanwhile, the long-established fallow deer has also been increasing in range and abundance, particularly in England, and concerns about its impact on farming and forestry, as well as amenity woodlands, are growing (Putman and Moore 1998).

6 Discussion

The years since 1975 have been dynamic ones for British mammals: perhaps they always have been in flux, and it is simply that the changes are now better recorded. For the carnivores and deer, the larger mammals in the countryside, the changes are generally towards increased ranges and populations, responses in all cases to reductions in historic times (often nineteenth century or earlier). For many of the smaller, more widespread species the trends seem to be downward, as a result more of habitat loss and change than of persecution. This is paradoxical, since the carnivores tend to depend on these smaller and widespread species as prey. The pernicious impact of introductions on native species is a clear and unwelcome trend; grey squirrels replacing red squirrels, mink eliminating water voles, Sika and feral cats introgressing with red deer and wildcats respectively. The elimination of the coypu might be seen as the one bright star in this particular firmament: however, another optimistic note is offered by the increasing use of reintroduction as a conservation tool. Dormice have been successfully reintroduced to two counties, Cheshire and Cambridgeshire, which they once inhabited. Otter populations in East Anglia and North Yorkshire have also been successfully strengthened, perhaps initiated, with released animals. Scottish Natural Heritage is actively considering the reintroduction of European beavers *Castor fiber*, and wild boar seem to have re-established themselves by escaping from specialist venison farms.

REFERENCES

Arnold, H.R. (1978) *Provisional Atlas of the Mammals of the British Isles.* Abbots Ripton: Biological Records Centre.

Arnold, H.R. (1984) *Distribution Maps of the Mammals of the British Isles.* Abbots Ripton: Biological Records Centre.

Arnold, H.R. (1993) *Atlas of Mammals in Britain.* London: HMSO.

Baker, S.J. (1990) Escaped exotic mammals in Britain. *Mammal Review*, **20**, 75–96.

Barlow, K.E. and Jones, G. (1996) *Pipistrellus nathusii* (Chiroptera: Vespertilionidae) in Britain in the mating season. *Journal of Zoology*, **240**, 767–773.

Barratt, E., Deaville, R., Burland, T.M., Bruford, M.W., Jones, G., Racey, P.A. and Wayne, R.K. (1997) DNA answers the call of pipistrelle bat species. *Nature*, **387**, 138–139.

Birks, J.D.S. and Kitchener, A.C. (1999) *The distribution and status of the polecat* Mustela putorius *in Britain in the 1990s*. London: Vincent Wildlife Trust.

Bright, P.W., Morris, P.A. and Mitchell-Jones, A.J. (1996) A new survey of the dormouse *Muscardinus avellanarius* in Britain, 1993–4. *Mammal Review*, **26**, 189–195.

Chanin, P.R.F. and Jefferies, D.J. (1978) The decline of the otter *Lutra lutra* L. in Britain: an analysis of hunting records and discussion of causes. *Biological Journal of the Linnean Society*, **10**, 305–328.

Chapman, N., Harris, S. and Stanford, A. (1994) Reeves' Muntjac *Muntiacus reevesi* in Britain: their history, spread, habitat selection, and the role of human intervention in accelerating their dispersal. *Mammal Review*, **24**, 113–160.

Clutton-Brock, J. and MacGregor, A. (1988) An end to medieval reindeer in Scotland. *Proceedings of the Society of Antiquaries of Scotland*, **118**, 23–35.

Clutton-Brock, T.H., Guinness, F.E. and Albon, S.D. (1982) *Red Deer: Behaviour and Ecology of Two Sexes*. Edinburgh: Edinburgh University Press.

Clutton-Brock, T.H. and Albon, S.D. (1989) *Red Deer in the Highlands*. Oxford: Blackwell Scientific.

Corbet, G.B. (1974) The distribution of mammals in historic times, in Hawksworth, D.L. (ed.) *The Changing Flora and Fauna of Britain*. London and New York: Academic Press, pp. 179–202.

Corbet, G.B. and Southern, H.N. (eds) (1977) *The Handbook of British Mammals*. 2nd edn. Oxford: Blackwell Scientific.

Corbet, G.B. and Harris, S. (eds) (1991) *The Handbook of British Mammals*. 3rd edn. Oxford: Blackwell Scientific.

Craik, J.C.A. (1997) Long-term effects of North American mink *Mustela vison* on seabirds in western Scotland. *Bird Study*, **44**, 303–309.

Cresswell, P., Harris, S. and Jefferies, D.J. (1990) *The History, Distribution, Status and Habitat Requirements of the Badger in Britain*. Peterborough: Nature Conservancy Council.

Daniels, M.J., Balharry, D., Hirst, D., Kitchener, A.C. and Aspinall, R.J. (1998) Morphological and pelage characteristics of wild living cats in Scotland: implications for defining the 'wildcat'. *Journal of Zoology*, **244**, 231–247.

Easterbee, N., Hepburn, L.V. and Jefferies, D.J. (1991) *Survey of the status and distribution of the wildcat in Scotland, 1983–1987*. Edinburgh: Nature Conservancy Council for Scotland.

Gosling, L.M. and Baker, S.J. (1988) Planning and monitoring an attempt to eradicate coypus from Britain. *Symposia of the Zoological Society of London*, **58**, 99–113.

Gosling, L.M. and Baker, S.J. (1989) The eradication of muskrats and coypus from Britain. *Biological Journal of the Linnean Society*, **38**, 39–51.

Harris, S. (1979) History, distribution, status and habitat requirements of the harvest mouse (*Micromys minutus*) in Britain. *Mammal Review*, **9**, 159–171.

Harris, S. (ed.) (1989) British mammals: past, present and future. *Biological Journal of the Linnean Society*, **38**, 1–118.

Harris, S., Morris, P., Wray, S. and Yalden, D. (1995) *A Review of British Mammals: population estimates and conservation status of British mammals other than cetaceans*. Peterborough: Joint Nature Conservation Committee.

Harting, J.E. (1880) *British Animals Extinct Within Historic Times*. London: Trübner.

Hutchings, M.R. and Harris, S. (1996) *The Current Status of the Brown Hare* (Lepus europaeus) *in Britain*. Peterborough: Joint Nature Conservation Committee.

Jones, G. and Barratt, E.M. (1999) *Vespertilio pipistrellus* Schreber, 1774 and *V. pygmaeus* Leach, 1825 (currently *Pipistrellus pipstrellus* and *P. pygmaeus*: Mammalia, Chiroptera): proposed designation of neotypes. *Bulletin of Zoological Nomenclature*, 56, 182–186.

Kitchener, A.C. and Bonsall, C. (1997) AMS radio-carbon dates for some extinct Scottish mammals. *Quaternary Newsletter*, 83, 1–11.

Langley, P.J.W. and Yalden, D.W. (1977) The decline of the rarer carnivores in Great Britain during the nineteenth century. *Mammal Review*, 7, 95–116.

Lloyd, H.G. (1983) Past and present distributions of red and grey squirrels in Britain. *Mammal Review*, 13, 69–80.

O'Connor, T.P. (1992) Pets and pests in Roman and medieval Britain. *Mammal Review*, 22, 107–113.

Putman, R.J. and Moore, N.P. (1998) Impact of deer in lowland Britain on agriculture, forestry and conservation habitats. *Mammal Review*, 28, 165–184.

Smal, C.M. and Fairley, J.S. (1984) The spread of the bank vole *Clethrionomys glareolus* in Ireland. *Mammal Review*, 14, 71–78.

Strachan, R. and Jefferies, D.J. (1993) *The Water Vole* Arvicola terrestris *in Britain 1989–1990: its distribution and changing status*. London: Vincent Wildlife Trust.

Strachan, R. and Jefferies, D.J. (1996) *Otter Survey of England 1991–1994*. London: Vincent Wildlife Trust.

Strachan, R., Jefferies, D.J., Burreto, G.R., Macdonald, D.W. and Strachan, R. (1998) The rapid impact of resident American Mink on water voles: case studies from Lowland England. *Symposia of the Zoological Society of London*, 71, 339–357.

Tapper, S. (1992) *Game Heritage*. Fordingbridge: Game Conservancy.

Temple, R. and Morris, P. (1998) The lesser white-toothed shrew on the Isles of Scilly. *British Wildlife*, 9, 94–99.

Walton, K.C. (1968) The distribution of the polecat *Putorius putorius* in Great Britain, 1963–67. *Journal of Zoology*, 155, 237–240.

Wilson, G., Harris, S. and McLaren, G. (1997) *Changes in the British Badger Population, 1988 to 1997*. London: People's Trust for Endangered Species.

Woodman, P., McCarthy, M. and Monaghan, N. (1997) The Irish Quaternary fauna project. *Quaternary Science Reviews*, 16, 129–159.

Yalden, D.W. (1999) *The History of British Mammals*. London: T. and A.D. Poyser.

Chapter 24

Fishes

Alwyne Wheeler

ABSTRACT

Since last addressing this topic and the related issue of the origin of the freshwater fishes in the British Isles (Wheeler 1974, 1977) there have been a number of changes in the fauna as well as changes in legislation which have affected fishes. These principally concern freshwater fishes, although the Wildlife and Countryside Act of 1981, and its amendments, has also had an impact on estuarine and marine species. The review of historical changes in the freshwater fish fauna (Wheeler 1974) omitted all marine species but in the present paper the terms of reference have been widened to include both freshwater fishes and those living in estuarine and marine conditions. The factors affecting these faunal assemblages are very different and pressures affecting their status differ. Thus, for example, freshwater fishes are exploited by anglers who mostly do not remove the fish on capture (the exceptions being anglers fishing for salmonids or game fish who usually kill large fish to eat). Despite this, a major pressure on freshwater fisheries is the continuing demand to add more fish to the population.

A positive change since the 1970s has been the reduction nationally of riverine and estuarine pollution following the establishment of the National Rivers Authority in 1989 and its transformation into the Environment Agency in 1995. The establishment of effective control of pollution in freshwaters has resulted in the partial restoration of the fish fauna and its improvement in most rivers, although it has to be emphasised that the restoration is only partial. Experience with the River Thames, which was severely polluted until the 1960s, shows that restoration of the fauna is a slow process which may take decades to complete if left to recover unassisted and may be very expensive if recovery has to be assisted (as with the attempted restoration of the salmon in the Thames).

Widening the scope of the topic to include freshwater fishes as well as marine and estuarine species introduces bias because of the unevenness of the coverage, marine fish being subject to commercial fisheries; the data are thus collected as a byproduct of these fisheries, while freshwater fish data are collected by selective methods biased in favour of angling species and estuarine species are very little studied at all.

1 Origin and status of freshwater fishes

This topic was addressed at some length by Wheeler (1974, 1977) and its broad conclusions have not been subsequently challenged, although some aspects of the scenario were open to debate. These conclusions were developed from the earlier presentations by Scharff (1899) and Regan (1911). The discussion stems from what was then understood of the history of glaciation in the British Isles, when at the greatest advance of the ice most of the land mass was ice-covered or endured peri-glacial conditions during the annual warm seasons or during short-term amelioration of the climate. Even in the peri-glacial areas there were insufficient open waters and temperatures remained close to freezing point so that fish, as poikilothermous animals, could not mature gonads – assuming the fish themselves could survive. As a result, it is postulated that any primary freshwater fishes that were present in the British Isles could not have survived into the Holocene (primary freshwater species are those which cannot tolerate exposure to sea water). Colonisation of Irish, Welsh and Scottish lakes after the retreat of the ice was effected by migratory fishes such as salmon, trout, charr and shads, or eutrophic species such as eel and three-spined stickleback, which are capable of living in salt water up to 30 ppm, especially at low temperatures. The later retreat of the ice exposed land bridges between eastern England and the European continent and (probably) southern England and France. Although these land bridges assisted passage by terrestrial and flying animals, they proved a barrier to aquatic organisms that had retreated into refugia in more southern or eastern parts of Europe and were therefore not immediately available as colonists into these newly unfrozen areas. In the case of the Irish fauna, sea level rises would have isolated freshwaters in Ireland from both France and Wales or Scotland and primary freshwater species would have been unable to cross into Irish freshwaters. The consequence of these glacial changes was that only secondary freshwater fishes (that is, those like eel and stickleback, which can tolerate full salinity levels), but not primary freshwater fishes (such as roach and pike), are truly native to Ireland. This is a challenge to accept, particularly for English anglers who visit Ireland to catch the abundant pike, bream, rudd and tench which were all introduced in historic times (probably mostly by English anglers or landowners).

Some freshwater fishes, such as the charr and the whitefishes, colonised rivers in the British Isles during the immediately post-glacial period as conditions allowed. The whitefishes (*Coregonus* spp.) must have been early colonisers while temperatures were low and the salinity of the sea was depressed by the melting of the ice, while the charr was probably a later arrival. Today, the charr is migratory in Norway and Iceland at about 64° north, entering rivers in late summer and spawning in lakes in early winter (often October) in deep water beneath the winter ice. Young fish stay in these deep lakes for up to five years before migrating to the sea. In the early Holocene, many of these populations became land-locked due to changes in sea level relative to the land following glaciation. There is no reason to suppose that this was a unique occurrence and it may have taken place on numerous occasions involving sibling taxa during the 8000 to 10 000 years through to the present.

The presence of freshwater fishes in England is more simply explained by reference to the connection between rivers during the period when a land bridge existed in the bed of the present North Sea and possibly in the eastern English Channel. Some evidence in support of this is provided by the differences between the fishes in a westward-flowing river (such as the River Severn) and a river which discharges to the North Sea (for example the Ouse), although the list of species given from the Severn in *The Species Conservation Handbook* (English Nature 1994) is greatly inflated by marine and introduced species. Although the Thames discharges to the North Sea and thus appears to be an eastern river, it is a special case as it lacked several of the critical species, notably spined loach, burbot and the silver bream, which although present must have been introduced. The distinctive absence of these species in the Thames is due to its early isolation from the North Sea rivers; rivers flowing into the northern North Sea were always isolated from colonisation through the North Sea, at first by ice sheets and later by the salinity of the water once the sea connection was effected. However, once the sea had flooded the North Sea basin, secondary freshwater fishes could colonise the eastern rivers of northern England and Scotland. The presence of primary freshwater fishes in Scotland and even in islands such as Orkney is entirely the result of human introduction.

Introduction and translocations within the British Isles and between their component parts are the most important feature affecting fish distribution.

2 Changes in legislation affecting fishes

Since 1973–74, there have been a number of changes in legislation affecting fishes which have in one sense clarified the legal position regarding the introduction of fish from abroad, although they have done little to decrease the level of fish movements into the country. Prior to the 1973 period introduction of fish depended on the permission of the riparian owner in the case of still waters and the controlling authority (variously river boards, river authorities or catchment boards, which were controlled by the Ministry of Agriculture). The responsible officers in these cases were Fisheries Officers. There seems to have been at least a notional control of these officers in each area. During this period several of the more undesirable movements of fish took place, such as the introduction of zander (a native European species, not found in Britain). Zander were introduced by the fisheries officer of the Great Ouse River Authority although advised by the fisheries office of the Ministry of Agriculture that this was undesirable. The only reason given for doing so was that the presence of zander would give anglers another large fish to catch. This period also saw increases in the distribution of the barbel into the western rivers of England, notably into the River Severn on the initiative of an angling newspaper, despite the protests of game fishermen.

From 1989 the formation of the National Rivers Authority resulted in a unified body for England and Wales which offered an integrated control of fisheries management, pollution control and drainage, and from 1995 the Environment Agency extended control over the whole aquatic environment. The Environment Agency now has responsibility for exercising control over the movement of fish for angling purposes and enforcing controls on the presence of disease or parasites on fish for which consents to move them from water to water are mandatory. However, the quantity of fish movements and changes in national law arising from the association of Great Britain with the European Union led to the abandonment of these well-intentioned conditions, except in the case of introductions to rivers. In place of mandatory 'health checks' on fish planned to be introduced from one water to another anglers, fishery owners and angling clubs were warned to ensure that fish for stocking came from reputable sources only.

The Wildlife and Countryside Act 1981 discourages the introduction of exotic species to UK waters, a position strengthened in late 1998 by the introduction of an Order requiring the licensing of non-native species coming into trade.

Despite the numerous attempts to control the movement of freshwater fishes in the British Isles, the overall effect has been that both exotic and native species have been very widely introduced and redistributed, particularly in areas where coarse angling is widely practised: effectively all of England, parts of Wales and even southern Scotland. The measure by which this can be judged is the speed with which exotic fish parasites have become established in large areas of the British Isles as the fish hosts have been transferred from one water to another without effective health checks being performed. Three specific cases can be cited, but there are others. The example given by Yeomans *et al.* (1997) is of the Asiatic tapeworm *Khawia sinensis*, first recognised in Britain in 1987, which has now been reported from practically the whole of the present range of its host, the carp (which is itself an introduced fish); this shows that the obligation to have health checks conducted before fish are introduced, under Section 30 of the Salmon and Freshwater Fishes Act 1975, has failed to stop the spread of this potential pathogen. The blood fluke *Sanguinicola inermis*, another parasite of the carp, was probably introduced to Britain in the 1950s–60s and was first recorded in the UK in 1977 (Kirk and Lewis 1994). It was classed as a category A parasite for which consent to move fish would be refused by the NRA, but it is now so widespread that there is no purpose in attempting to limit its further spread and it is no longer given this high category.

A similarly widespread introduced parasite involves the gill fluke *Ergasilus briani*, which was first reported in Britain in 1983 but is now widespread in southern England. An example of the ease with which this species can be spread occurred in the SSSI site of Epping Forest, Essex, when the gill fluke was found on fish in a small isolated pond which also contained a fauna consisting of carp, goldfish, rudd, crucian carp and their hybrids. A similar fish fauna was found in a pond at Epping, about 3 km to the north, and the anglers who fished the latter pond were known to be planning to remove some of the fish. It is believed that some fish had in fact already been transferred, complete with parasites, but without any authorisation.

The legal barriers controlling fish movements have proved ineffective in preventing the spread of exotic parasites and pathogens, and unauthorised introductions have continued. The introduction of the larger, high-value angling species such as the wels catfish *Silurus glanis* (a European species) and carp has continued unhindered. Compared with the 1970s commercial fisheries and fish dealers are numerous, and the relaxation of trade barriers and the enormous increase in cross-Channel traffic means that large numbers of non-native fish are available for purchase. Another loophole in the regulations is the abuse of the system that permits the import of exotic fishes as ornamental fish. Many of these only come to light with the publicity surrounding the capture of a large specimen, but there are numerous instances when fishes such as sterlet *Acipenser ruthenus* and sturgeon hybrids such as *A. ruthenus* × *A. guldenstati* have been unexpectedly found in still waters in Britain.

3 Changes in the freshwater fish fauna

Within the period 1973–99 several major changes affected freshwater fishes. The popularity of carp among anglers, largely because of the size they attain and their resistance to handling, resulted in the importation of large numbers, from European sources. Some imports are accompanied by health certificates verifying they are free from parasites and disease (depending where they were issued, these are sometimes of dubious status). Carp are also imported into Britain from Asia as ornamental fish; these often turn up in the wild as their owners tire of them as pets. Others are released into commercial fisheries.

Sturgeon are occasionally found in still waters and rivers. Most of these are believed to be *Acipenser rutheneus* or hybrids *A. rutheneus* × *A. guldenstati*; these are easily available from European sturgeon breeders and have been imported as ornamental fish. Some outgrow their accommodation and are released into the wild; some, however, are stocked into commercial angling lakes.

The wels catfish, a species native to the Danube basin but now widespread throughout Europe including the Iberian peninsula, is frequently introduced into commercial fisheries. It is a large species which can grow to a weight in excess of 45 kg (100 lb) and they are frequently stocked in commercial fisheries. It is imported by a licence issued by the Minstry of Agriculture, Fisheries and Food (MAFF) that requires the approval of English Nature. According to figures released by English Nature for the period April 1991 to December 1994, 19 licences were issued permitting the introduction of catfish. This level of introduction suggests that there is a steady level of interest but also that this fish might be relatively uncommon in Britain. However, according to the angling body the Catfish Conservation Group (1998), there are 171 waters that are known to contain the species, all but five in England. Other catfish species, notably the introduced North American channel catfish *Ictalurus punctatus*, are known to have been introduced to commercial fisheries.

The grass carp *Ctenopharyngodon idella*, is an Asiatic cyprinid that was initially of interest as a herbivore that does not breed in the conditions obtaining in British rivers. As it grows to a considerable size (35 kg in its native waters) it has attracted interest from fishery owners and there are a moderate number of applications to stock it. MAFF licences issued amounted to 64 in the period April 1991 to December 1994 (English Nature 1995).

Two other exotic fishes, both ornamental in interest, have become established within the last 25 years. The European fish sunbleak *Leucaspius delineatus* is now abundant in parts of Somerset and Hampshire. In England it was initially released in 1990 from a sales outlet for fish for garden ponds and was later spread by angling clubs whose members were under the impression that they were native bleak (Farr-Cox *et al*. 1996). It is probably now present in the River Wye, in addition to the Itchen and the Test, and in several lakes to which 'bleak' have been introduced.

The second is the clicker barb or sharp-nosed gudgeon *Pseudorasbora parva*, a member of the carp family originally from Japan, Korea and parts of China. This was accidentally imported to Romania along with other cyprinids introduced for fisheries purposes. Subsequently the clicker barb has spread as fish were moved around eastern Europe and it has been reported in lakes and rivers from Greece to France; it was imported to England by a fish dealer as an ornamental garden-pond fish in the Hampshire area. This stock was kept in ponds with orfe *Leuciscus idus* (another European species now well established in many waters in Britain) so that purchasers of orfe frequently get 'bonus' clicker barbs together with their orfe. It has been caught in several ornamental lakes in southern England, has appeared in the stock of other fish dealers and at least one has been found in the wild (Wheeler 1998).

While the introduction of exotic species is widely regarded as a major threat to the native fauna, redistribution of native species has possibly as serious implications. The barbel is now much more widely distributed than it was originally. A certain amount of redistribution had occurred before the 1970s (Wheeler 1974) and this has continued under the National Rivers Authority and the Environment Agency. The status of the barbel was described by Wheeler and Jordan (1990) who reported it from the Severn, the Bristol Avon, Somerset Frome and the River Chew in the south-west, and in northern England in the Swale, Wear, Aire, Calder and Don. Many of these were illegal introductions made by angling interests. More recently barbel have become widely available for stocking purposes, and as they are currently considered a desirable angling species are frequently introduced often in marginally suitable habitats and outside the original range of the species. In part the problem is caused by the duty of the Environmental Agency under the Water Resources Act (1991) 'to maintain, improve and develop freshwater fisheries', thus placing environmental concern for fish (as opposed to fisheries) in the hands of the influential angling lobby as opposed to conservationists.

The small percoid fish, the ruffe *Gymnocephalus cernuus*, has been introduced to a number of lakes in north Wales and Scotland as well as north-western England. It is a predator on bottom-living invertebrates, particularly chironomids, but in Scandinavia has been known to feed on the eggs of whitefishes (*Coregonus* sp.) in the appropriate season. All the *Coregonus* populations in Britain are regarded as threatened and the introduction of ruffe into these lakes may pose an additional threat to these already stressed species. The ruffe are said to have been introduced by anglers using them as live bait although there is no proof of this.

The zander, the introduction of which has already been discussed, has continued to spread in lowland rivers and lakes in England (as predicted by Wheeler and Maitland 1973).

4 Conservation measures involving freshwater fishes

Since 1973 there have been several studies of rare or threatened freshwater fishes in Britain and elsewhere in Europe. These have included commissioned reports for the Nature Conservancy Council (now English Nature) or European bodies such as those by Lelek (1987), Lyle and Maitland (1992) and Maitland and Lyle (1991) and represent useful status summaries of the species. Like many other commissioned reports they suffer from the lack of peer review and quickly become embedded in official literature as absolute truth. However, these reports have focused attention on the status of the species concerned. The taxa involved are Arctic charr, powan (= gwyniad or schelly), vendace and pollan, and burbot. Of these the charr has seriously declined in both numbers and populations as has the vendace, while the burbot is now extinct in Britain: claims that the burbot was ever common in Britain cannot be sustained, and the only objective evidence we have suggests it was uncommon in the eighteenth century in Fenland rivers which were the centre of distribution for the species in Britain (Forbes and Wheeler 1997). These authors suggest that this species has been in decline throughout the Holocene in Britain and changes in the environment, mostly human-induced, combined to cause its extinction.

5 Changes in the status of estuarine fishes

Prior to the 1950s, estuaries had in general suffered severely from pollution of many kinds, their flow was impeded by dams, weirs and navigation locks, and their water was polluted and sometimes anaerobic and even toxic. The situation has now greatly improved and fishes have returned to rivers from which they were either completely or partially absent. Several of the threatened fish taxa identified by Maitland and Lyle (1991) were estuarine: these included the sturgeon *Acipenser sturio*, allis shad *Alosa alosa*, twaite shad *A. fallax*, houting *Coregonus oxyrhynchus* and smelt *Osmerus eperlanus*. Of these the sturgeon and the houting were both rare vagrants and never known to have bred in British freshwaters; concern for their status was thus unnecessary.

The smelt was known from a number of rivers in the Solway area of south-west Scotland where a small fishery existed in the River Cree in the early 1980s but which has been greatly reduced recently (Lyle and Maitland 1997). Small populations existed in extensive low salinity areas, such as the Wash, Liverpool Bay and the Thames Estuary. The scarcity of the smelt in England was severely understated by Maitland and Lyle (1991) who cited a paper on the smelt's status in the Humber and the Tees as evidence for its status in eastern England, whereas smelt had been caught in small numbers on the Essex coast throughout the 1950s and 1960s and were abundant in the Thames in the 1960s and 1970s (Wheeler 1979). Since pollution in the Thames Estuary has been reduced, the smelt population has increased enormously. In the 1990s, this species was found in many of the rivers of East Anglia including the Great Ouse and in the Norfolk Broads; it has also been caught recently in Sussex rivers (R. Horsfield pers. comm.) and in the Tamar (personal observation) in the spring on a spawning migration. Its range and numbers have significantly increased since 1973. The smelt is known to breed in the Irish rivers Shannon, Fergus and Foyle (Quigley and Flanery 1996).

Twaite shad were caught more frequently in estuaries or on the coast in the 10 years to 1998, and there are possibly small breeding stocks in some rivers such as the Medway (Kent) the Tamar and on the Sussex coast as well as the well-known stock in the River Severn and other west-coast rivers. However, their breeding success is not known to have increased in the same way as that of the smelt. This may be a result of the difficulty inherent in catching and identifying young shad in the lower reaches of estuaries. Although they spawn in rivers, breeding by a small population would be difficult to confirm. However, it is certain that adult shad records in outer estuaries and on parts of the southern English coast have become more frequent in the last decade.

The allis shad is very infrequently caught in estuaries or on the coast, although occasionally spent females have been caught in the summer (in the Thames, on the Sussex coast and in the Tamar within the last decade). This may indicate that small numbers are attempting to breed but this is not confirmed. It is, however, always possible that these vagrant fish have crossed from French rivers as the species is reported to spawn in the Loire, Garonne, Dordogne and Adour (Quignard and Douchement 1991). A coastwise migration into British rivers by a fish which attains 60 cm total length is quite possible.

The amelioration of pollution which is general in the lower rivers and estuaries of the UK has led to improvements in the status of migratory fish in general. The salmon and sea trout, which were greatly diminished in numbers by pollution and obstructions, responded very quickly to improved conditions in estuaries. Salmon have been caught in a number of rivers such as the Don, where they were unheard of before, while the River Tyne now contains an appreciable run of salmon. Great efforts have been made to re-establish a salmon run in the River Thames by stocking with young fish, but so far it has only been partially successful: but then the Thames never was a particularly good salmon river and the present-day changes in rainfall and temperature are holding its recovery back. However, the spin-off from pollution control and the building of fish passes by the Environment Agency and the Thames Salmon Trust will improve the habitat over time for other migratory fish (for example eel and the shads) and freshwater fishes such as barbel *Barbus barbus*, chub *Leuciscus cephalus* and dace *L. leuciscus*.

6 Changes in the marine fish fauna

Compared with our knowledge of the conservation status of freshwater fishes knowledge of the status of marine fishes is very limited. Conservation can only be realistic in terms of species which can be assessed as to numbers in the shallow sea or on the shore, although statements of the need for conservation of such fish as basking shark *Cetorhinus maximus* and the skate *Raja batis* are made, the latter with some justification. The conservation requirements of rare British marine fishes were addressed in Report 1228 to the Nature Conservancy Council (Potts and Swabey 1991) which suggested in summary that 'Endangered species' included 10 taxa. Of these, the sturgeon and houting are present only as vagrants (and the latter has not been recorded since the beginning of the nineteenth century) and the allis and twaite shads have been discussed as estuarine species in this chapter. This leaves six taxa that are considered endangered, namely Couch's goby *Gobius couchi*, Steven's goby *G. gasteveni*, the giant goby *G. cobitis*, the black-faced blenny *Tripterygion delaisi*, the basking shark and the skate.

The three gobies and the black-faced blenny are all shallow water inshore species which are at the northern edge of their range in the British Isles. They are all small (although the 'giant' goby is only gigantic relative to other gobies – it reaches a total length of 27 cm), and cryptic in their lifestyle. Couch's goby appears to be the most restricted having been recognised only from the coastline of County Donegal, in the almost unique Lough Ine, County Cork and in the Helford estuary in Cornwall; Steven's goby has been recognised sparsely in the English Channel (in 35 m to 100 m depth), as well as in Madeira and the Canary Islands; the giant goby lives in high shore pools on rocky shores from the Isles of Scilly to Cornwall and South Devon (Wembury), but is abundant in the Channel Isles, the Brittany peninsula, south to Morocco and throughout the Mediterranean. There seems to be no record of spawning in the British Isles although fish of the year are common in the late summer along the Cornish and Devon coasts, but they are in much greater abundance in suitable habitats on the coasts of the Channel Isles and Brittany at this time of the year (Wheeler 1970, unpubl. data 1967). It has been suggested by Wheeler (1993) that recruitment of this species to the British fauna may depend to a considerable extent on immigration by young fish from the Channel Islands or the French coast. Conservation of the British population can only rely on protection of suitable shores from development or 'improvement' to the foreshore. The black-faced blenny has been found in the western English Channel (Dorset, the Channel Isles and Brittany) and the Atlantic coast of Spain and Madeira; it is believed to be conspecific with the Mediterranean form.

The very restricted distribution of these fishes appears to qualify them for inclusion in conservation discussions, but they are simply at the extreme end of their geographical range and all of them will be found in greater numbers when searches are made using appropriate methods and in suitable habitats. Conservation of marine fish should not be approached as a local nationalistic exercise: the inclusion of the two oceanic species (the basking shark and the skate) in conservation legislation is an exercise in sentiment rather than an actual attempt to preserve them. Both are large elasmobranchs, all of which are especially vulnerable to exploitation because of their low fecundity, the long development period of the young, and, where known, slow growth of subadults. Both are or have been commercially exploited and have shown a negative response to fisheries exploitation (as could be predicted for fish with these reproductive and developmental features).

The skate is said by Brander (1981) to have 'disappeared' from the Irish Sea and his study of MAFF ships' catches showed no skate caught over 10 years in the Irish Sea and none over 15 years at Menai Bridge, although fishermen are quoted as saying they were not uncommon in the late 1940s. Brander also cited the capture of 20 specimens exceeding 120 lbs weight in the 1950s and 1960s. The major evidence for the comparative rarity of the skate today and at the beginning of the twentieth century was Herdman and Dawson (1902) in which this species was said to be 'abundant in all parts, and is taken both by line and trawling all the year round on nearly all our fishing grounds'. Such generalisations are unsatisfactory and need rigorous examination of the evidence. They do, however, agree with general statements of the species elsewhere, for example Murie (1903) for the Thames Estuary. It seems therefore that of all British sea fish the skate may merit protection from over-exploitation: the same could be said, however, for most exploited elasmobranchs.

On the other hand, protection has recently been accorded to the basking shark, which is a pelagic fish living in boreal to warm temperate waters. It feeds on plankton and can be seen feeding with its back breaking the surface (hence basking shark). Its reproductive cycle is unknown (despite considerable work on its biology) but on occasions when fisheries have been established the sharks have quickly declined, doing so so quickly that it seems as if the sharks have quit the area rather than been locally exterminated. As a result of the activities of a pressure group in the UK this shark has now been protected from fishing. Within five months of protection being granted, schools of up to 500 sharks were reported in the western Channel; this number must have been approaching the greater part of the population of these huge sharks in northern European seas. If anything this example devalued the currency of conservation of marine fishes as there are many more worthy species which require protection.

7 The changing fauna

Since 1973 there have been many observed changes in the composition of the marine fish fauna. Most of them concern species which have increased their range towards the north or have simply become more common in the British Isles. These notes on such fish refer to species that have occurred in some number and not to the occurrence of a few isolated individuals.

Discussion of this subject has to proceed with caution, as in the last 25 years there have been many changes in fishing methods and grounds; this has resulted in unfamiliar fish appearing on the major fish markets which were not caught previously, or, if caught, would not have been landed. The extension of fishing areas and the use of large mesh surface nets has also brought on to the market species which were not caught off the British coast but long distances into the Atlantic. The species listed below are those which do genuinely appear to have increased in abundance. No attempt has been made to identify species which have significantly decreased in number as these are more difficult to identify.

- Bass *Dicentrarchus labrax*. A fish which has increased in numbers especially to the south and east of its range. Bass are now seasonally abundant as young fish in many British estuaries, and protection areas have been established in a number of areas to limit exploitation until the young fish have dispersed on their coastal migrations. The bass was very little exploited by fisheries 25 years ago; it is now a valuable commercial species which is much more abundant than formerly.
- Sole *Solea solea*. The sole is also much more common as a young fish in several British river mouths. This may be due to changes in climatic or meteorological conditions but could also be in response to the reduction of pollution, particularly in estuaries. The Thames is now recognised as an important nursery ground for sole and its impact will supplement the sole fishery in the North Sea.
- Trigger fish *Balistes capriscus*. In the period 1950 to 1973 the trigger fish was occasionally captured, mostly in crab-pots on the south and south-western coasts. Within the last two decades it has become much more common and is regularly caught on the southern Irish, Welsh and English coasts. Many are caught by anglers, others in baited pots and occasionally in trawls. From being a fish which was reported occasionally it is now regularly captured and is no longer considered to be worthy of special mention. A wide range of sizes are reported, including young specimens of about 15 cm, and it is possible that this species is now breeding in northern European seas.
- Smooth pufferfish *Sphoeroides pachygaster*. This species was first reported in northern Europe from Donegal Bay in 1984; it was later reported on a number of south-western localities. It has also been recorded from the Atlantic coast of southern Europe and the Mediterranean (and Adriatic) (Wheeler and van Oijen 1985).

Other species that could be described as representatives of the Lusitanian fauna have been captured in considerable numbers on the coasts of Britain, where formerly they were very uncommon or rare. These include the sea horses *Hippocampus ramulosus* and *H. hippocampus*, Couch's sea-bream *Pagrus pagrus* and the gilthead bream *Sparus aurata*. These, and other examples of fishes that have become increasingly frequently reported around the coasts of Britain, strongly indicate that there has been a major change in the marine fish fauna of northern Europe.

8 Postscript

Since this text was submitted in December 1999 there have been two major developments affecting the British fish fauna one concerning freshwater fishes and the other observed mostly in marine fishes.

The introduction of carp *Cyprinus carpio* to England and Wales in the last 25 years, and to a lesser extent Scotland, has been followed by a decline in the ecology of shallow lakes and slow-flowing rivers as the carp feed on soft vegetation and degrade the lake bed in search of food. The effect is serious depending on the size of the population of fish introduced. As carp can attain a considerable size (commonly growing to 4.5 kg (10 lb) in five years) depending on the initial density of stocking and the richness and type of vegetation, the biomass increases quickly. Many, possibly most, carp waters are stocked by angling club members or by fishery managers who have little concern for environmental impacts. Many carp waters thus become overstocked in terms of biomass.

The period from the 1950s to 2000 has seen an enormous increase in the number of carp in England and Wales and carp angling became very popular. The effects of inadequate control of carp are evidenced by many lakes with muddy clouded water, reductions in water plants and plant-eating waterfowl, and a consequent loss of aquatic invertebrates.

Despite the popularity of carp with anglers, relatively few have been introduced to Ireland; they had been stocked in only small numbers to late November 2000 when 96 carp, weighing 2–2.7 kg (4.5–6 lb) were imported and released into McNeils Ponds at Dromore in the area of Bambridge District Council (Anon. 2001). This is to the south-south-east of Lough Neagh within the flood plain of the rivers feeding it. Lough Neagh contains the major population of pollan *Coregonus pollan* in the British Isles (assuming those in Lough Erne, Fermanagh and the River Shannon lakes are conspecific (cf. Kottelat 1997). This single introduction may result in the extermination of these coregonine fishes in Ireland and changes in climate, notably winter low temperatures, will speed up these changes.

Several species never before been reported on the British coast have also been noted:

> *Seriola carpenteri* Mather 1971 Guinean amberjack. Caught by angler off Herm Island, Channel Islands, 7 September 2000. (Reports of other specimens caught off the French coast were received at the same period.)
> *Diplodus cervinus* (Lowe, 1841) Zebra sea bream. Netted off Portland, Dorset. December 2000. Reported to, and kept alive by, the Weymouth Sealife Centre.
> *Oblada melanura* (Linnaeus, 1758) Saddled bream. Caught by angler at St Austell Bay, Cornwall, 30 July 2000.

A number of other rarities have occurred in the last twelve months, but they had been reported elsewhere in British waters before.

REFERENCES

Anon. (2001) undated. Carp introduced to Northern Ireland. News Round-up. *Carp-talk Weekly*. Carp Fishing News, Newport, East Yorkshire, 13.

Brander, K. (1981) Disappearance of the common skate Raia batis from the Irish Sea. *Nature*, **290**, 48–49.

English Nature (1994) *Species Conservation Handbook*.

English Nature (1995) *Species Conservation Handbook*, vol. 2, *Fish*.

Farr-Cox, F., Leonard, S., and Wheeler, A. (1996) The status of the recently introduced fish *Leucaspius delineatus* (Cyprinidae) in Great Britain. *Fisheries Management and Ecology*, **3**, 193–199.

Forbes, I. and Wheeler, A. (1997) A comparison of late eighteenth century and present-day fishery surveys on the River Witham, Lincolnshire, UK. *Fisheries Management and Ecology*, **4**, 325–335.

Herdman, W.A. and Dawson, R.A. (1902) Fishes and Fisheries of the Irish Sea. *Lancashire Sea-fisheries Memoir*, **2**, 1–98.

Kirk, R.S. and Lewis, J.W. (1994) Sanguinicoliasis in cyprinid fish in the UK, in Pike, A.W. and Lewis, J.W. (eds) *Parasitic Diseases of Fish*. Tresaith, Dyfed: Salmara Publishing, pp. 101–117.

Kottelat, M. (1997) European freshwater fishes. An heuristic checklist of the freshwater fishes of Europe (exclusive of former USSR), with an introduction for non-systematists and comments on nomenclature and conservation. *Biologia, Bratislava*, **52** (Supplement 5), 1–271.

Lelek, A. (1987) *The Freshwater Fishes of Europe*, Vol. 9, *Threatened Fishes of Europe*. Wiesbaden: AULA.

Lyle, A.A. and Maitland, P.S. (1997) The spawning migration and conservation of smelt *Osmerus eperlanus* in the River Cree, southwest Scotland. *Biological Conservation*, **80**, 303–311.

Maitland, P.S. and Lyle, A.A. (1991) Conservation of freshwater fish in the British Isles: the current status and biology of threatened species. *Aquatic Conservation: Marine and Freshwater Ecosystems*, **1**, 25–54.

Maitland, P.S. and Lyle, A.A. (1992) Conservation of freshwater fish in the British Isles: proposals for management. *Aquatic Conservation: Marine and Freshwater Ecosystems*, **2**, 165–183.

Murie, J. (1903) *Report on the Sea Fisheries and Fishing Industries of the Thames Estuary*. London: Kent and Essex Sea Fisheries Committee.

Potts, G.W. and Swabey, S.E. (1991) *Evaluation of the Conservation Requirements of Rarer British Marine Fishes*. Final Report to the Nature Conservancy Council.

Quigley, D.T.G. and Flanery, K. (1996) Endangered freshwater fish in Ireland, in Kirchhofer, A. and Hefti, D. (eds) *Conservation of Endangered Freshwater Fish in Europe*. Basel: Birkhauser, pp. 27–34.

Quignard, J.P. and Douchement, C. (1991) *Alosa alosa*, in Hoestlandt, H. (ed.) *The Freshwater Fishes of Europe*, Vol. 2: *Clupeidae, Anguillidae*. Wiesbaden: AULA.

Regan, C.T. (1911) *The Freshwater Fishes of the British Isles*. London: Methuen.

Scharff, R.F. (1899) *The History of the European Fauna*. London: W. Scott.

Wheeler, A. (1970) Notes on a collection of shore fishes from Guernsey, Channel Islands. *Journal of Fish Biology*, **43**, 652–655.

Wheeler, A. (1974) Changes in the freshwater fish fauna of Britain, in Hawksworth, D.L. (ed.) *The Changing Flora and Fauna of Britain*. London: Academic Press, pp. 157–178.

Wheeler, A. (1977) The origin and distribution of the freshwater fishes of the British Isles. *Journal of Biogeography*, **4**, 1–24.

Wheeler, A. (1979) *The Tidal Thames: the History of a River and its Fishes*. London: Routledge and Kegan Paul.

Wheeler, A. (1993) The distribution of *Gobius cobitis* in the British Isles. *Journal of Fish Biology*, **43**, 652–655.

Wheeler, A. (1998) Ponds and fishes in Epping Forest, Essex. *The London Naturalist*, **77**, 107–146.

Wheeler, A. and Jordan, D.R. (1990) The status of the barbel. *Barbus barbus* (L.) (Teleostei, Cyprinidae), in the United Kingdom. *Journal of Fish Biology*, **37**, 393–399.

Wheeler, A. and Maitland, P.S. (1973) The scarcer freshwater fishes of the British Isles, I: Introduced species. *Journal of Fish Biology*, **5**, 49–68.

Wheeler, A. and van Oijen, M.J.P. (1985) The occurrence of *Sphoeroides pachygaster* (Osteichthyes-Tetraodontiformes) off north-west Ireland. *Zoologisches Mededelingen*, **59**, 101–107.

Yeomans, W.E., Chubb, J.C. and Sweeting, R.A. (1997) *Khawia sinensis* (Cestoda: Caryophyllidea) – an indicator of legislative failure to protect freshwater habitats in the British Isles? *Journal of Fish Biology*, **51**, 880–885.

Chapter 25

Tracking future trends
The Biodiversity Information Network

Keith Porter

ABSTRACT

Over the last 25 years there has been wide recognition of the value of information on wildlife status. Knowledge of distribution, rarity and population trends has enabled priority to be given to threatened or vulnerable species and habitats. These priorities are now expressed both through wildlife legislation and the Government commitment to the Biodiversity Action Plan. Without biological records, and a history of biological recording, none of this would have been possible. Our knowledge of the flora and fauna of Britain and Ireland is significantly better than 25 years ago, with an exponential increase of data for some groups. This trend is likely to continue as the demand for information grows and new inititiatives and technology enable us to access data more efficiently.

1 Introduction

The individual field recorder remains the bedrock of biological recording. Maintaining, or even increasing, the recording community is fundamental to realising the benefits of the new initiatives. We estimate that over 70% of all biological records are collected by 'volunteer' recorders and their role is now firmly recognised. The main issue which has emerged in the last 25 years is that the character of biological recording is changing. Some groups, such as butterflies or dragonflies, have more active recorders than ever before, while other groups are supported by small, or shrinking, numbers of recorders. The young people that are attracted to biological recording and natural science today do so without the benefit of formal taxonomic training and in circumstances where collecting, and the retention of reference specimens, is discouraged by society. The challenge for the next 25 years is to build on the progress made and ensure that the individual recorders are supported and further encouraged.

2 The importance of access to data

A knowledge of changes in the status of the biota is essential to nature conservation. The limitations of time and resources mean that species or habitats which are declining rapidly must be given the highest priority for conservation action. Equally, knowing that a species is rare encourages recorders to give it special attention. If information is not available on the trends in distribution or species number, then species or habitats will decline to the point of extinction. Over the last 25 years we have progressed from having a patchy knowledge of how much habitat existed, or how common a species was, to having a reasonably comprehensive view of the status of priority species or habitats; even so, the time lag between collecting data and producing the necessary trend data is still too long. The creation of habitat inventories takes many years and by the time we begin to estimate national extent or rates of decline many of the sites will already have been destroyed or degraded. Thus, despite the best efforts we are still only seeing retrospective views of trend for many habitats or species. The challenge for the immediate future is to improve our ability to detect, or predict, trends much faster, and thereby improve the chances of remedial action resulting in recovery of declines.

Data on status were not always a priority for nature conservationists. The establishment of Wicken Fen by the National Trust in 1899 as one of the first 'nature reserves' was driven by the knowledge that the area of fenland habitat was declining rapidly; the extent and decline of fenland was obvious to any observer and did not require justification with facts and figures. Similarly, the demise of several bird species, such as the white-tailed sea eagle (*Haliaeetus albicilla*) and the great auk (*Pinguinus impennis*), prompted legislation to protect vulnerable species. Declines and extinctions still occur today, the difference being that in the case of macro-organisms we usually know when something is declining and roughly how many populations or how much habitat remains.

3 Pressure for change

Pressure to establish views of the distribution and status of our flora and fauna has been a recurrent theme since the days of John Ray in the seventeenth century (Allen 1976). A primary aim of any naturalist or environmental scientist is to understand how their individual observation of a plant, animal, fungus, or rock type fits into the overall picture of distribution and status. The concept of rarity, and hence something 'special', would not exist if we did not know something about the frequency with which a given species was recorded. This concept of rarity was a major driver in the heyday of Victorian and Edwardian collecting, an impulse which precipitated some species to ultimate decline or extinction (Sheail 1976). This emphasis on rarity persists today with many of the 'priority' species listed in legislation or action plans being selected for inclusion on the basis of their rarity. The emphasis is changing, however, and shifting attention to species which are suffering declines in status, irrespective of their 'rarity'. Many species with special habitat requirements, such as the violet click beetle (*Limoniscus violaceus)*, have probably always been rare (Welch in Shirt 1987). Their populations may be small, but they are stable and need to be treated differently from species that are declining to potential extinction from once having been widespread or more abundant.

Tansley (1902) emphasised the need to harness the knowledge of amateur botanists as a way of securing a national picture of extent and status of species. This is a message that has been repeated many times since then and remains the driving force at the heart of current initiatives. The historical development of what has become known as biological recording is documented in Berry (1988).

In 1954 the Botanical Society of the British Isles launched a scheme to map the distribution of the British flora and produced the first *Atlas of the British Flora* in 1962 (Perring and Walters 1962). This was the realisation of Tansley's 1902 vision and provided a valuable baseline for measuring trends over the coming decades. In 1964 the data and mapping equipment were moved to the Nature Conservancy's office at Monks Wood and the Biological Records Centre (BRC) was established. In 1973 the ownership of the BRC changed when the Nature Conservancy was broken up to form the Nature Conservancy Council and the Institute of Terrestrial Ecology (ITE). The BRC was incorporated into the ITE at Monks Wood and this remains the position today. Thus the last 25 years have seen the establishment of the principle of using the enthusiasm and knowledge of voluntary recorders to compile a national picture, and this has operated very successfully: evidence of this is seen in the proliferation of distribution atlases (Harding 1989).

The model first envisaged for the BRC relied upon the central co-ordination of recording effort with the main products being atlases and their supporting databases; the purpose of the BRC was to provide a national picture of the known distribution of species, with the subsequent revision of atlases providing information on changes in distribution over time. This system relied upon a species' presence on a 10-km square scale, with the potential to add information on the number of records within each 10-km square through the use of scaled dots or symbols. In the past 25 years the BRC has perfected the art of atlas production and they are now widely used at local and national scales, a major use being to provide information on the status of a species and enable the development of *Red Data* books in the UK. The basic criterion for identifying species which should be given special attention has been the number of 10 km squares they occur in: species that occur in 15 or fewer squares are automatically *Red Data Book* (RDB) species, with five grades of priority allocated according to the degree of threat (Shirt 1987). Species which occur in 16–100 squares are regarded as 'notable' species and are given specific weightings when evaluating nature conservation priority. This approach has been extensively used in the selection of Sites of Special Scientific Interest (SSSIs) (Nature Conservancy Council 1989) and the evaluation of sites outside the designated series of SSSIs (Hawkswell 1997). Atlas data are a very crude measure of priority as their utility depends upon the extent to which a particular taxa has been recorded. For the less popular species groups, the draft atlases are largely indications of where recorders have been (Rich and Woodruff 1992; Rich and Smith 1996). This has been recognised by the BRC in calling their products 'provisional atlases'. These provisional atlases serve a valuable function in drawing attention to the gaps in coverage and thus encouraging recording in new areas.

With the growing enthusiasm for biological recording and the growth in the number of records, it became clear that a single central recording centre was inadequate to meet the demand. As a consequence the BRC has had to be highly selective in determining which atlas project to proceed with and explore better ways of securing records. In the early 1970s several conservationists were trying to establish a network of local record centres which would feed data into the BRC (e.g. Greenwood 1971; Perring 1971). This concept of local centres with a link to the national BRC was subject to a series of workshops and conferences throughout the 1970s and 1980s (Perring 1973, 1976; Copp 1984; Copp and Harding 1985; Berry 1988).

A consistent theme throughout this period was the recognition of local data being stored locally and a national summary being held by the BRC. Many local record centres were based within museums and were seen as a logical extension of a catalogue of museum collections. These catalogues were usually based on card indexes and data entry or extraction was a laborious task. The BRC approach streamlined data collection through an increasingly sophisticated system of record cards. The aim was to help recorders capture the relevant details as quickly and simply as possible. Early variants included cards which could be coded for direct entry to the new computers of the 1970s; by the late 1970s the BRC recording cards had been accepted as the standard way of making biological records (Heath and Scott 1977) and were familiar items to field botanists. The growing backlog of such record cards became a problem for local centres and the BRC in that transcription to a card index, or entry into the new computer databases, was a costly exercise. In addition, the BRC emphasised the need to check and validate the records to remove grid reference errors and errors of identification. This was a critical aspect of the BRC's work which set a standard of data quality that has stood the test of time. In transcribing data from field notebooks or literature it is inevitable that numbers and spellings will be miscopied: the phenomenon of butterfly colonies in the North Sea is well known to the early users of biological records! A major burden on the local–national record centre link was the double entry and checking of records, with the potential for mistakes made in transcription.

The BRC covers the United Kingdom, but atlases show records for the Republic of Ireland wherever these are available. The logical need is for a common view across the British Isles in order to see the full picture of species distribution. The position of Biological Recording in the Republic is evolving. The National Parks and Wildlife Service holds data for some taxa, whilst a few Societies collate their own records.

The BRC covers the United Kingdom, but atlases normally show records for the Republic of Ireland wherever records were available. The logical need is for a common view across Britain and Ireland in order to see the full picture of species distributions, particularly in relation to the rest of Europe. The position of biological recording in the Irish Republic is evolving: the government agency holds data for some taxa, while a few societies collate their own records.

The growing environmental awareness of the 1970s and 1980s increased the demand for information on the species and habitats in local areas. The development pressures of the 1970s, and the consequent impact on wildlife, increased the need for good, reliable data for use in planning inquiries or development control. The most significant change in the last 25 years has been the shift in emphasis on the use of biological records in development control; the atlases still provide the context, but greater detail is often required at a local level to inform planning decisions. This 'new' purpose for biological records enabled local centres to move away from their museum roots and seek funding from local authority planning departments. This shift is still taking place as new legislation and initiatives continue to emphasise the need for good local data as a basis for sound decisions and sustainable planning.

4 A watershed for biological recording

A key study of biological recording was published by the Department of the Environment (1995) on behalf of the Co-ordinating Commission for Biological Recording (CCBR). This body was established as a consequence of a debate initiated by the Linnean Society and the publication of *Biological Survey: need and network* (Berry 1988). The CCBR report, in two volumes, sets out the position of biological recording in the mid-1990s; the report was based on the response to a national questionnaire sent to all known naturalists and organisations involved in biological recording. Information was collected on a wide range of aspects of biological recording including the number, location and detail of records, systems used to store and manage records, uses to which they are put, data ownership and details of local record centres. This information was used to suggest a way forward in improving access to biological records. This valuable study was subsequently partly overshadowed by the publication of *Biodiversity: The UK steering group report* (HMSO 1995) but has been used to underpin the concept of a national biodiversity network. Some of the key findings are summarised in Table 25.1.

Table 25.1 Selected points adapted from *Biological Recording in the United Kingdom: present practice and future development* (Department of the Environment 1995).

Data holdings

In the UK there are over 60 million species records with over 42% relating to birds and 14% to vascular plants.

It is estimated that over 2000 organisations and 60 000 individuals are involved in collecting or holding biological records.

Volunteer recorders have collected over 70% of all taxon records and 36% of all habitat records. In contrast, the statutory conservation agencies have collected only 2% of taxon records and an estimated 13% of habitat records.

The growing awareness of biological records is reflected in the fact that 85% of all taxon records have been made since 1970, with 60% since 1980.

This increase in recording effort is even more pronounced for habitat records with 95% collected since 1970 and 79% collected since 1980.

Habitat data cover more than 3 million land parcels with at least 150 000 'sites' recognised.

The majority of records are still held on paper with only around 20% of respondents holding computerised data.

Key recommendations

Review the statutory requirements for biological records and thus secure support from users.

Establish biological recording in a formalised network.

Establish a biological record data standard to improve consistency of data.

Establish methods to control the quality of data and thus improve user confidence in the records.

Establish protocols for the compilation and content of databases.

Establish a dispersed national system for biological recording through local and national data centres.

Establish a management mechanism for a national system of biological recording.

5 Legislative change and biological recording

Pressure for change has been provided by new legislation over the past 25 years. Nature conservation has seen the introduction of a major piece of legislation in the form of the Wildlife and Countryside Acts (1981, 1985) and the European Directives on birds and habitats (CEC 1979, 1992). Among other measures, the Wildlife and Countryside Act builds on the National Parks and Access to the Countryside Act (1949) and Countryside Act (1968) to improve the safeguarding of Sites of Special Scientific Interest and endangered species. The consequence of this act was an extensive review of all existing SSSIs and a programme of 'renotification'. This exercise, and the supporting guidance, placed great emphasis on a sound understanding of the distribution of wildlife and the status of individual species and habitats. The BRC atlases formed the basis of species prioritisation through the recognition of RDB and 'notable' species and this information was then elaborated through access to detailed site records held in local centres.

The implications of the European Directives are still being realised, but will require the statutory agencies to have a more comprehensive and credible view of the distribution and trends in species and habitats than ever before. At the heart of this legislation is the concept of 'favourable conservation status' that aims to secure a sustainable position for all the priority features listed in the Directives' annexes. This implies that any given species or habitat needs to be present across its natural range in a way that ensures the continuation of that feature into the future. The identification of key sites for these features (Special Areas for Conservation and Special Protection Areas) will contribute to securing favourable conservation status, but for some species and habitats a significant proportion of the natural resource lies outside these special areas. In order to focus conservation actions and to report on progress towards favourable conservation status, up-to-date information is needed on the status of key species and habitats across the UK. Thus the new legislation has increased our need to support the establishment of a credible network of data centres across the UK. The international legislation places a responsibility on the UK government, and by implication the government departments involved, in delivering environmental schemes and action.

6 Biodiversity

While not a legislative requirement, the most significant driver for nature conservation and biological recording to emerge in the past 25 years has been the Convention on Biological Diversity enacted in Rio in 1992. This Convention has led directly to the production of the *UK Biodiversity Action Plan* (HMSO 1994, 1995; English Nature 1998) which establishes a set of national priorities for action. The national plans and targets are delivered through local Biodiversity Action Plans (BAPs) and rely upon local knowledge of the location of key species and habitats to deliver action.

The BAP has placed a great strain upon the national knowledge of some species and habitats and has used the best available information to define a set of national targets. Some of these targets may already be met for species which are actually more common than national datasets would suggest, for others the species may already be extinct. At a local scale the production of plans has likewise placed great strain on the local data sources and has revealed both strengths and weaknesses of the existing data sources. Overall, the BAP has further emphasised the need for effective data networks which allow up-to-date access to information. The biodiversity concept has considerable political backing and there is an expectation that government departments and relevant local authorities will include biodiversity and wider environmental aims into their policies and actions.

The high-profile emphasis on the BAP has revealed both the strengths and weaknesses of existing biological recording systems. This provides the opportunity for which we have waited since Tansley's prophetic words in 1902. We now have a requirement for access to up-to-date, quality data on the natural environment; such data are needed to support the delivery and monitoring of Government initiatives such as local Agenda 21, local BAPs and better development planning decisions. The same data at a higher level of detail are needed to inform government, and their agencies, of the impact of environmental policy and the effectiveness of environmental schemes.

7 Public perception and expectation

A further pressure for change has been the growing recognition of the need for public involvement and the educational value of biological data. The public's view of national and local museums has undergone a sea change; we have moved away from the dusty Victorian image of scientific collections towards interactive hands-on learning. This has precipitated financial crises for some museums and their record centres as the value of the scientific collection side of museums has been marginalised, and this will cause problems for biological recording in the future unless the value of the reference collection and retention of voucher specimens is recognised by data users.

Two linked changes have set the scene for a potential crisis in years to come. These changes have been the removal of taxonomy from most syllabuses of higher education, and the increasing pressures against 'collecting' biological specimens. A change in attitude to 'collecting' has affected the recruitment of new recorders. A straw poll of any group of active biological recorders over the age of 40 is likely to reveal that their formative years included some collecting activity as a stimulus to developing an interest in ecology or natural history. The shift away from recording through the establishment of reference collections has had two main impacts: the first is that for many taxa which require critical examination of taxonomic features for valid identification there is now a shortfall in the next generation of new recorders; the second is that many records now rely solely upon observations rather than voucher specimens. For some taxa, such as birds, butterflies or vascular plants, this is not usually a problem where the use of handbooks in the field can produce reliable records. For other groups a reliable identification requires critical examination and a proportion of all records will require checking before the record may be accepted. In groups such as the fungi where many species are imperfectly characterised and ones new to science are regularly found in the British Isles, records without vouchers have been viewed as so much waste paper (Dennis 1968). The retention of voucher specimens allows a record to be reassessed if a query arises, or a 'species' is found to form a complex of similar species. For example, a common hoverfly *Platycheirus clypeatus* was discovered to consist of a group of four species (Speight and Goedlin de Tefenau 1990). The records of *P. clypeatus* prior to 1990 will all be open to question where vouchers have not been kept.

Linked to the potential loss of new recorders is the change that has occurred in recent decades in the teaching of taxonomy in formal education. While taxonomy is seen as a relatively 'dry' subject within biological sciences, its teaching did provide an awareness of the rules and tools for establishing the correct identity of species. With the demise of such teaching, especially in higher education, and the moral pressures on not collecting, the budding entomologist or mycologist has to evolve without the benefit of these disciplines and freedoms. Similarly, such skills and knowledge are needed by those within conservation organisations who need to assess sites or confirm the existence of key species.

Thus the political incorrectness of collecting and the loss of taxonomic teaching have prejudiced the recruitment of new naturalists who should become the 'experts' of tomorrow at a time when we need more recording activity than ever before. We now have more active recorders in groups such as butterflies and dragonflies than 25 years ago, but for taxa that require more than a photographic handbook to confirm their identification the future is less certain. The changing attitude towards the 'child with a jam jar' is no doubt prompted by the public perception that collecting is bad because it leads to declines or even extinction. Equally, academia now has topics it considers more exciting to teach undergraduates than taxonomy; both views must be challenged and overcome if the increasing demands for biological records are to be met.

Any increase in recording activity will need to be supported by an appropriate community of specialists who can help validate the more difficult species records. This role is perhaps best supported through the national schemes and societies that can provide the co-ordination and support for specialists. Similarly, museums have a role to play in holding the voucher specimens in a safe and secure environment.

A major block to progress towards the co-operation and sharing of data is the misconception that data and information are worth money or influence. In the battle for survival and resources many local record centres or data holders have offered their services as consultants, with the expectation that large fees would result. The evidence of the CCBR report suggests that income from such consultancies is rarely sufficient to cover costs of providing the data, never mind subsidising the local record centre. A consequence of the consultancy era has been an unwillingness by some voluntary recorders to provide their records to such centres, and for the data centres to lose sight of the nature conservation objectives. In the mid-1990s the concept of open public access to data became a powerful force within politics and the statutory agencies were obliged to provide access to environmental data where it does not prejudice the delivery of nature conservation.

8 The changing role of recorders

Biological recorders of today have far more tools at their disposal to manipulate and summarise their data than in the early 1970s. The national atlas is now being supplemented by a wide range of local atlases and this can be a disincentive for recorders to send their records into a national BRC, or indeed any record centre! Until recently the thrill of seeing your dots on the national map and the feedback from the BRC were much valued by individual recorders. With modern computer technology an individual can now hold all their records and produce good quality atlases from their own home; the motivation for recorders to submit their records to a local or national centre may thus be changing.

A growing problem for biological recording is the proliferation of systems and approaches to recording, holding and providing access to records. A national overview of species or habitats requires a common basis for recording biological data. In modern parlance this means the establishment of standards which will ensure that maximum value can be derived from each record. Over the past 25 years these standards have been set by the BRC and have enabled them to provide quality products. Standards reflect the use to which data are put and one of the main purposes in the past was to produce the national picture; this meant that the grid reference of the record need be no more precise than a 10 km square. Similarly the scale adopted by many local centres was the tetrad, or 2 km square. Both these 'standards' served the purpose of showing the distribution of species at national and local scales. Such data are of little value when trying to manage or protect sites, however, and the time is now right to advocate a standard for biological records based upon the CCBR report and the exploration of user needs being undertaken as part of the national biodiversity network.

A biological record has a clear definition, in that it reflects a judgement made at a certain point in time: the identification of the species will be made according to the latest keys and taxonomy; the name of the locality will be recorded according to the understanding of the person making the observation. With manual transcription of records from recorder to local centre and then on to the BRC it is tempting to 'correct' the nomenclature or override someone's interpretation of the habitat. Once modified it may be very difficult to reconcile queries; indeed, many single records suddenly reappear as multiple records as each successive stage alters the detail of the original record. Modern technology, with increasing compatibility of systems, now allows us to enter the record once and use it many times without manual handling and the risk of 'helpful' changes. This approach does need to be supported by appropriate standards and a system of validation at source, however, and this places a clear responsibility on the originator of the record to ensure that appropriate confirmation of identification is obtained and that this is registered as part of the record.

Similarly the advent of the electronic networks enables access to a record no matter where it is held within the network. Thus all records for Kent do not need to be physically held in the Kent data centre, but can be held wherever the recorder enters the data into the network. This will revolutionise how biological recorders operate in the coming years, as this approach allows for the entry of data into local centres and for records to be collected and collated by national societies. The decision as to which option to adopt will depend upon a range of factors including the difficulty of identification and the access to the network.

9 The national biodiversity network

Arising from the key issues identified by the CCBR (Table 25.1), and the need for better access to biological data to help delivery of the BAP, a major initiative began in 1996. This started as two separate bids to the Millennium Commission relating to the establishment of an effective network of local record centres and securing better information on the natural environment. The bodies involved in separate bids realised that they were all developing parts of a much bigger picture and subsequently reshaped their bids into a single submission. This was subsequently rejected by the Millennium Commission in 1997, but the discussions which were prompted by pulling together the different bids and the realisation that working together was a sensible way forward had sown

Table 25.2 Programmes and projects within the national biodiversity network.

Co-ordinating development

Programme management: ensuring the projects are co-ordinated and supported.

Communication: ensuring that the NBN is promoted.

Membership and accreditation: to gain supporters for use and supply of data.

Standards for Biodiversity Information

Access terms: developing the 'rules' for users to allow data exchange and protection of copyright and ownership.

Data standards: to aid consistency of data collection, storage and use.

Recorder software: a software package that uses the data standards and is readily available to all recorders and users.

Checklist and identification standards: to improve consistency and quality of data.

Meeting information needs: ensuring that NBN supports the needs of users.

Access to biodiversity information

Linking national societies and schemes: helping these to increase participation in recording and understanding biodiversity.

Linking local record centres: securing a network of local data centres across UK.

Linking national biodiversity organisations: securing national data centres.

Education and public access: providing access to biodiversity information.

Using and applying biodiversity information

Network index and gateway: establishing a public online access to the network.

Modified from unpublished business plan for establishing the NBN 1998.

the seeds of the new concept of a national biodiversity network. The consortium which began to pull together a coherent proposal is a mixture of voluntary and statutory bodies, including the Wildlife Trusts, the Royal Society for the Protection of Birds, the National Federation of Biological Recorders and associated organisations, the Natural History Museum, the Natural Environmental Research Council, and the Joint Nature Conservation Committee (JNCC) on behalf of all the statutory conservation agencies.

By late 1997 the concept of a national biodiversity network (NBN) for the United Kingdom had progressed to proposing a series of linked projects. These are co-ordinated by the JNCC but led by the staff from the consortium members. These projects (Table 25.2) take a lead from the CCBR report and together should put in place the necessary framework of data centres, the electronic links between them, and a range of changes to the way we collect, maintain and access information on the natural environment.

The overall aim of the NBN is to provide access to information on the natural environment of the United Kingdom. This covers both the provision of access to existing data and ensuring that new data are recorded, held and maintained to standards that will maximise their use. The projects within the NBN build upon extensive consultation with data providers and users, and use the recommendations arising from the CCBR report. A key focus for the NBN is to enable an overview to be taken of progress with the BAP, both at local and national levels.

Key areas for action are to establish appropriate standards for biological records, improving access to biodiversity information and helping users to apply biodiversity information. All the NBN projects are interdependent and the overall objective can be met only through the delivery of all project areas.

The package of proposals which form the 'business plan' of the NBN was submitted to the Heritage Lottery Fund as a proposal for funding. The proposal focussed on a demonstration phase where a limited series of data centres are established and a small number of recorder groups and users are helped to use a data network, and then to progress to full implementation across the UK from 2004, once the benefits have been demonstrated. A modified submission was accepted for funding by the Heritage Lottery Fund in 2000 and will support the stronger involvement of volunteers in the NBN. Further support has also come in 2000 from Government with a package of measures being paid for by the Department of Transport, Environment and the Regions. Irrespective of the success of these first funding proposals, it is certain that a national biodiversity network will be established, the rate of implementation being dependent upon the commitment of users and the provision of funds.

10 Looking forward to the next 25 years

Given the changes in the last 25 years it is likely that advances in technology will continue to have a powerful influence on the way we capture and provide access to information on the status of species and habitats. The lessons from countries like the United States suggest that we need to become far more willing to provide open access to data on the natural environment; only then can we realise the full benefit of the collective effort of data gatherers.

The way forward will be painful, as we realise that change is needed in some established and respected approaches. National specialist societies, several of which for various reasons do not make their data available to the BRC, will want to review how they organise and maintain their information. The delays between making field observations and seeing the results in national summaries will be a thing of the past; data collected in the field will be immediately accessible once entered on a computer linked to the network. New standards will allow use of the data in a variety of ways, hence the notion of 'enter once, use many times'. The improvements in hand-held computers and less expensive geographic positioning systems will enable records to be collected in the field while data entry will be aided by online access to similar records or aids to identification.

The national biodiversity network programme includes a project to establish standard checklists of species and ensure that these are maintained and made accessible to all. This is a much-needed product required to help recorders tread through the maze of synonymy. The process set up to produce such a checklist will help to improve consistency of recording in the coming years and thus remove some of the anomalies of past records.

The major constraint on achieving this improved access to information is maintaining the community of biological recorders who provide the records on which the whole edifice stands. This brings us back full circle to the natural historian and the enthusiasm of voluntary recorders, especially in a situation of declining numbers of specialists in universities, museums and other systematic institutions. The lesson of the recent past is that we must find ways of encouraging a new generation of naturalists and carry forward the enthusiasm and expertise that has generated such a wealth of knowledge of our natural heritage.

REFERENCES

Allen, D.E. (1976) *The Naturalist in Britain*. London: Allen Lane.

Berry, R.J. (1988) *Biological Survey: Need and Network*. London: Linnean Society/PNL Press.

CEC (1979) *The Conservation of Wild Birds* [Council Directive 79/409/EEC]. Brussels: CEC.

CEC (1992) *The Conservation of Natural Habitats and Wild Fauna and Flora* [Council Directive 92/43/EEC]. Brussels: CEC.

Copp, C.J.T. (1984) Local record centres and environmental recording – where do we go from here? *Biological Curators Group Newsletter*, 3, 489–497.

Copp, C.J.T. and Harding, P.T. (eds) (1985) *Biological Recording Forum 1985* [Biological Curators' Group special report no. 4.]. Bolton: Biology Curators Group.

Dennis, R.W.G. (1968) *British Ascomycetes*. Lehre: J.Cramer.

Department of the Environment (1995) *Biological Recording in the United Kingdom: present practice and future development*, 2 vols. London: Department of the Environment.

English Nature (1998) *UK Biodiversity Group: Tranche 2 Action Plans*. Vol. 1 *Vertebrates and Vascular Plants*. Peterborough: English Nature.

Greenwood, E.F. (1971) North west biological field data bank. *Museums Journal*, 71, 7–10.

Harding, P.T. (1989) *Current Atlases of the Flora and Fauna of the British Isles*. Huntingdon: Biological Records Centre.

Hawkswell, S. (ed.) (1997) *The Wildlife Sites Handbook: 2nd edition*. Lincoln: The Wildlife Trusts.

Heath, J. and Scott, D. (1977) *Biological Records Centre – Instructions for recorders*, 2nd edn. Huntingdon: Institute of Terrestrial Ecology.

HMSO (1994) *Biodiversity: the UK Action Plan*. London: HMSO.

HMSO (1995) *Biodiversity: The UK Steering Group Report*, 2 vols. London: HMSO

Nature Conservancy Council (1989) *Guidelines for the Selection of Biological SSSIs*. Peterborough: Nature Conservancy Council.

Perring, F.H. (1971) The Biological Records Centre: a data centre. *Biological Journal of the Linnean Society*, 3, 237–243.

Perring, F.H. (1973) The Biological Records Centre, in G. Stansfield (ed.) *Centres for Environmental Records* [Vaughan papers in Adult Education, no. 18]. Leicester: Department of Adult Education, University of Leicester, pp. 4–13.

Perring, F.H. (1976) A biological records network for Britain, in P.T. Harding and D.A. Roberts (eds) *Biological Recording and a Changing Landscape*. Cambridge: National Federation for Biological Recording, pp. 3–5.

Perring, F.H. and Walters, S.M. (eds) (1962) *Atlas of the British Flora*. London: Botanical Society of the British Isles.

Rich, T.C.G. and Smith, P.A. (1996) Botanical recording, distribution maps and species frequency. *Watsonia*, **21**, 155–167.

Rich, T.C.G. and Woodruff, E.R. (1992) Recording bias in botanical surveys. *Watsonia*, **19**, 73–95.

Sheail, J. (1976) *Nature in Trust*. Glasgow: Blackie.

Shirt, D.B. (ed.) (1987) *British Red Data Books*. Vol.2. *Insects*. Peterborough: Nature Conservancy Council.

Speight, M.C.D. and Goeldlin de Tefenau, P. (1990) Keys to distinguish *Platycheirus angustipes*, *P. europaeus*, *P. occultus*, *P. ramsarensis* (Dipt. Syrphidae) from other *clypeatus* group species known in Europe. *Dipterist Digest*, **5**, 1–44.

Tansley, A.G. (1902) Research in British Ecology. *New Phytologist*, **1**, 84–85.

Chapter 26

Prospects for the next 25 years

David L. Hawksworth

ABSTRACT

Most changes in the wildlife of Great Britain and Ireland over the last 25 years have been the result of human intervention, whether directly or indirectly. Extinctions have been few, but the number now endangered or threatened is increasing in many groups. Although species new to science are still being discovered here, and some species thought to be extinct 25 years ago have been refound, the outlook is not good. Indeed, the next 25 years is likely to witness an increasing number of losses, although the implementation of Biodiversity Action Plans (BAPs) will safeguard some species. Monitoring the approximately 100 000 species present in these islands is a formidable task; enormous progress in recording has, however, been made by a burgeoning number of knowledgeable amateur recorders through an array of computerized mapping schemes. The Biodiversity Information Network is building on these foundations. There is a need for more long-term site studies, and a greater focus on habitats in deciding conservation priorities and actions which would help safeguard species in less well-studied groups. The major threat overall is agriculture, followed by introduced species, drainage, habitat management and air pollution. There is a need to strengthen the protection accorded to Sites of Special Scientific Interest (SSSIs), and to develop procedures to reduce the chance of harmful introductions. In many less well-known groups, a marked decline in the number of professional systematists will remain a barrier to an understanding of their occurrence and status.

1 Background

The contributions in this volume aim to provide an authoritative stock-check of the level of understanding and status of the wildlife of Great Britain and Ireland. Wildlife is used here in the broadest sense to encompass all groups of organisms, not only the more conspicuous, but also the most numerous and less studied. Periodic assessments by leading specialists are imperative in the light of current concerns for the protection of all forms of organisms.

The Linnean Society of London held a meeting in 1935 to examine the situation under the title 'Changes in the British fauna and flora during the past fifty years' (Anon. 1935); mammals, birds, insects, flowering plants, marine algae and fungi were considered. Subsequently, Edlin (1952) endeavoured to address the situation in mammals, birds, fish, amphibians, reptiles and plants. Stimulated by such works, a symposium to provide a broader overview was held by the Systematics Association in 1973; that involved specialists on 20 different groups of organisms (Hawksworth 1974). *The UK Action Plan* for biodiversity aimed to provide an overall assessment (Department of the Environment 1994), but included little specific information on particular categories of organism.

The production of the present volume was stimulated by the Linnean Society of London, which considered a reappraisal of the situation by specialists was opportune 25 years after the 1973 meeting and the new millennium was being entered. Dragon-flies and damselflies, mites and ticks, nematodes, protozoa, true bugs and their allies, and viruses, which were not covered in 1973, are included here. Contributions on aculeate *Hymenoptera*, arthropod ectoparasites of humans, beetles, amphibians and reptiles, seaweeds and spiders were not secured this time, nor were additional chapters on bacteria and marine invertebrates delivered. In the event that a parallel stock-take is planned for 2025, attempts to fill these lacunae should be sought.

2 State of knowledge

The advances in knowledge of the wildlife of Great Britain and Ireland since 1973 have been substantial, especially in the more charismatic and conspicuous groups. Even in less well-studied groups such as viruses, nematodes, mites, and fungi, the situation is immeasurably better than at that time. Nevertheless, the coverage remains patchy. For larger organisms in general, it is unusual to discover species new to science or to the area, whereas in the most speciose and microscopic groups knowledge is still in the exploratory phase with species being added to the known biota every year (see below).

The number of wild species actually present in Great Britain and Ireland is conse-quently still uncertain. The figure of 88 000 included in *The UK Action Plan* (Depart-ment of the Environment 1994) must be viewed only as an approximation in the absence of critical checklists for many groups, and the ease with which additional spe-cies can be added in less conspicuous ones. I suspect that the actual figure is unlikely to be less than 100 000 species, but even such a figure could well prove to have been found conservative by 2025.

3 Data capture

The dawn of the computer age, and particularly the availability of personal computers, has meant an increase in the way data can be captured and organized to a level that could not have been contemplated 25 years ago. The work of the Biological Records Centre, established in 1965, has been particularly commendable in this respect (Har-ding 1992), but some major data sets are still maintained independently, for example those for birds, lichens, and other fungi.

The need to integrate an increasing range of independent recording schemes was rec-ognized by a working party established by the Linnean Society of London (Berry 1988). Among other things, that working party recommended the establishment of a co-ordinating commission that would integrate regional and national centres and develop operational standards. This vision is now starting to be realized through the Biodiversity Information Network (Chapter 25 this volume) and augurs well for the future. The amount of information that will have accumulated by 2025, through the network and numerous other recording schemes, is likely to be scarcely believable by current standards.

However, a wealth of data backed by voucher specimens exists in the national and local collections which remains largely uncomputerized and inaccessible (Williams 1987). This is also true for published information scattered through books and an amazing array of international, national and local journals or newsletters, as well as unpublished site surveys and field notebooks. Advances in technology are likely to facilitate the incorporation of these types of data and aid the unravelling of now unused names. The inclusion of such back data will enable analyses of long-term trends and extinctions to be carried out more confidently than is currently possible.

4 Workforce

The issue of who will record changes differs according to the scientific understanding of the groups concerned. For vertebrates, butterflies, flowering plants, ferns, mosses and macrolichens there are sufficient manuals that empower a cadre of amateur naturalists to contribute to this task. But what of the species-rich groups such as bacteria, fungi, mites, nematodes, protozoa and viruses where few manuals exist? In such cases the need is for professional systematists. The UK Systematics Forum (1998) has endeavoured to compile a list of systematists active in the UK. That survey revealed that while there were 421 professional systematists, 185 worked on plants and vertebrates and disproportionately few on less well-known groups such as protozoa, fungi and invertebrates. Further, the professional systematic community is ageing and set on a course towards eventual extinction; the number in UK universities appears to be halving each decade (Claridge and Ingrouille 1992). The issue has even been addressed by a committee of the House of Lords (Anon. 1992), but no long-term prospects to alleviate this disturbing trend have emerged.

By 2025, recording will consequently be even more the domain of dedicated and enthusiastic members of specialist societies and local natural history and wildlife groups. That workforce will also be increasingly well equipped with any required specialist equipment such as microscopes and global positioning devices. More importantly, the amateur will be able to access a wealth of original scientific literature, keys, illustrations, maps, and other data directly over the internet. He or she will also have the ability to input data directly through links to local and national record centres, including the Biodiversity Information Network (Chapter 25).

The number of professional systematists in post will have declined, but the ranks of retired systematists continuing to work from home will have swelled as the 'over-45s' age group moves out of their current university and institutional positions. Sadly, many are unlikely to be replaced while current fashions persist and the long-term value of descriptive and monographic work is unrecognized through Science Citation Index evaluations (Valdecasas *et al.* 2000). In some groups, no professional systematists will exist in Great Britain and Ireland by 2025, unless there is a major outcry for support from the amateur workforce or some large-scale crisis involving human welfare or economics to which governments cannot but respond. This situation has been developing for many years, and for a variety of reasons (Hawksworth and Bisby 1988; Bisby and Hawksworth 1991), but the vision of Isely (1972) that taxonomists will need to be recreated after they have become extinct and their true value is recognized is approaching reality in many developed countries. A declining professional workforce will inevitably impact on the production of identification manuals and other aids; the number and standard of identification training courses; the recognition of species new to science, discovered here for the first time, or which have been introduced; and also lead to unverified and erroneous records of critical groups increasingly being included in databases

5 Synopsis of changes since 1973

The outcome of the 1973 meeting was one of optimism: there had been fewer losses and more gains in species, or increases in populations or ranges than decreases (Anon. 1973a, b). This conclusion was contrary to some scare-mongering prevalent at that time in the wake of *Silent Spring* (Carson 1962), which while it may have exaggerated some of the potential effects of agricultural chemicals in an overall UK context (Harding 1988), has to be seen against a backcloth of long-term decline in our wildlife (Vesey-Fitzgerald 1969; Ratcliffe 1984).

The contributions presented in this book do show the adverse effects of the ongoing agricultural revolution (see below), but again not of a sudden catastrophe. Indeed, it is gratifying overall to see the modest number of confirmed losses from the British and Irish biota reported over the last 25 years. For example, only seven more plants have become extinct (Chapter 2), whereas 8 bryophytes (Chapter 4) and 16 lichens (Chapter 7) thought to have been extinct in 1973 have actually been rediscovered. No mammals, breeding birds, butterflies or ferns appear to have been lost over the period.

The extent to which this view might be skewed by better known and recorded groups of organisms is unclear, but if they are accepted as surrogate indicators we can be confident that there has not been a substantial loss of species over this time. There is however no room for complacency as many groups were found to have numerous representatives in decline, and to have many restricted to only a few sites. Further, an alarming array of species in diverse groups, which have been introduced, are extending their ranges, sometimes at the expense of native species.

However, species do not exist in isolation but depend on each other in a variety of ways, for example as food sources, components of food-chains, mutualistic symbionts, or for camouflage. An analysis of various independent data sets on a global scale showed that on average each plant species had 15 organisms confined to it: three bacteria, 5.3 fungi, 10 insects, 0.5 nematodes and one virus (Hawksworth 1998). The loss of seven plant species from Great Britain may thus imply the extinction of 105 obligately associated organisms, many of which may not even have been recognized as present here.

Species new to science have continued to be discovered in speciose groups during the period; for example 17 new mosses and liverworts (Chapter 4), 65 new mites (Chapter 15), 23 nematodes (Chapter 14), and perhaps as many as 1150 fungi (Chapter 6).

Carnivorous mammals and deer have been recovering their former ranges in the islands (Chapter 23), and there has been a continued steady increase in the number of breeding bird species, about four being added each decade (Chapter 22). Nevertheless, while there has also been an 8% increase in mean bird populations since the mid-1970s, it is wetland birds that have fared particularly well while those of woodland and especially farmland have declined (Chapter 22). Conversely, many slugs, snails and flies preferring small areas of water have declined (Chapters 16 and 21). In the case of protozoa, it is the existence of appropriate habitats that seems to be the key; if the habitat is right the same assemblage of species develops, apparently worldwide (Chapter 12).

It is, however, important to recognize the difficulty of assessing the status of species in groups with few appropriately trained recorders. Furthermore, in some cases a recorder needs to be in the right place at the right time to witness a species, so sporadically fruiting organisms such as mushrooms may go undetected for gaps of a century or more (Chapter 5). When species do attain a higher profile, experience is that they are then searched for more avidly: The Devil's bolete mushroom *Boletus satanus* was thought to be endangered but is now confirmed as still present in 22 of 40 pre-1970 sites (Chapter 5). This experience has also been shared by hemipterists (Chapter 17).

6 Factors affecting change

6.1 Agriculture

The key changes in agriculture over the past three decades have been land drainage, hedgerow removal, introduction of new crops (such as oilseed rape and linseed), pasture 'improvement', increased agrochemical input, autumn rather than spring sowing, harvesting grass for silage, reduction of traditional rotations, and reduction in undersown leys (Fuller 2000). Collectively, it has been claimed that these factors have led to a loss of 10 million breeding individuals of farmland birds in the last 10 years, and that a second 'Silent Spring' may be upon us (Krebs *et al.* 1999). It must also be remembered that agricultural impacts can have significant effects on air and water quality many kilometres from the source (Skinner *et al.* 1997).

Hedge removal, which amounted to 140 000 miles between 1946–47 and 1974 in England and Wales (Ratcliffe 1984), has continued. This adversely affects a wide range of organisms from ferns to birds and lichens. The unploughed margins by hedges in fields are a particular haven for plants – provided that agricultural chemical use is limited. However, these are newly threatened by a European Court of Auditors ruling in 2000 that requires farmers to claim subsidies only for land actually under cultivation and does not permit boundary features more than 2 m wide. While it is to be hoped that this decision can be renegotiated, some farmers are already acting on it.

A reformed Common Agricultural Policy (CAP) is an ongoing aim of the statutory conservation agencies in Great Britain. By changing subsidy arrangements, and so farming practices, major benefits could result. The Agenda 2000 reforms made some progress, but while these do not yet go as far as to justify subsidies only where there are positive outcomes for wildlife, sustained lobbying may eventually prove effective.

The issue of Genetically Modified Organisms (GMOs) is currently particularly 'hot'. The potential threats to wildlife will differ from case to case, depending on what is being modified and by what kind of gene. Concerns include threats to birds, beneficial and harmless insects, soil organisms and gene transfer to native plants (especially by pollination). It is too early to predict possible long-term effects of GMOs on our wildlife, but pleasing to note that English Nature is represented on the panel approving the field trials announced in March 2000.

Conversely, the increased demand for organically produced farm products can only be beneficial to our wildlife, for example to birds, spiders, earthworms, butterflies and weeds (Krebs *et al.* 1999).

6.2 Fisheries

Marine fisheries have been an ongoing source of concern, with quotas legislated by the EEC in an attempt to maintain stocks. Fines can be substantial but these actions appear to be more effective in some cases than in others. Fish availability may also impact on some birds, and the Arctic tern failed to breed in Shetland in 1984–90 because of the overfishing of sand eels (Chapter 21). Pearl mussel overfishing has also occurred, partly accentuated by poor recruitment (Chapter 24). Other aspects of fisheries are discussed under Pollution and Introductions below.

6.3 Commercial collecting

Commercial collecting for restaurants focuses on a few species of mushrooms and snails (e.g. *Helix pomatia*; Chapter 21). Mushroom collecting had become a major source of concern, although its effects would be primarily on flies and other organisms associated with particular fruit-bodies, as well as limiting the capacity of species to spread. A *Code of Conduct* has now been prepared by a range of pertinent statutory and voluntary organizations which should lead to better practice in the future (English Nature 1998) – although its launch was not without controversy.

6.4 Habitat management

In woodlands, the decline in coppicing is an on-going matter of concern as its cessation causes reductions not only in plants but also butterflies (Chapter 18), and snails (Chapter 21). This may also be a key factor in the decline of woodland birds (Chapter 22). Where coppicing has been reintroduced, there are marked wildlife benefits and the practice clearly merits encouragement even where it is uneconomic.

The importance of maintaining ecological continuity in woodlands over periods of centuries was highlighted in several contributions to the 1973 symposium. It is gratifying that this is now accepted, and that old-forest indicator species are taken note of in criteria for the recognition of Sites of Special Scientific Interest (SSSIs). It may however take 300 years for some species to move into a 'newly established' woodland (Chapter 7). The Veteran Trees scheme, launched in 1997 to increase awareness of the importance of aged trees for invertebrates (especially spiders and beetles), larger fungi, and lichens in particular, was opportune and will also have long-term effects.

Grazing of pastures can adversely affect ferns and bryophytes, indeed the moss *Racomitrium lanuginosum* has been in decline since the 1950s (Chapter 4), but is needed by butterflies, some plants and associated larger fungi and lichens. EEC subsidies have favoured more extensive grazing, especially in Ireland. Over-grazing by sheep in upland habitats can also cause problems, in extreme cases removing all vegetation leading to erosion. The issue requires sensitive management related to conservation objectives that will vary from site to site.

6.5 Pollution

Water eutrophication affects a wide range of organisms, from diatoms and insect larvae through to aquatic plants and fish. It also affects mosses, lichens, and aquatic fungi although these have been less well studied. Sewage discharge is cited as the last cause of the loss of the dragonfly *Oxygastra curtisii* (Chapter 20). The increasing vigilance of the Environment Agency with respect to water quality is welcome, but their increased use of bioindicator organisms would be beneficial to wildlife generally. Reduction of water pollution in estuaries has already encouraged certain fish (Chapter 24). The phasing out of lead-shot is already encouraging wildfowl (Chapter 22).

The dramatic falls in ambient sulphur dioxide levels that have occurred in lowland Britain since the 1970s have had dramatic effects on lichens (Chapter 7), mosses (Chapter 4) and some plant pathogenic fungi (Chapter 6). Recolonization has been spectacular in major conurbations such as West Yorkshire and Greater London, but some mosses are still in decline (Chapter 4) as are some of the more sensitive lichens in rural areas (Chapter 7). An equilibration seems to be in progress, but one which is leading to extensions of range that are set to continue for the foreseeable future.

The effects of long-distance pollution, particularly as expressed in acid rain, continue to be a cause of concern for ferns, mosses and lichens and this situation is unlikely to be rectified without transnational action and changes in policies towards the burning of fossil fuels in tall-stack power stations.

6.6 Drainage

The drainage of wetland sites continues to adversely affect some ferns, flowering plants, butterflies, flies, slugs, and snails. There was a massive expansion in field under-drainage in the 1970s–80s (Chapter 16), the repercussions of which on wildlife as a whole have not been fully analysed.

Water abstraction can also have sometimes unforeseen and not easily reversible effects way downstream from abstraction points. This is an issue the Environment Agency needs to take note of when granting licences.

Commercial peat cutting adversely affects species such as dragonflies and damselflies that like small ponds (Chapter 20), and has led to the loss of one moss from Ireland (Chapter 4). Fortunately, new garden ponds are encouraging dragonflies, damselflies (Chapter 21) and certainly other invertebrates. The creation of larger artificial water-bodies is also encouraging wildfowl (Chapter 22).

6.7 Public pressure

Visitor pressure on the countryside as yet seems to have had little overall impact on our wildlife, although there are local instances of damage. For example, the loss of the moss *Andreae rothii* from west Cornwall is attributed to nitrogen oxide from tourists' cars (Chapter 4).

6.8 Introductions

Introduced species emerge as perhaps the greatest long-term threat to the wildlife of Great Britain and Ireland. The effects of the Dutch elm disease fungus epidemic in the mid-1970s were dramatic, but pathogens threatening native species continue to become newly established, such as the daisy rust (Chapter 6). Viruses can also transfer from crops or pets to native plants or animals respectively (Chapter 11). Vigilance has so far kept a few pests from becoming established, such as the gypsy moth that could attack our forests as badly as it has those of eastern North America (Chapter 18).

Introduced crayfish are now established in over 40 rivers, and the signal crayfish brought in a fungal infection that has caused a plague in native white-clawed crayfish (Chapter 13). The American flatworm *Dugesia tigrina* is also becoming well-established in river systems (Chapter 13). Exotic fish introduced for sport in some cases also thrive (Chapter 24), as do ornamental garden-pond plants (e.g. *Crassula helmsii*; Chapter 2), and diatoms that may even be moved from continent to continent through sea-going vessels and more locally through water pipelines (Chapter 10).

Accidental escapes of garden plants continue to be a problem (Chapter 2), especially when they enter sites of conservation importance. When such plants or inadvertently introduced weeds spread, they may also carry associated obligate organisms with them, such as tephretid flies, fungi and viruses.

Mosses *Orthodontium lineare* and *Campylopus introflexus* continue to spread (Chapter 4).

Red squirrels and water voles continue to be ousted and predated by grey squirrels and American minks respectively (Chapter 23) and will become increasingly endangered. The notorious Canada goose and ruddy ducks are among 14 introduced birds, and these can affect the status of native species (Chapter 22). However, introduced snails do not appear to affect the status of native species (Chapter 21), and neither do many of the introduced crickets and their allies (Chapter 19).

Deliberate introductions of native species into new sites are sometimes made as part of conservation programmes, but those of exotic species made by individual naturalists (e.g. of butterflies; Chapter 18) are a cause for concern. Researchers themselves are not always without blame, and accidently introduced the American invertebrate *Phagocata woodworthi* into Loch Ness (Chapter 13).

Action needs to be taken to minimize the risk of accidental or deliberate importation of organisms that could threaten our wildlife as pests, pathogens, or competitors, but this key issue has not been taken up as a part of the UK's Action Plan for biodiversity (Department of the Environment 1994). The issue is likely to become even more significant in future as the climate warms.

6.9 Climate change

It is difficult to speculate on the extent to which climate change may affect the wildlife of Great Britain and Ireland over the next 25 years, but evidence that warming is occurring is accumulating (Menzel and Fabian 1999). To date there are no more than slight indications of responses from native wildlife to global warming; for example the peaking of chanterelle (*Cantharellus cibarius*) fruiting in 1997–98 (Chapter 5), and increases in marine fish – including some new to our waters (Chapter 24). Additional breeding birds are expected (Chapter 22), and viruses and other associated organisms may move northwards with host plants extending their ranges. Escaped garden- and pond-plants may also have an increased chance for establishment, and the prospect of more introductions of pathogens or other damaging species has already been mentioned.

6.10 National and international legislation

The legal background to nature conservation has changed out of all recognition since 1973, and in ways that could scarcely be even dreamed of at that time. As this issue was not addressed at that time, an authoritative account of the framework developed in Great Britain has been presented in Chapter 1. Further discussions of the development of nature conservation in Britain are provided by Sheail (1976), Moore (1987) and Evans (1992).

In order to further safeguard our wildlife for the future, the immediate need is to strengthen the protection and positive management of the large number of Sites of Special Scientific Interest (SSSIs) now recognized; there were 4063 SSSIs in England alone by the end of July 1999. This will require revision of the Wildlife and Countryside Act 1981 to place greater obligations on owners and occupiers, and to empower the staff of the statutory agencies to enter and inspect existing and potentially notifiable sites. Increased penalties for the elimination or reduction of the scientific interest of notified sites will also be needed if some of the failures of the past are to be avoided for the future. A consultation document on proposed changes has been made available (Department of the Environment, Transport and the Regions 1998). The growth in the series of National Nature Reserves (NNRs) since 1973, some now managed by approved bodies other than the statutory agencies, is especially welcome; by the end of July 1999, there were 200 NNRs in England alone, which covered 81 764 ha.

The criteria for the notification of biological sites (Nature Conservancy Council 1989) have been supplemented by the Joint Nature Conservation Committee with detailed guidance for some particular groups of organisms and habitats – a process that requires further refinement and extension in the years ahead.

Schedules to the Wildlife and Countryside Act lists species that it is an offence to collect. These are added to periodically, and now include a considerable variety of organisms and not only conspicuous species; further information is included in several of the contributions to this book.

The legal framework is increasingly assuming an international dimension. The EC Habitats and Species Directive entered UK law in 1994, and a series of Special Conservation Areas (SACs) have been proposed and are currently undergoing a moderation process. Other sites are recognized under the Ramsar Convention 1973 for wetlands, or as Special Protection Areas (SPAs) for birds under a 1979 EC Directive.

One outfall of the Convention on Biological Diversity drawn up in 1992 has been the development of a suite of species recovery programmes for species recognized as endangered in *The UK Action Plan* for biodiversity (Department of the Environment 1994) and the subsequent reports of its steering committee. It is encouraging that these programmes are focusing on some of the less charismatic insects and fungi as well as on plants, birds, mammals and butterflies. Further information as to the species covered is included in the individual contributions to this volume.

7 Prognosis for the next 25 years

Most changes that have been seen in the wildlife of Great Britain and Ireland during the last 25 years have been a result of human intervention, whether intentional or accidental. While extinctions have been few in the last period, these are certain to accelerate as 2025 is approached as a result of the attrition of the number of sites where already endangered species occur. The proportion of species in some groups now considered as threatened is alarming, for example 40% of our butterflies (Chapter 18).

As populations become separated, isolated gene pools will inevitably develop their own identities over time, which may or may not be expressed phenotypically. The genetic dimension to conservation is rarely addressed by its practitioners, but it must always be remembered that protecting a 'species' may not equate to safeguarding the full range of its former genetic diversity. There may in any case be a danger in putting too great a proportion of the limited resources available into maintaining rarities. Some of these are at the edge of their southern or northern ranges in Great Britain and Ireland, but scarcely endangered elsewhere in Europe. There may be a need for a more holistic habitat-focused approach which will safeguard the little-known and unknown as well as the endangered, perhaps based on the Natural Areas maps launched in 1997. That naturalists in general may be more amenable to such an approach is evidenced by the success of the journal *British Wildlife*, launched in 1989, and which embraces notes on habitats and reserves as well as a broad spectrum of wildlife.

Monitoring the status of the perhaps 100 000 species in Great Britain and Ireland is unlikely to become realistic in the forseeable future. There could, however, be some advantage in monitoring a subset selected because of known sensitivities to particular pollutants or management regimes. This monitoring could be carried out nationally or at a few selected sites. In some of the more species-rich groups, there may also be a case for the accumulation of long-term fuller data sets in a few sites on the lines of the Rothamsted Insect Survey. Such suites of data should be seen as complementary to and not alternatives to the planned Biodiversity Information Network (Chapter 25).

The adoption of the precautionary principle, that is, setting limits at a level where the most vulnerable species will be safeguarded, has already played a crucial part in placing acceptable levels of air pollutants at figures where even the most sensitive organisms are unlikely to be affected (Ashmore and Wilson 1992). This is an encouraging model for other kinds of environmental disturbance, although the costs may not always be easy to justify (Rogers *et al.* 1997).

Habitat restoration – creative conservation – is a relatively new development in wildlife protection (Sheail *et al.* 1997). While such initiatives can favour particular species and communities, it is questionable whether habitats such as ancient woodlands and pastures will attain a reasonably full complement of the biota of natural sites within an acceptable time-frame. Such initiatives are commendable for particular purposes, but no alternative to the protection of natural sites.

What we can be sure of is more introductions, especially with global warming, many of which will adversely affect our wildlife as pathogens or competitors.

Overall, the future looks bleaker than the past. Further, if the current commitments of governments are to be maintained, public pressure over concern for the environment and wildlife must be sustained. This requires an ongoing effort by conservation scientists to strengthen the depth of understanding of the issues in society at large (Osborn

1997). Governments and their statutory agencies will also need to be increasingly proactive, rather than retroactive or reactionary. Then there will be hope that we can retain a wildlife of which we can be proud to pass into the care of the next generation.

REFERENCES

Anon. (1935) Changes in the British fauna and flora during the past fifty years. *Proceedings of the Linnean Society of London*, 148, 33–52.

Anon. (1973a) Scientist optimistic about survival of wildlife in Britain. *The Times* (7 May 1973), 4.

Anon. (1973b) Hope for Britain's wildlife, *Nature*, 243, 9.

Anon. (1992) *House of Lords Select Committee on Systematic Biology*. London: Her Majesty's Stationery Office.

Ashmore, M. R. and Wilson, R. B. (eds) (1992) *Critical Levels of Air Pollutants for Europe*. London: Department of the Environment.

Berry, R. J. (1988) *Biological Survey: Need and Network*. London: Polytechnic of North London Press.

Bisby, F. A. and Hawksworth, D. L. (1991) What must be done to save systematics?, in Hawksworth, D. L. (ed.), *Improving the Stability of Names: Needs and Options*. Königstein: Koeltz Scientific Books, pp. 323–336.

Carson, R. (1962) *Silent Spring*. New York: Houghton Miffin.

Claridge, M. F. and Ingrouille, M. (1992) Systematic biology and higher education in the UK, in *Taxonomy in the 1990s*. London: Linnean Society of London, pp. 39–46.

Department of the Environment (1994) *Biodiversity. The UK Action Plan*. London: Her Majesty's Stationery Office.

Department of the Environment, Transport and the Regions (1998) *Sites of Special Scientific Interest: Better Protection and Management*. London: Department of the Environment, Transport and the Regions.

Edlin, H. L. (1952) *The Changing Wildlife of Britain*. London: B. T. Batsford.

English Nature (1998) *The Wild Mushroom Pickers Code of Conduct*. Peterborough: English Nature.

Evans. D. (1992) *A History of Nature Conservation in Britain*. London: Routledge.

Fuller, R. J. (2000) Relationships between recent changes in lowland British agriculture and farmland bird populations: an overview, in Aebischer, N. J., Evans, A. D., Grice, P. V. and Vickery, J. A. (eds) *Ecology and Conservation of Lowland Farmland Birds*. Tring: British Ornithologists Union.

Harding, D. J. L. (ed.) (1988) *Britain Since 'Silent Spring': an Update on the Ecological Effects of Agricultural Pesticides in the UK*. London: Institute of Biology.

Harding, P. T. (ed.) (1992) *Biological Recording of Changes in British Wildlife* [Institute of Terrestrial Ecology Symposium no. 26.]. London: HMSO.

Hawksworth, D. L. (ed.) (1974) *The Changing Flora and Fauna of Britain*. London: Academic Press.

Hawksworth, D. L. (1998) The consequences of plant extinctions for their dependent biotas: an overlooked aspect of conservation science, in Peng, C.-I. and Lowry, P. P. (eds) *Rare, Threatened, and Endangered Floras of Asia and the Pacific Rim* [Institiute of Botany Monograph Series no. 16]. Taipei: Academia Sinica, pp. 1–15.

Hawksworth, D. L. and Bisby, F. A. (1988) Systematics: the keystone of biology, in Hawksworth, D. L. (ed.) *Prospects in Systematics*. Oxford: Clarendon Press, pp. 3–30.

Isely, D. (1972) The disappearance. *Taxon*, 21, 3–12.

Krebs, J. R., Wilson, J. D., Bradbury, R. B. and Siriwardena, G. M. (1999) The second Silent Spring? *Nature*, 400, 611–612.

Menzel, A. and Fabian, P. (1999) Growing season extended in Europe. *Nature,* 397, 659.

Moore, N. W. (1987) *The Bird of Time: the Science and Politics of Nature Conservation.* Cambridge: Cambridge University Press.

Nature Conservancy Council (1989) *Guidelines for Selection of Biological SSSIs.* Huntingdon: Nature Conservancy Council.

Osborn, D. (1997) Some reflections on UK environment policy, 1970–1995. *Journal of Environmental Law,* 9, 3–22.

Ratcliffe, D. A. (1984) Post-medieval and recent changes in British vegetation: the culmination of human influence. *New Phytologist,* 98, 73–100.

Rogers, M. F., Sinden, J. A. and De Lacy, T. (1997) The precautionary principle for environmental management: a defensive-expenditure application. *Journal of Environmental Management,* 51, 343–360.

Sheail, J. (1976) *Nature in Trust: the History of Nature Conservation in Britain.* Glasgow: Blackie.

Sheail, J., Treweek, J. and Mountford, J. O. (1997) The UK transition from nature preservation to 'creative conservation'. *Environment and Conservation,* 24, 224–235.

Skinner, J. A., Lewis, K. A., Bardon, K. S., Tucker, P., Catt, J. A. and Chambers, B. J. (1997) An overview of the environmental impact of agriculture in the UK. *Journal of Environmental Management,* 50, 111–128.

UK Systematics Forum (1998) *The Web of Life.* London: UK Systematics Forum.

Valdecasas, A. G., Castroviejo, S. and Marcus, L. F. (2000) Reliance on the citation Index undermines the study of biodiversity. *Nature,* 403, 698.

Vesey-Fitzgerald, B. (1969) *Vanishing Wild Life of Britain.* London: MacGibbon and Kee.

Williams, B. (1987) *Biological Collections. UK.* London: The Museums Association.

Index

LIBRARY
THE NORTH HIGHLAND COLLEGE
ORMLIE ROAD
THURSO
CAITHNESS
KW14 7EE

acari, *see* mites and ticks
acid rain 19, 121, 133, 158, 364; *see also* air
  pollution
aculeates 436
additions, *see* increases
afforestation, *see* forestry, woodlands
agricultural chemicals 133, 438; *see also*
  fertilisers, pesticides
agriculture 8, 17, 30–1, 42–3, 45, 63, 88,
  91, 111, 120, 133, 157, 172–3, 203–4,
  223–4, 250–1, 271, 291, 320, 337, 344,
  346, 349–50, 379, 382, 384–5, 403, 405,
  438–40
air pollution 42, 55, 65, 69, 83, 88–9, 121–2,
  129, 133–5, 138–9, 324, 364, 441
algae 89, 148–61, 350
aliens, *see* introductions
amateurs 104–5, 142, 235, 244, 282, 301,
  303, 337–8, 343, 351, 367, 375, 399,
  422–3, 433; *see also* work force
ammonia 133
amphibians and reptiles 14, 172, 436
anglers, *see* fishes
antibiotics 120, 252
aphids, *see* true bugs
aquatic organisms 30–1, 34, 36, 42, 44, 64,
  148–61, 170, 175–84, 188–206, 233,
  351–2, 354–5, 441; *see also* fish
arthropods, see under groups

bacteria 256–7, 436, 438
BAP, *see* Biodiversity Action Plan
bats 14, 401, 403
beetles 192, 195–7, 423, 436, 440
Berne Convention 16, 362
biocides 350; *see also* pesticides
biocontrol 256
Biodiversity Action Plan(s) 11, 13, 16, 18–9,
  44, 104, 118–9, 137, 139–42, 197–8, 205,
  242–3, 245–6, 252, 257–8, 279, 283,
  303–4, 309, 320, 324, 337, 345, 356–7,

361–2, 382, 387, 427–8, 430, 432, 436,
  442–3
Biodiversity Information Network 205, 243,
  253, 442–3, 436, 437, 444
bioindicators 70, 190, 217, 224
Biological Records Centre 424–5, 429–30,
  436
birds 6, 13, 18, 139, 171, 215–6, 223, 234,
  343, 352, 367–98, 420, 423, 428, 436,
  438–43
Bonn Convention 16
British Trust for Ornithology 369–71, 375
bryophytes, *see* mosses, liverworts and
  hornworts
BSE 250
bugs, *see* true bugs
building, *see* urbanization
butterflies and moths 14, 138–9, 195,
  300–25, 422, 428, 438, 440, 443–4

canals 198, 201
CAP, *see* Common Agricultural Policy
cars, *see* roads
cicadas, *see* true bugs
CITES 13, 15
climate 36, 39–40, 42, 45, 57–8, 60, 83, 92,
  109, 121–2, 148, 160–1, 172, 205, 257,
  312, 337, 341, 343, 351–2, 363, 385–6,
  442–4
coal mines, *see* mining
coast 30, 36, 38, 45, 64–5, 155, 160, 214,
  254, 257, 331, 332, 371–4, 380–1, 383,
  386, 417, 419
collecting 43, 88, 110, 138, 324, 377, 440
collections 23, 45, 64, 104, 231–2, 242, 273
Common Agricultural Policy 44, 321, 353,
  385, 439
coniferous forests and trees, *see* forestry,
  woodland
Conservation of Wild Creatures and Wild
  Plants Act 8, 14

Convention on Biological Diversity 13, 16, 116, 142, 197, 257, 345, 443
Convention on Trade in Endangered Species, *see* CITES
Council for the Protection of Rural England 15
Countryside Agency 12
Countryside Commission 10–11, 15
Countryside Council for Wales 10–11, 17, 141, 345, 369–70, 399
countryside stewardship 44
countryside survey 45
crickets, *see* grasshoppers, crickets and allied insects
crustacea, *see* freshwater invertebrates
cyanobacteria 134, 117, 128, 189–91

databases 105, 117, 128, 189, 191; *see also* mapping, recording
decreasing species 30, 34–6, 57–8, 63, 87–9, 96, 129–30, 133–4, 137, 142, 198–200, 234, 245–6, 265–8, 279, 289–92, 300, 305–12, 330, 344, 346–7, 357–61, 377–80, 392–8, 402–3, 405–6, 415–8
Department of the Environment, Transport and the Regions 7, 13, 15, 18
dermaptera, *see* grasshoppers, crickets and allied insects
diptera, *see* flies
diseases, *see* pests
ditches 64, 148, 252, 256, 271, 342
dragonflies and damselflies 189–90, 192, 195, 197, 200, 202, 340–53, 422
drainage 19, 42, 57–8, 63, 87, 203, 250–1, 291, 344, 347, 349–50, 406, 439, 441
Dutch elm disease 4, 19, 43, 118–21, 137, 253, 288, 312, 442

Earth Summit, *see* Convention on Biological Diversity
earthworms 440
endemics 32, 34, 39
English Nature 10–12, 17, 19, 140–1, 330, 333, 369–70, 399, 414, 439
Environment Agency 154, 191, 412, 415, 441
Environmental Protection Act 10
ephemoptera, *see* freshwater invertebrates
escapes, *see* introductions
European Habitat Directives 13, 15–16, 18, 44, 139, 142, 205, 257, 345, 356, 362, 427, 443
extinctions 32–4, 86–7, 91, 95, 110, 120, 130–2, 176, 199, 235, 245, 300–1, 305, 317, 322, 330, 337, 341, 344, 346, 349, 365, 376, 383, 400–1, 415, 438, 444

farming, *see* agriculture
felling, *see* forestry, woodland
ferns 50–7, 438–9, 441
fertilisers 42, 120, 133, 157, 291, 350; *see also* agricultural chemicals
fire 8, 57–8, 61, 69
fish 160–1, 199, 216, 383, 386, 410–20, 440–2
flies 88, 190, 192, 195–8, 240–58, 322, 428, 438, 441–2
flooding 38
flowering plants 13, 23–46, 50, 54–5, 428, 438, 441
Forest Enterprise 9–10
forestry 6, 8–99, 42, 55, 57–8; *see also* woodland
Forestry Authority 319
Forestry Commission 9, 17
freshwater, *see* aquatic organisms
Freshwater Biological Association 188–9, 244
freshwater invertebrates 188–206, 214, 420, 442
fungi 63, 103–11, 114–123, 215, 246, 428, 439–43; *see also* lichens
fungicides 384; *see also* pesticides
future changes, *see* prospects

genetic diversity 68, 134, 322, 334, 407
genetically modified organisms 385, 439
GMOs, *see* genetically modified organisms
grasshoppers, crickets and allied insects 328–38, 442
grassland 30–1, 36, 38–9, 42, 63, 105, 117, 119–20, 137, 250, 270, 289–90, 312, 321, 330, 334, 360, 364, 378
grazing 17, 42, 55, 61–3, 69, 87–8, 135, 137, 245, 252, 291, 320–1, 337, 360, 385, 406, 440

Habitat Directives, *see* European Habitat Directives
heathland 30, 32, 38, 60, 69, 137, 142, 250, 269, 334, 344
hedges 30–1, 53, 57–9, 65, 120, 312, 364, 406, 439
hemiptera, *see* true bugs
herbaria, *see* collections
herbicides 42, 61, 120, 133, 203, 252, 362; *see also* pesticides
Heritage Lottery Fund 12, 432
history of recording 21–4, 51–3, 78–9, 115–6, 126–7, 159–60, 177–8, 188–9, 211–4, 231–3, 240–3, 264, 281–2, 301, 342, 368, 400, 435

hornworts, *see* mosses, liverworts and hornworts
hybrids 50–1, 56, 58–60, 64, 413–4

increasing species 30, 38–42, 64, 89–90, 99–100, 106, 108–9, 129–31, 134–5, 150, 198, 200–2, 246–9, 265–8, 284–9, 292, 312–9, 330–5, 344, 347–8, 356–8, 360, 376–9, 384, 392–8, 403–7, 416, 418–20
insecticides, *see* pesticides
insects, see under common names for groups
International Biological Programme 5
introductions 4, 19, 30, 32, 38, 41–3, 45, 66–7, 85, 90–1, 97, 106–8, 119, 121, 130, 135, 158, 171, 192–3, 201, 222, 234, 250, 273–6, 279–81, 287–8, 319–20, 324, 335–7, 348–9, 358–9, 363–4, 377–8, 380–1, 392–8, 401, 412–5, 419–20, 442, 444
invertebrates, freshwater 188–206; see also under common names for groups
IUCN–The World Conservation Union 28, 91, 197, 245, 304

Joint Nature Conservation Committee 10–11, 13, 15–17, 117, 191–2, 244, 302, 369–70, 375, 231, 443

lakes 64, 152–61, 190; *see also* fish, freshwater invertebrates
landfill 12, 154
leaf-hoppers, *see* true bugs
legislation 1–19; see also under names of acts, conventions, etc.
lepidoptera, *see* butterflies and moths
lichens 55, 89, 126–43, 222, 436, 439, 441
Life Programmes 10, 19
liverworts, *see* mosses, liverworts and hornworts
local nature reserves 3, 5, 17

mammals 4, 14, 62, 111, 169, 171–2, 215–6, 234, 352, 399–407, 438, 443
mapping and recording 24–5, 29–31, 45, 53–4, 79–85, 104–5, 127–8, 184, 189, 205, 217, 220, 232, 240–3, 282, 301–4, 329, 332, 338, 342–3, 355, 368–75, 422–33, 436–7, 444
marine fish 417–9
meadows 31, 38; *see also* grasslands
migrants 16, 348–9, 373, 416–7; *see also* birds
mineral extraction, *see* mines, quarries
mines 58–9, 64, 66, 69, 150
mites and ticks 230–236, 436, 438
molluscs, *see* slugs and snails

moorland 9, 30–1; *see also* heathland
mosses, liverworts and hornworts 55, 78–100, 438, 440–2
moths, *see* butterflies and moths
museums, *see* collections
mushrooms, *see* fungi
myxomatosis 4–6, 172, 321, 360, 403

National Biodiversity Network, *see* Biodiversity Information Network
National Nature Reserves 3, 5, 12, 17, 91, 351–2, 443
National Parks 3, 5
National Rivers Authority 412–3, 415
National Trust 4, 17, 137
Natura 2000 18
Natural Environment Research Council 6, 219
Nature Conservancy Council 2, 4–14, 17, 242, 244, 256, 269, 334, 415, 417, 424
nematodes 139, 219–24, 413, 436, 438, 442
new species 89, 96, 103–4, 118, 211, 215, 222, 232–4, 438
NGOs, *see* Non-Governmental Organizations
nitrogen 43, 135, 151, 203
Non-Governmental Organizations 2, 7, 10, 12–13, 18–19, 26, 142, 257, 301, 352
Northern Ireland Council for Nature Conservation and the Countryside 10
numbers of species 28, 51, 90, 115, 117–8, 126, 130, 176–7, 194–6, 230, 239, 262, 329, 341, 375–6, 399, 436

odonata, *see* dragonflies and damselflies
oil 203, 254
orthoptera, *see* grasshoppers, crickets and allied insects
over-collecting, *see* collecting
ozone 43, 121, 135; *see also* air pollution

parasites, *see* pests
pathogens, *see* pests
peat 5, 9, 17, 43, 87, 254, 270–1, 320, 337, 344, 347, 350
pesticides 6, 122, 203, 223, 252, 383–4
pests 402; animal 88, 164–7, 169–71, 183–4, 193, 199, 214–5, 223, 230, 232, 234–5, 315, 317, 323, 413, 443; plant 43, 109, 118–21, 164–71, 211, 219–23, 230, 271, 319, 335; *see also* weeds
picoplankton 150
plankton, *see* cyanobacteria, diatoms, protozoa
plantations, *see* forestry, woodland
plant-hoppers, *see* true bugs

Plantlife 26, 104
ploughing 42, 63, 137, 291
pollution 43, 56–8, 69, 83, 91, 137, 148,
    150, 152–3, 155, 160, 223, 344, 350–1,
    416–7, 441; *see also* air pollution
ponds 34, 36, 42, 44, 64, 190–1, 204, 344,
    346, 349–50, 413, 442–3
prognosis, *see* prospects
prospects 45–6, 70, 91, 122, 142, 171–3,
    184, 205–6, 223–4, 235–6, 257–8, 324–5,
    337–8, 385–7, 432–3, 444–5
protozoa 175–84, 436, 439
public pressure 8, 138, 441

quarantine 171
quarries 66, 69, 108, 135, 254, 276, 340, 347

railways 59, 65
RAMSAR Convention 15, 443
recording, *see* mapping and recording
Red Databooks and Lists 28, 45, 54, 85–6,
    93–4, 105, 120, 139–42, 148, 196–7,
    199–200, 241–2, 244–5, 252, 257, 263,
    270, 283, 301–2, 329
reservoirs 150
rivers 254–5, 411, 416; *see also* aquatic
    organisms, fish
Rivpacs 190–1
roads and verges 8, 31, 38, 42, 57–9, 135,
    402
Royal Society for Nature Conservation 12, 15
Royal Society for the Protection of Birds 13,
    15, 17, 351–2, 369–70, 375, 382, 431

SACs, *see* Special Areas of Conservation
salt marshes 38, 160, 254, 257
sand dunes 34, 36, 109, 270, 278, 286
scale insects, *see* true bugs
scheduled species, *see* Wildlife and
    Countryside Act
Scottish Natural Heritage 10–11, 17, 140–1,
    369–70, 399, 407
Scottish Office 18
scrub 30, 121, 136
seashore, *see* coast
seaweeds 436
seeds 32
sewage 344
Sites of Special Scientific Interest 5, 8, 10, 12,
    17–18, 91, 119, 139, 150, 205, 224, 243,
    252–3, 255, 333, 345, 351–2, 424, 440,
    443
slugs and snails 138, 189, 192, 197–8, 215,
    246, 255–65, 438, 440–1
snails, *see* slugs and snails

Society for the Promotion of Nature
    Reserves 3–4, 12
soil 43, 56, 63, 111, 137, 184, 211–2,
    222–4, 230
SPAs, *see* Special Protection Areas
Special Areas of Conservation 16, 18–19,
    139, 255, 427, 443
Special Protection Areas 16, 18, 443
Species Recovery Programme 17, 29, 40–1,
    283, 330, 333–4
spiders 444
SSSI, *see* Sites of Special Scientific Interest
sulphur dioxide, *see* air pollution

taxonomists, *see* work force
ticks, *see* mites and ticks
trampling, *see* public pressure
true bugs, leaf- and plant-hoppers and their
    allies 195–7, 200, 262–97

uplands 30, 57–8, 61, 109
urbanization 91, 204, 337, 386

vascular plants, *see* ferns, flowering plants
VCOs, *see* Non-Governmental
    Organizations
viruses 164–73, 219, 256, 405, 436, 438,
    442–3
voluntary conservation organizations, *see*
    Non-Governmental Organizations

water abstraction 19, 44–5, 199, 203, 257,
    350, 411
water pollution, *see* pollution
water resources 204
weeds 30, 42, 440, 442
whiteflies, *see* true bugs
Wildlife and Countryside Act 7–8, 10, 29,
    44, 85, 89, 93–4, 139–41, 198, 205, 243,
    256, 283, 301, 304, 329–30, 345, 362,
    382, 400, 404, 412, 427, 443
Wild Life Conservation Special Committee
    3–4, 7
Wildlife Enhancement Scheme 17, 44
Wildlife Trusts 12, 17, 205, 431
woodland 30–1, 42–3, 55, 61–2, 108–9,
    111, 121, 129, 135–7, 142, 204, 223,
    248–9, 252–5, 320–1, 343, 350, 359,
    364, 380, 403, 406, 440
work force 26–27, 53, 105, 110, 118, 122,
    129, 142, 183, 191–2, 206, 216–7, 223,
    244, 258, 303, 343, 375, 399, 428–9,
    437–8; *see also* amateurs
World Wildlife Fund 12, 15

Systematics Association publications

1. Bibliography of key works for the identification of the British fauna and flora, 3rd edition (1967)†
Edited by G.J. Kerrich, R.D. Meikie and N.Tebble

2. Function and taxonomic importance (1959)†
Edited by A.J. Cain

3. The species concept in palaeontology (1956)†
Edited by P.C. Sylvester-Bradley

4. Taxonomy and geography (1962)†
Edited by D. Nichols

5. Speciation in the sea (1963)†
Edited by J.P. Harding and N. Tebble

6. Phenetic and phylogenetic classification (1964)†
Edited by V.H. Heywood and J. McNeill

7. Aspects of Tethyan biogeography (1967)†
Edited by C.G. Adams and D.V. Ager

8. The soil ecosystem (1969)†
Edited by H. Sheals

9. Organisms and continents through time (1973)†
Edited by N.F. Hughes

10. Cladistics: a practical course in systematics (1992)*
P.L. Forey, C.J. Humphries, I.J. Kitching, R.W. Scotland, D.J. Siebert and D.M. Williams

11. Cladistics: the theory and practice of parsimony analysis (2nd edition) (1998)*
I.J. Kitching, P.L. Forey, C.J. Humphries and D.M. Williams

* Published by Oxford University Press for the Systematics Association
† Published by the Association (out of print)

Systematics Association Special Volumes

1. The new systematics (1940)
Edited by J.S. Huxley (reprinted 1971)

2. Chemotaxonomy and serotaxonomy (1968)*
Edited by J.C. Hawkes

3. Data processing in biology and geology (1971)*
Edited by J.L. Cutbill

4. Scanning electron microscopy (1971)*
Edited by V.H. Heywood

5. Taxonomy and ecology (1973)*
Edited by V.H. Heywood

6. The changing flora and fauna of Britain (1974)*
Edited by D.L. Hawksworth

7. Biological identification with computers (1975)*
Edited by R.J. Pankhurst

8. Lichenology: progress and problems (1976)*
Edited by D.H. Brown, D.L. Hawksworth and R.H. Bailey

9. Key works to the fauna and flora of the British Isles and northwestern Europe, 4th edition (1978)*
Edited by G.J. Kerrich, D.L. Hawksworth and R.W. Sims

10. Modern approaches to the taxonomy of red and brown algae (1978)
Edited by D.E.G. Irvine and J.H. Price

11. Biology and systematics of colonial organisms (1979)*
Edited by C. Larwood and B.R. Rosen

12. The origin of major invertebrate groups (1979)*
Edited by M.R. House

13. Advances in bryozoology (1979)*
Edited by G.P. Larwood and M.B. Abbott

14. Bryophyte systematics (1979)*
Edited by G.C.S. Clarke and J.G. Duckett

15. The terrestrial environment and the origin of land vertebrates (1980)
Edited by A.L. Pachen

16. Chemosystematics: principles and practice (1980)*
Edited by F.A. Bisby, J.G. Vaughan and C.A. Wright

17. The shore environment: methods and ecosystems (2 volumes) (1980)*
Edited by J.H. Price, D.E.C. Irvine and W.F. Farnham

18. The Ammonoidea (1981)*
Edited by M.R. House and J.R. Senior

19. Biosystematics of social insects (1981)*
Edited by P.E. House and J.-L. Clement

20. Genome evolution (1982)*
Edited by G.A. Dover and R.B. Flavell

21. Problems of phylogenetic reconstruction (1982)
Edited by K.A. Joysey and A.E. Friday

22. Concepts in nematode systematics (1983)*
Edited by A.R. Stone, H.M. Platt and L.F. Khalil

23. Evolution, time and space: the emergence of the biosphere (1983)*
Edited by R.W. Sims, J.H. Price and P.E.S. Whalley

24. Protein polymorphism: adaptive and taxonomic significance (1983)*
Edited by G.S. Oxford and D. Rollinson

25. Current concepts in plant taxonomy (1983)*
Edited by V.H. Heywood and D.M. Moore

26. Databases in systematics (1984)*
Edited by R. Allkin and F.A. Bisby

27. Systematics of the green algae (1984)*
Edited by D.E.G. Irvine and D.M. John

28. The origins and relationships of lower invertebrates (1985)‡
Edited by S. Conway Morris, J.D. George, R. Gibson and H.M. Platt

29. Intraspecific classification of wild and cultivated plants (1986)‡
Edited by B.T. Styles

30. Biomineralization in lower plants and animals (1986)‡
Edited by B.S.C. Leadbeater and R. Riding

31. Systematic and taxonomic approaches in palaeobotany (1986)‡
Edited by R.A. Spicer and B.A. Thomas

32. Coevolution and systematics (1986)‡
Edited by A.R. Stone and D.L. Hawksworth

33. Key works to the fauna and flora of the British Isles and northwestern Europe, 5th edition (1988)‡
Edited by R.W. Sims, P. Freeman and D.L. Hawksworth

34. Extinction and survival in the fossil record (1988)‡
Edited by G.P. Larwood

35. The phylogeny and classification of the tetrapods (2 volumes) (1988)‡
Edited by M.J. Benton

36. Prospects in systematics (1988)‡
Edited by D.L. Hawksworth

37. Biosystematics of haematophagous insects (1988)‡
Edited by M.W. Service

38. The chromophyte algae: problems and perspective (1989)‡
Edited by J.C. Green, B.S.C. Leadbeater and W.L. Diver

39. Electrophoretic studies on agricultural pests (1989)‡
Edited by H.D. Loxdale and J. den Hollander

40. Evolution, systematics, and fossil history of the Hamamelidae (2 volumes) (1989)‡
Edited by P.R. Crane and S. Blackmore

41. Scanning electron microscopy in taxonomy and functional morphology (1990)‡
Edited by D. Claugher

42. Major evolutionary radiations (1990)‡
Edited by P.D. Taylor and G.P. Larwood

43. Tropical lichens: their systematics, conservation and ecology (1991)‡
Edited by G.J. Galloway

44. Pollen and spores: patterns of diversification (1991)‡
Edited by S. Blackmore and S.H. Barnes

45. The biology of free-living heterotrophic flagellates (1991)‡
Edited by D.J. Patterson and J. Larsen

46. Plant-animal interactions in the marine benthos (1992)‡
Edited by D.M. John, S.J. Hawkins and J.H. Price

47. The Ammonoidea: environment, ecology and evolutionary change (1993)‡
Edited by M.R. House

48. Designs for a global plant species information system (1993)‡
Edited by F.A. Bisby, G.F. Russell and R.J. Pankhurst

49. Plant galls: organisms, interactions, populations (1994)‡
Edited by M.A.J. Williams

50. Systematics and conservation evaluation (1994)‡
Edited by P.L. Forey, C.J. Humphries and R.I. Vane-Wright

51. The Haptophyte algae (1994)‡
Edited by J.C. Green and B.S.C. Leadbeater

52. Models in phylogeny reconstruction (1994)‡
Edited by R. Scotland, D.I. Siebert and D.M. Williams

53. The ecology of agricultural pests: biochemical approaches (1996)**
Edited by W.O.C. Symondson and J.E. Liddell

54. Species: the units of diversity (1997)**
Edited by M.F. Claridge, H.A. Dawah and M.R. Wilson

55. Arthropod relationships (1998)**
Edited by R.A. Fortey and R.H. Thomas

56. Evolutionary relationships among Protozoa (1998)**
Edited by G.H. Coombs, K. Vickerman, M.A. Sleigh and A. Warren

57. Molecular systematics and plant evolution (1999)
Edited by P.M. Hollingsworth, R.M. Bateman and R.J. Gornall

58. Homology and systematics (2000)
Edited by R. Scotland and R.T. Pennington

59. The Flagellates: unity, diversity and evolution (2000)
Edited by B.S.C. Leadbeater and J.C. Green

60. Interrelationships of the Platyhelminthes (2001)
Edited by D.T.J. Littlewood and R.A. Bray

61. Major events in early vertebrate evolution (2001)
Edited by P.E. Ahlberg

62. The changing wildlife of Great Britain and Ireland (2001)
Edited by David L. Hawksworth

* Published by Academic Press for the Systematics Association
† Published by the Palaeontological Association in conjunction with the Systematics Association
‡ Published by Oxford University Press for the Systematics Association
**Published by Chapman & Hall for the Systematics Association